The Economics of Sustainable Development

Gurudas Nulkar
Editor

The Economics of Sustainable Development

A Machine-generated Literature Overview

Springer

Editor
Gurudas Nulkar
Centre for Sustainable Development
Gokhale Institute of Politics and Economics
Pune, India

ISBN 978-981-99-7378-1 ISBN 978-981-99-7379-8 (eBook)
https://doi.org/10.1007/978-981-99-7379-8

The auto-summaries have been generated by a recursive clustering algorithm via the Dimensions Auto-summarizer by Digital Science handled by Subject Matter Experts and the external editor. The editor of this book selected which SN content should be auto-summarized and decided its order of appearance. Please be aware that the auto-summaries consist of original sentences, but are not representative of its original paper, since we do not show the full length of the publication. Please note that only published SN content is represented here, and that machine-generated books are still at an experimental stage.

This Springer imprint is published by the registered company Springer Nature Singapore Pte Ltd.
The registered company address is: 152 Beach Road, #21-01/04 Gateway East, Singapore 189721, Singapore

Preface

Disagreements between the proponents of quantitative and qualitative indicators of progress often arise in discussions on economic growth and social development. While economic expansion is quantifiable, progress in human and social realms is not. Subjectivity can creep into qualitative measures, and hence they are not the first choice for economists. In spite of this, there is increasing agreement that qualitative factors that supplement the Gross Domestic Product (GDP) must be employed in measuring national development. The United Nations Secretary General stated:

> Now is the time to correct a glaring blind spot in how we measure economic prosperity and progress. When profits come at the expense of people and our planet, we are left with an incomplete picture of the true cost of economic growth.
>
> United Nations Secretary-General Antonio Guterres, Our Common Agenda: Report of the Secretary-General

While the GDP is still a good measure of economic output, the qualitative indicators can include cultural, social, and economic contexts, that are essential for establishing the impact of development initiatives. Furthermore, they also help measure intangible outcomes like social inclusion, well-being, and social capital, that is challenging to quantify.

The Human Development Index, proposed by United Nations, is a composite measure of average achievement in key dimensions of human development: a long and healthy life, being knowledgeable and having a decent standard of living. When measured with this index, rapidly growing economies like India fare poorly on many counts.[1] This questions the rationale of "economic growth at any cost", and the reliance on GDP as a single measure of development.

The problems afflicting the planet today are beyond climate and environmental issues. Nor are they limited to a single country. The very survival of life on Earth is under jeopardy as a result of the rapid climate change. Human economic activity and the fixation with consumption-driven economic growth have led us into a position that triggers the melting of the polar ice caps, acidification of the oceans, loss of

[1] Human Development Index | Human Development Reports (undp.org).

biological variety, and spread of new diseases. These problems are threatening every nation, not just those who have contributed the most toward it. In this situation, the solutions are neither simple nor convenient. The technologies that we employ to solve the problems have either pushed the problem elsewhere or triggered newer problems.

Solutions to these challenges require a holistic understanding rather than a narrow perspective. Technological solutions are useful in the short run, but for a sustainable planet, we need take a systems approach. The interconnectedness of natural capital, livelihoods, pollution, employment, and other factors are not easy to unravel. Moreover, the current environmental problems are deeply connected to social issues. Problems of overconsumption and unfair distribution of natural resources are as much environmental as they are social. The challenges of inequality, poverty, climate change, rising pollution, dwindling resources, and vanishing biodiversity that have built up over the last few decades are now a serious threat to human survival. While the problems are global, solutions are often local, and require an interdisciplinary approach based on a clear understanding of the biogeography and socio-cultural characteristics.

Modern economics has not been successful in handling many of these issues. More resources are being diverted to space travel, entertainment, and the development of weapons instead of areas that contribute to human well-being. While the corporate world is busy in competition and expansion, the government is expected to use tax money to mitigate social and environmental issues. Greed bestowed by a regime of unrestrained freedom and the excessively competitive spirit bequeathed by the free market economy are no longer desirable traits for sustaining life on the planet. Sustainable development is now a non-negotiable condition for the future of humanity.

My intention in collating this book was to cover ground that would help readers make connections and gain a holistic understanding of sustainable development. This book covers a wide range of topics, from traditional knowledge and alternative livelihoods to challenges of intensive agriculture and farmer poverty. The topic of decoupling growth and resource consumption is introduced as an outcome of discussions on the expendable component of economic growth. The potential for ecological restoration and green jobs to improve human well-being is called into question. In the final chapter, we debate whether or not we need to continue using GDP as a measure, and argue the need to switch to a broader measure that reflects social progress and human well-being.

The compelling arguments in favor of sustainable development can lead to innovation and the development of low-carbon technologies, research in newer renewable energy sources, sustainable materials, and efficient production processes. When these are employed in a circular economy, they present an economic pathway to mitigate climate change in the pursuit of prosperity for humanity. Decentralization, diversification, resource efficiency, and community empowerment can improve the resilience of vulnerable communities and reduce the risks to national economies. By working together toward common goals, countries can address global issues more effectively and promote peace, stability, and shared prosperity.

The way forward is to balance economic growth, environmental protection, and social well-being, to ensure the long-term sustainability of our species on this planet. Economic development and environmental conservation are not mutually exclusive, but can be mutually reinforcing if we are able to change our dominant paradigm of consumption driven economic growth. A prudent use of natural resources is critical for our future generations and necessitates a careful consideration of the impact of economic activities on the natural resource base.

For those studying or engaging in research on sustainable development, I hope this book will be a valuable resource.

> When the study of the household (ecology) and the management of the household (economics) can be merged, and when ethics can be extended to include environmental as well as human values, then we can be optimistic about the future of humankind.
>
> Eugene Pleasants Odum (1993). "Ecology and our endangered life-support systems"

Pune, India
Gurudas Nulkar

About This Book

Auto-summaries can be generated by either an abstractive or extractive auto-summarization:

- An extraction-based summarizer identifies the most important sentences of a text and uses the original sentences to create the summary.
- An abstraction-based summarizer creates new text based on deep learning. New phrases are created to summarize the content.

The auto-summaries you will find in this book have been generated via an extractive summarization approach.

Each chapter was carefully edited by Dr. Gurudas Nulkar. The editors selected the papers which were then auto-summarized. The editors have not edited the auto-summaries due to the extraction-based approach and have not changed the original sentences. You will find the editors' reviews and guidance on the auto-summaries in their chapter introductions.

In machine-generated books, editors are defined as those who curate the content for the book by selecting the papers to be auto-summarized and by organizing the output into a meaningful order. Next to the thoughtful curation of the papers, editors should guide the readers through the auto-summaries and make transparent why they selected the papers.

The ultimate goal is to provide a current literature review of Springer Nature publications on a given topic in order to support readers in overcoming information overload and to help them dive into a topic faster; to identify interdisciplinary overlaps; and to present papers which might not have been on the readers' radar.

Please note that the selected papers are not used to train a LLM while the auto-summaries are created.

Contents

About the Editor

Prof. Gurudas Nulkar is Professor and Director of the Centre for Sustainable Development at the Gokhale Institute of Politics and Economics, Pune, India. He teaches undergraduate and graduate students, supervises Ph.D. students, and is engaged in empirical and applied research in the field of ecological economics and sustainable development. He has an experience of 19 years working in the industry, after which he shifted to academia where he has been teaching for the last 13 years. His doctoral research was in environmental sustainability within industrial ecosystems. This was followed by Post-doctoral Endeavour Fellowship of the Government of Australia. He has published four books, of which three are in Marathi language and one in English. His Marathi book *Anartha Shashtra* has received the Maharashtra State Literature Award in the category "C. D. Deshmukh Award for Economic Literature". His papers have been published in reputed journals and as research reports. Professor Nulkar has served as an expert on several committees and panels, including UN-ESCAP program, Government of Maharashtra, and Pune Divisional Commissioner's Office. He contributed to the Pune Metropolitan Region Development Authority Draft Plan as an ecological expert. He is a member of Pune International Centre, Energy, Environment and Climate Change, and a member of the Indian Society for Ecological Economics (INSEE). He has completed several projects funded by organizations such as World Resources Institute, Global Forest Watch, Centre for Environmental Education, UNICEF, and others.

Chapter 1
Intensive Farming and Its Effects on Natural Ecosystems

1.1 Introduction by Editor

Over 12,000 years ago humans discovered farming. This was to be a game-changer in human history as the surplus farm production bestowed food security on communities that adopted it. A reliable supply of food and settled lifestyle speeded up the transition from hunting-gathering to farming. As farming locked the cultivators to the land they tilled, this was to lay the foundation of urban growth. Perhaps the biggest advantage of farming was the enormous amount of surplus food that it could produce, and the free time it offered. Humans no longer had to spend the day gathering food. As agriculture diffused in societies, this allowed more time in advancing science, arts, and literature. This which was to change the way human societies lived and enjoyed the planet.

Agriculture proliferated around rivers and freshwater sources, and cities grew around them. Forests were cleared for farming and freshwater diverted for irrigation. Agriculture significantly raised the human impact on the natural environment. However, the primeval farm practices using human and animal muscle power could support the sparse population of the time. The food security and the settled lifestyle contributed to an increase in population. As time progressed, an increasing amount of land was diverted to farming. Around the year 1798, Thomas Malthus, the renowned English economist, made a prescient prediction in his seminal work "An Essay on the Principle of Population". According to him, human populations exhibited an exponential rise, whereas food production was constrained by arithmetic growth. During this period food production faced many challenges. Farm productivity was significantly reduced by droughts and pest infestations, resulting in numerous famines being experienced worldwide. In the 18th Century, severe famines occurred in Estonia, Finland, England, Africa, and the notorious Bengal famine in India resulted in a staggering death toll of over 10 million individuals. The Malthusian forecast, in conjunction with the law of diminishing returns, served as a catalyst for the adoption of more intensive agricultural practises.

G. Nulkar (ed.), *The Economics of Sustainable Development*,
https://doi.org/10.1007/978-981-99-7379-8_1

Around the year 1945, Norman Borlaug, an American agricultural scientist, was commissioned to work with farmers in Mexico to increase the yield of their wheat crop. Over the next 15 years, Borlaug created high-yielding, disease resistant varieties of wheat. These changed the fortunes of Mexican farmers and Borlaug's program came to be known as the 'green revolution'. Borlaug's varieties of wheat and rice helped countries like India and Pakistan to improve farm production and attain self-sufficiency in food. However, for a higher yield these seeds needed higher irrigation and had to be supplemented with chemical fertilisers. This was to pave the path of intensification of farming. Crop yields in the Green Revolution were boosted through specialization, increased scale of production units, minimized labour requirements, and maximized use of technological inputs (Zimdahl, 2006).

The Green Revolution triggered the worldwide production of chemical fertilisers and pesticides. As farms grew in size and yield, farm mechanisation developed. The Green Revolution placed 'farm yield' on a pedestal, while ignoring other agricultural conditions. This necessitated generous use of inorganic nitrogen and phosphorus fertilizers to push up productivity. Over time, the impacts of excessive chemical use manifested out in the soil. Agricultural intensification, characterized by the extensive use of synthetic fertilizers and pesticides, has had a detrimental impact on soil health and soil microbiology (Tripathi et al., 2020). While these farming practices have undoubtedly increased agricultural productivity, the environmental consequences have been severe. The excessive use of fertilizers and pesticides has led to soil degradation, enhanced greenhouse gas emissions, accumulation of pesticides, and a decline in the availability and quality of water. The focus on short-term production and profit has come at the expense of long-term environmental sustainability.

Acidification, nitrification, decline in organic matter, soil compaction, and erosion are seen on these farms. Inappropriate farming practices like excessive fertilization, improper use of pesticides, and the use of heavy machinery are some of the drivers of soil degradation (Thomas K. Alexandridis, 2018). Furthermore, the global food system has been a primary driver of biodiversity loss on the planet. Over the past 50 years, the conversion of natural ecosystems for crop production or pasture has been the principal cause of habitat loss, in turn reducing biodiversity (Tim G. Benton, 2021). As cropland expansion increases and farming intensification does not reduce, they will continue to drive biodiversity loss (Zabel, 2019).

The dairy industry contributes significantly to environmental stress, as is evidenced in the study from New Zealand. This industry is fraught with issues like nitrate contamination of drinking water, nutrient pollution in lakes, soil compaction, and greenhouse gas emissions. These costs are not internalised in the dairy industry and are contrary to the green image that the dairy industry claims.

One the one hand agricultural productivity has increased but on the other, the number of undernourished people in the world continued to rise in 2020. As per the report by Food and Agriculture Organisation (FAO) of the United Nations[1] (2021),

[1] The State of Food Security and Nutrition in the World 2021. https://www.fao.org/interactive/state-of-food-security-nutrition/2021/en/ Last accessed 18th June 2024.

nearly 810 million people in the world faced hunger in 2020, and an estimated 660 million people may still remain hungry by 2030.

Sustainable agriculture plays a crucial role in the overall framework of sustainable development and holds enormous potential to improve livelihoods and nutrition in rural communities, and promote equitable development. However, it will need to be mainstreamed in national development strategies and action plans.

The FAO outlines five principles of Sustainable Food and Agriculture strategy.[2] They are (i) boosting food system productivity (ii) safeguarding and upgrading natural resources (iii) enhancing resilience (iv) strengthening livelihoods (v) and governing. The principles offer a foundation for a road map that consists of twenty interconnected acts that can help nations prioritise and choose resources in order to hasten the shift to sustainable methods. To speed up the shift to more sustainable food systems, a variety of strategies, tactics, and instruments are used in these initiatives, which also rely on collaboration and involvement from various stakeholders.

In the absence of comprehensive reforms to the global food systems, it seems improbable that human activities will effectively mitigate the worst effects of climate change. This chapter examines the dynamics of eating habit alterations, the environmental ramifications of intensification practises, the dairy sector, and the escalating water footprint associated with agricultural activities.

References

Steffan-Dewenter, et al. (2007, March 20). Tradeoffs between income, biodiversity, and ecosystem functioning during tropical rainforest conversion and agroforestry intensification. National Academy of Sciences, 104(12), 4973–4978. https://doi.org/10.1073/pnas.0608409104.

Tripathi, S., Srivastava, P., Devi, R S., & Bhadouria, R. (2020, January 1). Influence of synthetic fertilizers and pesticides on soil health and soil microbiology. Elsevier BV, 25–54. https://doi.org/10.1016/b978-0-08-103017-2.00002-7.

Veldkamp, E., et al. (2023, January 24). Multifunctionality of temperate alley-cropping agroforestry outperforms open cropland and grassland. Nature Portfolio, 4(1). https://doi.org/10.1038/s43247-023-00680-1.

1.2 Machine Generated Summaries

Disclaimer: The summaries in this chapter were generated from Springer Nature publications using extractive AI auto-summarization: An extraction-based summarizer aims to identify the most important sentences of a text using an algorithm and

[2] A summary to assess synergies and tradeoffs among the twenty interconnected sustainable food and agriculture actions (2020).
https://openknowledge.fao.org/server/api/core/bitstreams/b06c9ea2-49ad-4116-b1f0-11f58ca8204a/content Last accessed 18th June 2024.

uses those original sentences to create the auto-summary (unlike generative AI). As the constituted sentences are machine selected, they may not fully reflect the body of the work, so we strongly advise that the original content is read and cited. The auto generated summaries were curated by the editor to meet Springer Nature publication standards. To cite this content, please refer to the original papers.

Machine generated keywords: farming, intensification, livestock, farm, production, yield, cost, northern, land use, expansion, adverse, develop country, demand, industry, half.

1.2.1 Sustainable Intensification Farming as an Enabler for Farm Eco-efficiency? [66]

This is a machine-generated summary of:

Weltin, Meike; Hüttel, Silke: Sustainable Intensification Farming as an Enabler for Farm Eco-Efficiency? [66].

Published in: Environmental and Resource Economics (2022).

Link to original: https://doi.org/10.1007/s10640-022-00718-6.

If you want to cite the papers, please refer to the original.

For technical reasons we could not place the page where the original quote is coming from.

Abstract-Summary

"Sustainable Intensification (SI) practices offer adopters exploiting improvement potentials in environmental performance of farming, i.e. enhance ecosystem functionality, while maintaining productivity."

"We conclude that the SI adopters determined the sample's system frontier and showed higher mean eco-efficiency, but that most farms in our sample did not fully exploit the improvement potentials in biodiversity as ecological outcome."

Introduction

"We base our evaluation on a utility maximization model, where farmers opt consciously for a SI technology, and on a two-output (economic and ecological) production model and a directional meta-frontier approach to measure eco-efficiency (cf. Beltrán-Esteve and Reig-Martínez [1]; O'Donnell and others [2])."

"We hypothesize that adopting SI offers an eco-efficient reduction of farms' improvement potentials and farms provide more maintenance in the form of ecological output at no economic cost."

"As a second aspect, we explore whether adopting the SI technology reduces farms' improvement potentials, i.e., the distances to the system frontier in the direction of the ecological output, and as a third aspect, we acknowledge that farms' performance in the chosen technology can impact the improvement potential and refer to this as the performance effect."

Conceptual Approach

"We base our analysis on the behavioural model of Chabé-Ferret and Subervie [3] to frame the decision to adopt SI measures for causal identification of farms' respective improvement potentials."

"The farm household's decision to adopt SI determines the possible improvement in the ecological direction by the respective group-specific frontier."

"Presuming that sequential preferences guide the farm household's decision to adopt SI measures (cf. Asmild and Hougaard [4]), the SI farm aims to reduce the improvement potential in the ecological direction compared to their non-SI reference situation."

"In the short-run, however, inefficiencies within the chosen technology may occur and farms may tolerate the adjustment costs of technology adoption in terms of output reductions (Ang and Oude Lansink [5]), although ignoring the possible inefficiencies of SI adopters would bias the retrieved improvement potentials (performance effect)."

"Farm B is eco-efficient within the non-SI technology but could exhibit eco-inefficiencies in the SI technology that prevent moving to a point B' that reduces the improvement potential."

Data and Empirical Approach

"Given the study region's urgencies to meet climate-protection and biodiversity goals by SI, our ecological outcome measures focus on on-farm and on-land diversity, and we selected six agronomic SI measures with potential contributions to biodiversity and thus ecosystem functionality of grassland and arable farming: (1) reduced tillage, (2) intercropping, (3) growing legumes, (4) integrated pest management, (5) grazing, and (6) extensive use of grassland."

"We use the Simpson diversity index to indicate on-farm diversity on cropped and non-cropped area, i.e. share of arable land, extensive grassland, and other grassland (cf. Van Eck and Koomen [6])."

"We weight the indicators for arable and grassland by the respective share of each land-use type on the farm in the composite indicator for on-land diversity to reflect the heterogeneity of farm types in the sample."

Results

"The meta-technology ratio that captures the eco-efficiency to the system frontier is 0.99 on average and is equal to 1 for 78 SI farms."

"Based on a Kolmogoroff–Smirnov test, we reject the null hypotheses that the distributions of eco-efficiency scores to the group-specific frontier and system frontier are identical for non-SI farms ($D = 0.42$; $p = 0.00$), but cannot reject them for SI farms ($D = 0.07$; $p = 0.89$)."

"The result is consistent with similar differences for all scenarios of the robustness analysis except for raising the weight of on-land diversity in the ecological outcome indicator (i.e., adopters are still more eco-efficient but the gap narrows to 0.04 score points; lowering the SI cut-off to the adoption of at least two SI measures yields the maximal difference of 0.14 score points)."

"Non-SI farms' average eco-efficiency score regarding the group-specific frontier is 0.74."

"For SI farms, 70% have a higher eco-efficiency score to the system frontier than their respective matched control non-SI farms."

Discussion

"For the majority of the sampled SI farms, adopting SI could in fact reduce the environmental improvement potential compared to their matched control references (i.e., eco-inefficiency in the ecological direction measured against the system frontier)."

"Results also lend support to our assumption of sequential preferences (i.e., farm managers could prioritize to become more efficient in an economic direction prior to adopting SI measures and improving in the ecological direction)."

"It is possible that our assumption explained why non-SI farms with traditional farming technology and rather high eco-efficiency scores in the environmental direction did not adopt SI, or that intentional inefficiencies (Hansson and others [7]) explained SI farms' larger distance to the frontier in an economic direction than in an ecological direction."

"Encouraging large-scale adoption of SI and sustainable farming requires understanding of origins of eco-inefficiencies and overall performance effects in economic and ecological direction."

Conclusion

"We proposed directional meta-frontier approach combined with matching to assess selected agronomic SI measures for improving ecological outcomes in agricultural land use."

"Eco-efficiency scores to the system frontier estimated by directional DEA, captured farms' ecological improvement potential."

"SI farms had higher eco-efficiency scores to the system frontier compared to farms staying with traditional farming at mean."

"To facilitate exploiting full ecological improvement potentials, we suggest smart mixes of policy measures including result-based support to offer incentives for large scale SI adoption and for improving within-technology efficiency."

1.2.2 How Dietary Transition Changed Land Use in Mexico [67]

This is a machine-generated summary of:

Tello, José; Garcillán, Pedro P.; Ezcurra, Exequiel: How dietary transition changed land use in Mexico [67].

Published in: Ambio (2020).

Link to original: https://doi.org/10.1007/s13280-020-01317-9.

If you want to cite the papers, please refer to the original.

For technical reasons we could not place the page where the original quote is coming from.

Abstract-Summary

"In Mexico, the shift in food consumption and production patterns, particularly in relation to animal source foods (ASF), has changed land use."

"We studied the consumption and production of ASF and change in agricultural land use in Mexico during the second half of the twentieth century and until 2013; using domestic and international data sources, our findings show an increasing proportion of farmed area devoted to the production of feed crops domestically, and also an increasing demand of farmed feed beyond national borders."

Introduction

"Despite the scientific and cultural interest of the traditional Mexican diet, the reality is that everyday food consumption in Mexico has changed drastically during the second half of the twentieth century and continues changing today."

"Flores and others [9], for example, reported that the emerging new dietary patterns significantly increased the risk of being overweight or obese compared with traditional Mexican diets."

"The role of sugar-sweetened beverages in the nutrition transition in Mexico and its health consequences has been widely analyzed (Aburto and others [10]; Sánchez-Pimienta and others 11), but the role of animal source foods (ASF) consumption and especially its ecological effects has been less explored."

"Our aim was to analyze the relationship between the temporal dynamics of consumption of livestock products as a result of the nutrition transition in the late twentieth century and its effects on land-use change due to demand of feed crops."

Materials and Methods

"We used annual data of production, imports, and exports of animal products for human consumption (i.e., animal protein such as beef, pork, and poultry, as well as milk and eggs) between 1961 and 2013 from FAOSTAT (FAO [8]) to build annual series of total consumption (production plus imports minus exports), and per capita consumption of livestock products."

"In the case of maize, we compiled data from different sources to estimate consumption and production (Appendixes S2 and S3)."

"For 1950–1960, we used published data of total harvested area and feed crop area from Barkin [12], and based on these data we estimated the harvested area of food crops."

"For 1961–2013, we compiled data from FAO [8] and SAGARPA [13] to estimate total harvested area in Mexico."

"We used available data of annual harvested land of sorghum (FAO [8]), soybean (FAO [8]), and assembled different sources to elaborate time series of harvested land of maize; alfalfa, oats and barley fodder; and cultivated grasses."

Results

"Consumption of commodities and products to feed animals (cultivated grasses, fodder, grains and oilseeds, and byproducts) rose 2157% between 1961 and 2013, from 5.3 to 119.3 million tons, at a mean annual rate of 6%."

"Consumption of feed grains, oilseeds, and byproducts increased 2533% between 1961 and 2013, from one to 27 million tons, at an annual mean rate of 6.3%."

"One of the best indicators of the impact of the dietary transition on Mexican agriculture is the proportion of farmed area devoted to the production of food for direct human consumption (food crops) relative to the area devoted to the cultivation of feed crops: The area devoted to the cultivation of food crops expanded during the mid-twentieth century, from 8.2 million ha in 1950 to 14.6 million ha in 1966 (an increase of 77%)."

"Total domestic production of maize increased from 6.2 to 22.7 million tons (363%) between 1961 and 2013, but the amount devoted to feed animals increased at a much faster rate, from 0.54 to 4.3 million tons (796%)."

Discussion

"The first large-scale impact on land use due to increasing livestock production in Mexico was the clearing of tropical forests to open space for cattle grazing in the 1970s and 1980s (Bray and Klepeis [14])."

"During the last 50 years, livestock production has clearly changed agricultural land-use patterns in Mexico."

"Considering Mexican agricultural yields in 2103, it would have been necessary to add 5.6 million ha (i.e., and additional 29.4% of the total harvested area) to produce domestically the total feed production of maize, soybeans, and sorghum that were imported that year from the USA."

"A conservative estimation of future demand, based on 2013 per capita consumption, shows that 26 881 million tons of livestock products may be consumed in Mexico by 2050, requiring an additional 5.2 million hectares of national agricultural area to produce feed crops equivalent to 27.1% of today's harvested area."

Conclusion

"Although the stabilization of extensive livestock production has reduced the rate of conversion of tropical forests into induced grasslands, in Mexico, the agricultural production for feed for barn and feedlot rearing of animals has increased the demand for new croplands, expanding the agricultural frontier, putting pressure over food crop fields, and requiring more water and other agricultural inputs like fertilizers, linked to large-scale pollution processes."

"Our analysis provides an insight at a national scale of land-use changes and large-scale environmental impacts induced by the increasing demand for ASF in the Mexican diet."

"Compared to the direct effects on human health, these environmental changes have been a less-discussed consequence of the transition that has occurred in the Mexican diet during the last half-century."

"We demonstrate that if these trends continue the environmental sustainability of Mexican food production will be seriously challenged."

1.2.3 The Environmental Costs and Benefits of High-Yield Farming [68]

This is a machine-generated summary of:

Balmford, Andrew; Amano, Tatsuya; Bartlett, Harriet; Chadwick, Dave; Collins, Adrian; Edwards, David; Field, Rob; Garnsworthy, Philip; Green, Rhys; Smith, Pete; Waters, Helen; Whitmore, Andrew; Broom, Donald M.; Chara, Julian; Finch, Tom; Garnett, Emma; Gathorne-Hardy, Alfred; Hernandez-Medrano, Juan; Herrero, Mario; Hua, Fangyuan; Latawiec, Agnieszka; Misselbrook, Tom; Phalan, Ben; Simmons, Benno I.; Takahashi, Taro; Vause, James; zu Ermgassen, Erasmus; Eisner, Rowan: The environmental costs and benefits of high-yield farming [68].

Published in: Nature Sustainability (2018).

Link to original: https://doi.org/10.1038/s41893-018-0138-5.

If you want to cite the papers, please refer to the original.

For technical reasons we could not place the page where the original quote is coming from.

Abstract-Summary

"High-yield farming raises other concerns because expressed per unit area it can generate high levels of externalities such as greenhouse gas emissions and nutrient losses."

"Such metrics underestimate the overall impacts of lower-yield systems."

"We develop a framework that instead compares externality and land costs per unit production."

"We apply this framework to diverse data sets that describe the externalities of four major farm sectors and reveal that, rather than involving trade-offs, the externality and land costs of alternative production systems can covary positively: per unit production, land-efficient systems often produce lower externalities."

"Our conclusions are limited: remarkably few studies report externalities alongside yields; many important externalities and farming systems are inadequately measured; and realizing the environmental benefits of high-yield systems typically requires additional measures to limit farmland expansion."

Main

"In terms of supply, farming at high yields (production per unit area) has considerable potential to restrict humanity's impacts on biodiversity."

"Detailed field data from five continents and almost 1,800 species from birds to daisies [15–20] reveal that so many depend on native vegetation that for most the impacts of agriculture on their populations would be best limited by farming at high yields (production per unit area) alongside sparing large tracts of intact habitat."

"Measured per unit area of farmland, the production of such externalities is sometimes greater in high- than lower-yield farming systems [21, 22], potentially weakening the case for land sparing."

"While expressing externalities per unit area can help identify local-scale impacts [23], it systematically underestimates the overall impact of lower-yield systems that occupy more land for the same level of production [24]."

"To be robust, assessments of externalities also need to include the off-site effects of management practices, such as crop production for supplementary feeding of livestock, or off-farm grazing for manure inputs to organic systems [24–26]."

A Novel Framework for Comparing System-Wide Costs

"We argue that comparisons of the overall impacts of contrasting agricultural systems should focus on the sum of externality generated per unit of production [16] (paralleling measures of emissions intensity in climate change analyses)."

"We develop a more general framework, and apply it to a diversity of data on some major farm sectors, farming systems and environmental externalities."

"We used both literature searches and consultation with experts to find paired yield and externality measurements for contrasting production systems in each sector."

"To be included, data had to be near-complete for a given externality—for example, most major elements of GHG emissions or N losses had to be included, and if systems involved inputs (such as feeds or fertilizers) generated off-site we required data on the externality and land costs of their production."

Findings for Four Sectors

"Few studies measured paired externality and yield information, many reported externalities in substantially incomplete or irreconcilably divergent ways, and we could find no suitable data at all on some widely adopted practices."

"The type of data available differed across these combinations (which we view as a useful test of the flexibility of our framework)."

"For one combination, the most extensive data we could find was from a long-term experiment at a single location."

"In livestock systems, all data we could find showed positive covariation between land costs and externalities."

Incorporating Land Use

"As a final analysis, we examined the additional externalities resulting from the different land requirements of contrasting systems."

"To generate the same quantity of agricultural product, low-yield systems require more land, allowing less to be retained or restored as natural habitat."

"This is in turn likely to increase GHG emissions and soil loss, and alter hydrology—although we could find only enough data to explore the first of these effects."

"These findings thus confirm recent suggestions [27, 28] that high-yield farming has the potential, provided land not needed for production is largely used for carbon sequestration, to make a substantial contribution to mitigating climate change."

Conclusions, Caveats and Knowledge Gaps

"A multi-dimensional understanding of the environmental effects of alternative production systems is integral to delivering sustainable intensification, more field measurements linking yield with a broader suite of externalities across a much wider range of practices and sectors are urgently needed."

"Many low-yield systems impose high costs in other ways too and, although certain yield-improving practices have undesirable impacts (for example, organic fertilization of paddy rice increasing CH_4 emissions; see also ref."

"Although most systems we examined are relatively high-yielding, other recent work suggests that positive associations (rather than trade-offs) among environmental and land costs may, if anything, be more likely in lower-yielding systems [29]."

"Some high-yield practices we did not examine, such as the heavy use of pesticides in much tropical fruit cultivation [30], are likely to increase externality costs per unit production."

Methods

"Systems generated significant co-products (wheat and rapeseed from rotational rice, beef from dairy), we allocated land and externality costs to the focal product in proportion to its relative contribution to the gross monetary value of production per unit area of farmland (from focal and co-product combined) [31]."

"GHG costs were expressed in tonnes of CO_2 equivalents (CO_2e) per tonne of rice grain, and included CH_4 and N_2O emissions for growing and fallow seasons (with the latter where necessary based on mean values from refs [32–35]), and embodied emissions from N fertilizer production (Yara emissions database; F. Brendrup, personal communication)."

"For rotational systems, we adjusted the land and GHG costs of rice production downwards by multiplying them by the proportional contribution of rice to the gross monetary value of production per unit area of farmland from rice and co-product combined (using mean post-2000 prices from ref."

1.2.4 Environmental Effects of Intensification of Agriculture: Livestock Production and Regulation [69]

This is a machine-generated summary of:

Chakravorty, Ujjayant; Fisher, Donna K.; Umetsu, Chieko: Environmental effects of intensification of agriculture: livestock production and regulation [69]

Published in: Environmental Economics and Policy Studies (2007).

Link to original: https://doi.org/10.1007/bf03353963.

If you want to cite the papers, please refer to the original.

For technical reasons we could not place the page where the original quote is coming from.

Abstract-Summary

"This article deals with the relationship between industrialization of agriculture and the environment in developing countries."

"We develop a simple economic framework to demonstrate the effect of location on intensification of industrial activity in farming, and discuss this issue in the context of urbanization and economic growth in developing countries."

"Policy implications of the model are discussed in light of the experience of developed countries in regulating livestock pollution and other externalities."

"We argue that environmental problems from agricultural industrialization in developing countries may pose major challenges."

1.2.5 New Zealand Dairy Farming: Milking Our Environment for All Its Worth [70]

This is a machine-generated summary of:

Foote, Kyleisha J.; Joy, Michael K.; Death, Russell G.: New Zealand Dairy Farming: Milking Our Environment for All Its Worth [70]

Published in: Environmental Management (2015).

Link to original: https://doi.org/10.1007/s00267-015-0517-x.

If you want to cite the papers, please refer to the original.

For technical reasons we could not place the page where the original quote is coming from.

Abstract-Summary

"There have been major increases in dairy production in New Zealand."

"Intensified dairy farming thus incurs considerable environmental externalities: impacts that are not paid for directly by the dairy farmer."

"This is the first nationwide assessment of some of the environmental costs of the recent increase of dairy intensification in New Zealand."

"At the higher end, the estimated cost of some environmental externalities surpasses the 2012 dairy export revenue of NZ$11.6 billion and almost reaches the combined export revenue and dairy's contribution to Gross Domestic Product in 2010 of NZ$5 billion."

"For the dairy industry to accurately report on its profitability and maintain its sustainable marketing label, these external costs should be reported."

"This assessment is in fact extremely conservative as many impacts have not been valued, thus, the total negative external impact of intensified dairying is probably grossly underestimated."

Introduction

"This intensification has been associated with environmental effects that are external to the farm, termed externalities."

"The current dominant economic system of valuation [using Gross Domestic Product (GDP)] does not adequately incorporate the costs of environmental impacts or the true value of natural resources and services (Bertram [36]; Costanza and others [37]; Daily and others [38]; Daly [39]); hence costs are externalized (Pretty and others [40])."

"New Zealand is not exempt from this process and the dairy industry is a classic example of unaccounted for externalities left for the general public to pay."

"The dairy industry does not pay for all its environmental pollution and the growth of dairy farming has occurred with little balanced economic evaluation or even awareness of the true environmental impacts and costs."

"This study identified some of the environmental impacts and externalities produced from the New Zealand dairy industry and the costs associated with mitigating these."

Trends in Dairy Production and Inputs

"The cost of imported fertilizers for dairy farming was estimated using the proportion of each fertilizer used on dairy farms [around 70% of nitrogen and 34% of phosphorus (Statistics New Zealand [41])], equalling around NZ$503 million in 2012."

"Despite dairy being New Zealand's predominant export industry, there has been no national level analysis on the environmental impacts and costs of dairy farming in New Zealand."

"We included economic measures relating to dairy impacts of the cost to remove nitrogen from water exceeding drinking standards; the cost to remove nutrients from lakes; the cost to remedy soil compaction; the potential price of a tax on GHG emissions; and the potential loss of dairy export revenue from degradation of New Zealand's 'clean green image'."

"Carbon prices in New Zealand were \$25/t CO_2-e (Stroombergen and others [42]); at this price, dairy emissions would cost NZ$480 million."

Summary of Environmental Costs from Dairy Impacts

"If costs from biodiversity loss and ecosystem degradation were calculated, this total would be much higher."

"Only costs estimated on a national scale were included in the total."

"Although cost estimates are available for remediation at a small scale (e.g., a lake catchment), it was not possible to extrapolate for the entire country because of the large differences in cost between lakes, and is thus omitted from the total."

"Costs may become a reality in the future if further degradation requires mitigation or if New Zealander's decide they would like higher environmental standards."

Conclusion

"There needs to be a more holistic conversation in New Zealand of whether the dairy industry is actually beneficial for the country: economically, environmentally, and socially."

"This is not at present a cost for the dairy industry, but an estimate of the potential external cost to the public of New Zealand from dairy farming—from having a degraded environment or paying to clean up."

"If the dairy industry is to continue to expand and intensify, accurate reporting of the real costs needs to be used in the evaluation of the true value of this industry."

1.2.6 Population Growth and Land Use Intensification in a Subsistence-Based Indigenous Community in the Amazon [71]

This is a machine-generated summary of:

Sirén, Anders Henrik: Population Growth and Land Use Intensification in a Subsistence-based Indigenous Community in the Amazon [71].

Published in: Human Ecology (2007).

Link to original: https://doi.org/10.1007/s10745-006-9089-y.

If you want to cite the papers, please refer to the original.

For technical reasons we could not place the page where the original quote is coming from.

Abstract-Summary

"Shifting cultivation practiced by indigenous peoples living at low population densities in tropical forests has often been described as sustainable and compatible with conservation."

"As many indigenous peoples in tropical forests currently experience rapid demographic growth, this raises the question to what extent their agricultural activities actually contribute to deforestation."

"During the last seven decades, people have increasingly begun to clear fallows instead of old-growth forest to farm."

"Whereas these changes may seem negligible in terms of deforestation, they do cause hardships to the local people, because of increasing walking distance to old-growth forest, and problems with weeds, pests, and decreasing soil productivity when farming after reclearing fallows."

Introduction

"At least until recently, shifting cultivation for subsistence needs has been the main direct impact of the indigenous people on forest cover."

"Shifting cultivation in tropical forests typically involves clearing the forest in order to farm for a few years until nutrients in the soil are exhausted or weed infestation becomes too severe, and then the land is fallowed, letting the forest grow back again (Beckerman [43]; Clay [44])."

"Fallowing the land and allowing the forest to grow up restores soil fertility and reduces populations of pests, pathogens, and weeds over time."

"Such shifting cultivation as well as other "traditional" land use by indigenous peoples has often been described as sustainable and compatible with conservation (e.g., Dufour [45]; Sponsel [46]; Schwartzman and others [47])."

"As populations have increased, shifting cultivators have often reduced the extent of old-growth forest and gradually decreased the length of the fallow period, which has led to decreased productivity (Boserup [48])."

Study Area

"This is done on favorable sites such as sandy soils or sometimes when clearing old-growth forest or when there is a hurry to get the crops planted."

"There are also two major types of maize—traditional runasara, particularly suitable as a food crop because the ears dry slowly and asynchronously, allowing a prolonged period of harvesting of green ears, and the fairly recently introduced morocho, which provides higher yields and dries fast and quite synchronously, facilitating a one-time harvest in order to store the maize as poultry feed."

"On hill fields on red soils, however, only the guinia type of plantain is planted, mostly along the low-lying edges."

"Clearing of old-growth forest within your area of use rights gives exclusive ownership not only to the field itself, but also to the fallow that grows up afterwards."

Materials and Methods

"Given these limitations, I made five separate estimates of the rate of demographic change, based on four different methods and sets of data."

"Almost all fields had a life span of about two and a half years before they were abandoned and got invaded by regrowth, I estimated the annual rate of clearing of forest and fallows by dividing the area of cultivated land by 2.5."

"Based on the age of the fallow at the last clearing of each land use plot, I also calculated the age distribution of fallows in the landscape."

"Although the remote sensing analysis only allowed estimation of the area of fallows younger than 20 years of age, during fieldwork I found fallows up to 65 years old."

"As the field data analysis provided an estimate of the age distribution of fallows in the landscape, this allowed me to calculate also the area of fallows between 20 and 65 years old."

Results

"During fieldwork, people sometimes showed me fields that had recently been made by clearing old-growth forest quite close to settlements."

"The annual rate of clearing of forest and fallow for establishing fields was 53 ha."

"Out of this, an estimated 18 ha was cleared in old-growth forest."

"In relation to the total area of fields and fallows, this corresponds to a rate of expansion of the intervened area of about 0.4% per year."

"In relation to the total area of old-growth forest in the study area, on the other hand, this corresponds to a deforestation rate of 0.015% per year."

"This suggests that, except for a brief increase in the mid 1980s when many people established cattle pastures, the absolute rate of old-growth forest clearing has changed little since the establishment of the community in spite of population growth."

Discussion

"The Sarayaku community has adapted to increasing competition for land by adopting firmer rules regulating the rights to land for farming (Sirén [49])."

"It may be possible to strengthen rules further, encouraging permanent agriculture on alluvial plains and/or discouraging the type of competition for land that leads to shortened fallow periods or expansion into old-growth forest."

"Whatever the direct cause, such fissions reduce competition for land for cultivation, and such competition for resources may be an underlying cause of social conflict."

"This would facilitate access to old-growth forest, and thus alleviate land scarcity while also causing loss of old-growth forest."

"Increasing scarcity of land to farm is a serious socioeconomic problem for the Sarayaku people."

"Increased access to salaried off-farm employment could also lead to increased consumption of imported foods, reducing the need for land to farm (cf. Kaimowitz and Angelsen [50])."

1.2.7 Seasonal and Regional Limits to Growth of Water-Intensive Crop Farming in Bangladesh [72]

This is a machine-generated summary of:

Akanda, M. Aminul Islam: Seasonal and regional limits to growth of water-intensive crop farming in Bangladesh [72]

Published in: Sustainable Water Resources Management (2018).

Link to original: https://doi.org/10.1007/s40899-018-0263-3.

If you want to cite the papers, please refer to the original.

For technical reasons we could not place the page where the original quote is coming from.

Abstract-Summary

"Irrigation-led expansion of crop farming raised demand for groundwater in Bangladesh."

"Out of total 76.5 bm^3 water consumption for crop farming in the early 2010s, 34.1 bm^3 was supplied through irrigation of which 79.8% was from groundwater and that would reach at 100.5, 46.6 and 88.2%, respectively in 2029–30."

"Pulses, oilseeds, vegetables and potato were identified as feasible crops on the basis of groundwater demand, dietary-balanced self-sufficiency and domestic resource cost ratio."

"Any reduction of cereal and sugarcane areas over projections and its adjustment with feasible crops would save about one and a half billion m^3 of groundwater in dry season."

Introduction

"Rice farming was reported to suffer from water scarcity during winter season (Mirza [51])."

"Akanda [52] carried out a research on water requirement for crop farming and potentials of crop diversification in Bangladesh."

"There was a scope of further study with seasonal projections for all types of crops, new method of water accounting and regional potentials for diversification."

"This study was conducted with specific objectives of (1) estimating seasonal demand for water and pressures on groundwater for present and future farming, (2) relocating future cropped areas coherent to food security, groundwater demand and cost factors and (3) exploring strategies to cope with landscape and groundwater scenario among hydrological regions in Bangladesh."

"Regional land type, water sources and groundwater scenario were used to evaluate demand for diversification projects among hydrological regions."

Materials and Method

"The TCA calculation as its bottom-up summing was not accurate because of probable seasonal maladjustment with ceiling for forecasted net cropped area (NCA)."

"The researchers here used data on daily soil water balance, actual use of water and actual crop yield to calculate water demand per unit of production."

"The footprint data were available for all categories of crops but were not calculated with water consumption data for all crops grown in Bangladesh."

"The demand for water and irrigation was calculated for each category of crops grown in different crop seasons."

"The DSF-adjusted cropped area, calculated as the multiple of readily available cropped area with DSF, would ensure production for desirable intake with full self-sufficiency."

"The required cropped area for 2029–30 at 2015–16 level of availability ($CAPA_{2030}$) was calculated as the ratio of required production to yield in 2029–30."

Results and Discussion

"Areas under all non-cereal crops would increase from 2.82 to 3.93 million ha (18.7–24.4% of TCA) during 2010–2030, where vegetable and potato farming would expand largely."

"Crop farming demanded 0.09, 0.23 and 33.77 bm^3 of irrigation water during summer, monsoon and winter seasons, respectively, in the early 2010s, which would increase to 0.15, 0.91 and 45.16 bm^3 in 2029–30."

"Over-exploitation of groundwater made it necessary to come out of water-intensive winter farming for NC and NW regions and their adjacent areas of SW and NE regions."

"The EH and NW regions demanded the highest level of diversification efforts as per the DDE index, which was followed by the SW, NC, NE, SE and SC regions."

"A special diversification project was ongoing in some areas of the NW, SW and SC regions and similar was identified necessary for the EH, NC and NE regions."

Conclusion

"The present study aimed to identify a cropping pattern compatible to future accessibility of groundwater and to explore regional potentials of diversification with feasible crops."

"Rice, other cereals, cash crops, potatoes and sugarcane were identified as water-intensive crops for demanding groundwater over 4500 m^3 per ha during the winter."

"The demand for groundwater was accounted for 4718 m^3/ha for the NW, 4331 m^3/ha for the NC and 4261 m^3/ha for the SW regions, which could be reduced through more expansion of feasible crops during the winter."

"Crop diversification efforts were most demanded for the EH and NW regions followed by the SW and NC regions."

"Based on the DDE index, it would be effective to undertake special projects for expansion of feasible crops in the potential EH, NC and NW regions."

1.2.8 Farming Intensification and Environmental Justice in Northern Ireland [73]

This is a machine-generated summary of:

Gladkova, Ekaterina: Farming Intensification and Environmental Justice in Northern Ireland [73].

Published in: Critical Criminology (2020).

Link to original: https://doi.org/10.1007/s10612-020-09488-3.

If you want to cite the papers, please refer to the original.

For technical reasons we could not place the page where the original quote is coming from.

Abstract-Summary

"This article employs an environmental justice perspective to scrutinize farming intensification in a community affected by this phenomenon."

"Based on findings from semi-structured interviews with local residents, the article describes their exposure to environmental risks as a result of farming intensification."

"The article concludes that farming intensification in Northern Ireland is marked by procedural environmental injustice and should be addressed by downscaling levels of production and reforming decision-making processes."

Introduction

"The strategy focused on farming expansion in Northern Ireland, allowing the country's agri-food industry to keep pace with the global trends in meat production governance."

"The rising number of intensive pig farms is likely to continue to have serious impacts on human health and to exacerbate further the existing environmental problems in Northern Ireland."

"The environmental justice paradigm appears to be a suitable lens for evaluating the impact of pig farming intensification on a rural community in Northern Ireland because it illuminates not only the distribution of environmental benefits and burdens but also the extent of individual and community recognition and participation (Schlosberg [53]) in relation to farming intensification."

"The article concludes with a call for not only reforming the environmental decision-making forums in Northern Ireland but, more importantly, for reversing pig farming intensification and changing the dominant mode of production."

Green Criminology and Environmental Justice

"Proponents and scholars of environmental justice have considered questions surrounding procedural justice, including individual and community recognition in the decision-making process and political participation."

"Many environmental justice advocates are also concerned with social, cultural, and political processes of environmental decision-making (Chakraborty [54]; Holifield and others [55]), thereby echoing the call of prominent environmental justice scholars to explain the causes of and dynamics leading to the inequitable distribution of environmental advantages and disadvantages (Bullard [56]; Fraser [57]; Pulido [58]; Young [59])."

"Procedural environmental justice is conceptualized as fairness in the processes of decision-making—reflecting an understanding of "who has the power to make decisions, the kinds of decisions that are made, in whose interests they are made, and how social practices based on these decisions are materially organized" (White [60]: 56)."

"The notion of capabilities, thus, links distributional and procedural components of environmental justice, while considering a broader set of conditions necessary for a full human functioning (Schlosberg [53])."

Methodology

"This article utilized a case study research design, and its empirical basis originated from semi-structured interviews with the residents affected by farming intensification in the Newtownabbey area in County Antrim north of Belfast in Northern Ireland."

"The participants interviewed were "the 'right persons' to study in terms of the 'theory'" (Brannen [61]: 16) in that out of the total possible number of resident participants, they were the ones affected directly by the existing farms and/or the ones who participated in the decision-making processes behind farming intensification."

"Government officials interviewed in the course of this research were involved directly in decision-making regarding farming intensification."

"During the interviews, the resident participants were asked about their views on the impact of pig farming intensification, on the natural environment and cultural identity of the area, on their participation in the decision-making process around new farm developments, and about the recognition of their views in that process."

Distributional Environmental Injustice and Farming Intensification

"Already, water contamination has taken place in the area: "There have been episodes of water pollution from the existing farm and the farmer has been convicted in court"

(COM002), and the respondents expressed their uncertainty about the future of local water quality."

"The production of risk from water pollution is linked to the rise in intensive farms in Newtownabbey area, but at the same time, it is part of the collective production of environmental risk from farming intensification in Northern Ireland and elsewhere (Curran [62])."

"The distribution of environmental risks (Curran [63]) from water pollution is uneven insofar as the residents living in close proximity to existing farms are exposed to a higher amount of risk, and the proliferation of farms in the area is likely to exacerbate it."

"The high concentration of environmental burdens from existing pig farms in the area is affecting the community's quality of life: "everywhere they are spreading the slurry, the fields are surrounded by villages."

Procedural Environmental Injustice and Farming Intensification

"Before discussing procedural environmental justice in the context of the Newtownabbey area in County Antrim, it is necessary to describe the opportunities for individuals and communities that exist within the planning framework in Northern Ireland to contribute to the decision-making regarding the construction of new farms."

"Local residents were included formally in the decision-making process and could provide input by commenting on the farm planning application and engaging with the local council."

"Of "getting heard," local residents have to engage with the council and, in some cases, the government agencies involved in the decision-making around pig farm planning applications."

"In terms of procedural environmental injustice, local residents' formal participation in the decision-making does not result in the redistribution of power needed to challenge the dominant logic of farming intensification."

Conclusion

"The environmental justice paradigm employed in this article demonstrates how an already disproportionate exposure of some rural communities to environmental harms from farming may be amplified in the future as more farms are built in the process of farming intensification legitimized by the GfG agri-food strategy in Northern Ireland."

"This process is marked by recognitional and procedural environmental injustice, where the inability of local communities to "counter-argue the prevailing consensuses that sustain their situation" (Bustos and others [64]: 292) fuels farming intensification even further."

"The prevailing consensus in the political economy of capitalism prioritizes the economic benefits from farming intensification over the social interests of affected communities, as well as the integrity of local ecological systems."

"Such decisions depend on challenging the dominant mode of meat production to transform the "not in my backyard" argument into "not in anyone's backyard" (Faber 65: 252) and re-orient profit-driven farming toward the concerns of environmental and social sustainability."

Bibliography

1. Beltrán-Esteve M, Reig-Martínez E (2014) Comparing conventional and organic citrus grower efficiency in Spain. Agric Syst 129:115–123
2. O'Donnell CJ, Rao DP, Battese GE (2008) Metafrontier frameworks for the study of firm-level efficiencies and technology ratios. Empir Econ 34(2):231–255
3. Chabé-Ferret S, Subervie J (2013) How much green for the buck? Estimating additional and windfall effects of French agro-environmental schemes by DID-matching. J Environ Econ Manag 65(1):12–27
4. Asmild M, Hougaard JL (2006) Economic versus environmental improvement potentials of Danish pig farms. Agric Econ 35(2):171–181
5. Ang F, Oude Lansink A (2017) Decomposing dynamic profit inefficiency of Belgian dairy farms. Europ Rev Agric Econ 45(1):81–99
6. Van Eck JR, Koomen E (2008) Characterising urban concentration and land-use diversity in simulations of future land use. Ann Reg Sci 42(1):123–140
7. Hansson H, Manevska-Tasevska G, Asmild M (2018) Rationalising inefficiency in agricultural production—the case of Swedish dairy agriculture. Europ Rev Agric Econ 1–24
8. FAO. 2017. FAOSTAT statistics database. Retrieved July 1, 2017, from http://www.fao.org/faostat/en/#data .
9. Flores, M., N. Macias, M. Rivera, A. Lozada, S. Barquera, J. Rivera-Dommarco, and K.L. Tucker. 2010. Dietary patterns in Mexican adults are associated with risk of being overweight or obese. The Journal of Nutrition 140: 1869–1873. https://doi.org/10.3945/jn.110.121533 .
10. Aburto, T.C., L.S. Pedraza, T.G. Sánchez-Pimienta, C. Batis, and J.A. Rivera. 2016. Discretionary foods have a high contribution and fruit, vegetables, and legumes have a low contribution to the total energy intake of the Mexican population. The Journal of Nutrition 146: 1881–1887. https://doi.org/10.3945/jn.115.219121 .
11. Sánchez-Pimienta, T.G., C. Batis, C.K. Lutter, and J.A. Rivera. 2016. Sugar-sweetened beverages are the main sources of added sugar intake in the Mexican population. Journal of Nutrition 146: 1888–1896. https://doi.org/10.3945/jn.115.220301 .
12. Barkin, D. 1981. The use of agricultural land in Mexico. Problemas del Desarrollo 12: 59–85 (In Spanish).
13. SAGARPA. 2016. SIACON statistics database. Retrieved September 10, 2016, from https://www.gob.mx/siap/acciones-y-programas/produccion-agricola-33119 (In Spanish).
14. Bray, D.B., and P. Klepeis. 2005. Deforestation, forest transitions, and institutions for sustainability in Southeastern Mexico, 1900–2000. Environment and History 11: 195–223. https://doi.org/10.3197/096734005774434584 .
15. Phalan, B., Onial, M., Balmford, A. & Green, R. E. Reconciling food production and biodiversity conservation: land sharing and land sparing compared. Science 333, 1289–1291 (2011).
16. Balmford, A., Green, R. & Phalan, B. Land for food & land for nature? Daedalus 144, 57–75 (2015).

17. Hulme, M. F. et al. Conserving the birds of Uganda's banana-coffee arc: land sparing and land sharing compared. PLoS ONE 8, e54597 (2013).
18. Kamp, J. et al. Agricultural development and the conservation of avian biodiversity on the Eurasian steppes: a comparison of land-sparing and land-sharing approaches. J. Appl. Ecol. 52, 1578–1587 (2015).
19. Dotta, G., Phalan, B., Silva, T. W., Green, R. & Balmford, A. Assessing strategies to reconcile agriculture and bird conservation in the temperate grasslands of South America: grasslands conservation and agriculture. Conserv. Biol. 30, 618–627 (2016).
20. Williams, D. R. et al. Land-use strategies to balance livestock production, biodiversity conservation and carbon storage in Yucatán, Mexico. Glob. Change Biol. 23, 5260–5272 (2017).
21. Matson, P. A., Parton, W. J., Power, A. G. & Swift, M. J. Agricultural intensification and ecosystem properties. Science 277, 504–509 (1997).
22. Tilman, D., Cassman, K. G., Matson, P. A., Naylor, R. & Polasky, S. Agricultural sustainability and intensive production practices. Nature 418, 671–677 (2002).
23. Didham, R. K. et al. Agricultural intensification exacerbates spillover effects on soil biogeochemistry in adjacent forest remnants. PLoS ONE 10, e0116474 (2015).
24. Seufert, V. & Ramankutty, N. Many shades of gray – the context-dependent performance of organic agriculture. Sci. Adv. 3, e1602638 (2017).
25. Kirchmann, H., Bergström, L., Kätterer, T., Andrén, O. & Andersson, R. in Organic Crop Production – Ambitions and Limitations (eds Kirchmann, H. & Bergström, L.) 39–72 (Springer, Dordrecht, 2008).
26. Madhusudan, M. D. The global village: linkages between international coffee markets and grazing by livestock in a South Indian wildlife reserve. Conserv. Biol. 19, 411–420 (2005).
27. Gilroy, J. J. et al. Optimizing carbon storage and biodiversity protection in tropical agricultural landscapes. Glob. Change Biol. 20, 2162–2172 (2014).
28. Lamb, A. et al. The potential for land sparing to offset greenhouse gas emissions from agriculture. Nat. Clim. Change 6, 488–492 (2016).
29. Poore, J. & Nemecek, T. Reducing food's environmental impacts through producers and consumers. Science 360, 987–992 (2018).
30. Bravo, V. et al. Monitoring pesticide use and associated health hazards in Central America. J. Int. J. Occup. Environ. Heal. 173, 1077–3525 (2011).
31. Guinée, J. B., Heijungs, R. & Huppes, G. Economic allocation: examples and derived decision tree. Int. J. Life Cycle Assess. 9, 23–33 (2004).
32. Shang, Q. et al. Net annual global warming potential and greenhouse gas intensity in Chinese double rice-cropping systems: a 3-year field measurement in long-term fertilizer experiments. Glob. Change Biol. 17, 2196–2210 (2011).
33. Liu, Y. et al. Net global warming potential and greenhouse gas intensity from the double rice system with integrated soil–crop system management: a three-year field study. Atmos. Environ. 116, 92–101 (2015).
34. Chen, Z., Chen, F., Zhang, H. & Liu, S. Effects of nitrogen application rates on net annual global warming potential and greenhouse gas intensity in double-rice cropping systems of the Southern China. Environ. Sci. Pollut. Res. Int. 23, 24781–24795 (2016).
35. Yao, Z. et al. Improving rice production sustainability by reducing water demand and greenhouse gas emissions with biodegradable films. Sci. Rep. 7, 39855 (2017).
36. Bertram G (2013) Green border control: issues at the environment/economy border. Simon Terry Associates Research, Wellington
37. Costanza R et al (1997) The value of the world's ecosystem services and natural capital. Nature 387:253–260
38. Daily GC et al (1997) Ecosystem services: benefits supplied to human societies by natural ecosystems Issues. Ecology 2:1–16
39. Daly H (2005) Economics in a full world. Sci Am 293:100–107
40. Pretty JN et al (2000) An assessment of the total external costs of UK agriculture. Agric Syst 65:113–136

41. Statistics New Zealand (2013a) 2012 Agricultural census tables. Statistics New Zealand. http://www.stats.govt.nz/browse_for_stats/industry_sectors/agriculture-horticulture-forestry/2012-agricultural-census-tables.aspx
42. Stroombergen A, Schilling C, Ballingall J (2009) Economic modelling of New Zealand climate change policy. Report to Ministry for the Environment. NZIER and Infometrics, New Zealand. https://www.climatechange.govt.nz/emissions-trading-scheme/building/reports/economic-modelling/economic-modelling-of-new-zealand-climate-change-policy.pdf
43. Beckerman, S. (1987). Swidden in Amazonia and the Amazon rim. In Turner, B. L. and Brush, S. B. (eds.), Comparative Farming Systems, Guilford, New York, pp. 55–94.
44. Clay, J. W. (1988). Indigenous Peoples and Tropical Forests, Cultural Survival, Cambridge, Massachusetts.
45. Dufour, D. L. (1990). Use of Tropical Rainforests by Native Amazonians. Bioscience 40(9): 652–659.
46. Sponsel, L. E. (1992). The environmental history of Amazonia: Natural and human disturbances, and the ecological transition. In Steen, H. K. and Tucker, R. P. (eds.). Changing Tropical Forests: Historical Perspectives on Today's Challenges in Central & South America, Forest History Society/IUFRO Forest History Group, pp. 233–251.
47. Schwartzman, S., Moreira, A. and Nepstad, D. (2000). Rethinking Tropical Forest Conservation: Perils in Parks, Conservation Biology 14(5): 1351–1357.
48. Boserup, E. (1965). The Conditions of Agricultural Growth: The Economics of Agrarian Change Under Population Pressure, Allen & Unwin, London, UK.
49. Sirén, A. H. (2006). Natural Resources on Indigenous Peoples' Land in Amazonia: A Tragedy of the Commons? The International Journal of Sustainable Development and World Ecology.
50. Kaimowitz, D. and Angelsen, A. (1998). Economic Models of Tropical Deforestation: A Review, CIFOR. Bogor, Indonesia.
51. Mirza MMQ (1997) Hydrological changes in the Ganges system in Bangladesh in the post Farakka period. Hydrol Sci 42(5):613–631
52. Akanda AI (2010) Rethinking cop diversification under changing climate, hydrology and food habit in Bangladesh. J Agric Environ Int Dev 102(1&2):3–23
53. Schlosberg, D. (2007). Defining environmental justice: Theories, movements, and nature. New York: Oxford University Press.
54. Chakraborty, J. (2017). Focus on environmental justice: New directions in international research. Environmental Research Letters, 12(3), 1–4.
55. Holifield, R., Walker, G., & Chakraborty, J. (Eds.). (2018). The Routledge handbook of environmental justice. New York: Routledge.
56. Bullard, R. (1993). Confronting environmental racism: Voices from the grassroots. Boston: South End Press.
57. Fraser, N. (1998). Social justice in the age of identity politics: Redistribution, recognition, and participation. WZB Discussion Paper, Wissenschaftszentrum Berlin für Sozialforschung (WZB) (pp. 98–108). Berlin.
58. Pulido, L. (1996). Environmentalism and social justice: Two chicano struggles in the southwest. Tucson, AZ: University of Arizona Press.
59. Young, I. (1990). Justice and the politics of difference. Princeton, NJ: Princeton University Press.
60. White. (2008). Crimes against nature. Environmental criminology and ecological justice. Cullompton, Devon, UK: Willan Publishing.
61. Brannen, J. (2012). Expert voices. In S. E. Baker & R. Edwards (Eds.), How many qualitative interviews is enough. National Centre for Research Methods Review Discussion Paper (pp. 16–17). Available at http://eprints.ncrm.ac.uk/2273 .
62. Curran, D. (2016). Risk, power, and inequality in the 21st century. Basingstoke, Hampshire, UK: Palgrave Macmillan.
63. Curran, D. (2018). Environmental justice meets risk class: The relational distribution of environmental bads. Antipode, 50(2), 298–318.

64. Bustos, B., Folchi, M., & Fragkou, M. (2017). Coal mining on pastureland in Southern Chile; challenging recognition and participation as guarantees for environmental justice. Geoforum, 84, 292–304.
65. Faber, D. (2008). Capitalizing on environmental injustice. Lanham, MD: Rowman & Littlefield Publishers Inc.
66. Weltin, Meike; Hüttel, Silke Sustainable Intensification Farming as an Enabler for Farm Eco-Efficiency?. Environmental and Resource Economics (2022). https://doi.org/10.1007/s10640-022-00718-6
67. Tello, José; Garcillán, Pedro P.; Ezcurra, Exequiel How dietary transition changed land use in Mexico. Ambio (2020). https://doi.org/10.1007/s13280-020-01317-9
68. Balmford, Andrew; Amano, Tatsuya; Bartlett, Harriet; Chadwick, Dave; Collins, Adrian; Edwards, David; Field, Rob; Garnsworthy, Philip; Green, Rhys; Smith, Pete; Waters, Helen; Whitmore, Andrew; Broom, Donald M.; Chara, Julian; Finch, Tom; Garnett, Emma; Gathorne-Hardy, Alfred; Hernandez-Medrano, Juan; Herrero, Mario; Hua, Fangyuan; Latawiec, Agnieszka; Misselbrook, Tom; Phalan, Ben; Simmons, Benno I.; Takahashi, Taro; Vause, James; zu Ermgassen, Erasmus; Eisner, Rowan The environmental costs and benefits of high-yield farming. Nature Sustainability (2018). https://doi.org/10.1038/s41893-018-0138-5
69. Chakravorty, Ujjayant; Fisher, Donna K.; Umetsu, Chieko Environmental effects of intensification of agriculture: livestock production and regulation. Environmental Economics and Policy Studies (2007). https://doi.org/10.1007/bf03353963
70. Foote, Kyleisha J.; Joy, Michael K.; Death, Russell G. New Zealand Dairy Farming: Milking Our Environment for All Its Worth. Environmental Management (2015). https://doi.org/10.1007/s00267-015-0517-x
71. Sirén, Anders Henrik Population Growth and Land Use Intensification in a Subsistence-based Indigenous Community in the Amazon. Human Ecology (2007). https://doi.org/10.1007/s10745-006-9089-y
72. Akanda, M. Aminul Islam Seasonal and regional limits to growth of water-intensive crop farming in Bangladesh. Sustainable Water Resources Management (2018). https://doi.org/10.1007/s40899-018-0263-3
73. Gladkova, Ekaterina Farming Intensification and Environmental Justice in Northern Ireland. Critical Criminology (2020). https://doi.org/10.1007/s10612-020-09488-3

Chapter 2
Farmer Poverty in India—Drivers and Challenges

2.1 Introduction by Editor

Farmer poverty is a pressing issue that transcends geographical boundaries, affecting small and marginal farmers across the world including developed countries like United States and developing ones like India. This chapter delves into the challenges faced by farmers, that are not isolated to a single region but rather a global phenomenon. In the United States, for instance, small farmers often struggle with economic viability. Some of them pursue alternative sources of income to supplement their farm incomes (Hazell et al., 2010).

One of the primary drivers of farmer poverty is the rise in input costs, which erodes the profit margins of these farmers (Ahmad et al., 2020). As the costs of seeds, fertilizers, and other agricultural inputs continue to increase, small and marginal farmers find it increasingly difficult to maintain their livelihoods, often resorting to diversifying their efforts towards off-farm activities. Furthermore, the issue of low price realization is another significant challenge facing farmers globally. Farmers often struggle to receive fair prices for their produce, as they are frequently at the mercy of intermediaries and powerful market forces beyond their control. This dynamic is exacerbated by the increasing globalization of agricultural trade, which can expose small and marginal farmers to the volatility of international markets (Havinal, 2020).

Consider the Indian economy, which has been fairly buoyant over the last decade. After the COVID-19 pandemic, the economy has rebounded with an exponential growth rate of 9.1 percent in the 2021–22 financial year. However, the growth has hardly been equitable. The contribution of agriculture sector to the GDP has been on a steady decline since 1951. The nation produced 315.7 million tons of food grains in 2021–22, that ensured food sufficiency. However, over 190 million citizens are yet undernourished. Free foodgrains have been distributed to about 81.4 crore beneficiaries under the National Food Security Act (Ministry of Finance, 2023). Agriculture and allied sectors are the largest source of livelihoods in India. This sector employs over 44% of the total employed population, and 70 percent of rural

G. Nulkar (ed.), *The Economics of Sustainable Development*,
https://doi.org/10.1007/978-981-99-7379-8_2

households still depend primarily on agriculture for their livelihood. However, the contribution of this sector to the GDP is a little over 18% (Ministry of Agriculture & Farmers Welfare, 2023). This indicates where the bulk of poverty lies in India.

Nevertheless, the agricultural industry's contribution to the Gross Domestic Product (GDP) is not as worrisome as the distressing increase in farmer suicides. According to data from the National Crime Records Bureau, there has been a significant increase in the incidence of suicides during the mid-1990s. This topic has been the subject of extensive scholarly investigation, some of which has resulted in policy measures and governmental initiatives. However, the issue of farmer suicides appears to persist without resolution. Farmer poverty is influenced by a range of issues, including limited land holdings, fluctuations in prices, the presence of middlemen cartels, inadequate infrastructure, and the occurrence of droughts, among others. India's irrigation sector is one of the largest in the world in terms of irrigated area, volume of investment and number of large dams including under construction (Narayanamoorthy, 2022). However, they provide irrigation to less than 48 per cent of land under cultivation (Directorate of Economics and Statistics, 2020). The changing climate has brought about a rise in frequency of natural disasters. Cyclones and droughts are experienced throughout the year.

Despite the implementation of various governmental initiatives such as Minimum Support Prices for specific crops, interest subsidy, loan waivers, and the recent establishment of the National Agriculture Market (eNAM)—an electronic trading platform that links existing agricultural markets (APMC mandis) to establish a unified national market for agricultural commodities—the incidence of suicides in the agricultural sector remains persistent.

In order to facilitate India's progress, it is imperative to boost the financial condition of the nation's small and marginal farmers. This necessitates overcoming the influence of partisan politics and committing substantial investment in infrastructure. This chapter presents a compilation of scholarly studies pertaining to the phenomena of farmer suicides, working conditions experienced by farm labourers, occurrences of natural disasters, and the issue of ground water shortages. It is anticipated that readers would acquire an understanding of the various facets associated with the challenge of farmer poverty in India. We present studies from India that examine the profitability of small farming, drivers of farmer suicides, issues with tea plantation labour force, and impacts of natural disasters on farm incomes.

Agriculture is an economy that relies on renewable resources. For developing countries, this is the growth that holds a promise of reducing income inequality, and hence makes up a critical sector in sustainable economic growth.

References

Ahmad, M I., Oxley, L., & Ma, H. (2020, April 14). What Makes Farmers Exit Farming: A Case Study of Sindh Province, Pakistan. Multidisciplinary Digital Publishing Institute, 12(8), 3160–3160. https://doi.org/10.3390/su12083160.

Havinal, R. (2020, August 7). The Role and Potential of Information Technology in Agricultural Development. International Research Publication House, V9(07). https://doi.org/10.17577/ijertv9is070663.

Hazell, P., Poulton, C., Wiggins, S., & Dorward, A. (2010, October 1). The Future of Small Farms: Trajectories and Policy Priorities. Elsevier BV, 38(10), 1349–1361. https://doi.org/10.1016/j.worlddev.2009.06.012.

2.2 Machine Generated Summaries

Disclaimer: The summaries in this chapter were generated from Springer Nature publications using extractive AI auto-summarization: An extraction-based summarizer aims to identify the most important sentences of a text using an algorithm and uses those original sentences to create the auto-summary (unlike generative AI). As the constituted sentences are machine selected, they may not fully reflect the body of the work, so we strongly advise that the original content is read and cited. The auto generated summaries were curated by the editor to meet Springer Nature publication standards. To cite this content, please refer to the original papers.

Machine generated keywords: India, farmer, Indian, poverty, disaster, seed, household, wheat, rural, situation, rate, poor, variety, labour, region.

2.2.1 The Economic Situation of the Indian Farmers: Some Findings from the NSSO 59th Round [78]

This is a machine-generated summary of:

Bhalla, G. S.: The economic situation of the Indian farmers: some findings from the NSSO 59th round [78].

Published in: Journal of Social and Economic Development (2022).

Link to original: https://doi.org/10.1007/s40847-022-00201-4.

If you want to cite the papers, please refer to the original.

For technical reasons we could not place the page where the original quote is coming from.

Abstract-Summary

"Many aspects of Indian farming are discussed that include farmers' desire not to continue with farming, input usage scenario, adoption of risk mitigation strategies such as crop insurance, membership in cooperatives etc. Importantly this lecture provides an all India scenario as well as presents state level diversities."

Introduction

"This euphoria is also accompanied by the fact that despite high growth of the GDP and per capita income, the agricultural sector has experienced a significant deceleration in growth after the initiation of economic reforms in 1991."

"The growth rate of agricultural export accelerated after the devaluation of the rupee in 1991 and these grew rapidly till 1996–97."

"Most important, income terms of trade of agriculture have not shown any visible improvement after the introduction of economic reforms."

"This is serious matter since the entire rationale of economic liberalisation was to end discrimination against agriculture with a view to improving its terms of trade."

"It also comes out that employment situation in agriculture has deteriorated and the rate of growth of employment has visibly slowed down during 1993–94 to 199–00 as compared with 1987–88 to 1993–94."

Situation Assessment Survey of Farmers

"These are Consumption Expenditure of Farmer Households (495), Some Aspects of Farming (496), Income Expenditure and Productive Assets of Farmer Households (497), Indebtedness of farming households (498), and Access to Modern Technology for Farming (499)."

"In all the reports, farmer households have been defined as those households who were engaged in farming activities (which include cultivation of field or horticultural crops, growing of trees or plants such as rubber, cashew, coconut, pepper, coffee, tea, etc.; animal husbandry, bee keeping, vermiculture, sericulture etc. At the all-India level, 60% of the rural households belonged to farmer households."

"The highest percentage of farmer households of 77% was reported by UP and the lowest of 35% by Tamil Nadu."

"The reports throw light on the consumption expenditure of farmer households."

Some Aspects of Farming

"It studies eight aspects of farming namely: educational level of farmers; awareness of important technical and institutional developments in agriculture; farmers' liking for farming; farmers' membership of registered farmers' organisations and self-help groups; crop insurance; services obtained from farmers' cooperatives usual source of seeds; and practice of seed replacement."

"At the all-India level, only 4% of farmer households reported to have ever insured their crops."

"Almost 30% farmer households at the all-India level reported using veterinary services during the last kharif season, compared to 22% during the rabi season."

'At the al-India level, only 2% of farmers' households said that the facility was available, 52% reported that the facility was not available, and the rest 47% said that they did not know about it."

Land Used for Different Kinds of Farming

"The use of land for orchards and plantations, dairy, goats etc., poultry and fishery and bee keeping is very small indeed."

"The share of farmed land under fishery was highest in West Bengal (between 2.5 and 3%), followed by Andhra Pradesh (nearly 2% reported during rabi season), and then Assam."

"The share of "other farming", which includes farming of goats, sheep, pigs, poultry, and all other animal farming except dairy, fishery, sericulture and vermiculture, is much higher for Scheduled Tribes than for the other social groups."

"The size class " < 0.01 hectare", which normally means "homestead land only", devoted 68–69% of farmed land to dairy farming and only about 14% to cultivation."

"Of all irrigated lands, 50% during the kharif season and 60% during the rabi season were irrigated by tubewells."

Energy Used for Farming and Other Activities

"As far as sources of energy for various operations are concerned, it comes out that for most of the operations bullock power is being replaced by diesel or electricity."

"The agricultural advanced states use more of diesel or electricity for most operations and the poorer states still are relatively more dependent on bullock power."

"At the all-India level, 47% of farmer households reporting energy used for ploughing during the last 365 days used diesel tractors for ploughing, while 52% used animals."

"Seventy per cent of households in this class reported the use of diesel tractors."

"Diesel and electricity have also emerged as important sources of energy for irrigation."

Harvesting

"Almost 59% of households reporting use of energy for harvesting crops during the last 365 days used diesel-powered machines, 38% used animal power and only 3% used electric power."

"There was a direct relationship between the size of holding and the frequency of use of diesel power for harvesting."

"The incidence of use of diesel power increased from 47% in the lowest size class to 65% in the class 4–10 hectares, and to 80% in the class "10.01 + "."

"The use of diesel-powered threshers, too, increased with land possessed; from 51% for the class " < 0.01" it reached 66–67% in the highest two size classes—"4–10 ha" and "10.01 + "."

Access to Modern Technology: Sources of Information

"The state wise data shows that in Gujarat the main sources for accessing information are input dealers, other progressive farmers and extension workers, in that order."

"Percentage of farmer households accessing information through 'other progressive farmers' was highest in Andhra Pradesh (34%), followed by Gujarat (30%) and West Bengal (25%)."

"Percentage of farmer households accessing information through 'input dealers' was highest in West Bengal (36%), followed by Andhra Pradesh (30%) and Gujarat (24%)."

"The 'improvement in quality and reliability of information' was the primary concern to the farmer households for different sources."

"Farmer households were asked to give suggestions for improvement in extension services for different sources of information accessed by them."

"The 'improvement in quality and reliability of information' was the primary concern to the farmer households for different sources available to them."

Consumption and Poverty Among Farmer Households

"It is possible to use these two sources to obtain comparative results for total rural households and for farmer households."

"Poverty estimates in India are based on the Household Consumer Expenditure Surveys conducted by the NSSO."

"The SAS data also give details about monthly state wise monthly per capita expenditure (MPCE) of persons belonging to farmer households over MPCE classes for all the states of India."

"A comparison with the results for 1999–00 shows that the incidence of poverty for all rural households declined from 27.09 to 23.99% but for farmer households the incidence was higher at 30.73%."

"It comes out that in all the major states of India, the incidence of poverty among farmer household is significantly higher than in all rural households."

"Among states, the lowest incidence of poverty among farmer households was in Jammu and Kashmir and Punjab."

2.2.2 *"Who Will Stand Up for Us?" the Social Determinants of Health of Women Tea Plantation Workers in India [79]*

This is a machine-generated summary of:

Rajbangshi, Preety R.; Nambiar, Devaki: "Who will stand up for us?" the social determinants of health of women tea plantation workers in India [79].

Published in: International Journal for Equity in Health (2020).

Link to original: https://doi.org/10.1186/s12939-020-1147-3.

If you want to cite the papers, please refer to the original.

For technical reasons we could not place the page where the original quote is coming from.

Abstract-Summary

"Workers are dependent on plantation estates for a range of basic services under the 1951 Plantation Labour Act and have been subject to human rights violations."

"Ad hoc reports related to poor health outcomes exist, yet their determinants have not been systematically studied."

"This study in Assam, situated in Northeast India, sought to understand the Social Determinants of Health (SDH) of women plantation workers with an aim to offer directions for policy action."

"As part of a larger qualitative study, 16 FGDs were carried out with women workers in three plantations of Jorhat district covering permanent and non-permanent workers."

"Determinants at structural, intermediary and individual levels were associated with health."

"Poverty and poor labour conditions, compounded by the low social position of women in their communities, precluded their ability to improve their economic situation."

"Health services were found wanting and social networks were strained even as women were a critical support to each other."

"These factors impinged on use of health services, diet and nutrition as well as psychosocial stress at the individual level."

"Notwithstanding this, to have heard from women workers themselves has been an important step in visibilizing and building accountability for action on the health and SDH of women in plantation estates."

Background

"While poor conditions of plantation workers and non-compliance of PLA in Assam have been documented by several national and international organizations [1–3], far less is known about the health and well-being of women workers."

"What is known is that more than 50% of the labour force in plantation estates is women as it is believed that "soft hands and nimble fingers of women" are suited for plucking tea leaves [4]."

"Knowledge of the high burden of maternal deaths (237 per 100,000 live births in Assam as against 130 per 100,000 live births national) [5] in the state motivated the health department to intervene and provide health services in plantation estates."

"There remains a broader lack of understanding of women worker's conditions and their lives and rights – their socioeconomic position, conditions of work, access to health and education - the Social Determinants of Health (SDH)."

Methodology

"The selection of tea plantation estates was purposive and based on consultation with district health officials: one government tea plantation and two private sector plantation estates were chosen."

"The participant selection strategy for focus group discussions (FGDs) was developed by a team of researchers and involved criterion sampling designed to reflect particular working conditions and entitlements."

"Sub-groups were selected based on years of service in tea plantations."

"The following groups were formed (a) women who had worked in the tea plantations for less than a year; (b) women who had worked in the tea plantation for 1–5 years; and (c) women who had worked in the tea plantation for over 5 years."

"On acceptance to participate, groups were formed based on our selection criteria."

"Each recorded group discussion and interview was transcribed by a professional transcriber to obtain verbatim transcripts."

Results

"Poverty was linked to a clear asymmetry in the knowledge and social position of women workers relative to others in their family, community, and work, across plantation types."

"Women from government plantations did not report a membership of or leadership in worker unions."

"Women workers, irrespective of the type of contract in private plantations, reported that they were entitled to receive a total of 6 kg ration - 3 kg rice and 3 kg wheat flour after 12 working days."

"Women also noted that most non- permanent workers in plantations did not have a ration card."

"Permanent workers mentioned that husbands of women workers were not entitled while wives and children of male permanent workers had access to health services in plantations."

"In plantations, women would delay seeking health care because cost of being sick was substantial: a day's wage for non-permanent workers and half a day wage for permanent workers."

Discussion

"In the plantation sector in Munnar, Kerala, poor working conditions and emoluments were among the root causes for a plantation strike in 2015 by women workers whose demands were ignored by the local trade union [6]."

"These approaches to address women's health in plantations are typically piecemeal, episodic, and do not catalyse the agency of workers themselves, and thus are neither taking a holistic approach to act on social determinants of health (SDH) nor one that can cumulatively do so."

"This study revealed that the conditions in which women workers are working and living cannot be ignored if we truly want to improve their health."

"To improve women's health in plantations, action on the SDH must be taken."

"The government must enforce PLA provisions requiring tea plantation estates to improve the daily living and working condition of workers."

Conclusion

"This paper, focusing on structural, intermediary and individual SDH has shown the range of factors that require action to improve the health of women plantation workers in Assam."

"To have heard from women themselves has been an important step in visibilizing and building accountability for action on the health and social determinants of health of women workers in tea plantation estates in Assam."

"There is a need to respond to the question posed by these workers that: who will stand up for them?".

2.2.3 Deconstructing Indian Cotton: Weather, Yields, and Suicides [80]

This is a machine-generated summary of:

Gutierrez, Andrew Paul; Ponti, Luigi; Herren, Hans R; Baumgärtner, Johann; Kenmore, Peter E: Deconstructing Indian cotton: weather, yields, and suicides [80].

Published in: Environmental Sciences Europe (2015).

Link to original: https://doi.org/10.1186/s12302-015-0043-8.

If you want to cite the papers, please refer to the original.

For technical reasons we could not place the page where the original quote is coming from.

Abstract-Summary

"Hybrid cotton was introduced in the 1970s with increases in fertilizer and in insecticide use against pink bollworm that caused outbreaks of bollworm."

"Hybrid Bt cotton, introduced in 2002 to control bollworm and other lepidopteran pests, is grown on more than 90% of the cotton area."

"Biological modeling of the pre-1970s cotton/pink bollworm system was used to examine the need for Bt cotton, conditions for its economic viability, and linkage to farmer suicides."

"Use of Bt cotton seed and insecticide in rainfed cotton is questionable."

"Bt cotton may be economic in irrigated cotton, whereas costs of Bt seed and insecticide increase the risk of farmer bankruptcy in low-yield rainfed cotton."

"Annual suicide rates in rainfed areas are inversely related to farm size and yield, and directly related to increases in Bt cotton adoption (i.e., costs)."

"High-density short-season cottons could increase yields and reduce input costs in irrigated and rainfed cotton."

Background

"In India, before the widespread use of insecticide, the key pest of cotton was the cosmopolitan stenophagous semi-tropical pink bollworm (Pectinophora gossypiella Saunders; hereafter pink bollworm (PBW))) and also generalists such as jassids (Amrasca biguttula Ishida), spotted bollworm (Earias vitella Fabricius), and defoliators such as Spodoptera litura (Fabricius) [7, 8]."

"Before the introduction of Bt cotton, insecticide was used to control pink bollworm, and as has occurred worldwide wherever PBW infests cotton, insecticide use caused ecological disruption due to the destruction of generalist natural enemies resulting in the resurgence of PBW, outbreaks of formerly secondary pests, insecticide resistance, and adverse ecological and human health effects [9, 10]."

"Prior to the 1970s, when high insecticide use began, bollworm and whitefly were minor pests in Indian cotton, but by 2002, when Bt cotton was introduced, 75% of insecticide use in cotton was for bollworm control [11]."

"The inescapable conclusion is that Bt cotton was introduced to India to solve a bollworm problem created by insecticide use [8, 11]."

Results

"As occurs in other areas of the world, insecticide use to control PBW in India induces outbreaks of bollworms, defoliators, whiteflies, and other pests [12–14]."

"Predicted yields and lack of sizeable PBW populations (not shown) accord qualitatively with long-run field studies on rainfed cotton at Nagpur, MH, roughly 100 km to the NW of Yavatmal [15, 16]."

"We first assess rainfed cotton production across the agro-ecological zones of India and then focus on the states of AP, GJ, KA, and MH."

"As a metric of weather-related risk, we define high-risk areas as having average yields < 500 kg ha^{-1} and CV $> 50\%$."

"Using this metric, GJ has relatively few high-risk areas, KA has a large number, and AP and MH are intermediate."

"The relative level of risk for each state (i.e., GJ.

Discussion

"Insecticides continue to be used in Bt cotton in India, and as in China productivity effects of Bt cotton and pesticide use likely depend on the action of natural control agents with the profitability of damage control measures increasing with the severity of ecosystem disruption due to insecticides [17]."

"In India and Pakistan, sucking insect pests had been of minor concern in cotton but are now increasing in Bt cotton and contribute to yield losses and increased insecticide use [18, 19]."

"Absent resistance to Bt toxin [20], quality Bt cotton provides excellent control of PBW [21] and good control of bollworm and would appear to have a role in high-yielding irrigated long-season cotton in India where PBW is a chronic problem and where the marginal cost of the technology is low."

Conclusions

"In the light of this more thorough holistic agro-ecological analysis, recommendations by international agricultural economists [22, 23] and some Indian government official to implement the Bt technology in cotton (and other crops) as a solution for pest problems become questionable."

"When insecticides became available for control of PBW and other pests, outbreaks of highly damaging bollworms (and other pests) were induced, and F1 Bt hybrid cotton that discouraged seed saving was introduced for control."

"Suicides in rainfed areas of south-central India are inversely related to farm size and yields and directly related to area of Bt cotton adoption, or more likely the combined high costs of Bt seed and insecticide."

"Short-season high-density cotton is a viable solution in both irrigated and rainfed cotton reducing the need for the Bt technology."

Methods

"Physiologically based demographic models (PBDM) of cotton and PBW (see [24–26]) are used to explain the underlying effects of weather and planting density on yield and PBW dynamics at the local, state, and national levels under irrigated and rainfed conditions prior to the introduction of insecticides and Bt cotton."

"The effects of one and two toxin Bt cottons on pest dynamics, crop damage, and development of resistance in PBW and other cotton pests have been modeled [24, 27] but were not required in this analysis The landscape of India, excluding Andaman and Nicobar Islands, was divided into 2855 lattice cells of 38 × 38 km."

"Using the weather data to run the model, simulations were made for all of lattice cells of India, but smaller time- and geographic-scale studies were also made."

"The cotton-PBW model was assumed equilibrating during the first year, and hence, the data were not used in computing the summary statistics."

2.2.4 Farmer-Suicide in India: Debating the Role of Biotechnology [81]

This is a machine-generated summary of:

Thomas, Gigesh; De Tavernier, Johan: Farmer-suicide in India: debating the role of biotechnology [81].

Published in: Life Sciences, Society and Policy (2017).

Link to original: https://doi.org/10.1186/s40504-017-0052-z.

If you want to cite the papers, please refer to the original.

For technical reasons we could not place the page where the original quote is coming from.

Abstract-Summary

"Indian Biotech opponents have attributed the increase of suicides to the monopolization of GM seeds, centering on patent control, application of terminator technology, marketing strategy, and increased production costs."

"The contentions of the biotech opponents, however, have been criticized for a lack of transparency in their modus operandi i.e. the use of methodology in their argumentation."

"This article will review and evaluate the multifaceted contentions of both opponents and defenders."

"The underlying thesis of biotech opponents becomes all the more significant when analysed vis-à-vis the contention of the globalization critics that there has been a political and economic marginalization of the Indian farmers."

Introduction

"India has witnessed around 300,000 farmer-suicides over the past two decades (Philpott, 28; Mishra, 29)."

"Biotech opponents attribute the majority of these suicides to the monopolization of the cotton seed sector."

"Holding transgenic seeds responsible, they dubbed them "Seeds of Suicide, Seeds of Slavery, and Seeds of Despair" (Shiva and Jalees, 30, p. 281; Shiva, 31)."

"The Acharya report attributes the increase of farmer-suicides after 2002 to Bt."

"We intend to critically evaluate the pros and cons, focusing on the veracity of the alleged link between GM cotton cultivation and farmer-suicide."

"Before we start doing this, we describe a succinct history of GM cotton and the opposition against it, and evaluate the significance of the contentions of the biotech opponents by positioning their criticism within the larger debate on the deleterious impact of economic liberalization on Indian farmers."

A Succinct Chronology of Bt. Cotton in India

"Monsanto requested authorities to conduct field trials."

"In 1995, Mahyco was permitted to conduct field trials with local varieties backcrossed with Monsanto's imported 100 g gene."

"The Genetic Engineering Approval Committee (GEAC) of the Department of Environment, Forest and Wild Life (DOE) - India's highest authority on large-scale experimentation and commercialization – was not satisfied with the DBT's observations on bio-safety because the trial was allegedly conducted during times of low pest presence; therefore, on 19 June 2001, it ordered a repetition of the field trials."

"This "proprinquinity" between Mahyco and the regulatory agencies engendered skepticism about the effectiveness of GEAC's monitoring of MMB's field trials (Scoones, 32) - a good example of Quis Custodiet Ipsos Custodes?"

"This public outrage coincided with MMB 's field trials, and the failure of that cotton crop aggravated the public's fear of allowing terminator technology in India."

The Campaign Against gm Cotton

"Media drama, the KRRS also demanded a 5-years ban on GM cotton seeds (Scoones, 33)."

"The longest opposition came from Vandana Shiva and her associations, Research Foundation for Science and Technology (RFSTE) and Navdanya, through activist writings, public interest litigations (PIL) and agitation-events, e.g., seed tribunals attended by farmers from Europe, the U.S. and India (Scoones, 32).'

"Ever since Shiva's PIL, courts were inundated with similar cases filed under the Freedom of Information Act, like the petition to release trial results in Delhi alleging the Gujarat government's connivance with Mahyco-Monsanto, and the petition in Mumbai demanding compensation for crop loss under the Plant Variety and Protection Act by farmers' movements in Vidarbha (Scoones, 32)."

"When Monsanto advertised the benefits of GM cotton, Shiva's Navdanya and other anti-GM organizations launched the Monsanto Quit India Movement to raise public awareness on the negativities of GM crops."

The Seed Monopolisation Debate

"Shiva asserts the driving force behind farmer-suicide swirls around Monsanto's monopolization of the cotton seed sector by patent control, removing alternatives, the increase in production costs, and the fear for crop failure."

"Farmers have replicated the transgenic seeds because, unlike the hybrid seeds produced with two distinct parental lines, these Bt."

"The impact of other factors together with higher prices farmers must pay for the transgenic seeds can put farmers in a more precarious position than their non-Bt."

"While MMB's Bollgard cotton seeds were advertised as resisting bollworms, discouraging farmers from using pesticides, the opposite seems to be true because bollworms have now become Bt."

"The average expense for 3 years on seeds and pesticides/acre for Bt was ₹1557 and ₹2571 respectively for MMB Bt cotton vs ₹2571 and ₹2776 for non-Bt farmers."

"With the rapid adoption of transgenic cotton by farmers, seed firms frantically attempted to collaborate with MMB (Ramaswamy and others 34, 34)."

Is There a Connection Between Monsanto's bt. Seed and Farmer-Suicide?

"Srijit Mishra's study on this topic, between 1995 and 2001, shows farmer-suicide rates in Andhra Pradesh, Gujarat, Haryana, Punjab, Tamilnadu, Karnataka and Rajasthan to be unstable, while increasing in Maharashtra and Madhya Pradesh."

"His study shows that the suicide rate for male farmers increased from 1995 to 2001, i.e., before the Bt."

"Ian Plewis' calculation of the suicide rate from 1996 to 2011 in the nine cotton producing states shows a constant suicide rate in Madhya Pradesh, Karnataka and Tamilnadu, a varying rate in Rajasthan, a steady rise in Andhra Pradesh and Haryana, a reduction in Gujarat until 2004 with an increase in Punjab and an increase in Maharashtra until 2005 followed by a reduction."

"Fifthly, the non-cotton cultivating state Kerala is also infamous for a high number of farmer-suicides (Sadanandan, 35; Münster, 36)."

"This demonstrates there is no necessary link between the cultivation of GM cotton and farmer-suicide."

What Is the Relevance of Critics' Claims Abouttransgenic Technology?

"The underlying thesis of the globalization critics and the biotechnology opponents is the same: India's agrarian crisis and farmer-suicides are revealing the political and economic marginalization of the agrarian sector (Vasavi, 37)."

"Anoop Sadanandan's study shows that 1/3 of agricultural land is irrigated, illustrating the excessive dependence of Indian farmers on rainfall (Sadanandan, 35)."

"Instead of focusing on matters significant to farmers like resource reallocation, strategies to ensure a sufficient price for their products, and crisis intervention, governments accorded increasing emphasis to India's industrialization, thus aggravating the marginalization of the India's farmers (Vasavi, 37)."

"After several suicides in January 2011 in Madhya Pradesh, the state's agriculture minister downplayed the distress as being a consequence of the farmers' past shortcomings of destroying land fertility via excessive chemical application (Vasavi, 2012)."

Conclusions

"The chief components of the allegations by the biotech opponents on the link between MMB's seed monopolisation and farmer-suicide are: the firm's thwarting of seed-saving and exchange via patency and use of terminator technology; the clever destruction of farmers' seed variety; the increased production cost; the insufficient crop-yield and financial returns."

"Irrigation presents added problem for those farmers who must invest heavily in seeds and pesticides."

"cotton is not disputed, the national and inter-state data do not support transgenic cotton as an overall boon to the Indian farmers."

"farmer facing an unfavourable expenditure-yield ratio can find himself in a more precarious position than a non-Bt."

"cotton notwithstanding, Shiva's direct link between MMB's seed monopolisation and farmer-suicide stand tall."

"Despite the high number of farmer-suicides, India's governments failed to sufficiently address the crisis, attributing these suicides to personal weaknesses vs. systemic problems."

2.2.5 *Understanding Farmers' Suicidal Ideation: A Structural Equation Modeling Study in Maharashtra, India [82]*

This is a machine-generated summary of:

Swami, Deepika; Dave, Prashant; Parthasarathy, Devanathan: Understanding farmers' suicidal ideation: a structural equation modeling study in Maharashtra, India [82].

Published in: Climatic Change (2020).

Link to original: https://doi.org/10.1007/s10584-020-02935-8.

If you want to cite the papers, please refer to the original.

For technical reasons we could not place the page where the original quote is coming from.

Abstract-Summary

"India is witnessing an increasing rate of farmers' suicide, which has been attributed to indebtedness by most of the studies."

"Indebtedness can be attributed to multiple factors, including individual, household, community, climatic, institutional, and market-oriented factors."

"The relationship of these adaptation measures toward indebtedness and farmers' suicide was found to be missing in the existing literature, which would otherwise provide valuable information regarding how to deal with farmers' suicide."

"Considering the compounding effect of these factors on farmers' suicide, we evaluated and quantified the relationship of individual, community, climatic, credit, and market-oriented factors with adaptation and indebtedness, which were subsequently related to suicidal ideation of farmers."

"SEM results show that individual traits, i.e., age, experience, and illness of farmers; community-oriented factors, i.e., provision of lakes, wells, water tanks; credit factor, i.e., moneylenders; perception of climate change; and market factors are the primary factors contributing toward farmers' suicidal ideation."

"Adaptation was found to play a vital role in reducing the indebtedness and suicidal thoughts of the farming community."

Introduction

"There are multiple other factors such as socio-economic condition, indebtedness, availability of credits, climatic, market, etc. These are either derived or dependent upon the above-mentioned factors, which also play a highly significant role in increasing the rate of farmers' suicide in India."

"For the Vidarbha region of Maharashtra (India), indebtedness and economic factors such as unavailability of basic amenities and assets including toilets, drinking water, mobile phones, and lack of transportation facilities were considered the leading causes of farmers' suicide (Mishra 38)."

"This study is the first of its kind which comprehensively evaluates the effect of multiple stressors such as individual characteristics, i.e., age, illness; economic status, i.e., income, provision of assets; community resources, i.e., wells, lakes, water tank; climatic, i.e., drought, floods, heat waves, storms; crop failure, institutional schemes, indebtedness, and adaptation factors on farmers' suicidal tendency for India using structural equation modeling (Bollen 39)."

Methodology

"Only, a few farmers (30–35 farmers out of 400) could answer regarding the measures of suicidal ideation, duration and intensity of suicidal thoughts, and thereby owing to this limitation during the current study, we could only use one measure of this variable, i.e., "whether they are getting any recurring suicidal thoughts from past 5 years or not", which was responded by all 400 farmers in the study region."

"All the adaptation variables used in the model are coded as 0 and 1, i.e., whether farmers adopted that particular adaptation measure (coded as 1) or not (coded as 0)."

"Our variable of interest, i.e., suicidal ideation is a binary variable (0/1) where zero indicates when the farmer has never thought of committing suicide in the past five years and one when the farmer continuously thinks about committing suicide."

Descriptive Statistics

"Most of the farmers are young, i.e., 26 to 35 years old (~44%), and ~ 36% are 36 to 50 years."

"Our sample is skewed toward small and marginal farmers, and most of them are 26 to 50 years old."

"Basic amenities consist of toilet, drinking water, electricity, house, etc. Approximately 85% of farmers possess a higher number of basic amenities, indicating that most of the farmers have better conditions concerning assets and basic amenities."

"Water tanker cannot be used for irrigation purposes because farmers hardly have the money to buy it for drinking purposes."

"We found that at least two strategies, i.e., organic farming and use of fertilizers were followed by almost all the farmers."

"This indicates that only very few farmers are following all eight adaptation strategies."

Results and Discussion

"The primary purpose of this study was to evaluate the comprehensive effect of adaptation, indebtedness, and other variables on suicidal ideation of farmers."

"The results of the study highlight that the latent variables such as credit and market factors reduce the likelihood of adopting adaptation measures, while the increase in age, experience in farming, provision of assets, community oriented resources, and perception regarding climate change increase the likelihood of adopting adaptation measures by the farming community in Maharashtra, India."

"Indebtedness is also affected by multiple other factors such as age, experience, income, education of farmers, etc. Therefore, the next sub-section will focus on understanding the effect of adaptation and other multiple stressors on indebtedness."

"This is because adaptation leads to an increase in crop productivity (Alauddin and Sarker 40; Below and others 41; Deressa and others 42) that helps to generate more income to the farmers, which reduces the likelihood of indebtedness among them."

Conclusions

"The present work identified the influence of individual, assets, community, climatic, credit, and market oriented factors on suicidal ideation of farmers via quantifying their impacts on adaptation and indebtedness."

"Findings reveal that opting for multiple adaptation strategies eases the indebtedness, resulting in reduced suicidal thoughts among farmers."

"Apart from indebtedness and adaptation, age, income, community oriented resources, and perception regarding climate change play a vital role in farmers' suicidal ideation."

"Our study found that indebtedness is a highly significant factor for farmers' suicidal ideation."

"Market oriented factors, i.e., crops sold at lower price, buying high yielding variety, and traditional seeds lead to a higher rate of indebtedness among farmers that aggravates their suicidal thoughts."

"The government should buy the crops directly from farmers along with timely payment to them, which will motivate them toward adaptation along with reducing their indebtedness."

2.2.6 The Potential of Small Ruminant Farming as a Means of Poverty Alleviation in Rural Southern India [83]

This is a machine-generated summary of:

Lalljee, Sehr Vanya; Soundararajan, C.; Singh, Yogendra Deep; Sargison, Neil D.: The potential of small ruminant farming as a means of poverty alleviation in rural southern India [83].

Published in: Tropical Animal Health and Production (2018).

Link to original: https://doi.org/10.1007/s11250-018-1686-4.

If you want to cite the papers, please refer to the original.

For technical reasons we could not place the page where the original quote is coming from.

Abstract-Summary

"Twenty-six small ruminant landless farmers in three villages in the Kanchipuram District of the state of Tamil Nadu were surveyed with regard to their sheep farming practice and production indices, with the preliminary aim of evaluating the potential of small ruminant farming in alleviating poverty in parts of rural in southern India."

"The small ruminant farmers reared mostly indigenous Madras Red sheep as a means of generating primary or supplementary income."

"When calculating the annual farm profits without taking into consideration the opportunity cost of labour, 83% of annual sheep production cycles over a 4-year period added to household incomes."

"Further, 23% of the instances that accounted for the opportunity cost of labour, household income was raised above the Indian Government's defined poverty line solely through small ruminant farming."

"While small ruminant farming generated income in most instances, the way it is practiced creates opportunities for simple changes in husbandry and management that could make it more efficient in poverty alleviation."

Introduction

"There is a growing population of landless farmers in southern Indian state of Tamil Nadu (TN), who use small ruminants as a means of generating of income."

"Small ruminants have played a role in providing income and dietary protein to poor, small-scale producers and their families (Iniguez 43)."

"Global markets for small ruminant products have expanded due to a growing urbanised human population, creating opportunities for landless producers in emerging agricultural economies (Kosgey and others 44)."

"The aim of this study was to undertake a preliminary evaluation of the efficacy of small ruminant farming as a means of alleviating poverty through income generation in landless households in part of the Kanchipuram District, TN."

Materials and Methods

"The cost effectiveness of small ruminant farming was determined using expenditure and returns to calculate annual flock profits."

"The purchase cost of the animals was allocated as an expense in the year of the instance."

"Only those costs associated with new introduction of animals were included in our calculations of profitability over the 4-year period."

"To further investigate the viability of small ruminant farming, we modelled farm profitability using median values for the key variables such as numbers for ewes and rams purchased in the first year of operations."

"Expenses in the initial year were calculated as the purchase cost of the animals, herd maintenance costs and the opportunity cost of labour."

"For these models, the purchase cost of the animals was expensed in the year of acquisition due to a lack of unambiguous data on longevity."

Results

"Comparison of variable costs, including the opportunity cost of labour and miscellaneous expenses, against the returns received from the sale of lambs, old ewes and rams showed that 58 of 103 (56%) of the instances over a 4-year period across the 26 farmers surveyed made a profit and thus added to household income."

"Including the opportunity cost of labour in the profit calculations, 35 of the 45 loss-making instances were in the production cycle years of 2015 and 2016, during and 1 year after the period when the flooding occurred."

"The landless farmers could break even within the first year of starting small ruminant farming with a flock size of 262 ewes and the best observed values for birth rates, lamb losses, ewe losses and selling prices."

Discussion

"Equally, all of the data used in the study were primary, and the credibility of information was supported by the participatory manner in which it was collected, involving our visiting and observing the landless farmers as they engaged with their sheep flock routines."

"When the opportunity cost of labour was removed from our calculations, 25 of the 26 sheep flocks generated a net profit over a 4-year study period, further demonstrating the potential of small ruminant farming in poverty alleviation."

"Our study shows the intelligence and agribusiness acumen of TN small ruminant landless farmers and implies a willingness to improve the production efficiency and profitability of their sheep flocks."

"The management practices and production indices identified when interviewing farmers in the Kanchipuram District provide a pragmatic starting point for engagement in planned animal health management and education to improve the productivity of their sheep flocks (Sargison 45)."

2.2.7 Agrarian Crisis in Wheat-Producing Region of India: Analysis of Profitability and Costs [84]

This is a machine-generated summary of:

Jain, Ankur; Sheekha, Neela Madhaba: Agrarian crisis in wheat-producing region of India: analysis of profitability and costs [84].

Published in: Journal of Social and Economic Development (2022).

Link to original: https://doi.org/10.1007/s40847-022-00186-0.

If you want to cite the papers, please refer to the original.

For technical reasons we could not place the page where the original quote is coming from.

Abstract-Summary

"It has resulted in many farmers committing suicides and leaving farming as a profession as agriculture has become less profitable."

"Under this context, the present study examines the trends in profitability of wheat cultivation in major wheat-growing states of India."

"There is intelligible relationship between wheat profitability and farmers committing suicides in all the selected states except Punjab where other socio-economic and psychological factors are important."

"There is an urgent need to increase public spending in agriculture and farmers should be given incentives for ensuring agriculture profitability to farmers as well as sustainability of food security in India besides removing the agrarian distress of crop failure and as well as realizing the goal of doubling of farmers' income."

Introduction

"The performance of trend in the productivity growth of wheat is overwhelming; the contribution of agriculture to the Gross Value Added (GVA) was reported 16.5 per cent in 2019–20 which was 16.1 per cent in 2018–19 (Government of India 46–20)."

"Dev and Rao (47) examined the effectiveness of agricultural price policies, for rice and wheat crops, in ensuring profitability and food security to the farmers by studying rice and wheat prices and found the similar trends in the significant growth rate in the yield of wheat crop in Madhya Pradesh, Punjab and all India from 1981–82 to 2006–07."

"Under this context, the paper aims to examine the trends in profitability of wheat cultivation in major wheat-growing states of India in the groaning of rising cost of cultivation and inadequate profit from crop cultivation and agricultural crisis."

Methodology

"India has experienced crisis in such as decline in rate of growth, rising number of farmer's suicides, declining prices of several crops and a widening gap between the agriculture and non-agriculture sector."

"This paper studies crisis in agriculture sector particularly wheat-producing region during 1970–71 to 2016–17 by dividing the whole study period of 47 years in Agrarian Crisis period Pre1990s and Agrarian Crisis period post-1990s."

"The study employs secondary data pertaining to farmers suicides since 1995 was collected from National Crime Record Bureau (NCRB) whereas the wheat crop statistics, i.e. production, area and cost of cultivation for the studies year is collected from Commission for Agriculture Cost and Prices (CACP), Ministry of Agriculture and Farmers Welfare (MOFAW)."

"The profitability of wheat is calculated by taking the difference between value of output of wheat and cost C2 or C3 in respective years."

Agrarian Crisis in India

"Rural India has faced agrarian distress that results in farmer suicides in many states."

"It is also observed that introduction of Agricultural Debt Waiver and Debt Relief Scheme of 2008 reduced the numbers of farmer suicides in the state."

"Narayanamoorthy (48) reported widespread indebtedness due to low margin in agricultural production which is reported as the main cause of farmer suicides in the country."

"Mishra (49) pointed out that farmers commit suicide for poor returns and absence of non-farm opportunities due to multiple risks in terms of low yield, price fluctuation, input, technology and credit in the agricultural production."

"The number of farmer suicide has fallen since 2000 except in Punjab state."

"The crop failure and inability to pay farm equipment loans are found to be the two major causes of farmer's suicide."

Growth Rate and Profitability Analysis

"If cost C3 is considered for profitability calculation then farmers in Punjab accrue losses in wheat cultivation during 1970–71 to 1990–91."

"In the agrarian crisis post-1990s, 2 times losses accrued to wheat farmers in Haryana, 1 time in Punjab and 3 times in Uttar Pradesh on the basis of cost C2."

"The profitability of wheat in Haryana during the whole period under study has increased when cost C2 is considered, and reverse happened in the Madhya Pradesh and Uttar Pradesh wheat farmers."

"Further, the study infers that taking cost C2 as the basis of profitability in the entire period, the wheat cultivation was profitable 80 per cent times in Haryana, more than 70 per cent in Punjab, more than 90 per cent times in Rajasthan, more than 45 per cent in Uttar Pradesh and while in case of Madhya Pradesh farmers accrued losses in wheat cultivation in all the years taken into consideration."

Conclusions and Policy Suggestions

"In Haryana and Rajasthan, wheat profitability is 66.66 per cent and 100 per cent, respectively, taking cost C2 post-1990s and there is decline in farmer committing suicides which shows clear relationship between profitability and farmer suicides as with increase in profitability there is a decline in farmers committing suicides."

"In Uttar Pradesh, the profitability of wheat is 50 per cent and there is decline in farmer's suicides marginally."

"In Madhya Pradesh, farmers are accruing losses 100 per cent and there is also increase in farmer suicides which also shows indirect relationship between profitability in wheat cultivation and farmer's suicides."

"In Punjab, the profitability in wheat cultivation is 83.33 per cent post-1990s along with increase in farmer's committing suicides."

"The new research may also use primary survey to estimate the number of farmer's suicide at all India level and attempt to forecast the profitability of wheat crop as well as other crops amid rising cost of cultivation."

2.2.8 Unravelling the Association Between the Impact of Natural Hazards and Household Poverty: Evidence from the Indian Sundarban Delta [85]

This is a machine-generated summary of:

Hajra, Rituparna; Szabo, Sylvia; Tessler, Zachary; Ghosh, Tuhin; Matthews, Zoe; Foufoula-Georgiou, Efi: Unravelling the association between the impact of natural hazards and household poverty: evidence from the Indian Sundarban delta [85].

Published in: Sustainability Science (2017).

Link to original: https://doi.org/10.1007/s11625-016-0420-2.

If you want to cite the papers, please refer to the original.

For technical reasons we could not place the page where the original quote is coming from.

Abstract-Summary

"Natural hazards, such as fluvial or coastal floods, impact poorer and more vulnerable communities greater than more affluent communities."

"Quantitative assessments of how natural hazards affect vulnerable communities in deltaic regions are still limited, hampering the design of effective management strategies to increase household and community resilience."

"Drawing from Driving Forces–Pressure–State–Impact–Response (DPSIR), we quantify the associations between household poverty and the likelihood of material and human loss following a natural hazard using new survey data from 783 households within Indian Sundarban Delta community."

"The results suggest that the poorest households are significantly more likely to endure material and human losses following a natural hazard and repeated losses of livelihood make them more vulnerable to future risk."

Introduction

"In the US, economically poor people have been found to be more vulnerable to natural hazard events due to less preparedness and lower ability to recover from livelihood losses (Fothergill and Peek 50)."

"The present study aims to assess the associations between household poverty and the likelihood of material and human loss following a natural hazard in the islands of the Indian Sundarban Delta (ISD)."

"Although earlier studies investigated the impacts of natural hazards on livelihoods in nearby coastal islands of Bangladesh Sundarban Delta and low-elevation deltas in other geographical setup (Hossain and others 51; Shamsuddoha and Chowdhury 52; Fothergill and Peek 50; Lal and others 53), there is limited research examining the connections between vulnerability to natural hazards and poverty in the ISD."

"Our study aims to fill this gap by examining on how economic poverty affects losses from natural hazards at a household level in the ISD and offer a number of concrete policy recommendations to reduce vulnerability to natural hazards."

Socio-economic Vulnerability and Impacts of Natural Hazards in the Indian Sundarban Delta

"The Indian Sundarban Delta (ISD) region extends from 21°30′N to 22°40′48″N latitude and from 88°1′48″E to 89°04′48″E longitude."

"Both climatic and non-climatic events adversely affect the livelihoods of the people of the Indian Sundarban Delta (ISD)."

"The ISD remains one of the most under-developed places in India, with 34% of the 4.6 million people who inhabit the islands living below the poverty line (Hazra and others 54)."

"Climate change, induced sea-level rise, changing rainfall patterns, and changes in the frequency and intensity of extreme weather events have had significant impacts on the islanders of ISD (WWF 55)."

Conceptual Framework

"The DPSIR framework helps illustrate a key relationship between households' socio-economic vulnerability and the likelihood of loss following an extreme event or natural hazard."

"Natural hazards put a disproportionate amount pressure on the state of poor households, resulting in livelihood loss and material poverty."

"The UN Commission on Sustainable Development (UNCSD 56) attempted to formulate a new Driving Forces–State–Response (DSR) framework, which includes not only the social, political, economic, and demographic pressures in the PSR model, but also pressures that result from the natural system."

"While the former PSR model only focused on the anthropogenic pressure and response, the DPSIR model is more comprehensive and incorporates natural driving forces (Bowen and Riley 57; Poveda and Lipsett 58)."

Analytical Approaches

"The outcome variable used for the analysis is the binary variable measuring whether or not a household suffered a loss as a result of a natural hazard event."

"As highlighted previously, the conceptual framework of this study draws from DPSIR framework highlighting the association between household poverty (state) and the likelihood of material and human loss (impact) influenced by natural hazards (driver pressure)."

"The results of the unadjusted model show that for the richest households, the likelihood of suffering a loss following a natural hazard is significantly lower than for the poorest households ($p < 0.01$)."

"Controlling for other variables included in the model, the odds of a loss following a natural hazard for the richest households are 0.52 times the odds of the poorest households."

"Model 5 uses the same poverty measure but additionally controls for the type of natural hazards and household-level characteristics."

Discussion and Policy Implications

"The main objective of this paper was to examine the significance and strength of the associations between household poverty, as a key aspect of vulnerability, and the effect of natural hazards on livelihoods, measured by material and human loss."

"The results of the study confirmed that the poorest households are most likely to suffer from deteriorating livelihoods following a natural hazard."

"Similar to other studies (Hazra and others 54; Myers and others 59), our results show that migration is positively associated with the probability of loss following a natural hazard."

"Based on the conceptual framework of this study, we found positive association between economic poverty and material loss following natural hazard."

"Reduction of poverty may help lower the impact of loss on the economic condition of a household due to natural hazards."

"Development strategies, including efforts to improve income opportunities among the poor, are necessary to combat material losses from natural hazards."

Natural disasters and economic losses: controlling external migration, energy and environmental resources, water demand, and financial development for global prosperity [86]

This is a machine-generated summary of:

Abbas Khan, Khawar; Zaman, Khalid; Shoukry, Alaa Mohamd; Sharkawy, Abdelwahab; Gani, Showkat; Sasmoko, None; Ahmad, Jamilah; Khan, Aqeel; Hishan, Sanil S.: Natural disasters and economic losses: controlling external migration, energy and environmental resources, water demand, and financial development for global prosperity [86]

Published in: Environmental Science and Pollution Research (2019).

Link to original: https://doi.org/10.1007/s11356-019-04755-5.

If you want to cite the papers, please refer to the original.

For technical reasons we could not place the page where the original quote is coming from.

Abstract-Summary

"The objective of the study is to examine the impact of natural disasters on external migration, price level, poverty incidence, health expenditures, energy and environmental resources, water demand, financial development, and economic growth in a panel of selected Asian countries for a period of 2005–2017."

"The results confirm that natural disasters in the form of storm and flood largely increase migration, price level, and poverty incidence, which negatively influenced country's economic resources, including enlarge healthcare expenditures, high energy demand, and low economic growth."

Introduction

"It is the real fact that the natural disasters have put a serious threat to the economies of the world especially developing countries in terms of economic losses and physical damages."

"Since the last three decades, natural disasters episodes substantially increase worldwide, which caused an intimidating impact on the different economies of the world specifically the developing countries."

"One of the major disasters in the history is the tsunami of December 2004, which caused a high level of destruction with an earthquake of magnitude 9.0 effecting more than 18 countries from Southeast Asia to Southern Africa."

"The tsunami killed about 250,000 people in a single day and destroyed the large number of houses, which results in more than one million homeless people, and also posed a negative impact on economic activity causing millions of dollars loss by affecting tourist and fishing industries (Cavallo and others 60)."

Literature Review

"Loayza and others (61) investigated the impact of natural tragedies on development distinctly through disaster in economic segment and found that flood disasters can have a constructive development effect in certain areas, simple disasters do not, which need fair examination in the study area."

"The results reveal that natural disasters mitigation is connected with higher rates of human capital accumulation, rises in total factor productivity plus economic development, while it negatively effects on rate of return to physical capital."

"The motivation of the study is to amalgamate different socio-economic and environmental factors affected by natural disasters, to analyze economic losses of a country."

"The stated studies confirmed the damaging effects of natural disasters on socio-economic and environmental factors in different economic settings, while this study accessed vulnerabilities of Asian countries account of high natural disaster episodes to proposed sound economic policies to reduce its intensity and damages with quick recovery team for preventing high economic and human losses."

Data and Methodology

"The following variables are used in the study, including international migrant stock in numbers, consumer price index in annual %, poverty headcount in percentage, GDP per capita in constant 2010 US$, health expenditures as % of GDP, energy demand in kg of oil equivalent per capita, water supply as % of population access, broad money supply as % of GDP, the number of people affected by flood and storm, mineral depletion as % of GNI, FDI inflows as % of GDP, trade openness as % of GDP, industrialization as % of GDP, agriculture value added as % of GDP, fertilizer as Kg per hectare of arable land, cereal yield as Kg per hectare, real interest rate in %, mortality rate per 100 live birth, and carbon dioxide emissions as metric tons per capita."

Results and Discussion

"The results are confirmed by the previous studies of Reinhart and Rogoff (62), Sayek (63), etc. The result clearly shows that there is a positive relation between storm and poverty as increase in the storms results in rise in the poverty level, which tends to show that storms cause destruction in the infrastructure of the society including industrial zones and agriculture sector that lowers the income level across countries."

"The results are confirmed by the studies of Jaber and Probert (64), Karekezi (65), Pachauri and Spreng (66), etc. The per capita income relates positively with poverty which shows an increase in per capita causes an increase in poverty level, which account for high income inequality across countries."

"The result is confirmed by the previous studies of Pielke Jr and others (67), Yang (68), etc. The result found the positive relationship between storm and health expenditures in a panel of Asian countries, which implies that an increase in storms intensity causes large-scale destruction, which ultimately leads to increase healthcare expenditures in affected areas."

Conclusions and Recommendations

"Natural disasters led to the increase in poverty incidence in a country, while energy demand and health expenditures supported the poor through providing opportunity to do work with safety hands."

"The results proclaimed that natural disasters substantially decrease country's per capita income via the channel of energy infrastructure destructions across countries."

"The government should take the following steps to improve the disaster management process: build technically sound emergency management systems to handle the unfortunate situations, the Government should enhance the research activities to carry out new scientific methods to overcome the natural vulnerabilities, create awareness among the masses to get understanding about the importance of clean environment to stop manmade disasters, landslide and flood level areas always remain in danger during the disaster episodes; thus, there is a need of proper planning for reconstruction homes, apartments, businesses, etc., as per safety standards, tree planting plays an important role in changing the environment."

2.2.9 *Effects of Drought and Flood on Farmer Suicides in Indian States: An Empirical Analysis [87]*

This is a machine-generated summary of:

Parida, Yashobanta; Dash, Devi Prasad; Bhardwaj, Parul; Chowdhury, Joyita Roy: Effects of Drought and Flood on Farmer Suicides in Indian States: An Empirical Analysis [87].

Published in: Economics of Disasters and Climate Change (2018).

Link to original: https://doi.org/10.1007/s41885-018-0023-8.

If you want to cite the papers, please refer to the original.

For technical reasons we could not place the page where the original quote is coming from.

Abstract-Summary

"The study examines the effects of drought and flood on farmer suicides using state-level panel data from 17 Indian states for the period 1995–2011."

"The empirical estimates based on unconditional fixed effect Negative Binomial model show that while drought significantly increases farmer suicides, flood has no direct impact on the same."

"The results also show that incidence of farmer suicides is higher in cotton producing states of India because these states experience frequent drought conditions."

"Our findings reveal that states with high levels of rural poverty experience a higher number of farmer suicides as a result of frequent occurrence of droughts and moderate floods."

Introduction

"Subsequent studies done in Indian context have indicated various reasons such as excessive economic liberalization, low import tariffs, growing disparities between agriculture and non-agriculture sectors, withdrawal of state support system, government's poor extension services and dumping of agricultural goods in global markets as the main determinants of increasing farmer suicides in the recent years. (Mishra 69; Sridhar 70; Vaidyanathan 71; Jeromi 72)."

"In view of the increasing incidence of farmer suicides, the present study empirically examines the effect of drought and flood, and cropping pattern in states and incidence of rural poverty on farmer suicides using a state-level panel data from 17 Indian states over the period 1995 to 2011."

"The study examines if the states with higher incidence of rural poverty experience a higher number of farmer suicides as a result of frequent droughts and floods."

Overview of Farmer Suicides in Indian States

"By Basu and others (73), more than 0.3 million farmers have committed suicides across various states of India."

"We analyze the trend of farmer suicides in India for the period 1995 to 2011."

"We have used state-wise farmer suicide data available from NCRB."

"The focus is primarily on the 17 major states of India, where the problem of farmer suicides is more acute."

Exploring the Linkage Between Farmer Suicides and Vulnerability of a Region Towards Natural Disasters

"Maharashtra state, which ranks next to Karnataka in terms of average number of farmer suicides, has the highest percentage of drought prone area and a relatively low flood prone area."

"It is observed that the top 5 states (Karnataka, Maharashtra, Kerala, Andhra Pradesh and Madhya Pradesh) which have seen the highest incidence of farmer suicides also feature among the top five states with the highest percentage of drought prone area."

"This has two important implications: (i) the states which have experienced a relatively larger number of farmer suicides contain a higher percentage of drought prone area and a lesser percentage of flood prone area and (ii) the less poor states have witnessed a higher number of farmer suicides."

"The northern states such as Himachal Pradesh, Punjab and Uttar Pradesh are relatively well irrigated states, which consist of a relatively lower percentage of drought prone area and a concomitant lower number of farmer suicides."

Data Sources and Econometric Identification

"State-wise cotton and wheat production information is obtained from the state-wise area production and yield statistics, Directorate of Economics and Statistics, Ministry of Agriculture and Farmer Welfare, GoI. The data on RHCR was compiled from the "Report of the Expert Group to Review the Methodology for Measurement of Poverty", Planning Commission, GoI. In this empirical analysis, we could not use 'state-wise total area affected by flood' as one of the explanatory variables because it combines low, moderate and even high flood magnitudes based on total area affected."

Empirical Results; Effects of Drought and Flood on Farmer Suicides

"In Model-1, the coefficient of drought dummy is positive and statistically significant, implying that the frequent occurrence of drought significantly increases farmer suicide in Indian states due to crop failure."

"Another interesting finding is that the coefficient of moderate flood magnitude dummy is positive and significant in Model-3 and Model-4 which implies that the states with frequent occurrence of moderate flood witness higher occurrence of farmer suicides."

"Further, in Model-2 to Model-5, the interaction variable (Cotton producing state dummy*drought dummy) is positive and significant, which implies that the probability of farmer suicide is higher in cotton producing states because those states experience frequent occurrence of droughts."

"The coefficients of (RHCR*drought dummy) and (RHCR*moderate flood magnitude dummy) are positive and significant in Model-3 implying that one unit change in rural poverty increase farmer suicides by 2.89% and 3.52% respectively as a result of frequent occurrence of droughts and moderate floods respectively."

Conclusion and Discussion

"The study examines the effect of extreme weather events (mainly floods and droughts) on farmer suicides in Indian states over the period 1995 to 2011."

"The estimates based on unconditional FE Negative Binomial model reveal that drought has significantly increased the incidence of farmer suicides across Indian states."

"The study also shows that agricultural wage earnings of male workers significantly reduce farmer suicides due to increasing income securities of rural households."

"The result also shows that the incidence of farmer suicides is higher in cotton producing states because these states experience frequent drought conditions."

"Further, our findings reveal that states with high level of rural poverty experience a higher number of farmer suicides as a result of frequent occurrence of droughts and floods."

"Our results show that higher agricultural wage income of male workers and availability of alternative employment opportunities in agriculture and allied sectors can potentially arrest the rising incidence of farmer suicides."

2.2.10 Groundwater Irrigation Reduces Overall Poverty but Increases Socioeconomic Vulnerability in a Semiarid Region of Southern India [88]

This is a machine-generated summary of:

Fischer, Chloé; Aubron, Claire; Trouvé, Aurélie; Sekhar, Muddu; Ruiz, Laurent: Groundwater irrigation reduces overall poverty but increases socioeconomic vulnerability in a semiarid region of southern India [88].

Published in: Scientific Reports (2022).

Link to original: https://doi.org/10.1038/s41598-022-12814-0.

If you want to cite the papers, please refer to the original.

For technical reasons we could not place the page where the original quote is coming from.

Abstract-Summary

"The development of irrigation is generally considered an efficient way to reduce poverty in rural areas, although its impact on the inequality between farmers is more debated."

"Assessing the impact of water management on different categories of farmers requires resituating it within the different dimensions of the local socio-technical context."

"Our results show that the differentiation of farm trajectories since the 1950s has been linked with the development of groundwater irrigation, interplaying with their initial assets, and the evolution of the national and local contexts."

"This extensive understanding of the agrarian context allowed us to formulate and assess the potential of different ways forward, including irrigation technology, change in cropping or livestock systems, land tenure, and value added distribution."

Introduction

"Apart from the question of the differentiated effects of the depletion of water resources on households, the subject of the links between irrigation and poverty is not new."

"The works that articulate these different dimensions and assess the effects of the development of irrigation on different social categories, remain relatively rare [75]."

"By studying the trajectories of farms in the region over time, it is for example possible to identify the mechanisms that give the households who first developed irrigated cropping an advantage over those who turned to irrigation later and to assess the resulting income differences that exist today."

"A combined approach to these different dimensions (farm trajectories, capital assets, effects on work and employment) seems to be particularly useful when it comes to understanding the impact of the development of irrigation on agricultural households, and assessing the extent to which it reduces poverty and ensures equity."

Methodology

"This second concept—the production system—refers to "a group of farms with the same range of resources (same area, same level of equipment, same workforce) in similar socio-economic contexts, with a similar crops and livestock mix" [76]."

"Highlighting the differentiation process of these groups of farms, and thus the current diversity of production systems within the agrarian system, is an essential step in the research in comparative agriculture."

"The agro-ecological zoning of the study area and the reconstitution of its agrarian dynamics led to a first understanding of the current diversity of farms, which helped select farmers for the next round of interviews."

"In-depth understanding, we then built archetypes representing the highlighted diversity of production systems (6), cropping systems (8) and livestock farming systems (2) in the area."

Results

"They either sold a part or their land or livestock or took out heavy loans at high interest rates (between 3 and 5% per month), essentially from farmers who had invested in irrigation in the previous period."

"As labour demand increased with the development of irrigated lands, small rainfed farmers were able to compensate the reduction of their farm size with more job opportunities, which also helped the increasing number of landless labourers to ensure their livelihoods."

"With the reduction of straw availability in the area, induced by the development of crops that do not provide straws (turmeric, vegetables, marigold, sunflower), it became increasingly difficult to maintain dairy cows and some of the rainfed farmers cultivating marigold in the west abandoned livestock farming."

Discussion

"To create employment, the irrigated cropping systems must however permit an increase in labour productivity, which is confirmed here: the value added per workday is between 1.2 and 7 times higher for irrigated cropping systems in comparison to rainfed crops."

"More than the increase in productivity per hectare, it is the increase in labour productivity and income that explains the farmers' huge shift to irrigated cropping systems when they have the means."

"With the shift from sugarcane to new irrigated crops such as vegetables, turmeric, ginger and banana, farmers using irrigation in the study area have clearly increased the risks they face in marketing their products as they have to sell them to private buyers in the market, and, unlike for sugarcane, there is no policy regulating their prices [77]."

Conclusion

"The comparative agriculture framework used in this study, by documenting farming system dynamics and diversity in the Gundlupet Taluk, showed how the differentiation of farm trajectories over time was determined by the development of groundwater irrigation."

"We showed that the development of groundwater irrigation clearly increased wealth in the region, in particular by increasing dramatically labour productivity, allowing even very small farms to generate decent livelihoods."

"We also showed that groundwater irrigation did not reduce inequality, in particular due to the high burden faced by most of the small irrigators, and, a more original result of this study, due to the unfair repartition of the value added between land owners and salaried workers."

"While technological solutions to improve the efficiency of water management should be encouraged, they are unlikely to reduce both the social inequality and the pressure on the groundwater resource, and might even worsen the problem, if implemented without considering the socio-economic context."

2.2.11 Agriculture in India's Twenty-First Century [89]

This is a machine-generated summary of:

Vyasulu, Vinod: Agriculture in India's twenty-first century [89].

Published in: Journal of Social and Economic Development (2022).

Link to original: https://doi.org/10.1007/s40847-022-00218-9.

If you want to cite the papers, please refer to the original.

For technical reasons we could not place the page where the original quote is coming from.

Abstract-Summary

"While agricultural production has increased in selected crops and in some regions in India, environmental issues like salinity, water logging, leaching of soil, and contamination of rivers have become serious in these regions."

"This lecture takes the position that all these are the unintended results of our agricultural policy."

A Tribute

"He has made contributions in econometrics and agricultural economics."

"He was a member of the Commission on Agricultural Costs and Prices in Delhi at the time of his unfortunate death in a train accident."

"In instituting an endowment lecture in his memory, ISEC is honouring the memory and legacy of one who has unobtrusively contributed much to this institution."

"This lecture has four parts: Agriculture in India—the changing economic structure Some paradoxes, puzzles and consequences A way forward."

Agriculture in the Economy

"Agriculture has always been of utmost importance in India; it is here that food and essential raw materials, like jute and cotton, are produced."

"An important aspect of agricultural development has been the fact that India today is self-sufficient in foodgrains."

"The maps shown, based on research at the Geospatial Lab of the Indian Institute for Human Settlements depict the possible role agriculture changing in space in twenty-first century India."

"There have been major changes that suggest that India's twenty-first century will be very different."

"In 1951 India was predominantly rural Many areas far away from urban markets A 'country of villages' that was essentially inward looking By 2011, this has changed."

"The rural area, is however, around 95% of India's total land."

Paradoxes, Puzzles and Consequences

"An increase in government spending helps to increase effective demand and bring labour back into employment."

"New technology has released a great deal of labour from agriculture, but the other sectors have not been able to absorb this freed labour."

"In labour scarce economies like the US, efforts are made consciously to develop technologies that use, relatively, less labour than capital for a given output."

"These economies develop labour saving, capital using technologies."

"If such technologies are used in economies where the factor endowments are different, output may indeed increase, but employment is unlikely to."

"India has made efforts to deal with this problem of surplus agricultural labour."

"The Green Revolution in rice and wheat came about as a result of using new technologies along with a complementary package of inputs like credit, on good watered lands."

A Way Forward

"From a health point of view, the ill effects of 'superior crops' encouraged by government policy should no longer be allowed to continue."

"The focus must be on traditional crops in arid areas."

"We must grow crops that help good health—nurture a strong link between health policy and agriculture policy."

"Alternative ways of providing food security must be developed and implemented at local levels."

"Is Cuba an example of sustainable organic agriculture?".

"After the collapse of the Soviet Union, and with the continuing US blockade, Cuba had no option but to follow a policy of low energy organic agriculture and the country has succeeded in making the transition and feeding its people."

"There will be a saving in energy costs and some storage costs as food will be produced and largely consumed locally; movement of food will come down and transport costs reduce."

The Agency Question

"We need a functioning federal governance structure."

"We need a suitable administrative Form to enable state and local governments to fulfil these functions."

"It is not suited to the requirements and functions of Federal Governance in India's twenty-first century."

"In local governments, the elected body must function as the executive, interpreting State policy in the local context, implementing programmes, ensuring horizontal integration, monitoring progress, making course corrections, etc. In this it needs its own staff to implement its programmes."

"The term Federal Governance is not used in the Indian Constitution, but nevertheless, thanks to Dr B.R. Ambedkar, India's governance is in essence federal India is a union of states, the union and state governments have their own spheres of responsibility, enshrined in the lists in the 7th schedule."

"This existing structure of public administration, designed for a colony, has to change to a genuine federal form with local governments having their staff and resources."

Bibliography

1. Columbia Law School Human Rights Institute. "The More Things Change..." The World Bank, Tata and Enduring Abuses on India's Tea Plantations [Internet]. Columbia Law School Human Rights Institute; 2014 Jan. Available from: https://web.law.columbia.edu/sites/default/files/microsites/human-rights-institute/files/tea_report_final_draft-smallpdf.pdf.
2. Goddard S. Tea break: a crisis brewing in India. ActionAid; 2015.
3. Baishya D. History of tea industry and status of tea garden Workers of Assam. Int J Appl Res. 2016;2(9):552–6.
4. Sarkar K. Wages, mobility and labour market institutions in tea plantations: The case of West Bengal and Assam [internet]. Centre for development studies; 2015. Available from: http://cds.edu/wp-content/uploads/2015/07/46_Kingshuk.pdf.
5. Registrar General of India. Sample Registration System: Special Bulletin on Maternal Mortality in India 2014–16. Delhi: Office of the Registrar General of India. Vital Statistics Division; 2018.
6. Levy S. Munnar Plantation Strike, 2015: A case study of Keralan female tea workers' fight for justice [internet]. Independent study project (ISP) collection; 2015. Available from: https://digitalcollections.sit.edu/isp_collection/2624.
7. Kranthi KR, Russell DA. Changing trends in cotton pest management. In: Peshin R, Dhawan AK, editors. Integrated pest management: innovation-development. Springer; 2009. p. 499–541.
8. ICAC, International Cotton Advisory Committee. Factors influencing the use of pesticides in cotton in India. Washington DC, USA: Report From the Expert Panel on Social, Environmental and Economic Performance of Cotton Production (SEEP); 2010.
9. van den Bosch R. The pesticide conspiracy. Berkeley: University of California Press; 1978.
10. DeBach P, Rosen D. Biological control by natural enemies. 2nd ed. Cambridge, UK: Cambridge University Press; 1991.
11. Kranthi KR, Russell D, Wanjari R, Kherde M, Munje S, Lavhe N, et al. In-season changes in resistance to insecticides in Helicoverpa armigera (Lepidoptera: Noctuidae) in India. J Econ Entomol. 2002;95(1):134–42.
12. Stone ND, Gutierrez AP. A field oriented simulation of pink bollworm in southwestern desert cotton. Hilgardia. 1986;54:1–24.
13. von Arx R, J B, V D. A model to simulate the population-dynamics of Bemisia tabaci Genn. (Stern., Aleyrodidae) on cotton in the Sudan Gezira. Z Angew Entomol. 1983;96(4):341–63.
14. Gutierrez AP, Falcon LA, Loew W, Leipzig PA, van-den Bosch R. An analysis of cotton production in California: a model for acala cotton and the effects of defoliators on its yields. Environ Entomol. 1975;4(1):125–36.
15. Blaise D. Cropping and forage systems/crop ecology/organic farming yield, boll distribution and fibre quality of hybrid cotton (Gossypium hirsutum L.) as influenced by organic and modern methods of cultivation. J Agron Crop Sci. 2006;192:248–56.
16. Aggarwal PK, B HK, V VM, Rani S, Bala A, Biswa A et al. Quantification of yield gaps in rain-fed rice, wheat, cotton and mustard in India: global theme on agro ecosystems. Report no. 43. International Crops Research Institute for the Semi-Arid Tropics, Patancheru, Andhra Pradesh, India. 2008. http://oar.icrisat.org/2335/. Accessed 25 March 2015.

17. Pemsl DE, Gutierrez AP, Waibel H. The economics of biotechnology under ecosystem disruption. Ecol Econ. 2008;66(1):177–83.
18. Nagrare VS, Kranthi S, Biradar VK, Zade NN, Sangode V, Kakde G, et al. Widespread infestation of the exotic mealybug species, Phenacoccus solenopsis (Tinsley) (Hemiptera: Pseudococcidae), on cotton in India. Bull Entomol Res. 2009;99:537–41.
19. Kranthi KR. Cotton production systems—need for a change in India. Cotton Statistics & News. 2014;38(16 December 2014):4–7. http://caionline.in/newsletters/issue_38_161214.pdf. Accessed 27 March 2014.
20. Fabrick JA, Ponnuraj J, Singh A, Tanwar RK, Unnithan GC, Yelich AJ, et al. Alternative splicing and highly variable cadherin transcripts associated with field-evolved resistance of pink bollworm to Bt cotton in India. PLoS One. 2014;9(5), e97900.
21. Carrière Y, Ellers KC, Sisterson M, Antilla L, Whitlow M, Dennehy TJ, et al. Long term regional suppression of pink boll worm by Bt cotton. Proc Natl Acad Sci U S A. 2003;100:1519–23.
22. Qaim M, Zilberman D. Yield effects of genetically modified crops in developing countries. Science. 2003;299(5608):900–2. https://doi.org/10.1126/science.1080609.
23. Naik G, Qaim MAS, Subramanian A, Zilberman D. Bt cotton controversy: some paradoxes explained. Econ Polit Weekly. 2005;40(15):1514–7.
24. Gutierrez AP, Ponsard S. Physiologically based demographics of Bt cotton–pest interactions: I. Pink bollworm resistance, refuge and risk. Ecol Model. 2006;191:346–59.
25. Gutierrez AP, d'Oultremont T, Ellis CK, Ponti L. Climatic limits of pink bollworm in Arizona and California: effects of climate warming. Acta Oecol. 2006;30:353–64. https://doi.org/10.1016/j.actao.2006.06.003.
26. Gutierrez AP, Pizzamiglio MA, Dos Santos WJ, Tennyson R, Villacorta AM. A general distributed delay time varying life table plant population model: cotton (Gossypium hirsutum L.) growth and development as an example. Ecol Model. 1984;26(3):231–49.
27. Gutierrez AP, Adamczyk JJ, Ponsard S, Ellis CK. Physiologically based demographics of Bt cotton-pest interactions II. Temporal refuges, natural enemy interactions. Ecol Model. 2006;191(3–4):360–82.
28. Philpott Tom. No, GMOs Didn't Create India's Farmer Suicide Problem, But.... 2015. http://www.motherjones.com/tom-philpott/2015/09/no-gmos-didnt-create-indias-farmer-suicide-problem. Accessed 24 Aug 2016.
29. Mishra S. Farmers' Suicides in India, 1995–2012: Measurement and Interpretation. London: Asia Research Centre; 2014.
30. Shiva V, Jalees K. Seeds of Suicide: The Ecological and Human Costs of Seed Monopolies and Globalisation of Agriculture. New Delhi: Navdanya; 2006.
31. Shiva Vandana. Seed Monopolies, GMOs and Farmer Suicides in India: A Response to Nature. 2013c. http://www.navdanya.org/blog/?p=744.
32. Scoones I. Regulatory Manoeuvres: The Bt. Cotton Controversy in India. Sussex: Institute of Development Studies; 2003.
33. Scoones I. Contentious Politics, Contentious Knowledges: Mobilising Against GM Crops in India, South Africa and Brazil. Sussex: Institute of Development Studies; 2005.
34. Ramaswamy B, et al. The spread of illegal transgenic cotton varieties in India: Bio-safety regulation, monopoly, and enforcement. World Dev. 2012a;40(1):177–88.
35. Sadanandan A. Political economy of farmer suicide: financial reforms, credit crunches and farmer suicide in India. J Dev Areas. 2014;48(4):287–307.
36. Münster D. Farmers' suicides and the state in India: conceptual and ethnographic notes from wayanad, Kerala. Contrib Indian Sociol. 2012;46(1&2):181–208.
37. Vasavi Ainahalli R. Suicides and the making of India's agrarian distress. South Afr Rev Sociol. 2009;40(1):94–108.
38. Mishra S (2006) Suicide of farmers in Maharashtra Indira Gandhi Institute of Development Research. Mumbai, Indira Gandhi Institute of Development Research, Mumbai
39. Bollen KA (2008) Fixed and random effects in panel data using structural equations models. California Center for Population Research, University of California. https://escholarship.org/uc/item/3sr461nd

40. Alauddin M, Sarker MAR (2014) Climate change and farm-level adaptation decisions and strategies in drought-prone and groundwater-depleted areas of Bangladesh: an empirical investigation. Ecol Econ 106:204–213. https://doi.org/10.1016/j.ecolecon.2014.07.025
41. Below TB, Mutabazi KD, Kirschke D, Franke C, Sieber S, Siebert R, Tscherning K (2012) Can farmers' adaptation to climate change be explained by socio-economic household-level variables? Glob Environ Chang 22:223–235. https://doi.org/10.1016/j.gloenvcha.2011.11.012
42. Deressa TT, Hassan RM, Ringler C, Alemu T, Yesuf M (2009) Determinants of farmers' choice of adaptation methods to climate change in the Nile Basin of Ethiopia. Glob Environ Chang 19:248–255. https://doi.org/10.1016/j.gloenvcha.2009.01.002
43. Iniguez, L., 2011. The challenges of research and development of small ruminant production in dry areas. Small Ruminant Research, 98, 12–20
44. Kosgey, I.S., Rowlands, G.J., van Arendonk, J.A.M., Baker, R.L., 2008. Small ruminant production in landless and pastoral/extensive farming systems in Kenya. Small Ruminant Research, 77, 11–24
45. Sargison, N., 2017. Chapter 11. Maintaining sheep flock health: an overview. In: J. Greyling (ed), Achieving Sustainable Production of Sheep. Burleigh Dodds, Cambridge. Pages 221–243. ISBN: print 978–1–78676–084–5; ePub 978–1–78676–086–9; ePDF 978–1–78676–087–6.
46. Government of India (2019) Agricultural statistics at a Glance: 2019–20, Ministry of Agriculture, New Delhi. www.agricoop.nic.in
47. Dev SM, Rao NC (2010) Agricultural price policy, farm profitability and food security. Econ Polit Wkly 45(26–27):174–182
48. Narayanamoorthy A (2018) Post harvesting requirements for reducing the gap between what consumers pay and farmers receive. Indian J Agric Econ 73(3):421–430
49. Mishra S (2006) Farmers' suicides in Maharashtra. Econ Pol Wkly 41(16):1538–1545
50. Fothergill A, Peek LA (2004) Poverty and disasters in the United States: a review of recent sociological findings. Nat Hazards 32(1):89–110
51. Hossain MA, Reza MI, Rahman S, Kayes I (2012) Climate change and its impacts on the livelihoods of the vulnerable people in the southwestern coastal zone in Bangladesh. In: Climate change and the sustainable use of water resources (pp. 237–259). Springer, Berlin
52. Shamsuddoha M, Chowdhury RK (2007) Climate change impact and disaster vulnerabilities in the coastal areas of Bangladesh. COAST Trust, Dhaka
53. Lal PN, Singh R, Holland P (2009) Relationship between natural disasters and poverty: a Fiji case study. UN International Strategy for Disaster Reduction (ISDR), the World Bank, Pacific Island Applied Geoscience Commission (SOPAC)
54. Hazra S, Das I, Samanta K, Bhadra T (2014) Impact of climate change in Sundarban area West Bengal, India. School of Oceanographic Studies, Earth Science & Climate Book 9326/ 17.02.00. Report submitted to Caritus India, SCiAF
55. WWF (2010) Sundarbans: future imperfect climate adaptation report
56. UNCSD (1997) From theory to practice: indicators for sustainable development. United Nations Division for Sustainable Development New York, United Nations
57. Bowen RE, Riley C (2003) Socio-economic indicators and integrated coastal management. Ocean Coast Manag 46(3):299–312
58. Poveda CA, Lipsett MG (2011) A review of sustainability assessment and sustainability/ environmental rating systems and credit weighting tools. J Sustain Dev 4(6):36–52
59. Myers CA, Slack T, Singelmann J (2008) Social vulnerability and migration in the wake of disaster: the case of Hurricanes Katrina and Rita. Popul Environ 29(6):271–291
60. Cavallo A, Cavallo E, Rigobon R (2014) Prices and supply disruptions during natural disasters. Rev Income Wealth 60(S2):449–471
61. Loayza NV, Olaberria E, Rigolini J, Christiaensen L (2012) Natural disasters and growth: going beyond the averages. World Dev 40(7):1317–1336
62. Reinhart MC, Rogoff MK (2003) FDI to Africa: the role of price stability and currency instability (no. 3–10). International Monetary Fund, Washington D.C.
63. Sayek S (2009) Foreign direct investment and inflation. South Econ J 76(2):419–443

64. Jaber JO, Probert SD (2001) Energy demand, poverty and the urban environment in Jordan. Appl Energy 68(2):119–134
65. Karekezi S (2002) Poverty and energy in Africa—a brief review. Energy Policy 30(11–12):915–919
66. Pachauri S, Spreng D (2004) Energy use and energy access in relation to poverty. Econ Polit Wkly 39(3):271–278
67. Pielke RA Jr, Rubiera J, Landsea C, Fernández ML, Klein R (2003) Hurricane vulnerability in Latin America and the Caribbean: normalized damage and loss potentials. Nat Hazard Rev 4(3):101–114
68. Yang D (2008) Coping with disaster: the impact of hurricanes on international financial flows, 1970–2002. J Econ Anal Policy 8(1):1–45
69. Mishra S (2006) Farmers' suicides in Maharashtra. Econ Polit Wkly 41(16):1538–1545
70. Sridhar V (2006) Why do farmers commit suicide? The case of Andhra Pradesh. Econ Polit Wkly 41(16):1559–1565
71. Vaidyanathan A (2006) Farmers' suicides and the agrarian crisis. Econ Polit Wkly 41(38):4009–4013
72. Jeromi PD (2007) Farmers' indebtedness and suicides: impact of agricultural trade liberalization in Kerala. Econ Polit Wkly 42(31):3241–3247
73. Basu D, Das D, Misra K (2016) Farmer suicides in India. Econ Polit Wkly 51(21):61–65
74. Singh, D. Groundwater markets and institutional mechanism in Fragile environments. In Water Resources, Sustainable Livelihoods and Eco-System Services (eds Chopra, K. R. et al.) 311–340 (Concept Publishing Company, New Delhi, 2003).
75. Bjornlund, V. & Bjornlund, H. Understanding agricultural water management in a historical context using a socioeconomic and biophysical framework. Agric. Water Manag. 213, 454–467. https://doi.org/10.1016/j.agwat.2018.10.037 (2019).
76. Cochet, H. & Devienne, S. Fonctionnement et performances économiques des systèmes de production agricole: Une démarche à l'échelle régionale. Cah. Agric. 15(6), 578–583 (2006).
77. Mukherjee, A. et al. Groundwater systems of the Indian sub-continent. J. Hydrol. Reg. Stud. 4, 1–14 (2015).
78. Bhalla, G. S. The economic situation of the Indian farmers: some findings from the NSSO 59th round. Journal of Social and Economic Development (2022). https://doi.org/10.1007/s40847-022-00201-4
79. Rajbangshi, Preety R.; Nambiar, Devaki "Who will stand up for us?" the social determinants of health of women tea plantation workers in India. International Journal for Equity in Health (2020). https://doi.org/10.1186/s12939-020-1147-3
80. Gutierrez, Andrew Paul; Ponti, Luigi; Herren, Hans R; Baumgärtner, Johann; Kenmore, Peter E Deconstructing Indian cotton: weather, yields, and suicides. Environmental Sciences Europe (2015). https://doi.org/10.1186/s12302-015-0043-8
81. Thomas, Gigesh; De Tavernier, Johan Farmer-suicide in India: debating the role of biotechnology. Life Sciences, Society and Policy (2017). https://doi.org/10.1186/s40504-017-0052-z
82. Swami, Deepika; Dave, Prashant; Parthasarathy, Devanathan Understanding farmers' suicidal ideation: a structural equation modeling study in Maharashtra, India. Climatic Change (2020). https://doi.org/10.1007/s10584-020-02935-8
83. Lalljee, Sehr Vanya; Soundararajan, C.; Singh, Yogendra Deep; Sargison, Neil D. The potential of small ruminant farming as a means of poverty alleviation in rural southern India. Tropical Animal Health and Production (2018). https://doi.org/10.1007/s11250-018-1686-4
84. Jain, Ankur; Sheekha, Neela Madhaba Agrarian crisis in wheat-producing region of India: analysis of profitability and costs. Journal of Social and Economic Development (2022). https://doi.org/10.1007/s40847-022-00186-0
85. Hajra, Rituparna; Szabo, Sylvia; Tessler, Zachary; Ghosh, Tuhin; Matthews, Zoe; Foufoula-Georgiou, Efi Unravelling the association between the impact of natural hazards and household poverty: evidence from the Indian Sundarban delta. Sustainability Science (2017). https://doi.org/10.1007/s11625-016-0420-2

86. Abbas Khan, Khawar; Zaman, Khalid; Shoukry, Alaa Mohamd; Sharkawy, Abdelwahab; Gani, Showkat; Sasmoko, None; Ahmad, Jamilah; Khan, Aqeel; Hishan, Sanil S. Natural disasters and economic losses: controlling external migration, energy and environmental resources, water demand, and financial development for global prosperity. Environmental Science and Pollution Research (2019). https://doi.org/10.1007/s11356-019-04755-5
87. Parida, Yashobanta; Dash, Devi Prasad; Bhardwaj, Parul; Chowdhury, Joyita Roy Effects of Drought and Flood on Farmer Suicides in Indian States: An Empirical Analysis. Economics of Disasters and Climate Change (2018). https://doi.org/10.1007/s41885-018-0023-8
88. Fischer, Chloé; Aubron, Claire; Trouvé, Aurélie; Sekhar, Muddu; Ruiz, Laurent Groundwater irrigation reduces overall poverty but increases socioeconomic vulnerability in a semiarid region of southern India. Scientific Reports (2022). https://doi.org/10.1038/s41598-022-12814-0
89. Vyasulu, Vinod Agriculture in India's twenty-first century. Journal of Social and Economic Development (2022). https://doi.org/10.1007/s40847-022-00218-9

Chapter 3
Alternative Farming for Sustainable Development

3.1 Introduction by Editor

The topic of alternative farming practices for sustainable development has gained significant attention in recent years as global concerns over soil nutrient loss, environmental degradation, climate change, and the limitations of industrial agriculture have been widely researched. Traditional farming practices in countries like China, India, and the USA have long incorporated principles of sustainability, and these approaches hold valuable lessons for modern agriculture (Altieri, 1999; Bachev, 2016).

As seen in the earlier chapter, intensive farming that has dominated the agricultural sector since the 1960s, has contributed to a range of environmental and social problems, including soil degradation, deforestation, water scarcity, and climate change (Sulaiman & Misnan, 2022). Excessive use of chemical inputs, such as fertilizers and pesticides, as well as heavy machinery, have led to the loss of organic matter, pollution, and soil compaction. In contrast, alternative agriculture offers a more holistic approach that emphasizes the conservation and regeneration of natural resources, biodiversity, and ecosystem services (Batie & Taylor, 1989; Altieri, 1999; Sulaiman & Misnan, 2022; Abubakar & Attanda, 2013). Traditional farming systems in many countries have long recognized the importance of maintaining biodiversity, soil nutrition by using sustainable practices. This was done to ensure the long-term productivity of the land. These traditional systems often incorporate techniques such as crop rotation, intercropping, and the use of organic fertilizers, which help to maintain soil fertility and minimize the need for external inputs.

There is ample evidence from various regions to suggest that the widespread adoption of alternative farming practices could have significant positive impacts, including improved profitability for farmers, higher yields, and enhanced environmental quality (Sulaiman & Misnan, 2022; Altieri, 1999; Abubakar & Attanda, 2013; Batie & Taylor, 1989). By supporting the conservation and regeneration of natural resources, alternative farming can play a crucial role in mitigating the effects of climate change

G. Nulkar (ed.), *The Economics of Sustainable Development*,
https://doi.org/10.1007/978-981-99-7379-8_3

and building resilience in agricultural systems (Altieri, 1999; Sulaiman & Misnan, 2022; Babikir et al., 2015; Batie & Taylor, 1989).

In the preceding chapter, we examined studies that elucidate the challenges linked to intensive agriculture. This chapter explores potential alternatives to conventional agriculture. The subject of sustainable agriculture is subject to varying interpretations among individuals. This chapter summarises various studies that explore agricultural methodologies with a potential to replace existing practises. This involves the examination of many streams, such as organic farming, agroecology, and permaculture. Nonetheless, the fundamental premise revolves around the capacity of alternative approaches to address the interconnected challenges posed by the degradation of soil, loss of biodiversity, and the decline in farmer incomes. All of the various alternative types of agriculture exhibit shared characteristics, such as the use of organic nutrition, the elimination of chemical inputs, the implementation of biological controls, adherence to natural cycles, reduction of machinery usage on farms, and other similar practises. Alternatives to conventional agriculture does not mean going back to preindustrial practices, but employing traditional methods with modern research and appropriate technologies (Shristi Bhandari, 2022).

Several studies establish and define sustainable agriculture practices and systems (SAPSs). Among the SAPSs, only 5—crop rotation, agroforestry, rainwater harvesting, mulching and precision are seen across the country, and they do not go beyond 5 per cent of net sown area. It is thus evident that sustainable agriculture is not yet mainstreamed in India (Gupta, 2021). The resource intensive methods employed by Indian farmers has increased water stress in the country. Desertification and land degradation also pose major threats to agriculture (Food and Agricultural Organisation, India).

This chapter looks at successful alternative practices across Thailand, Canada, France and Cuba. Studies related to seed conservation, participative agriculture, multi-cropping and crop rotation, organic farming, vertical farms, and soil conservation offer insights to what could be regional pathways to improve sustainable practices. The intention is to offer readers the various possibilities in alternative forms of agriculture. Traditional agricultural practices have been an integral part of food production in India since ages. These practices have the potential to mitigate the adverse effects of climate change with spatial and sequential diversity (Patel, 2020). However, the caveat is that choices must be made considering the agro-climatic situation, food and cultural diversity, and availability of natural resources in the location. In agriculture, there is no one-size-fits-all.

In the holy grail of sustainable development, the long-term ecological impacts of intensive agriculture must be urgently mitigated and replaced by more sustainable alternative forms of farming. This must be done without compromising the food security. This chapter intends to broaden the meaning of sustainable agriculture, and examine the benefits and costs of each one.

References

Abubakar, M., & Attanda, M. (2013, December 20). The Concept of Sustainable Agriculture: Challenges and Prospects. IOP Publishing, 53, 012,001–012,001. https://doi.org/10.1088/1757-899x/53/1/012001.

Altieri, M A. (1999, June 1). The ecological role of biodiversity in agroecosystems. Elsevier BV, 74(1–3), 19–31. https://doi.org/10.1016/s0167-8809(99)00028-6.

Babikir, O M., Muchina, S., Sebsibe, A., Bika, A., Kwai, A., Agosa, C., Obahai, G., & Wakhusama, S. (2015, June 10). Agricultural Systems in IGAD Region—A Socio-Economic Review. https://doi.org/10.5772/60012.

Bachev, H. (2016, January 1). Unpacking Sustainability of Farming Organizations. OMICS Publishing Group, 05(03). https://doi.org/10.4172/2162-6359.1000335.

Batie, S S., & Taylor, D B. (1989, December 1). Widespread adoption of non-conventional agriculture: Profitability and impacts. Cambridge University Press, 4(3–4), 128–134. https://doi.org/10.1017/s0889189300002952.

Sulaiman, N K., & Misnan, S H. (2022, September 1). Environmental Sustainability through Agriculture: Perspectives of Extension Agents on Adoption of Sustainable Practices. IOP Publishing, 1082(1), 012,024–012,024. https://doi.org/10.1088/1755-1315/1082/1/012024.

3.2 Machine Generated Summaries

Disclaimer: The summaries in this chapter were generated from Springer Nature publications using extractive AI auto-summarization: An extraction-based summarizer aims to identify the most important sentences of a text using an algorithm and uses those original sentences to create the auto-summary (unlike generative AI). As the constituted sentences are machine selected, they may not fully reflect the body of the work, so we strongly advise that the original content is read and cited. The auto generated summaries were curated by the editor to meet Springer Nature publication standards. To cite this content, please refer to the original papers.

Machine generated keywords: organic, farming, organic farming, pesticide, agriculture, water, soil, fertilizer, farm, farmer, chemical, industrial, network, agricultural production, producer.

3.2.1 The Challenge of Feeding the World While Conserving Half the Planet [125]

This is a machine-generated summary of:

Mehrabi, Zia; Ellis, Erle C.; Ramankutty, Navin: The challenge of feeding the world while conserving half the planet [125].

Published in: Nature Sustainability (2018).

Link to original: https://doi.org/10.1038/s41893-018-0119-8.

If you want to cite the papers, please refer to the original.

For technical reasons we could not place the page where the original quote is coming from.

Abstract-Summary

"Amid widespread concerns about biodiversity loss, a single clear conservation message is engaging leading conservationists: the proposal to give half the surface of the Earth back to nature."

Main

"We do this by assessing Half-Earth conservation strategies designed to maximize the protection of key biodiversity areas and existing wild lands while minimizing crop calorie losses caused by displacing arable agriculture."

"Our results demonstrate clear potential for trade-offs between Half-Earth and agricultural production that are strongly mediated by the approach taken and the spatial scale of the land units used for conservation planning."

"The demands of future human populations and losses of pasture-fed animal calories are not considered here, nor did we conduct an explicit spatial prioritization for maximizing the coverage of species ranges or connectivity (which may prioritize more productive lands than those given back in our analysis), so our findings should be seen as conservative with respect to Half-Earth's potential negative impacts on agriculture."

"The trade-offs between nature and agriculture are potentially large, but depend strongly on the strategies used to achieve Half-Earth, highlighting the possible strengths of shared landscape approaches to expanding conservation with lower trade-offs for agricultural production."

Methods

"We then ranked all pixels for this new land-class variable in the following order: protected areas > key biodiversity areas > non-agricultural lands (forests, wetlands, shrublands, savannahs, grasslands and barren land covered in snow or ice) > agricultural lands (pasture > cropland-pasture > cropland) > urban, and then ordered each pixel within each rank in order of increasing calorific production based on the data from ref [1].”

“We assessed agricultural losses by identifying the proportion of pasture land, cropland and calories included in the pixels summing to 50% of the total area (nature-only landscapes approach) or sub-pixel areas summing to 50% of the total area (shared landscapes approach) within each of the focal geographic boundaries (global, country and ecoregion) included in the analysis.”

“Under the shared landscapes approach, we maintained a priority ranking as: protected areas > key biodiversity areas > non-agricultural areas > non-food-calorie (feed/biofuel/other)-producing land and pasture land > food-calorie-producing land > urban, so that we prioritized giving back land area producing non-food calories and pasture lands before food-calorie-producing croplands.”

3.2.2 Converting to Organic Farming as a Way to Enhance Adaptive Capacity [126]

This is a machine-generated summary of:

Bouttes, Maëlys; Darnhofer, Ika; Martin, Guillaume: Converting to organic farming as a way to enhance adaptive capacity [126].

Published in: Organic Agriculture (2018).

Link to original: https://doi.org/10.1007/s13165-018-0225-y.

If you want to cite the papers, please refer to the original.

For technical reasons we could not place the page where the original quote is coming from.

Abstract-Summary

“Farmers are facing an increasingly turbulent context, driven by volatile markets, shifting policies, and new societal expectations.”

"Insights into farmers' adaptive capacity are helpful to understand how farmers perceive various options, and what they think is necessary to enable their farms to persist through uncertain times."

"We conducted interviews with 20 dairy farmers who were starting their conversion to organic farming in Aveyron, France."

"The analysis showed that the interviewed farmers perceived organic farming as less risky, especially given stable prices and positive consumer perception."

"They expected organic farming to increase their autonomy, especially regarding feed, thus reducing their farm's exposure to volatile input prices."

"The interviewed farmers expected that the higher prices of organic milk would enable them to reduce the number of cows, thus reducing their workload."

"The interviews thus showed that their decision to convert to organic farming was driven by their expectation that it would enhance their adaptive capacity, thus enabling them to better face current turbulences and future changes in the broader context."

Introduction

"At the farm level, adaptive capacity depends on the farm's organization, e.g., the diversity of activities and marketing channels, and on the farmer him/herself, e.g., through his/her monitoring of the context and engaging in learning processes (Darnhofer and others [2])."

"While a number of studies that assess adaptive capacity (e.g., Astigarraga and Ingrand [3]; David and others [4]) focus on the activities at the farm level, this paper focuses on farmers' perception."

"Indeed, organic farming may be perceived by farmers as offering levers to adapt their farms in the face of rising market turbulences (Milestad and Darnhofer [5])."

"We explore whether farmers expect that converting to organic farming will improve their ability to face changes."

"Our aim is to better understand what aspects of organic farming they perceive as increasing their adaptive capacity."

Material and Methods

"The area was selected, as it allowed us to study a group of farmers concurrently engaging in the conversion to organic farming."

"To achieve this aim, the dairy was actively encouraging dairy farmers to convert to organic farming by offering attractive contracts and by encouraging the advisory services of the Chamber of Agriculture to facilitate the conversion process."

"More broadly, the French government also promoted organic farming through its agri-environment program, offering direct payments to all dairy farmers during the first years after conversion."

"To collect farmers' perceptions regarding their conversion to organic farming, we used semi-structured interviews."

"In the analysis, our aim was not to assess the level of adaptive capacity of the individual farms but to understand how the interviewed farmers perceived organic farming as enabling them to enhance their adaptive capacity."

Results and Discussion

"In the context of our study, it was important to understand both the risks that the interviewed farmers perceived in relation to either continuing as conventional farmers or converting to organic farming (Xu and others [6])."

"A conversion to organic farming was widely perceived by the interviewed farmers as a welcome professional challenge."

"In the interviews, the emotional burden of feeling trapped in the conventional system was often mentioned, and farmers expected that converting to organic farming would contribute to improving their work satisfaction."

"The interviewed farmers felt increasingly apprehensive about using agrichemicals, especially given the reports in the media on environmental problems related to conventional farming: "I was never motivated to use the sprayer, we know very well that it is harmful" (farmer 12); "I think that the mentality is changing, even those who don't convert to organic farming think about reducing their use of chemicals" (farmer 9)."

Conclusion

"While these are important considerations, this study shows that the subjective, personal satisfaction of the farmer also plays a crucial role, especially if organic farming is perceived by the farmer as enabling him/her to achieve his/her professional ambitions."

"It might thus be helpful to present organic farming as a way that enables farmers to combine traditions, societal demands, and personal ambitions, rather than just as a set of production practices."

"In their assessment of organic farming, the interviewed farmers integrated very different elements, such as expected development of markets, regulations, demands by consumers and citizens, perception by peers, impact on emotional health, likely technical challenges, learning requirements, and the richness of the collective learning process."

"If organic farming enhances this adaptive capacity, this may be an important aspect to highlight to farmers considering a conversion, as many family farmers aim not so much to maximize their profits or yields on the short term, than to ensure the persistence of their farm over the long term."

3.2.3 Modeling Sustainable Food Systems [127]

This is a machine-generated summary of:

Allen, Thomas; Prosperi, Paolo: Modeling Sustainable Food Systems [127].

Published in: Environmental Management (2016).

Link to original: https://doi.org/10.1007/s00267-016-0664-8.

If you want to cite the papers, please refer to the original.

For technical reasons we could not place the page where the original quote is coming from.

Abstract-Summary

"Building sustainable food systems has become a predominating endeavor aiming to redirect our food systems and policies towards better-adjusted goals and improved societal welfare."

"Food systems are complex social-ecological systems involving multiple interactions between human and natural components."

"The systemic nature of these interdependencies and interactions calls for systems approaches and integrated assessment tools."

"This paper proposes a conceptual model that articulates crucial vulnerability and resilience factors to global environmental and socio-economic changes, postulating specific food and nutrition security issues as priority outcomes of food systems."

"By acknowledging the systemic nature of sustainability, this approach allows consideration of causal factor dynamics."

Introduction

"Food systems act as complex social-ecological systems, involving multiple interactions between human and natural components."

"We present an additional approach that considers vulnerability and resilience as the operating concepts to model the systemic factors that lead to final food system outcomes such as food and nutrition security."

"The proponents of the "Sustainable Diet" agenda, a closely related concept highlighting the role of consumers in defining sustainable options, provide in particular a food and nutrition security-orientated perspective on the question of the sustainability of food systems (FAO/Bioversity [7]; Johnson and others [8])."

"Transforming the abstract concept of sustainability into descriptive objectives, the authors propose in this paper a conceptual model that articulates crucial vulnerability and resilience factors to global environmental and socio-economic changes in the Mediterranean region, postulating specific food and nutrition security issues as priority food system outcomes."

Identifying the Fundamental Sustainability Properties of the Food System

"A crucial challenge towards sustainability of food systems is the management of dynamics originating from both global and internal changes, and their different synergistic impacts on systems' outcomes."

"There are several illustrations of approaches analyzing food systems for their vulnerability and resilience to global socio-economic and biophysical changes in order to explore sustainability, and highlighting key system processes and characteristics (Ericksen [9]; Darnhofer and others [10]; Allouche [11])."

"Ericksen [9] argues that the vulnerability approach "frames the consequences of environmental change for food systems in the context of socioeconomic and political change so as to understand the synergistic effects of the multiple stresses that interact with food systems, sometimes making these systems vulnerable".

"Foran and others [12] state that "The social-ecological system considers the human–environment interface as a coupled 'system' where socio-economic and biophysical drivers of change interact to influence activities and outcomes, of the food system, that subsequently influence drivers of changes in a feedback loops dynamic".

Formalizing the Food System as a Dynamic System

"This requires examining how the multicausality of dynamic processes within complex systems such as the food system could help understand changes over time towards food security."

"Modeling dynamic systems is about representing mathematically the dynamics between the inputs and outputs of the system of interest."

"When modeling input–output systems, in addition to an observed set of variables internal to the system that can be levers of action, external drivers e can enter the model as inputs (Ionescu and others [13])."

"To capture something of the internal dialectic of a system, we suggest fixing some crucial external variables, or drivers of changes e, and observing how these affect our system outcome of interest: food and nutrition security."

"To assess the sustainability of the food system, we need to understand what might affect its processes, to what extent the drivers of change impact the food system's outcomes, and how actors respond to these pressures."

Application: Addressing Context-Specific Issues

"It also requires simultaneously identifying the main drivers of change as the food system-specific issues of concern that the drivers are likely to affect (Schröter and others [14])."

"Several global and regional drivers of change affect the structure and processes of food systems, putting context-specific food and nutrition security outcomes at risk."

"Based on an extensive literature review and discussions conducted throughout two focus group sessions made up of seven experts, four critical food and nutrition security issues and four drivers of change were identified at a sub-regional level."

"Food consumption patterns are changing both in terms of total amount and composition."

"These sets of characteristics are indicating how changes in water, biodiversity, food prices, and food consumption patterns are transmitted through the food system, including the sequencing of events and the scale of interactions; how the food system is sensitive to these changes; and the adaptive capacity of the food system."

Discussion

"We propose the analysis and assessment of the sustainability of food systems using the concepts of vulnerability and resilience."

"Exposure, sensitivity, and resilience provide the concepts to identify the system's properties that shape causal pathways towards food system outcomes."

"Focusing on a number of external forces and highlighting systemic internal dialectic, the vulnerability/resilience model allows a dynamic analysis of some specific issues of the food systems and provides direction for policy-makers."

"Food systems are responsible for diverse environmental, economic, and social outcomes, and introducing these may have been more in line with the generally accepted understanding of sustainability."

"The second step is to understand what and how global or regional changes, either socio-economic or environmental, might be transmitted through the activities to impact the outcomes, because food systems' complexity means that impacts may not always be felt directly."

Conclusion

"Developing policy to ensure sustainable food security is a tremendous challenge that requires a comprehensive and integrated analytical approach."

"A vulnerability and resilience approach is suggested here as a possible framework to capture the food system as a whole, think prospectively and identify the system elements that policy can leverage."

"Vulnerability and resilience of food systems can have multiple sources, and these sources may interact to generate unexpected responses (SCAR [15])."

"As sustainability and food security become increasingly central, vulnerability/ resilience will be among the principles that will drive the reformulation of research, as well as policies (Brunori and Guarino [16])."

3.2.4 Economics of Vertical Farming in the Competitive Market [128]

This is a machine-generated summary of:

Moghimi, Faraz; Asiabanpour, Bahram: Economics of vertical farming in the competitive market [128].

Published in: Clean Technologies and Environmental Policy (2023).

Link to original: https://doi.org/10.1007/s10098-023-02473-8.

If you want to cite the papers, please refer to the original.

For technical reasons we could not place the page where the original quote is coming from.

Abstract-Summary

"This paper offers the first quantitative model in the environmental economics and policy literature that evaluates the economic prospect of vertical farming systems in a competitive market setting."

"Our framework identifies the principal factors to assess the economic and risk aversion potential of vertical farming and utilize a decision model quantify the trade-off between the two alternative farming practices."

"The model is utilized to evaluate the competitive economic prospect of vertical farming in seven locations with heterogeneous climate and economic conditions within the USA."

"We leverage these results to evaluate the current and future prospect of the vertical farming industry."

Introduction

"This study quantitatively models vertical farming's performance and economic prospect with respect to conventional farming under different conditions in a competitive marketplace setting."

"We propose a decision model framework to evaluate both the profit and risk prospects of vertical farming with respect to conventional farming."

"We leverage a series experiment results from our framework to develop an evidence-based discussion regarding the vertical farming industry and business practice from a macro perspective."

"There have been no attempts in the prior literature to develop models and frameworks that evaluate the economics of vertical farming with respect to conventional agriculture under different conditions in a competitive marketplace."

"To the best of our knowledge this is the first data-driven case study conducted on this scale to evaluate the economics of vertical farming across various markets in the USA."

Literature Review

"Our model's results withstand a variety of sensitivity robustness checks, illustrating the low sensitivity impact of our assumptions. (2) This allows us to lead an empirically guided discussion regarding the principal factors influencing the economic prospects of vertical farming, which adds to the current literature."

"We believe that especially climate-driven risk aversion is an extremely important element in determining the sustainability of a farming practice, especially in an era where climate change has significantly increased the climate-driven risk in agriculture. (5) Our fifth contribution to the literature is to propose decision framework specifically designed to identify the markets and locations in which vertical farming would be the most competitive. (6) Finally, to the best of our knowledge, this is the first case study that examines the economic prospect of vertical farming within different US-based locations on such scale."

The Quantitative Framework

"We leverage the regional subsidy adjusted insurance rates as a proxy for farming production risk: In this segment, we propose a profit margin indicator model that would provide the relative economic appeal of a vertical farming business venture with respect to traditional agriculture in a given marketplace."

"Let us define minmaxscalar() (statisticshowto, 2015) for a data set as: Using the minmax scaling tool let us define profit decision indicator (PDI) and risk decision indication (RDI) as follows: We know the higher the value of RDI(i) and PDI(i) the more appealing a location is for implementing a vertical farming operation."

"To that end, let us define w1 and w2 that would represent that relative importance of PDI(i) and RDI(i), respectively, where Now let us define the final location appeal (LA(i)) function based on the previous indicators: The LA(i) function would provide a score ranging between 0 and 1 that would indicate the relative appeal of a location for implementing vertical farming based on the relative economics, regional risks, and investor's preferences."

Case Study

"In this segment, we utilize the framework from the previous segment to conduct a series of experiments, examining the appeal of implementing vertical farming in different locations under different scenarios."

"This analyses seeks to investigate the framework's results for two hypothetical location alternatives that are, in theory, the most desirable and the least desirable places to implement a vertical farming business."

"Consistent with our hypothesis, we find that the system successfully recognizes the extreme alternatives as the best and worst relative location to implement vertical farming."

"That said, we validate our decision model framework in the following step: Production technology sensitivity test (4.4.2): We find that the model would provide robust results under possible universal errors in estimating industry benchmarks or implementing superior or inferior forms of vertical farming technology by a stakeholder."

Future of Vertical Farming

"We find that the savings are quite slim in land and water costs, where vertical farming holds the advantage."

"A few factors that might help vertical farming's case in the future are discussed."

"The water prices, as of now, are still too low to encourage vertical farming's minimal water usage advantage and make a significant difference in the cost of production comparisons."

"If fertile land prices happened to increase in the future, whether due to population growth or legislation preventing deforestation, this would heavily help vertical farming gain an edge over the traditional practice."

"Plus, considering the current high production costs and low variety of produced crops, the insurance and subsidy rates are probably not the priority problem blocking vertical farming's progress."

"If at any point in the future, the insurance rates increase due to global climate change or USDA's choice to reduce insurance subsidies, this would probably favor the vertical farming industry."

Conclusion and Future Work

"We provide a comprehensive quantitative framework that can be utilized as a business/economic analytics tool to evaluate vertical farming in a competitive marketplace."

"We provide the first account of examining the economics and risks of vertical farming on a relative basis with respect to conventional farming."

"Our study identifies several factors and economic trends that strengthen (hinder) the adoption of vertical farming in the marketplace."

"We showcase our framework by examining the economics of vertical farming across 7 heterogeneous markets across the USA, where we can identify the best and least promising markets for vertical farming as well as the underlying reasons for a given market's varying economic appeal."

"As the vertical farming industry grows, it will attract more interested parties to its ecosystem, investors, policymakers, analysts, scholars, etc. Our proposed framework is flexible enough to help them in some way or form."

Appendix Section

"Hydroponics is a soilless practice of growing crops in a medium of liquid and nutrients."

"Similar to hydroponic systems, aquaponics grow crops by using a medium of liquids and nutrients."

"Aquaponic systems use the nutrient-rich water of the fish tanks as the medium to grow crops."

"Unlike hydroponics or aquaponics, aeroponic system nourish plants without a liquid medium using nothing but nutrient-charged mist."

"Aeroponic farming requires less water and nutrient compared to hydroponics and aquaponics."

3.2.5 Organic Farming for Sustainable Agricultural Production. A Brief Theoretical Review and Preliminary Empirical Evidence [129]

This is a machine-generated summary of:

Dima, S.J.; Odero, A.N.: Organic Farming for Sustainable Agricultural Production. A Brief Theoretical Review and Preliminary Empirical Evidence [129].

Published in: Environmental and Resource Economics (1997).

Link to original: https://doi.org/10.1023/a:1026472410031.

If you want to cite the papers, please refer to the original.

For technical reasons we could not place the page where the original quote is coming from.

Abstract-Summary

"The basic premise of this paper is that modern agricultural production is not sustainable in view of the high cost of chemical inputs (notably fertilizers and pesticides) used and their detrimental effects on the environment."

"The results of the experiment carried out at the Chepkoilel campus farm, Moi University, Kenya to demonstrate the differences between the yields and returns to organic manure and chemical fertilizers show that high yields are obtained from the use of organic manure."

"The high yields are also shown to be obtained in subsequent years after the application of the organic manure."

"Maximum yields are obtained from a combination of organic manure and chemical fertilizers."

3.2.6 From Industrial Production to Biosensitivity: The Need for a Food System Paradigm Shift [130]

This is a machine-generated summary of:

Dyball, Robert: From industrial production to biosensitivity: the need for a food system paradigm shift [130].

Published in: Journal of Environmental Studies and Sciences (2015).

Link to original: https://doi.org/10.1007/s13412-015-0323-z.

If you want to cite the papers, please refer to the original.

For technical reasons we could not place the page where the original quote is coming from.

Abstract-Summary

"Urban consumers in affluent cities are typically divorced from the landscapes and farmers that produce their food."

"Most food is made available to these consumers via global retail systems, operating within an overarching paradigm of industrial commodity production."

"Potentially, this vulnerability could be reduced if urban food consumption was linked to regional production, but for many cities, the volumes of food required do not match regional output."

"Canberra, Australia, could not consume all its regional production and so is in food surplus."

"Copenhagen, Denmark, could probably meet its needs from its region but chooses to reduce its food-producing land area and focus production on high-value meat products from pigs fed on imported low-value grains."

"Despite their differing food procurement strategies, producers and consumers in all three cases remain co-dependent upon each other and vulnerable to the processes being driven by the industrial paradigm."

"This paradigm shift would reduce food choice and convenience and likely increase cost, so what would motivate consumers to support it?"

"Pragmatically, these values are most likely to arise from consumers interacting with local food systems."

"It is argued, the primary value of local food systems lies not in the absolute volumes of food that they produce but in their educative capacity to foster a shift to a biosensitive paradigm."

"This new paradigm could extend concern to all food-producing landscapes and farmers, wherever on the planet they were located."

Introduction

"Despite this, urban communities need to rediscover that they and the national and international producers of their food are co-dependent on each other and that both are vulnerable to systems of production that treat food merely as an industrial commodity."

"Despite benefiting from the cheap end product that this system delivers, first-world urban consumers need to develop a sensitivity to the means by which their food is made available to them and to extend that concern to all farmers and their environments that are producing that food."

"Local consumers certainly could demand sustainably and ethically produced local food products, and farmers and business could respond to satisfy that demand."

"Demands for food originating in cities drive changes in the state of the landscapes of origin and in the wellbeing of the rural communities that manage them for food production (Seto and others [17])."

Conclusion

"As the case studies demonstrate, many urban communities are likely to secure their food from a range of local and more remote landscapes, with the question of 'which is better' dependent on a range of social and environmental contexts specific to the city in question."

"Despite the likely ongoing global dependency for food in affluent cities, there are grounds for advocating engagement with, and support for, local food production and retail systems."

"By fostering consumer awareness of their relationship with, and concern for, local food systems, political support can be generated for minimum social and environmental standards."

"With a shift to a biosensitive paradigm, citizens would come to value their active engagement with food systems that they cared about and would expect as normal everyday practice that their food systems were just and sustainable."

3.2.7 Agricultural Transitions in the Context of Growing Environmental Pressure Over Water [131]

This is a machine-generated summary of:

Gasteyer, Stephen P.: Agricultural transitions in the context of growing environmental pressure over water [131].

Published in: Agriculture and Human Values (2008).

Link to original: https://doi.org/10.1007/s10460-008-9137-x.

If you want to cite the papers, please refer to the original.

For technical reasons we could not place the page where the original quote is coming from.

Abstract-Summary

"With the growth of population pressure and concern about water quality and quantity, demands are growing to alter the relationship of agriculture to water in both these locations."

"Boulder, CO: Westview Press, 1993) to frame discussions of water and agriculture in the upper Mississippi River watershed, particularly Iowa."

"The author concludes that contested views of agriculture and countryside, as well as differing views of how agriculture must change to adapt to growing water concerns, will shape coalitions that will ultimately play a significant role in shaping the future of agriculture."

Introduction

"Social concern over issues of water quality and quantity are demanding change in the very agricultural systems that emerged out of the technological reconfiguration of hydrologic systems."

"This paper will build on that literature by nesting the movements towards a changing structure of agriculture due to concern about water quality in broader social processes."

"I hope to shed light on the strength of the water-quality lobby within society; the extent to which local agricultural production is being impacted by larger societal trends and local-, state-, and national-level social movements; the role of technical information in shaping the discussion; and the role of culture in shaping these discussions."

"Whether agriculture is practiced in the form of large capitalist, industrial production systems, or smaller, family held, traditional production systems, social mobilization around issues of water quality and quantity are changing the context for agriculture."

"New concerns about the quality of water in the Midwest and quantity of water in the West have the possibility of dramatically changing the constraints on agricultural production."

Background and Context

"Competing demands for water, because of expanding urban interests as well as growing concern about water quality and habitat, are creating a new level of debate over the resources available to and the direction of agriculture."

"These actors plus self-proclaimed sustainable farmers, conventional farmers, and farm interests are playing a role in the reshaping of agriculture through changing the relationship of agriculture to water."

"These debates unfold in the context of changing technology, both for delivery of water resources and farming, and are contingent on changing economics, markets, and a changing (and often internally contradictory) role for government."

"This context will set the stage for a brief outline of the actor networks and advocacy coalitions engaging in the debates over agriculture and their activities in the context of growing concern about water quality and availability."

A Brief Overview of the History of Agriculture and US Society

"Starting in the middle of the nineteenth century, the US government underwrote increasing numbers of projects aimed at the expansion of settlement and agricultural production through changing nature."

"Starting in the early part of the twentieth century, an array of government subsidies, from price deficiency payments to Acreage Reserve Programs, encouraged farmers to focus on the production of a few commodities (Becker [19]; Jagger and Harwood [20])."

"While government research, extension, and infrastructure encouraged farmers to grow and industrialize to fit into the US capitalist system, rhetoric, both by farmers and social/political leaders, continued to convey the agrarian image of the independent, hard-working small farmer."

"His movement was aimed more at reform—particularly government curbs on the monopoly control of the railroad robber barons so that farmers could more fairly compete in the capitalist system (Danbom [21])."

The Role of Environmental and Conservation Concerns

"By the early 1980s, erosion on these lands sparked concern among scientists and agricultural activists about environmental damage and future productivity."

"Using discourse analysis of testimony from the Agriculture Committee hearings on the 1985 Farm Bill, Glenna [22] demonstrates how members of the alliance that pushed the conservation provisions of the 1985 Farm Bill through Congress operated from fundamentally different rationalities."

"Agricultural scientists, environmental activists, and sustainable farming advocates initially proposed legislation aimed at addressing the systematic problems of soil fertility by replacing commodity-based productivity requirements with conservation-based incentives."

"The farm crisis of the 1980s provided fuel for a growing popular and institutional environmental movement have steadily moved toward presenting an alternative vision of agriculture in the 1990s."

"Urbanizing populations, both in the Midwest and the West, have developed into specific political interest groups—both environmental and simply urban—who do not see their interests in increasing agricultural production or in an agrarian lifestyle that does not involve connections to ecology."

Theoretical Framework for the Case Analysis

"Each of these interests has different actors associated with it, and they are all acting as globalization sets a context of change in rural areas and expands the potential networks for all actors at the level of the locality."

"The advocacy coalition framework (ACF) of Sabatier and Jenkins-Smith [23] allows a better understanding of mobilization and the development of political networks and policies."

"ANT and ACF can be combined to provide a framework for analyzing the actors in ongoing disputes over water and the role of agriculture in Iowa."

"ANT provides the framework for identifying the different actors, both human and biological, that are creating the context that is changing agriculture."

"ACF then provides a framework for understanding how the coalitions are being mobilized to advocate for change within current the policy structure."

Iowa, Water Quality, and Hypoxia

"And the first part of the twenty-first century, Iowa has increasingly been confronted with a need to address water-quality concerns."

"The efforts to protect water quality and mitigate water-quality concerns have, ultimately, led to a discussion about the structure and practice of farming."

"Iowa has experienced the same trends as the rest of American agriculture, with increasing acres per farmer and decreasing numbers of farmers."

"Tracing the nutrient loads upstream on the Mississippi River, they indicate that this load in all probability came from the heavy row-crop farming states, namely Missouri, Iowa, Illinois, and Minnesota (the upper Mississippi River basin)."

"An association between agricultural practices and water quality, combined with generalized concern about the future of farming, has led to processes designed to renegotiate the relationship between water and the production of agricultural products."

Mobilization Process

"While administering federal programs, such as the EPA 319 local watershed initiative grants, the government of Iowa was unwilling to take an active role in mitigating water-quality problems if that meant implicating agriculture as the polluter."

"Iowa Farm Bureau Federation and other commodity groups have supported an increasing number of watershed groups and initiatives in recent years, in an effort to build support for the notion that it is possible to maximize commodity production while protecting water quality."

"Initiatives by the governor and congressional representatives to secure resources for programs that appeal to production farmers encourage adoption of conservation practices and arguably protect water quality, but do nothing to encourage alternative agricultural production."

"Iowa Farm Bureau reports that there is significant interest by farmers in implementing water-quality protection and soil conservation practices, as evidenced by the fact that there are far more farmers wanting to sign up for government conservation and wetland restoration programs than resources available to enroll farmers (Iowa Farm Bureau [24])."

Actors, Networks, and Coalitions

"Researchers focused on soil conservation and water quality at Iowa State University (actor 11), along with associated colleagues in government agencies such as US Geologic Survey and moderate non-governmental organizations (NGOs), such as the Soil and Water Conservation Society (SWCS) (actor 12), worked at documenting BMPs that farmers could implement to protect soil and water quality."

"The administration of Iowa governor Vilsack (actor 14), elected in 1999, developed initiatives aimed at addressing water-quality concerns from agriculture."

"Large watershed initiatives such as the RRWP (actor 17), which was supported by commodity organizations and the DMU (actor 10), aimed to address water-quality issues, such as nitrogen loading in drinking source water, through landscape mitigation efforts such as buffer strips."

"Major actors within this coalition eventually supported and provided funding for initiatives to mitigate water-quality impacts even as they continued assert a production orientation."

Impacts

"Iowan farmers also had 514,680 acres enrolled in these conservation programs as of 2006 and were first in the nation in the number of acres set aside as riparian buffers, grassed waterways, contour buffer strips, field borders, and other buffers on private farmlands—reducing soil erosion, protecting water quality, and stabilizing stream banks (IDALS n.d.)."

"Such local initiatives are capable of mobilizing state and national resources to develop nutrient and land management plans that can have the impact of simultaneously prescribing methods for minimizing water-quality impairment and increasing awareness of the relationships between agriculture and water quality (Corey et al. [25])."

"Others project that farmers will begin to move into the production of perennials, such as switchgrass, as part of the ethanol boom, which may have a positive effect on water quality (Jordan et al. [26])."

"This, in part, explains the simultaneous interest in expanding corn production acres to fuel ethanol initiatives, increasing acres in alternative farming, and pursuing water-quality initiatives."

Conclusion

"The point here is that social processes that link local to state, national, and international actors determine which direction agriculture is likely to take in response to concerns about water quality."

"Important actors in this coalition instead shifted focus to conservation measures to mitigate impacts of existing production practices on water quality while not interfering with the basic production system or structure."

"This paper has demonstrated how, in Iowa, the response to local, state, national, and international water concerns resulting from agriculture led to the simultaneous implementation of conservation programs meant to mitigate environmental impacts while accepting a conventional agricultural paradigm and in growing support for alternative agriculture scenarios."

"The competing coalitions at the extremes balance each other, and the mitigation coalition is increasingly empowered to mobilize farmers and others to mitigate production impacts on water quality at the watershed, state, and regional level."

[Section 11]

"It is impossible to talk about the history of human civilization without talking about water. (Postel [27], p. 3) The importance of water for life on earth can hardly be underestimated."

"Too much water at the same place is a problem, often a disaster."

"Water management is one of the oldest tasks to be taken on by government. (Schrama [28], p. 3) It is a paradox of modern democratic societies that the state is considered both the cause and the cure of injustice. (Walton [29], p. 1) [T]he social organization of agricultural production is centered in nature as its name implies: agri (meaning "field") and culture. (Mann [30], p. 28, italics in original)."

3.2.8 Transformative Participation in Agrobiodiversity Governance: Making the Case for an Environmental Justice Approach [132]

This is a machine-generated summary of:

Coolsaet, Brendan: Transformative Participation in Agrobiodiversity Governance: Making the Case for an Environmental Justice Approach [132].

Published in: Journal of Agricultural and Environmental Ethics (2015).

Link to original: https://doi.org/10.1007/s10806-015-9579-2.

If you want to cite the papers, please refer to the original.

For technical reasons we could not place the page where the original quote is coming from.

Abstract-Summary

"This paper makes the case for an environmental justice approach to the practice and study of participation and effectiveness in agrobiodiversity governance."

"By exploring the approach through empirical examples of participation in biodiversity and environmental governance, a research framework is outlined, attempting to bridge normative and practical approaches to environmental justice, and tested on two cases of agrobiodiversity governance in Western Europe."

Introduction

"While the case for increased participation is based on its potential for more environmental effectiveness, a decade of empirical research on participation in environmental governance has produced mixed results and failed to establish causal links between participation and higher environmental quality (among many others see Young and others [31]; Carr and others [32]; Newig and Fritsch [33]; Richards and others [34]; Beierlee and Konisky [35]).”

“Considering these two points, if we are to strengthen participation as a policy goal, an objective that seems to be firmly embedded in today’s environmental policy-making, the question to address therefore is not if participation in environmental governance produces effective results, but under which conditions it does.”

“Through the comparison of two effective agrobiodiversity initiatives (i.e. of which the outcome improves agrobiodiversity), this paper analyzes the justice-relevant governance conditions underlying the success, using environmental justice as an analytical framework.”

“Part three of the paper briefly describes the research methods and the studied cases and part four then analyses how justice-relevant conditions to participation may explain the environmental effectiveness of the cases.”

Linking Participation, Justice and Effectiveness in Environmental Governance

“Since the institutionalization of public participation for sustainable development by the Rio Declaration and the Aarhus Convention, a triple shift has been observed in dealing with participatory approaches in biodiversity and environmental governance (Rauschmayer and others [36]; Rowe and Frewer [37]).”

“Environmental justice then can provide a framework to analyze participation in environmental governance (Newell [38], 238 cited in Sikor [39]).”

“Rethinking participation through this three-dimensional environmental justice framework, with its focus on socio-material empowerment and democracy, allows shifting the focus from participation as a technical tool to participation as a justice-relevant political process (Carr and others [32]; Suiseeya and Caplow [40]), thereby developing what has been termed a “post-participation approach” (Reed [41], 2418).”

“Conceived as the consequence of a decision-making process (i.e. ex-post), environmental justice here is used as an ex-ante framework: the justice-relevant governance arrangements described below are seen as enabling conditions to (parity of) participation, rather than a result of it (Young [42]).”

Operationalizing Parity of Participation in a Farming Context

“According to Fraser, parity of participation is conditioned by three interacting dimensions: an economic dimension, a cultural dimension and a political dimension.”

“The economic dimension is straightforward: participation does not just happen, it requires financial and human resources.”

“These resources may not be equally accessible to participants, especially in times of exacerbating economic inequality (Piketty [43]), conditioning their ability for social interaction.”

"Fraser's economic dimension links up with the concept of distributive justice: overcoming disparities in the access of participants to the resources for participation (money, people, land, ...) requires a (re)distribution of available resources."

"Fraser's economic redistribution is operationalized as the material conditions facilitating participation of farmers, including measures fighting economic exploitation, marginalization and deprivation such as, for instance, fair(er) pricing, financial support, human resources, and access to land."

"According to Fraser, an equitable participation requires the recognition of social and cultural differences of the participants."

Methodology and Cases

"The next section hence analyses how justice-relevant conditions of participation in two different cases have contributed to the improvement of agrobiodiversity."

"Although the two cases are very different—the first one is a pig-breeders association in southwestern Germany, while the second one is seed exchange network in southwestern France; they thus have different production systems, different farming practices, and produce different agricultural goods—they share the same outcome, i.e. an improvement of their agrobiodiversity."

"Through a least-similar case comparison, justice-relevant conditions shared by both cases are identified, which explain why, despite a contrasting configuration, these cases both have an effective outcome."

"I am using a normative rationale as an analytical framework to identify justice-relevant conditions which have contributed to environmental effectiveness."

"Although the swine is still considered to be in danger, the community counts over 1400 farmers breeding the Swabian-Hall swine and the network was broadened with local cattle and poultry breeders."

Findings: Linking Justice-Relevant Conditions and Effectiveness

"The distribution of material means conditions the effectiveness of a decision-making process: participatory governance is unlikely to produce effective results when some participants are able to dominate others in terms of resources, be they human, financial or structural (Fung and Wright [44]), or when farmers are being exploited or economically marginalized."

"In participatory plant breeding approaches, such as in the French case, the participation of land-owning farmers is a necessary condition to allow for the development of environment-specific breeding techniques (Sperling and others [45])."

"The same is true for our two cases, where specific participants have been included in the process in order to improve farming practices or to redefine supply and demand."

"While farmers keep their autonomy and define their own farming practices, the inclusion of external participant is based on the awareness that "the lack of skilled practitioners able to facilitate participatory processes is a major limiting factor to sustainable development" (Tippet and others [46], 24)."

Conclusion

"Not only methodologically, by focusing on the conditions through which improved outcomes occur, but also politically, by adopting a post-participation approach that repoliticizes the study and practice of participation."

"It is argued that environmental justice, consisting of both political theories and empirical realities, can serve as an analytical framework, which, when put in relation with empirical data on participation and outcome-effectiveness in environmental governance, can be used to identify the determinants of a positive outcome."

"This practical approach to environmental justice allows studying participation from a rights-based perspective, searching for common ground between the empowerment of participants and environmental effectiveness."

"The governance processes in our cases depart from the common approaches of participation in which farmers are given agency and/or voice within existing structures, and involve a genuine engagement with power and politics to bring about socio-ecological transition."

3.2.9 Cultivating Greater Well-Being: The Benefits Thai Organic Farmers Experience from Adopting Buddhist Eco-spirituality [133]

This is a machine-generated summary of:

Kaufman, Alexander Harrow; Mock, Jeremiah: Cultivating Greater Well-being: The Benefits Thai Organic Farmers Experience from Adopting Buddhist Eco-spirituality [133].

Published in: Journal of Agricultural and Environmental Ethics (2014).

Link to original: https://doi.org/10.1007/s10806-014-9500-4.

If you want to cite the papers, please refer to the original.

For technical reasons we could not place the page where the original quote is coming from.

Abstract-Summary

"Little is known about whether farmers' values change as they make the shift from conventional farming to organic farming."

"We found that as Thai farmers adopted organic methods, they developed an eco-consciousness."

"In comparing members of a Buddhist temple-based organic farmer support group with members of a non-profit organic agriculture support group, we found that the temple-based group adopted deeper Buddhist eco-spiritual values while the non-profit group remained more commercially oriented."

"Farmers who adopted Buddhist eco-spiritual values enjoyed greater financial and health benefits from making the organic shift."

"A higher proportion of the temple-based group eliminated their debts after making the shift ($p > 0.001$)."

"A higher proportion of members in the temple-based group felt that their physical health improved after becoming an organic farmer ($p = 0.003$)."

"We conclude that for Thai farmers who are devoutly committed to Buddhism, temple-based organic support programs provided an effective pathway to improving their finances, health and well-being."

"For Thai farmers generally, a fundamental adoption of eco-conscious values is integral to experiencing deeper non-commercial benefits from organic farming."

Introduction

"Over the same two decades, opponents of conventional farming have been promoting organic agriculture as a safer alternative that builds farmer self-sufficiency and restores biodiversity (Hecht [47]; McNeely and Scherr [48])."

"These facts show that the vast majority of farmers in LMICs have not adopted organic farming."

"We can also assume that farmers who recognize the potential benefits of organic farming but have not switched, do not believe that organic farming is feasible."

"The situation in Thailand is much the same; in 2011 only 0.18% of agricultural land was farmed organically."

"From the perspective of advocating the adoption of organic farming, it is important to know why some farmers switch to organic farming."

"What influences farmers to shift to organic farming?"

"What benefits do organic farmers perceive from making the shift?"

"What types of farmers perceive the greatest benefits from shifting to organic farming?"

Background

"Domestic and international demand for organic food products has attracted some Issan smallholder farmers to organic agriculture (Roitner-Schobesberger and others [49])."

"A small group of farmers at the Dharma Garden Temple created an organic rice program based on the Five Precepts (Kaufman [50])."

"Moral Rice farmers have followed organic agriculture methods, worked in closely-knit groups, and monitored each others' progress in living according to the Five Precepts."

"Throughout Thailand, including Issan, Green Net Cooperative/Earth Net Foundation (Green Net) has provided assistance to organic farmers with the following objectives: Green Net Cooperative serves as a marketing channel for small-scale organic farmers, incorporating fair-trade principles in its marketing activities by combining organic agriculture and fair-trade as its core policies."

"Boonman and Anpim's [51] study in Yasothon Province suggested that rural farmers would stand to benefit from organic rice production methods aided by agricultural chemicals."

Methods

"The farmers we recruited to participate in the study were members of one of two organic farming support networks: the Dharma Garden Temple or Green Net."

"We tried to recruit conventional farmers to participate in the first part of this study as a comparison group, but conventional farmers were reluctant to respond to questions about their environmental values and their decisions to not farm organically."

"In Phase 3, we asked participants structured questions about organic farmers' environmental values, spirituality and organic farming practices."

"This approach resulted in a sample of 50 organic farmers for the qualitative interviews in Phase 2."

"Using sampling methods based on grounded theory, theoretical saturation was reached after analysis of in-depth interviews 50 farmers (45 organic rice farmers and five conventional rice farmers)."

Findings

"Participants reported that recently BAAC has supported some training programs on organic farming methods conducted by experts from the Dharma Garden Temple and Green Net."

"Although many Green Net farmers considered themselves devout Buddhists, Dharma Garden Temple farmers mentioned Buddhism's Five Precepts more frequently as a reason for adopting organic farming methods."

"One participant explained the importance of Buddhist values: Without the Five Precepts in a farmer's heart, they can't be a true organic farmer because they might easily break the rules (Dharma Garden Temple farmer)."

"Many Dharma Garden Temple farmers linked the Five Precepts to the use of organic farming methods: I believe organic farmers must believe in themselves to produce organic food in order to protect nature."

"Regardless of membership in Dharma Garden Temple or Green Net, the amount of debt the participants carried was significantly correlated with self-perceived improvements in health status after shifting to organic farming (Spearman's ρ 0.46, $p > 0.001$)."

Discussion

"Participants' values changed based on the organizations they belonged to, their observations, their understanding of the natural environment, and their experiences with organic farming."

"For participants in both groups, health concerns were their main motivation for shifting to organic farming."

"In the words of one participant who switched to organic farming but later purportedly died of complications from exposure to agro-chemicals during the time when he was a conventional farmer: I see nature all over my farm, green, animals, and it brings me peace and happiness."

"Our research is consistent with findings in Curry [52] showing that participants' preexisting spiritual values influenced the degree to which they were open to adopting organic agriculture methods, particularly among members of the Dharma Garden Temple."

"A longitudinal study was not feasible because the farmers who participated in this study shifted to organic methods at different points in time."

Conclusion

"Our findings raise a question: Is it essential for Thai conventional farmers to adopt eco-spiritual values to become organic farmers?".

"If the answer is yes, then our study suggests that among Thai smallholder conventional farmers a lack of interest or resistance to adopting eco-spiritual values may be the main barrier that keeps them from considering a shift to organic farming."

"If this is the case, then it may be difficult for many Thai conventional farmers to enjoy the benefits of a reduction in debt and improved health that come with the shift to organic farming."

"Our study shows that whether others make the shift will depend largely on the influence of organic farmer support groups like Dharma Garden Temple and Green Net."

3.2.10 Sustainable Agriculture in Canada and Cuba: A Comparison [134]

This is a machine-generated summary of:

Hiranandani, Vanmala: Sustainable agriculture in Canada and Cuba: a comparison [134].

Published in: Environment, Development and Sustainability (2009).

Link to original: https://doi.org/10.1007/s10668-009-9223-2.

If you want to cite the papers, please refer to the original.

For technical reasons we could not place the page where the original quote is coming from.

Abstract-Summary

"Over the last 50 years, most of the world's agriculture has transitioned into industrial agriculture that requires greater inputs of fossil-fuel energy, water, synthetic pesticides and fertilizers, which have created substantial harmful effects on air, soil, water and biodiversity."

"Sustainable farming that uses less chemicals and fossil-fuel energy and emphasizes localized production and consumption has come to be viewed as an eco-friendly alternative to modern agriculture."

"This paper will examine the concept of sustainable agriculture and compare and contrast its practice in Canada and Cuba."

Introduction

"Industrial agriculture, with its focus on increasing crop yields, has introduced new crop varieties that require greater inputs of fossil-fuel energy, water, synthetic pesticides and fertilizers compared to traditional varieties (Picone and Van Tassel [53])."

"While these transformations have generated unprecedented increases in world food production, modern agriculture has created substantial harmful effects on air, soil, water and biodiversity (Kimbrell [54]; Maynard and Nault [55]; Miller and Hackett [56])."

"Given the huge ecological footprint of industrial agriculture, environmentalists have advocated better management practices, such as sustainable agriculture, which emphasizes farming practices that protect soil, water and biodiversity."

"Despite the harmful effects of energy- and chemical-intensive agriculture, about 80% of the world's food supply is still produced by industrialized agriculture and 20% by environmentally friendly agriculture."

"The objectives of this essay are (1) to describe the extent, forms and key policies influencing sustainable farming practices in Canada and Cuba and (2) to compare and contrast sustainable agriculture in Canada and Cuba and examine its effectiveness in the two countries."

Sustainable Farming: An Alternative to Industrial Agriculture

"At best, sustainable agriculture has been loosely defined as "agricultural technologies and practices that maximize the productivity of the land while seeking to minimize damage both to valued natural assets (soils, water, air and biodiversity) and to human health (farmers and other rural people, and consumers)" (Pretty [57], p. 171)."

"Unlike industrial agriculture, the sustainable approach is multifunctional—it not only produces food, but also contributes to various public goods such as clean water, wildlife, carbon sequestration in soils and biodiversity."

"Sustainable agriculture encompasses a wide range of technologies and practices including water harvesting, soil and water conservation, rotational grazing, crop rotation, composting, using livestock and farmyard manures, irrigation scheduling and management, restoration of degraded or abandoned land, urban agriculture, integrated pest management, polyculture and organic farming. (Miller and Hackett [56]; Pretty [57])."

"Organic farming is a variant of sustainable agriculture that uses little or no chemical pesticides and fertilizers (Miller and Hackett [56])."

Sustainable Agriculture in Canada

"Farmers, governments and other stakeholders in Canadian agricultural industry have become increasingly cognizant of the need to integrate environmental factors into farm policies."

"Agriculture and Agri-Food Canada's 2007–2009 Strategy for Sustainable Development integrates sub-programs on biodiversity, soil protection, water use and quality, and development of agri-environmental indicators into Canada's Agricultural Policy Framework (Agriculture and Agri-Food Canada [58]; Janzen and others [59])."

"The report emphasizes that Agriculture and Agri-Food Canada recognizes the importance of sustainable agriculture, whose goal is not only to value and protect the environment and to safeguard the health and well-being of Canadians but also to help the industry prosper economically."

"In June 1999, the Government of Canada introduced a national standard for organic farming, which constituted the first real federal government intervention directed specifically to this sector of the agricultural industry (Forge [60])."

Sustainable Agriculture in Cuba

"Another remarkable Cuban achievement is the growth of urban agriculture that turned unused city plots into large vegetable gardens for feeding urban people."

"The earlier food shortages and the rise in food prices transformed urban agriculture into a very profitable activity for Cubans."

"Havana's 2002 strategic plan gives urban agriculture its own zoning category, for the first time thereby protecting this form of farming (Rosset [61])."

"The net result of the changes in the Cuban farming sector is that currently the country has the most sustainable agricultural production in the world (Levins [62]; Sullivan [63])."

"In less than a decade, Cuban agriculture revolutionized itself from high-input, non-sustainable, monoculture to a diversified, organic and sustainable model demonstrating that chemical-intensive industrial agriculture is not the only way to produce food."

"The important lessons from Cuba that can be applied elsewhere are agroecology, fair prices, land reform, local production, urban agriculture (Rosset [61]), investment in human development, education and R &D, and making agriculture a viable occupation (Tharamangalam [64])."

Key Differences and Similarities Between Canadian and Cuban Sustainable Farming

"Cuba has transitioned almost completely to sustainable agriculture, while only 1.5% of Canada's agricultural output is produced using organic agriculture (Miller and Hackett [56])."

"In Canada, organic farming is a profit-oriented industry, urban agriculture is not as widespread as in Cuba; however, organic agriculture is increasingly practiced and the growing number of community-supported agriculture initiatives (CSAs) is a positive aspect in Canadian agriculture sector."

"While Cuban farmers are world leaders in large-scale organic production and in the development of urban agriculture, Canadian farmers are world leaders in sustainable management practices and have expertise in processing, packaging, labeling and marketing (Holm [65])."

"The review of literature pertaining to sustainable agriculture and organic farming in Canada undertaken for this paper revealed that the economic, rather than the environmental dimension, is the main consideration in the move toward organic agriculture."

"While sustainable agriculture is considered utopian vision in Canada, it is government policy and widespread agricultural practice in Cuba."

Concluding Remarks

"Moving toward sustainable agriculture is crucial for improving environmental quality as well as human health."

"This paper described sustainable agriculture, and compared and contrasted this alternative farming approach in Cuba and Canada."

"Although Canada has made considerable progress toward sustainable agriculture since the 1990s, most reforms remain piecemeal."

"Cuba has been lauded as the first country to adopt sustainable agriculture on a nationwide scale."

"It is imperative to note that the differences in the scenario of sustainable agriculture in Canada and Cuba reflect the two diverging viewpoints in sustainable agriculture."

"While sustainable agriculture is largely seen as utopian vision in some quarters in Canada, it is government policy and widespread agricultural practice in Cuba."

"Current industrial agricultural techniques in many countries, including Canada, are dependent on expensive technology and chemical inputs that are environmentally unsustainable over the long term."

3.2.11 Agroecology and Organic Farming Foster Soil Health by Promoting Soil Fauna [135]

This is a machine-generated summary of:

Domínguez, Anahí; Escudero, Héctor Javier; Rodríguez, María Pía; Ortiz, Carolina Elizabeth; Arolfo, Romina Vanesa; Bedano, José Camilo: Agroecology and organic farming foster soil health by promoting soil fauna [135].

Published in: Environment, Development and Sustainability (2023).

Link to original: https://doi.org/10.1007/s10668-022-02885-4.

If you want to cite the papers, please refer to the original.

For technical reasons we could not place the page where the original quote is coming from.

Abstract-Summary

"The industrial paradigm is the current predominant agricultural model in the Pampean region of Argentina."

"We aimed to assess if alternative farming systems foster soil health-indicated by soil fauna-compared to industrial systems in the Pampean region of Argentina."

"We sampled soil meso and macrofauna in fields under agroecology (AEC), organic farming (ORG), industrial agriculture (IND) and semi-natural grasslands (GRA)."

"Our results suggest a better development of nutrient cycling and soil structure formation in the alternative systems than in the industrial; together with a better preservation of soil health, crucial for achieving sustainability."

"We think that recognizing the negative impact of the current model on soil health is a necessary step to promote a change toward agricultural systems that are more socially and environmentally sustainable, in Argentina and also in other Latin American countries that face similar situations."

Introduction

"Alternative farming systems in Argentina represent a very small proportion compared to industrial agriculture."

"Alternative farming systems propose to increase agricultural sustainability through the promotion of soil biological communities to enhance soil ecosystem processes crucial for preserving soil health (Altieri and others [66]; Tittonell [67])."

"Agricultural management practices, either industrial or alternative, affect soil fauna abundances and diversity compared to undisturbed soils."

"There is evidence that organic farming can favor some taxa of soil organisms compared to industrial agriculture (e.g., Bengtsson and others [68]; Domínguez and others [69]; Domínguez & Bedano [70]; Hole and others [71]; Mäder and others [72]; Ostandie and others [73]; Werner & Dindal [74]).”

“The development of field research in regards to the effect of different crop farming systems on soil meso and macrofauna is notoriously slow in the Pampean region.”

“In this article we aimed to assess if alternative farming systems foster soil health -indicated by soil fauna- compared to industrial systems in the Pampean region of Argentina.”

Materials and Methods

“For macrofauna sampling, in each site 5 soil monoliths of 25 × 25 × 20 cm (separated by at least 20 m from each other), were dug and hand-sorted for extracting all the macroinvertebrates visible to the naked eye from 0 to 10 cm and 10–20 cm soil depth (ISO [75]).”

“In 33 sites (GRA, ORG and IND) the whole meso and macrofauna (including earthworms) communities were sampled, totaling 410 samples for every group.”

“Earthworms were also sampled in another 54 sites (GRA, ORG, IND and AEC) with the same methodology but for 0–20 cm soil depth.”

“The mesofauna community that inhabit litter was analyzed together with the macrofauna community that inhabit the interface litter-soil (litter plus 0–10 cm soil depth).”

“The effect of agricultural systems on earthworm abundance and biomass (in 0–20 cm soil depth) and on soil meso- and macrofauna abundances (for each separate soil depth) was analyzed using generalized linear mixed models (GLMM).”

Results

“In litter, springtails and oribatid mites had similar abundances between ORG and IND.”

“Mesostigmatid mites instead were significantly more abundant in ORG than in IND ($p < 0.05$).”

“Astigmatid mites, though a minority in the community, were as abundant in ORG as in GRA and significantly less abundant in IND ($p < 0.05$).”

“Soil prostigmatid mites were significantly more abundant in ORG than in IND ($p < 0.05$).”

“In the first depth, spiders and larvae insects were significantly more abundant in ORG compared to IND ($p < 0.05$).”

“Ants, termites, hemipterans and orthopterans were negatively affected by ORG and IND compared to GRA ($p < 0.05$).”

“In the second depth beetles were significantly more abundant in ORG than in GRA and IND ($p < 0.05$).”

Discussion

"The increased earthworm abundance and biomass found in AEC and also in ORG suggest that the mentioned ecological processes to which they contribute have a better performance in the alternative systems than in the industrial one, and it is likely that this would contribute to improve soil health and sustainability of the agroecosystem."

"This is probably related to an increase in the availability of trophic resources in organic systems (through pastures, higher residue inputs to the soil, and sometimes spontaneous vegetation) as well as to lower competition with oribatid mites for the organic resources."

"Both groups are important regulators of soil carbon cycle and their abundances in ORG system suggest a strengthening of litter decomposition, soil aggregation processes and soil organic carbon stabilization altogether with earthworm activity."

"Higher abundances of potworms and millipedes were found in IND compared to ORG, this could reflect a reduction in the competition for the organic resources with earthworms (Cochran [76]; Snyder and others [77])."

Conclusions

"Alternative farming is more effective in preserving soil health than the currently dominant industrial paradigm."

"From a soil ecology perspective, our results confirm some early warnings of David Pimentel about the environmental consequences of the industrial agricultural model."

"It is necessary to look at this evidence as part of a wider context since this knowledge may be not comfortable (sensu Rayner [78]) because it defies the current agricultural paradigm."

"To analyze the uncomfortable nature of the knowledge provided here as well as the influence of governments, industrial farmers' associations, chemical companies, and media in concealing environmental issues related to the current industrial agricultural paradigm needs further research."

3.2.12 Commoning the Seeds: Alternative Models of Collective Action and Open Innovation Within French Peasant Seed Groups for Recreating Local Knowledge Commons [136]

This is a machine-generated summary of:

Mazé, Armelle; Calabuig Domenech, Aida; Goldringer, Isabelle: Commoning the seeds: alternative models of collective action and open innovation within French peasant seed groups for recreating local knowledge commons [136]

Published in: Agriculture and Human Values (2020).

Link to original: https://doi.org/10.1007/s10460-020-10172-z.

If you want to cite the papers, please refer to the original.

For technical reasons we could not place the page where the original quote is coming from.

Abstract-Summary

"Our empirical analysis emphasizes, using the most recent advances in the IAD/SES framework, the distributed and collaborative knowledge governance in a French peasant seed network as a key driver for reintroducing cultivated agrobiodiversity and on-farm seed conservation of ancient and landrace varieties."

"These inclusive peasant seed groups developed alternative peer-to-peer models of collaborative peasant-led community-based breeding and grassroots innovations in the search for more resilient population varieties."

"Our results highlight the various models of collective action within the network and discuss the organizational tradeoffs of opting out of peasant seed activities and recreating a shared collective knowledge base on the benefits of restoring cultivated agrobiodiversity."

"It helps us better understand how modern peasant seed groups function as epistemic communities which contributes to envisioning alternative agricultural systems."

Introduction

"We analyze, based on recent theoretical advances on knowledge commons (Hess and Ostrom [79, 80]; Hess [81]; Frischmann and others [82]), the role of the distributed and collaborative knowledge governance adopted by a French peasant seed network, the "Réseau Semences Paysannes" (RSP) and its "Participatory Wheat Breeding Group" (hereafter PWBG), on small grain cereals."

"We also demonstrate that peasant seed initiatives provide particularly relevant empirical cases to develop a better understanding of knowledge commons, in both their material and immaterial dimensions, and to address the specific cognitive, social and organizational framings at stake in individual and collective learning in the context of knowledge restoration (Douglas [83]; Hess and Ostrom [79]; North [84]; Hess [81])."

"Our study highlights the capacity of local peasant seed initiatives to develop innovative research collaborations and to more effectively blend the peasant's experiential and situated knowledge with the most recent scientific advances in population genetics (Enjalbert and others [85]; Goldringer and others [86]; Demeulenaere and others [87])."

Analytical Framework

"In the recent literature, a growing number of studies have considered the role of alternative seed systems as "commons", e.g., shared collective resources by local place-based peasant communities, but also as involving new innovative forms of civic engagement and stewardship towards a larger public (Wirz and others [88]; Girard and Frison [89]; Wolter and Sievers-Glotzbach [90]; Mazé and others [91])."

"The study of traditional natural and agrarian commons has been partially disconnected from studies on knowledge and cultural commons (Hess and Ostrom [80]; Hess [81]; Madison and others [92]; Frischmann and others [82])."

"Knowledge and the set of cognitive representations (or mental models) and underlying values supporting the social processes of individual and collective learning are also key aspects of the sustainability of SES (Hess and Ostrom [79]; Ostrom [93]; Pahl-Wostl [94])."

"When considered as a generative practice, coproducing, creating and preserving shared common knowledge requires the active participation and engagement of individuals, raising specific problems of collective action, called here "knowledge dilemma", different from other social dilemmas."

Methodology and Data

"Our analysis concentrates on the PWBG of the RSP and its 14 collective groups as well as several individual peasants involved in 2018."

"Our methodology builds upon the most recent theoretical and analytical developments within the IAD/SES framework, allowing a broader approach to the social and knowledge aspects of collective action and to the adaptive governance of SES (Ostrom [93]; Hess [81]; McGinnis and Ostrom [95]; Cole and others [96])."

"The IAD/SES framework was used here as a diagnostic guideline to analyze the formal and informal rules-in-use within the PWBG groups and the political and institutional context framing their action arenas (Ostrom and Cox [97]; Ostrom and Basurto [98])."

"Specific attention was paid to the political and institutional factors, the socialization processes and valuation models, as well as the missions and resources (human, financial, and material) influencing the specific "rule configurations" and models of collective action (Ostrom [99])."

Results and Analysis

"Through the RSP activities, individual knowledge acquisition and re-creation were harnessed via collective knowledge exchanges and familiarization with the practical experiences of other peasants in handling landraces and ancient varieties, as well as restoring cultivated agrobiodiversity."

"Whereas early core PWBG members were attracted by the idea of dynamic seed management and on-farm conservation of landraces and ancient varieties as a strategy to reintroduce crop genetic diversity in the fields, others also rapidly engaged in innovative breeding strategies and considered "new" or "interesting" seed types, such as cross-breeds, heterogeneous crop populations, "seed mixtures", or even foreign varieties (Rivière [100])."

"By acknowledging the central role of peasants as knowledge holders and key actors in the dynamics of on-farm conservation of agrobiodiversity, the PWBG proposed a shift from traditional approaches to research on plants and seeds towards more collaborative, dynamic and egalitarian approaches (Enjalbert and others [85]; Goldringer and others [86]; Demeulenaere [101])."

A diversity of Models of Collective Action and Self-Governance

"We analyze the specific models of collective action and collaborative seed governance adopted by local peasant seed groups participating in the PWBG as they influenced their capacity to develop their seed activities and their sustainability over time."

"The importance for the PWBG of stimulating interpersonal seed and knowledge exchanges as social practices of commoning is reflected in the priority given by the organization to annual or semiannual "open days", which allow individual peasants to meet and participate in different seed activities."

"According to our interviews, in most PWBG groups, their key seed activities are handled by a core group of more experienced peasants (1 to 50 for the larger groups), whereas seed exchanges involve a larger number of peasants."

Comparing Two Seed Governance Models: Pétanielle and Graine de Noé

"This locally centralized governance model involves (i) more formalized rules and conventions; (ii) a regional and centralized platform for seed conservation and multiplication (with a capacity of 150 plots) and a centralized information and data management system managed by the facilitator in accordance with the generic PWBG default guidelines defined at the national level; (iii) the mandatory centralization of seed exchange through the association, which provides sanitary protections to prevent wheat bunt as specified by the legislation; and (iv) technical follow-up of seed exchanges through the redistribution of seed based on peasant's demands and on principles of equity."

"New PWBG groups tend to adopt some of the formal aspects of Model 2, but Model 1, with its highly decentralized and distributed governance, is essential for certain seed activities, such as in situ conservation, propagation and breeding, as well as for promoting a collaborative and inclusive approach based on strong ethical values and principles of knowledge sharing and horizontality among participants that is at the core of the social practices of seed commoning."

Discussion and Conclusion

"Our study adds to this emerging literature in European countries on peasant seed communities by analyzing self-governance under less favorable conditions than those found in small, homogeneous communities whose right to self-govern is recognized by outside authorities (Wirz and others [88]; Koller and others [102])."

"In their works, Hess and Ostrom [80] acknowledged the need for a more inclusive approach to the study of knowledge commons in which knowledge is not viewed as a finite pool of resources but as an ongoing collective process of joint knowledge co-creation and innovation."

"By presenting the diversity of seed practices and their models of collective action, our study provides a different perspective on local peasant seed groups and the roles of knowledge co-creation and dissemination in collaborative and scientifically grounded breeding to restore and maintain crop genetic agrobiodiversity in European landscapes."

3.2.13 Direct Effects of Climate Change on Productivity of European Aquaculture [137]

This is a machine-generated summary of:

Cubillo, Alhambra Martínez; Ferreira, João G.; Lencart-Silva, João; Taylor, Nick G.H.; Kennerley, Adam; Guilder, James; Kay, Susan; Kamermans, Pauline: Direct effects of climate change on productivity of European aquaculture [137].

Published in: Aquaculture International (2021).

Link to original: https://doi.org/10.1007/s10499-021-00694-6.

If you want to cite the papers, please refer to the original.

For technical reasons we could not place the page where the original quote is coming from.

Abstract-Summary

"Aquaculture managers and industry must take into account the impact of climate change on production and environmental quality to ensure that sector growth is sustainable over the coming decades, a key requirement for food security."

"The potential effects of climate change on aquaculture range from changes to production capacity in existing cultivation areas to changes in the areas themselves, which may become unsuitable for particular species, but also suitable for new species."

"This work presents a modelling approach used to predict the potential effects of climate change on aquaculture, considering six key finfish and shellfish species of economic importance in Europe: Atlantic salmon (Salmo salar), gilthead seabream (Sparus aurata), sea bass (Dicentrarchus labrax), Pacific oyster (Crassostrea gigas), blue mussel (Mytilus edulis) and Mediterranean mussel (Mytilus galloprovincialis)."

"Climate projections for present-day conditions; mid-century (2040–2060) and end-of-century (2080–2100) were extracted from regionally downscaled global climate models and used to force bioenergetic models."

"Projected temperature changes will have variable effects on growth depending on the species and geographic region."

"Most of the marine finfish simulated were projected to have decreased feeding efficiency in both mid-century and end-of-century climate scenarios."

"Bivalve shellfish showed a decreasing trend with respect to most productivity parameters as climate change progresses, under both emission scenarios."

Introduction

"Anthropogenic-induced warming of aquatic habitats (marine and freshwater), as well as altered precipitation patterns, is expected to have direct and indirect effects on aquaculture such as changes in disease occurrence, physiology, growth performance and/or feed conversion efficiency and reproduction of farmed fish and shellfish (Catalán and others [103]; Reid and others [104]; Dahlke and others [105])."

"In order to assess the effects of climate change on cultivated organisms, models must be applied to predict (a) the changes in the driving variables that determine the performance of aquaculture farms and (b) the consequences of those changes on the performance of the farms using three key performance indicators (KPI): production, environmental effects and profitability."

"We selected and applied a set of models to address both the driving forces and the response of aquaculture farms to the direct effects of different climate change scenarios."

"Develop and implement a multi-model framework to address direct effects of climate change on growth performance and aquaculture production; 2."

Material and Methods

"The regionally downscaled climate models provided projections for daily mean temperature, salinity, chlorophyll and other biogeochemical variables, which were used as drivers by individual growth models and local-scale production models."

"For the finfish species, a selection procedure was devised to choose the pair of years which would capture the range of effects of future temperature variability, the key driver for the finfish models, within each 20-year time slice."

"The validated FARM model was then used to simulate different climate change scenarios, using mid- and end-of-century projections from regionally downscaled climate models (NORWECOM for Norwegian salmon and POLCOMS-ERSEM for the other species and locations), under two different carbon concentration scenarios: RCP 4.5 and RCP 8.5—RCP 4.5 for Norway (Kay and others [106])—which are further downscaled to the farm level."

Results

"The productivity of Mediterranean mussel aquaculture in Portugal was projected to remain similar or improve slightly under both emission scenarios when compared to the present-day situation; the worst results were predicted for the end-of-century high-emission scenario."

"At mid-century, blue mussel bottom culture in the Netherlands was similar to present-day but was predicted to decrease in production (and increase in economic uncertainty) at the end-of-century scenario."

"73% of days per year in the North Sea region were predicted to be in the optimal temperature range for blue mussel growth."

"Only 51% of days in the Mediterranean region were predicted to be in the optimal temperature range for blue mussel growth, and this was predicted to reduce to 47% and 43% under RCPs 4.5 and 8.5, respectively."

Discussion

"The projected temperature-related changes of climate change observed in this study had a different impact on each aquaculture species, with gilthead seabream in sea cages in the western Mediterranean and offshore suspended Mediterranean mussel culture in SW Portugal as the most and least vulnerable, respectively."

"The impact of direct effects of climate change such as ocean warming will mostly depend on (i) the thermal tolerance window and adaptation capacity of each species and (ii) how close the tolerance and optimal temperature windows for each species are from the present-day temperature range at each farm."

"Due to the coarse spatial resolution of downscaled climate model predictions, the modelling results of the present study should be interpreted with caution, particularly with respect to shellfish farming, although they do provide useful insights into the levels of spatial variability we might expect in species' responses to climate change in coming decades, and the importance of considering local-scale impacts (Montalto and others [107])."

Conclusions

"In order to contribute to this goal, our study applied a multi-model framework, including atmospheric models, water circulation, biogeochemistry and physiology and population dynamics of cultivated species, to explore and predict impacts of future climate change projections on the productivity of the European aquaculture sector."

"This multi-model approach helps understand to what extent the direct effects of climate change will present challenges (or opportunities) for different aquaculture species and geographical regions."

"The findings of this study are relevant for medium-term aquaculture planning, as the local impact of climate-related changes on physiology and growth will create species-specific risks and opportunities in terms of growth efficiency, productivity and profitability."

"Key indicators obtained by modelling growth, production and environmental effects in response to climate change will help quantify the challenges ahead."

3.2.14 Reductions in Water, Soil and Nutrient Losses and Pesticide Pollution in Agroforestry Practices: A Review of Evidence and Processes [138]

This is a machine-generated summary of:

Zhu, Xiai; Liu, Wenjie; Chen, Jin; Bruijnzeel, L. Adrian; Mao, Zhun; Yang, Xiaodong; Cardinael, Rémi; Meng, Fan-Rui; Sidle, Roy C.; Seitz, Steffen; Nair, Vimala D.; Nanko, Kazuki; Zou, Xin; Chen, Chunfeng; Jiang, Xiao Jin: Reductions in water, soil and nutrient losses and pesticide pollution in agroforestry practices: a review of evidence and processes [138].

Published in: Plant and Soil (2019).

Link to original: https://doi.org/10.1007/s11104-019-04377-3.

Copyright of the summarized publication:

If you want to cite the papers, please refer to the original.

For technical reasons we could not place the page where the original quote is coming from.

Abstract-Summary

"Agroforestry systems combining trees with crops or pastures have been widely used to reduce water, soil, and nutrient losses and associated water pollution from agricultural lands in both temperate and tropical regions."

"This paper synthesized the available evidence on the reduction in surface runoff, soil erosion, nutrient, and pollutant losses (e.g., herbicides, pesticides, and antibiotics) to quantify the effectiveness of agroforestry systems on water quality improvement based on published studies."

"Agroforestry systems reduced surface runoff, soil, organic carbon, and related nutrient losses by 1–100%, 0–97%, –175–92%, and –265–100%, respectively, with average values of 58%, 65%, 9%, and 50%, respectively."

"A comprehensive science-based review is needed to generalize agroforestry design and site adaptability for water and soil conservation where climatic, geographical, ecological, and socio-economic conditions are relatively similar in the world."

Introduction

"Both pecan (Carya illinoenis)–cotton (Gossypium hirsutum) alley cropping and silvopastoral systems in heavily fertilized soils in Florida, US, are more efficient at reducing nutrient leaching (e.g., N and P) than treeless pastures (Nair and Graetz [108]; Nair and others [109])."

"This paper synthesized the literature on the efficacy of agroforestry practices worldwide in terms of reducing water, soil, nutrient losses and pesticide transport."

"Besides, we described and assessed the water, soil, nutrient, and pesticide losses associated with the following major agroforestry practices: (1) silvoarable agroforestry, which includes alley cropping (synonymous to hedgerow intercropping) and agroforests combined with shaded-perennial crop systems (e.g., cacao, coffee and tea); (2) silvopastoral systems, which constitute an integration of trees, forage (pasture) and livestock production; (3) linear tree plantings including windbreaks, shelterbelts, and living fences (Muschler [110]; Iwasaki and others [111]); and (4) riparian and upland buffer strips that include at least one woody (ligneous) perennial plant."

Materials and Methods

"We chose the following six indicators of agroforestry related to erosion and nutrient loss reduction, for the meta-analyses: (i) runoff, defined as the flow of water, which can also occur as a result of excess stormwater, over the Earth's surface (in mm) (also known as overland flow); (ii) soil erosion, defined as the soil displacement caused by the flow of surface runoff (in t ha^{-1}); (iii) organic carbon (C) loss, defined as the variation in C stock associated with leaching and water erosion (kg ha^{-1}); and (iv–vi) nutrient losses of N, P, and K, defined as the nutrient losses caused by leaching, runoff and soil losses (in kg ha^{-1} or mg L^{-1})."

"The change in a service indicator in a case study, namely, the reduction in service loss (M, in %) was calculated as follows: where I_0 and I_{AF} are the mean indicators (i–vi, except iii) measured in control plots and agroforest plots, respectively."

"This study is therefore based on the effect of agroforestry on the reduction of services loss expressed in proportion values (comparison between agroforestry relative to control as baseline)."

Estimation of Reductions in Water, Soil, and Nutrient Losses

"Such agroforestry practices are considered to effectively control water, soil and related nutrient losses, in both temperate regions and humid tropical regions (Nair [112])."

"Compared to the use of single pastures, silvopasture can reduce water and soil nutrient losses more effectively due to environmental benefits from trees in such types of agroforestry systems (Nair and Graetz [108])."

"The removal efficacy of soil and nutrients by agroforestry buffer strips is affected by various factors such as size, slope, age, water content, and climate conditions (Lee and others [113]; Otto and others [114])."

"No significant difference was observed in terms of reducing soil loss among the four agroforestry systems although the following order of efficacy was determined: riparian and upland buffers (69%) > silvopastoral systems (68%) > silvoarable agroforestry (66%) > linear tree plantings (55%)."

"In comparison with mono-cropping systems, agroforestry systems effectively controlled soil erosion and related nutrient loss and water contamination."

Reductions in Pesticides, Herbicides and Other Pollutants

"Various protective agroforestry practices have been developed to reduce the effects of pollutants on water resources (Pavlidis and Tsihrintzis [115]; Dollinger and others [116])."

"Silvoarable agroforestry reduced pollutants by 41% (range of 7% to 100%)."

"Both of these agroforestry systems could reduce pollutants by up to 100%."

"Several studies have mainly focused on the trapping efficacy of buffer strips, while few researchers have investigated the function of silvoarable agroforestry, silvopastoral systems and linear tree plantings in terms of reducing pesticides, herbicides and other pollutants."

Water-Related Biophysical Drivers in Agroforestry Ecosystems

"Due to greater resistance to surface flow afforded by the presence of litter or buffer vegetation strips and the "safety net" afforded by the presence of (deep-rooted) trees (Pestemer and others [117]; Hoffman and others [118]; Udawatta and others [119]; Schultz and others [120]), these mitigation processes are characterized by enhanced infiltration and increased trapping efficiency for water, sediment and nutrients."

"On the other hands, some studies have also pointed out that some trees, shrubs, and forested buffer strips in agroforestry systems may be too young to have a detectable impact on the reduction in soil and nutrient losses due to the lack of effective ground surface cover (Schmitt and others [121])."

"Besides, the trees transpiration is considerably greater than agricultural crops; thus, the soil of inter-cropped agroforestry sites will tend to absorb relatively large amounts of water due to its lower water availability, reducing the leaching losses (van Noordwijk and Lusiana [122])."

"To reduce the competition for soil water and nutrients, the tree and crop species that present different rooting patterns should preferably be considered in agroforestry practices."

Conclusions and Outlook

"Besides, considering the significance of agroforestry systems in reducing water and soil nutrient losses, our review demonstrates that further research should also focus on: (1) addressing the impact of fertilization and pesticides management on the efficiency of agroforestry in reducing nutrient and pollutant loss and subsequent contamination of water streams; (2) clarifying the specific mechanisms of soil organisms involved in reducing water, soil and related nutrient losses, including a range of direct and indirect processes, such as improving soil properties and promoting biochemical cycles; (3) demonstrating the effect of biodiversity on pest management in agroforestry systems, as there may be a negative or positive impact with different species; (4) assembling systematic knowledge of desirable agroforestry design and determining site adaptability in reconciling with policy-makers and local environmental conditions; (5) enhancing the study of above- and belowground tree-crop interactions, such as resource competition and species coexistence in agroforestry ecosystems; and (6) highlighting agroforestry practices for biodiversity maintenance and mitigation of climate change on a global scale rather than on a site scale."

3.2.15 Increasing the Adaptive Capacity of Organic Farming Systems in the Face of Climate Change Using Action Research Methods [139]

This is a machine-generated summary of:

Bloch, Ralf; Knierim, Andrea; Häring, Anna-Maria; Bachinger, Johann: Increasing the adaptive capacity of organic farming systems in the face of climate change using action research methods [139].

Published in: Organic Agriculture (2015).

Link to original: https://doi.org/10.1007/s13165-015-0123-5.

If you want to cite the papers, please refer to the original.

For technical reasons we could not place the page where the original quote is coming from.

Abstract-Summary

“Adapting agricultural systems to the effects of climate change requires farm-specific innovations and adaptive measures.”

“Such innovative measures were developed for organic farming systems in the federal state of Brandenburg, Germany by a regional innovation network using an action research approach.”

“The goal of the presented approach was to address climate change-related farm-specific problems, develop specific adaptation measures and learn new approaches such as proactive risk management and strategic adaptive behaviour.”

“In a cyclical process of analysis, planning, action and reflection, the network of researchers and organic farmers repeatedly used SWOT (strengths, weaknesses, opportunities and threats) analyses to structure the transdisciplinary innovation and adaption process.”

“The organic farmers identified as main weaknesses the water and nitrogen supply likely to be worsened by climate change; then, farm-specific adaption measures were identified and tested by conducting on-farm experiments at six organic farms.”

“In adapting regions to climate change, these kinds of stakeholders will play a decisive role.”

Introduction

"To help develop such strategies, within the German Adaptation Strategy to Climate Change (GAS), seven transdisciplinary joint research projects were carried out in seven model regions throughout Germany from 2008 to 2014 (KLIMZUG projects)."

"The main purpose was to build and strengthen regional networks (including the scientific community, government, industry and society) in order to increase the adaptive capacity of the model regions and thus minimise their vulnerability to the effects of climate change."

"This paper will present results from the KLIMZUG joint research project 'Innovation Network of Climate Change Adaptation Brandenburg Berlin (INKA BB)' by the example of one of the sub-projects entitled 'Organic Farming'."

"INKA BB was based on an action research (AR) approach, which brought together scientists and practitioners and was used to generate climate innovations in 24 sub-projects in the model region, Brandenburg Berlin, from 2009 to 2014."

Materials and Methods

"Of the farmers who took on these farms and carried out the conversion process, most were young entrepreneurs who are highly educated (e.g. agricultural engineers) and not originally from Brandenburg."

"SWOT analysis is a widely used tool originating in business development, used in the strategic development process to help businesses and other organisations to direct and adapt internal capabilities to external conditions that cannot be changed (Piercy and Giles [123])."

"In INKA BB, SWOT analysis helped to determine the strategic direction of the sub-projects by having the participating stakeholders critically reflect on the measures from their different viewpoints."

"The stakeholders who participated in the sub-project 'Organic Farming' were organic farmers, agricultural consultants, representatives from organic farming associations and scientists from the Leibniz Centre for Agricultural Landscape Research (ZALF) and the Eberswalde University for Sustainable Development (HNEE)."

"Twenty-six stakeholders participated in the AR process, half of them farmers."

Results—Case Study 'Organic Farming'

"Because tillage-based adaptive measures played a key role in all the experiments, the participants all agreed to implement a ring-cutter that was new to organic farming in Brandenburg."

"Parallel to the on-farm experiments, corresponding small-plot trials were implemented for a deeper long-term comparison of the plough and the reduced tillage with the ring-cutter in the organic cropping system experiment at the experimental station of ZALF."

"These include the following: Using the ring-cutter to regulate perennial clover-grass stands (Experiment I Gut Wilmersdorf) The plough-less late sowing of forage maize in a double-cropping system (Experiment II, Landgut Pretschen) The stubble-sowing method for catch crops (Trial III, Ökodorf Brodowin) These measures can be used to diversify farming methods, make more effective use of time periods with challenging weather conditions and minimise risk in growing crops, especially forage crops."

Discussion

"The exchange of knowledge took placed in the AR process in the following ways: Multilateral exchange between all stakeholders in the network via regular stakeholder workshops (SWOT analyses, annual discussion of on-farm and small-plot experimental results) Regular bilateral exchange between scientists and farmers as they carried out on-farm experiments Occasional multilateral exchange between stakeholders in the network, new partners and interested members of the public at Field Days, seminars and trade events (knowledge transfer events) The different channels and forms for communication made possible a continual dialogue between the stakeholders, creating a basis for individual and collective learning processes."

"Through the AR process, a certain 'de-ideologisation' of the stakeholders took place: in the on-farm trials and small-plot on-station experiments, it became clear that, despite the use of innovative technology (ring-cutter and cultivator-drill), the positive effects of reduced tillage were outweighed by the negative (e.g. weed growth, reduced yields, soil compaction)."

Conclusions

"The way in which each farmer perceive the problems specific to their concerns regarding climate change impacts illustrates that regional adaptation strategies for agriculture should focus on the farm level and should take into account the specifics of the cropping system, the farm-specific management practices and the socioeconomic conditions."

"The AR process has proven as particularly suitable to ensure the cross-fertilisation of expertise between all stakeholders and thus expand their knowledge about real-world climate change adaption options."

"As innovators – and knowledge bearers and multipliers – organic farms are predestined to shape regional processes of change and development (Schäfer and others [124])."

"In adapting regions to climate change, they can thus take on a decisive role, as this case study confirms."

Bibliography

1. Cassidy, E. S. et al. Environ. Res. Lett. 8, 34015 (2013).

2. Darnhofer I, Bellon S, Dedieu B, Milestad R (2010) Adaptiveness to enhance the sustainability of farming systems. A review. Agron Sustain Dev 30:545–555. https://doi.org/10.1007/978-94-007-0394-0_4
3. Astigarraga L, Ingrand S (2011) Production flexibility in extensive beef farming systems. Ecol Soc 16:7. https://doi.org/10.5751/ES-03811-160107
4. David C, Mundler P, Demarle O, Ingrand S (2010) Long-term strategies and flexibility of organic farmers in southeastern France. Int J Agric Sustain 8:305–318. https://doi.org/10.3763/ijas.2010.0497
5. Milestad R, Darnhofer I (2003) Building farm resilience: the prospects and challenges of organic farming. J Sustain Agric 22:81–97. https://doi.org/10.1300/J064v22n03
6. Xu Q, Huet S, Poix C, Boisdon I, Deffuant G (2018) Why do farmers not convert to organic farming? Modeling conversion to organic farming as a major change. Nat Resour Model. https://doi.org/10.1111/nrm.12171
7. FAO, Bioversity (2012) Sustainable diets and bioversity: directions and solutions for policy, research and action. FAO, Rome
8. Johnson JL, Fanzo J, Cogill B (2014) Understanding sustainable diets: a descriptive analysis of the determinants and processes that influence diets and their impact on health food security and environmental sustainability. Adv Nutr 5:418–429
9. Ericksen P (2008) Conceptualizing food systems for global environmental change research. Global Environ Change 18:234–245
10. Darnhofer I, Fairweather J, Moller H (2010) Assessing a farm's sustainability: insights from resilience thinking. Int J Agric Sustain 8:186–198
11. Allouche J (2011) The sustainability and resilience of global water and food systems: Political analysis of the interplay between security resource scarcity political systems and global trade. Food Policy 36:S3–S8
12. Foran T, Butler JR, Williams LJ, Wanjura WJ, Hall A, Carter L, Carberry PS (2014) Taking complexity in food systems seriously: an interdisciplinary analysis. World Dev 61:85–101
13. Ionescu C, Klein RJ, Hinkel J, Kumar KK, Klein R (2009) Towards a formal framework of vulnerability to climate change. Environ Model Assess 14:1–16
14. Schröter D, Polsky C, Patt A (2005) Assessing vulnerabilities to the effects of global change: an eight step approach. Mitig Adapt Strat Glob Change 10:573–595
15. SCAR (2008) 2nd Foresight Exercise: New challenges for Agricultural Research: Climate change food security rural development agricultural knowledge systems. Standing Committee on Agricultural Research (SCAR), Brussels
16. Brunori G, Guarino A (2010) Security for whom? Changing discourses on food in Europe in times of a global food crisis. In: Lawrence G, Lyons K, Wallington T (eds) Food security, nutrition and sustainability. Earthscan, London, pp 41–60
17. Seto KC et al. (2012) Urban land teleconnections and sustainability. Proceedings of the National Academy of Sciences 109:7687–7692
18. Thompson, P.B. 1997. Agrarian values: their future place in US agriculture. In Visions of American agriculture, ed. P.B. Thompson, 17–30. Ames, IA: Iowa State University Press.
19. Becker, G.S. 1986. The fundamentals of domestic commodity price support programs. Washington, DC: Congressional Research Service.
20. Jagger, C., and J. Harwood. 1997. Ms. Farmer's wheat program adventures—before and after the 1996 Farm Bill: a case study for the classroom. Review of Agricultural Economics 20 (1): 259–272.
21. Danbom, D.B. 1997. Past visions of American agriculture. In Visions of American agriculture, ed. P.B. Thompson, 3–16. Ames, IA: Iowa State University Press.
22. Glenna, L. 1999. Systematic constraints to ecological well-being: The case of the 1985 Food Security Act. Rural Sociology 64 (1): 133–157.
23. Sabatier, P.A., and H.C. Jenkins-Smith. 1999. Advocacy coalition framework, an assessment. In Theories of the policy process, ed. P.A. Sabatier, 117–166. Boulder, CO: Westview Press.
24. Iowa Farm Bureau. 2006. Wetlands remain very popular with landowners. http://www.ifbf.org/fullarticle.aspx?artid=23602 . Accessed 16 September 2007.

25. Corey, K.G., L. Wright-Morton, E.A. Bitto, and T. Terrill. 2006. Partnerships in public connections in solving water-quality issues in Camp Creek. Poster presented at the 2006 US Department of Agriculture CSREES Annual Conference. http://www.usawaterquality.org/conferences/2006/posters/Corey.pdf . Accessed 31 March 2007.
26. Jordan, N., G. Boody, W. Broussard, J.D. Glover, D. Keeney, B.H. McCown, G. McIsaac, M. Muller, H. Murray, J. Neal, C. Pansing, R.E. Turner, K. Warner, and D. Wyse. 2007. Sustainable development of the agricultural bio-economy. Science 316 (5331): 1570–1571.
27. Postel, S. 1999. Pillar of sand: Can the irrigation miracle last? New York, NY: W.W. Norton & Company.
28. Schrama, G.J.I. 1998. Introduction. In Drinking water supply and agricultural pollution: Preventive action by the water supply sector in the European Union and the United States, ed. G.J.I. Schrama, 3–18. Dordrecht, The Netherlands: Kluwer Academic Publishers.
29. Walton, J. 1992. Western times and water wars: State, culture, and rebellion in California. Berkeley, Los Angeles, Oakland, CA: University of California Press.
30. Mann, S.A. 1990. Natural obstacles to agrarian capitalism. In Agrarian capitalism in theory and practice, ed. S.A. Mann, 28–46. Chapel Hill, NC: University of North Carolina Press.
31. Young, J. C., Jordan, A., Searle, K. R., Butler, A., Chapman, D. S., Simmons, P., & Watt, A. D. (2013). Does stakeholder involvement really benefit biodiversity conservation? Biological Conservation, 158, 359–370.
32. Carr, G., Blöschl, G., & Loucks, D. P. (2012). Evaluating participation in water resource management: A review. Water Resources Research, 18, W11401.
33. Newig, J., & Fritsch, O. (2009). Environmental governance, participatory, multi-level—And effective? Environmental Policy and Governance, 19, 197–214.
34. Richards, C., Blackstock, K., & Carter, C. (2004). Practical approaches to participation. Socio-Economic Research Group (SERG) Policy Brief 1.
35. Beierlee, T. C., & Konisky, D. M. (2001). What are we gaining from stakeholder involvement? Observations from environmental planning in the Great Lakes. Environmental and Planning C, 19, 515–527.
36. Rauschmayer, et al. (2009). Examining processes or/and outcome? Evaluation concepts in European governance of natural resources. Environmental Policy and Governance, 19(3), 2009.
37. Rowe, G., & Frewer, L. J. (2004). Evaluating public-participation exercises: A research agenda. Science, Technology and Human Values, 29(4), 512–556.
38. Newell, P. (2007). Trade and environmental justice in Latin America. New Political Economy, 12(2), 237–259.
39. Sikor, T. (2013). The justices and injustices of ecosystem services. London/New York: Earthscan.
40. Suiseeya, K. R. M., & Caplow, S. (2013). In pursuit of procedural justice: Lessons from an analysis of 56 forest carbon project designs. Global Environmental Change, 23, 968–979.
41. Reed, M. S. (2008). Stakeholder participation for environmental management: A literature review. Biological Conservation, 141, 2417.
42. Young, I. M. (1990). Justice and the politics of difference. Princeton: Princeton University Press.
43. Piketty, T. (2014). Capital in the twenty-first century. Cambridge: Harvard University Press.
44. Fung, A., & Wright, E. O. (2003). Thinking about empowered participatory governance. In A. Fung & E. O. Wright (Eds.), Deepening democracy. Institutional innovations in empowered participatory governance. London: Verso.
45. Sperling, et al. (2001). A framework for analyzing participatory plant breeding approaches and results. Euphitica, 122, 439–450.
46. Tippet, J., Handley, J. F., & Ravetz, J. (2007). Meeting the challenges of sustainable development—A conceptual appraisal of a new methodology for participatory ecological planning. Progress in Planning, 67, 9.
47. Hecht, S. (1987). The evolution of agroecological thought. In M. Altieri (Ed.), The scientific basis of alternative agriculture (pp. 1–27). London: Intermediate Technology Publications.

48. McNeely, J., & Scherr, S. J. (2003). Ecoagriculture: Strategies to feed the world and save biodiversity. Washington, DC: Island Press.
49. Roitner-Schobesberger, B., Darnhofer, I., Somsook, S., & Vogl, C. R. (2008). Consumer perceptions of organic foods in Bangkok. Food Policy, 33, 112–121.
50. Kaufman, A. (2012). Organic farmers' connectedness with nature: Exploring Thailand's alternative agriculture network. Worldviews, Global Religions, Culture and Ecology, 16, 154–178.
51. Boonman, C., & Anpim, U. (2006). The possibility of Homali rice production in organic farming systems as an alternative farming career with poverty alleviation potential for lower-northeastern farmers: A case study of Yasothon Province. Bangkok: Thai Research Foundation. (in Thai).
52. Curry, J. (2000). Community worldview and rural systems: A study of five communities in Iowa. Annals of American Geography, 90(4), 693–712.
53. Picone, C., & Van Tassel, D. (2002). Agriculture and biodiversity loss: Industrial agriculture. In N. Eldredge (Ed.), Life on earth: An encyclopedia of biodiversity, ecology, and evolution (pp. 99–105). Santa Barbara, California: ABC-CLIO.
54. Kimbrell, A. (2002). Introduction. In A. Kimbrell (Ed.), The fatal harvest reader: The tragedy of industrial agriculture (pp. xi–xiv). Sausalito, California: Foundation for Deep Ecology.
55. Maynard, H., & Nault, J. (2005). Big farms, small farms: Strategies in sustainable agriculture to fit all sizes. Agricultural Institute of Canada. Retrieved February 17, 2008, from http://www.aic.ca/about/pr_docs/AIC_discussion_paper_Final_ENG.pdf .
56. Miller, G. T., & Hackett, D. (2008). Living in the environment (1st Canadian ed.). Ontario: Thomson Nelson.
57. Pretty, J. (2002). Regenerating agriculture. In V. Desai & R. B. Potter (Eds.), The companion to development studies (pp. 170–175). London: Arnold.
58. Agriculture and Agri-Food Canada. (2006). Making progress together: Sustainable development strategy 2007–2009. Ottawa: Author. Retrieved February 17, 2008, from http://www4.agr.gc.ca/resources/prod/doc/policy/environment/pdfs/sds/sds4_e.pdf .
59. Janzen, H. H., Desjardins, R. L., Asselin, J. M. R., & Grace, B. (1998). The health of our air: Towards sustainable agriculture in Canada. Ottawa: Agriculture and Agri-Food Canada. Retrieved February 17, 2008, from http://www.agr.gc.ca/nlwis-snite/pub/ha_sa/pdf/intro_e.pdf .
60. Forge, F. (2004). Organic farming in Canada: An overview. Ottawa, Canada: Library of Parliament, Parliamentary Information and Research Service. Retrieved December 15, 2008, from http://www.parl.gc.ca/information/library/PRBpubs/prb0029-e.htm .
61. Rosset, P. (2000). Cuba: A successful case study of sustainable agriculture. In F. Magdoff, J. B. Foster, & F. H. Buttel (Eds.), Hungry for profit: the agribusiness threat to farmers, food and the environment (pp. 203–213). New York: Monthly Review Press.
62. Levins, R. (2005). How Cuba is going ecological. Capitalism, Nature, Socialism, 16(3), 7–26.
63. Sullivan, R. E. (2000, July 13). Cuba producing, perhaps, 'cleanest' food in the world. Earth Times News Service. Retrieved February 17, 2008, from http://www.globalexchange.org/countries/americas/cuba/sustainable/susAgriculture/earthTimes071300.html .
64. Tharamangalam, J. (2008) Can Cuba offer an alternative to corporate control over the world's food system. Retrieved October 15, 2009, from http://www.globaljusticecenter.org/articles/report_cubafood.html .
65. Holm, W. (2006). Canada-Cuba farmer to farmer project. Project rationale. Retrieved February, 20, 2008, from http://www.farmertofarmer.ca/rationale.html .
66. Altieri, M. A., Nicholls, C. I., & Montalba, R. (2017). Technological approaches to sustainable agriculture at a crossroads: An agroecological perspective. Sustainability, 9(3), 349. https://doi.org/10.3390/su9030349
67. Tittonell, P. (2014). Ecological intensification of agriculture—sustainable by nature. Current Opinion in Environmental Sustainability, 8, 53–61. https://doi.org/10.1016/j.cosust.2014.08.006 .

68. Bengtsson, J., Ahnström, J., & Weibull, A. C. (2005). The effects of organic agriculture on biodiversity and abundance: A meta-analysis. Journal of Applied Ecology, 42(2), 261–269. https://doi.org/10.1111/j.1365-2664.2005.01005.x
69. Domínguez, A., Bedano, J. C., Becker, A. R., & Arolfo, R. V. (2014). Organic farming fosters agroecosystem functioning in Argentinian temperate soils: Evidence from litter decomposition and soil fauna. Applied Soil Ecology, 83, 170–176. https://doi.org/10.1016/j.apsoil.2013.11.008
70. Domínguez, A., & Bedano, J. C. (2016). Earthworm and enchytraeid co-occurrence pattern in organic and conventional farming: Consequences for ecosystem engineering. Soil Science, 181(3/4), 148–156. https://doi.org/10.1097/SS.0000000000000146
71. Hole, D. G., Perkins, A. J., Wilson, J. D., Alexander, I. H., Grice, P. V., & Evans, A. D. (2005). Does organic farming benefit biodiversity? Biological conservation, 122(1), 113–130. https://doi.org/10.1016/j.biocon.2004.07.018 .
72. Mäder, P., Fliessbach, A., Dubois, D., Gunst, L., Fried, P., & Niggli, U. (2002). Soil fertility and biodiversity in organic farming. Science, 296(5573), 1694–1697. https://doi.org/10.1126/science.1071148 .
73. Ostandie, N., Giffard, B., Bonnard, O., Joubard, B., Richart-Cervera, S , Thiéry, D., & Rusch, A. (2021). Multi-community effects of organic and conventional farming practices in vineyards. Scientific Reports, 11(1), 1–10. https://doi.org/10.1038/s41598-021-91095-5
74. Werner, M. R., & Dindal, D. L. (1990). Effects of conversion to organic agricultural practices on soil biota. American Journal of Alternative Agriculture, 5(1), 24–32. https://doi.org/10.1017/S0889189300003192 .
75. ISO. (2011). ISO 23611–5. Soil quality-sampling of soil invertebrates-part 5. Sampling and extraction of soil macro-invertebrates. International Organization for Standardization.
76. Cochran, V. L., Sparrow, S. D., & Sparrow, E. B. (1994). Residue effects on soil micro-and macroorganisms. Managing Agricultural Residues,163–184.
77. Snyder, B. A., Callaham, M. A., & Hendrix, P. F. (2011). Spatial variability of an invasive earthworm (Amynthas agrestis) population and potential impacts on soil characteristics and millipedes in the great smoky mountains National Park, USA. Biological Invasions, 13, 349–358. https://doi.org/10.1007/s10530-010-9826-4
78. Rayner, S. (2012). Uncomfortable knowledge: The social construction of ignorance in science and environmental policy discourses. Economy and Society, 41(1), 107–125.
79. Hess, Charlotte, and Elinor Ostrom. 2003. Ideas, artefacts and facilities: Information as a common-pool-resource. Law and Contemporary Problems 66: 111–145.
80. Hess, Charlotte, and Elinor Ostrom. 2007. Understanding knowledge as a commons: From theory to practice. Cambridge: MIT Press.
81. Hess, Charlotte. 2012. The unfolding of the knowledge commons. St Antony's International Review 8 (1): 13–24.
82. Frischmann, Brett, Michael Madison, and Katherine Strandburg. 2014. Governing knowledge commons. Oxford: Oxford University Press.
83. Douglas, Mary. 1986. How institutions think. Syracuse: Syracuse University Press.
84. North, Douglas. 2005. Understanding the process of economic change. Princeton: Princeton University Press.
85. Enjalbert, Jerome, Julie Dawson, Sophie Paillard, Bénédicte Rhoné, Yves Rousselle, Thomas Matthieu, and Isabelle Goldringer. 2011. Dynamic management of crop diversity: From an experimental approach to on-farm conservation. Compte Rendu Biologies 334 (5–6): 458–468.
86. Goldringer, Isabelle, Jerome Enjalbert, Pierre Rivière, and Dawson J. Julie. 2012. Recherche participative pour des variétés adaptées à une agriculture à faible niveau d'intrants et moins sensibles aux variations climatiques. Pour 213 (1): 153–161.
87. Demeulenaere, Élise, Pierre Rivière, Alexandre Hyacinthe, et al. 2017. La sélection participative à l'épreuve du changement d'échelle. Trajectoire d'une collaboration entre paysans sélectionneurs et généticiens de terrain. Natures Sciences Sociétés 25 (4): 335–346.

88. Wirz, Johannes, Peter Kunz, and Ueli Hurter. 2017. Seed as a commons. Breeding as a source for real economy, law and culture. Switzerland: Goetheanum and Fund for Crop Development.
89. Girard, Fabien, and Christine Frison. 2018. The commons, plant breeding and agricultural research challenges for food security and agrobiodiversity. Abingdon: Routledge.
90. Wolter, Hendrik, and Stefanie Sievers-Glotzbach. 2019. Bridging traditional and new commons: The case of fruit breeding. International Journal of the Commons 13 (1): 303–328.
91. Mazé, Armelle, Aida Calabuig Domenech, and Isabelle Goldringer. 2021. Restoring cultivated agrobiodiversity: The political ecology of knowledge networks between peasant seed groups in France. Ecological Economics 179: 106821.
92. Madison, Michael, Brett Frischman, and Katherine Strandburg. 2010. Constructing Commons in the cultural environment. Cornell Law Review 95: 656–709.
93. Ostrom E (2009) A general framework for analyzing sustainability of social-ecological systems. Science 325:419–422
94. Pahl-Wostl, Claudia. 2009. A conceptual framework for analysing adaptive capacity and multi-level learning processes in resource governance regimes. Global Environmental Change 19: 354–365.
95. McGinnis, Mickael, and Elinor Ostrom. 2014. Social-ecological system framework: Initial changes and continuing challenges. Ecology and Society 19 (2): 30.
96. Cole, Daniel, Graham Epstein, and Mickael McGinnis. 2019. Combining the IAD and SES frameworks. International Journal of the Commons 13 (1): 244–275.
97. Ostrom, Elinor, and Mickael Cox. 2010. Moving beyond panaceas: A multi-tiered diagnostic approach for social-ecological analysis. Environmental Conservation 37: 451–463.
98. Ostrom, Elinor, and Xavier Basurto. 2011. Crafting analytical tools to study institutional change. Journal of Institutional Change 7 (3): 317–343.
99. Ostrom, Elinor. 2011. Background on the institutional analysis and development framework. Policy Studies Journal 39 (1): 2–27.
100. Rivière, Pierre. 2014. Méthodologie de la sélection décentralisée et participative: un exemple sur le blé tendre. Thèse de doctorat en génétique: Université Paris XI.
101. Demeulenaere, Elise. 2014. A Political ontology of seeds: The transformative frictions of a farmer' movement in Europe. Journal of Global and Historical Anthropology 69: 45–61.
102. Koller, Beste, Béla Bartha, Riccardo Bocci, María Carrascosa, Pierre Riviére, and Régine Andersen. 2018. Community seed banks in Europe. EU Project DIVERSIFOOD.
103. Catalán IA, Auch D, Kamermans P, Morales-Nin B, Angelopoulos NV, Reglero P, Sandersfield T, Peck MA (2019) Critically examining the knowledge base required to mechanistically project climate impacts: a case study of Europe's fish and shellfish. Fish and Fisheries 1-17 https://doi.org/10.1111/faf.12359
104. Reid GK, Gurney-Smith HJ, Flaherty M, Garber AF, Forster I, Brewer-Dalton K, Knowler D, Marcogliese DJ, Chopin T, Moccia RD, Smith CT (2019) Climate change and aquaculture: considering adaptation potential. Aquac Environ Interact 11:603–624
105. Dahlke FT, Wohlrab S, Butzin M, Pörtner H-O (2020) Thermal bottlenecks in the life cycle define climate vulnerability of fish. Science (80-) 369:65–70. https://doi.org/10.1126/science.aaz3658
106. Kay S, Andersson H, Catalán IA, Drinkwater KF, Eilola K, Jordà G, Ramirez-Romero E (2018) Projections of physical and biogeochemical parameters and habitat indicators for European seas, including synthesis of Sea Level Rise and storminess. Deliverable Report 1.3 CERES. Available at https://ceresproject.eu/ceres-results-and-solutions/ . Accessed Jan 2020
107. Montalto V, Helmuth B, Ruti PM, Dell'Aquila A, Rinaldi A, Sarà G (2016) A mechanistic approach reveals non linear effects of climate warming on mussels throughout the Mediterranean Sea. Clim Chang 139:293–306. https://doi.org/10.1007/s10584-016-1780-4
108. Nair VD, Graetz DA (2004) Agroforestry as an approach to minimizing nutrient loss from heavily fertilized soils: the Florida experience. Agrofor Syst 61:269–279. https://doi.org/10.1023/B:Agfo.0000029004.03475.1d
109. Nair VD, Nair PKR, Kalmbacher RS, Ezenwa IV (2007) Reducing nutrient loss from farms through silvopastoral practices in coarse-textured soils of Florida, USA. Ecol Eng 29:192–199. https://doi.org/10.1016/j.ecoleng.2006.07.003

110. Muschler RG (2016) Agroforestry: Essential for sustainable and climate-smart land use? In: Tropical forestry handbook. Springer, Berlin Heidelberg, pp 2013–2116
111. Iwasaki K, Torita H, Abe T, Uraike T, Touze M, Fukuchi M, Sato H, Iijima T, Imaoka K, Igawa H (2018) Spatial pattern of windbreak effects on maize growth evaluated by an unmanned aerial vehicle in Hokkaido, northern Japan. Agrofor Syst 23:1–13. https://doi.org/10.1007/s10457-018-0217-7
112. Nair PKR (2012) Climate change mitigation: A low-hanging fruit of agroforestry. In: Nair PKR, Garrity D (eds) Agroforestry-The future of global land use. Springer, Netherlands, pp 31–67
113. Lee KH, Isenhart TM, Schultz RC (2003) Sediment and nutrient removal in an established multi-species riparian buffer. J Soil Water Conserv 58:1–8 http://www.jswconline.org/content/58/1/1.short
114. Otto S, Vianello M, Infantino A, Zanin G, Di Guardo A (2008) Effect of a full-grown vegetative filter strip on herbicide runoff: maintaining of filter capacity over time. Chemosphere 71:74–82. https://doi.org/10.1016/j.chemosphere.2007.10.029
115. Pavlidis G, Tsihrintzis VA (2018) Environmental benefits and control of pollution to surface water and groundwater by agroforestry systems: a review. Water Resour Manag 32:1–29. https://doi.org/10.1007/s11269-017-1805-4
116. Dollinger J, Lin CH, Udawatta RP, Pot V, Benoit P, Jose S (2019) Influence of agroforestry plant species on the infiltration of S-Metolachlor in buffer soils. J Contam Hydrol 225:103498. https://doi.org/10.1016/j.jconhyd.2019.103498
117. Pestemer W, Radulescu V, Walker A, Ghinea L (1984) Residual effects of chlorotrazine herbicides in soil at 3 Romanian sites. I. Prediction of the persistence of simazine and atrazine. Weed Res 24:359–369 https://eurekamag.com/research/001/449/001449279.php
118. Hoffman DW, Gerik TJ, Richardson CW (1995) Use of contour strip cropping as a best management practice to reduce atrazine contamination of surface water. In: Proceedings of the second international conference IAWQ specialized conference on diffuse pollution. International Association on Water Quality, Brno and Prague, Czech Republic, pp 595–596
119. Udawatta RP, Krstansky JJ, Henderson GS, Garrett HE (2002) Agroforestry practices, runoff, and nutrient loss: a paired watershed comparison. J Environ Qual 31:1214–1225. https://doi.org/10.2134/Jeq2002.1214
120. Schultz RC, Isenhart TM, Colletti JP, Simpkins WW, Udawatta RP, Schultz PL (2009) Riparian and upland buffer practices. In: Garrett HE (ed) North American agroforestry, an integrated science and practice, 2nd edn. American Society of Agronomy, Madison, USA, pp 163–218
121. Schmitt TJ, Dosskey MG, Hoagland KD (1999) Filter strip performance and processes for different vegetation, widths, and contaminants. J Environ Qual 28:1479–1489. https://doi.org/10.2134/jeq1999.00472425002800050013x
122. van Noordwijk M, Lusiana B (1998) WaNuLCAS, a model of water, nutrient and light capture in agroforestry systems. Agrofor Syst 43:217–242. https://doi.org/10.1023/A:1026417120254
123. Piercy N, Giles W (1989) Making SWOT analysis work. Mark Intell Plan 7:5–7. https://doi.org/10.1108/EUM0000000001042
124. Schäfer M, Nölting B, Engel A (2009) Organic agriculture as a new player in sustainable regional development? Case studies of rural areas in Eastern Germany. Int J Agric Resour Gov Ecol (IJARGE) 8:158–179
125. Mehrabi, Zia; Ellis, Erle C.; Ramankutty, Navin The challenge of feeding the world while conserving half the planet. Nature Sustainability (2018). https://doi.org/10.1038/s41893-018-0119-8
126. Bouttes, Maëlys; Darnhofer, Ika; Martin, Guillaume Converting to organic farming as a way to enhance adaptive capacity. Organic Agriculture (2018). https://doi.org/10.1007/s13165-018-0225-y
127. Allen, Thomas; Prosperi, Paolo Modeling Sustainable Food Systems. Environmental Management (2016). https://doi.org/10.1007/s00267-016-0664-8
128. Moghimi, Faraz; Asiabanpour, Bahram Economics of vertical farming in the competitive market. Clean Technologies and Environmental Policy (2023). https://doi.org/10.1007/s10098-023-02473-8

129. Dima, S.J.; Odero, A.N. Organic Farming for Sustainable Agricultural Production. A Brief Theoretical Review and Preliminary Empirical Evidence. Environmental and Resource Economics (1997). https://doi.org/10.1023/a:1026472410031
130. Dyball, Robert From industrial production to biosensitivity: the need for a food system paradigm shift. Journal of Environmental Studies and Sciences (2015). https://doi.org/10.1007/s13412-015-0323-z
131. Gasteyer, Stephen P. Agricultural transitions in the context of growing environmental pressure over water. Agriculture and Human Values (2008). https://doi.org/10.1007/s10460-008-9137-x
132. Coolsaet, Brendan Transformative Participation in Agrobiodiversity Governance: Making the Case for an Environmental Justice Approach. Journal of Agricultural and Environmental Ethics (2015). https://doi.org/10.1007/s10806-015-9579-2
133. Kaufman, Alexander Harrow; Mock, Jeremiah Cultivating Greater Well-being: The Benefits Thai Organic Farmers Experience from Adopting Buddhist Eco-spirituality. Journal of Agricultural and Environmental Ethics (2014). https://doi.org/10.1007/s10806-014-9500-4
134. Hiranandani, Vanmala Sustainable agriculture in Canada and Cuba: a comparison. Environment, Development and Sustainability (2009). https://doi.org/10.1007/s10668-009-9223-2
135. Domínguez, Anahí; Escudero, Héctor Javier; Rodríguez, María Pía; Ortiz, Carolina Elizabeth; Arolfo, Romina Vanesa; Bedano, José Camilo Agroecology and organic farming foster soil health by promoting soil fauna. Environment, Development and Sustainability (2023). https://doi.org/10.1007/s10668-022-02885-4
136. Mazé, Armelle; Calabuig Domenech, Aida; Goldringer, Isabelle Commoning the seeds: alternative models of collective action and open innovation within French peasant seed groups for recreating local knowledge commons. Agriculture and Human Values (2020). https://doi.org/10.1007/s10460-020-10172-z
137. Cubillo, Alhambra Martínez; Ferreira, João G.; Lencart-Silva, João; Taylor, Nick G.H.; Kennerley, Adam; Guilder, James; Kay, Susan; Kamermans, Pauline Direct effects of climate change on productivity of European aquaculture. Aquaculture International (2021). https://doi.org/10.1007/s10499-021-00694-6
138. Zhu, Xiai; Liu, Wenjie; Chen, Jin; Bruijnzeel, L. Adrian; Mao, Zhun; Yang, Xiaodong; Cardinael, Rémi; Meng, Fan-Rui; Sidle, Roy C.; Seitz, Steffen; Nair, Vimala D.; Nanko, Kazuki; Zou, Xin; Chen, Chunfeng; Jiang, Xiao Jin Reductions in water, soil and nutrient losses and pesticide pollution in agroforestry practices: a review of evidence and processes. Plant and Soil (2019). https://doi.org/10.1007/s11104-019-04377-3
139. Bloch, Ralf; Knierim, Andrea; Häring, Anna-Maria; Bachinger, Johann Increasing the adaptive capacity of organic farming systems in the face of climate change using action research methods. Organic Agriculture (2015). https://doi.org/10.1007/s13165-015-0123-5

Chapter 4
Traditional Knowledge and Sustainable Livelihoods

4.1 Introduction by Editor

Traditional knowledge encompasses community-based systems that rely on the management and utilisation of natural resources. It is about the cultural heritage of indigenous communities, and how they have coped with food security and resource scarcity over generations. Such knowledge evolves with an intimate understanding of natural cycles, resource availability and is linked to the biogeographic conditions that communities have occupied over generations. Social scientists believe that such customs and practices can offer valuable lessons in resource use, land management, conservation, and scientific research. Especially in the area of biodiversity conservation, scientists acknowledge that traditional ecological knowledge can provide vital insights, where humans and nature have coevolved over millennia (Zs. Molnár, 2021). In Ethiopia, for example, indigenous knowledge has sustained landscapes and livelihoods for millennia (Zenebe Mekonnen, 2021).

Rooted in a deep knowledge of nature, seasons, and environmental conditions, traditional knowledge systems of indigenous communities across the globe, have the potential to unlock sustainable pathways for managing natural resources. These generations-old practices, cumulatively add up to a deep understanding of local ecosystems, agricultural techniques, and resource management strategies that can offer valuable insights for achieving the Sustainable Development Goals (SDGs).

Across India, Africa, South America, China, and other continents, indigenous peoples have developed holistic approaches to managing their environments, utilizing agroforestry, soil conservation, and other sustainable practices that have stood the test of time (Ahearn et al., 2019). In many cases, these traditional knowledge systems have enabled communities to sustain changing climatic conditions, demonstrating their resilience and adaptability (Sharma & Kumari, 2021). Unfortunately, as rural development has increasingly favored modern, technology-driven solutions, the importance of indigenous knowledge has often been overlooked or undervalued (Maunganidze, 2016). In this study, Maunganidze (2016) suggests "rural people's assets and

G. Nulkar (ed.), *The Economics of Sustainable Development*,
https://doi.org/10.1007/978-981-99-7379-8_4

knowledge have lost importance and no longer function as economic pillars for rural and national development". This process has not only threatened the preservation of traditional practices, but also the very identity and cultural integrity of indigenous communities. The gradual shift from traditional lifestyles to the market-linked economy has deepened the loss of ecological knowledge.

To address this challenge, concerted efforts must be made to recognize, document, and integrate traditional knowledge systems into sustainable development frameworks. By fostering this dialogue and embracing the wealth of traditional knowledge, policymakers and development practitioners can unlock innovative solutions that are rooted in local contexts and aligned with the principles of sustainability (Maunganidze, 2016; Pilgrim et al., 2008; Sharma & Kumari, 2021; Ahearn et al., 2019).

India has several laws that protect the traditional knowledge of indigenous communities. The Geographical Indications Act of 1999, Protection of Plant Varieties and Farmer's Rights Act of 2001 and the Biodiversity Act of 2002 are some examples. However, there are many formidable barriers yet to be addressed. Many indigenous communities are not yet recognized under the purview of Scheduled Tribes. Infrastructure development, rapidly expanding in tribal areas, has ousted millions of people from their ancestral homes, and this has forced them to take up modern lifestyles. This has contributed to the loss of traditional knowledge, and cultural expressions, language and traditional indigenous farming practices amongst others (Chakrabarty SP, 2021). The Indian Ayurvedic pharmaceutical industry has benefited from tribal sources of indigenous knowledge of the medicinal plants. They are being used for healing and curing diseases for better living (Manna, 2021).

The preservation of traditional knowledge begins with the protection of indigenous communities and their habitats. Furthermore, indigenous languages facilitate the transmission of knowledge to the next generation, and thus their languages need to be accorded protection too. Today, around 2,700 languages are estimated to be in danger of disappearing forever (Traditional knowledge—an answer to the most pressing global problems?, 2019).

Many studies across the world have emphasized the need to integrate indigenous knowledge with scientific knowledge in order to develop more effective and culturally-appropriate climate change adaptation strategies (Makondo & Thomas, 2018; Ahearn et al., 2019). The Intergovernmental Panel on Climate Change has recognized the relevance of traditional knowledge on a global scale. This is because indigenous communities have developed practices over centuries that integrate a deep understanding of the natural world, local ecological conditions, and sustainable resource management practices that have been refined over generations (Ahearn et al., 2019).

It has been noted that attempts to adapt to climate change do not always take into account indigenous knowledge. Indigenous peoples' rights are mentioned in the Paris Agreement, and it recognizes the importance of traditional knowledge in reducing the detrimental effects of climate change.

In this chapter, we explore the ways in which traditional knowledge systems can be put to use in the modern era. Studies from Colombia, Amazon, Africa and the Himalayan regions illustrate how such practices can offer ways to integrate livelihoods with conservation efforts. The chapter cuts across disciplines of agroforestry, farming, soil management, and seed conservation to establish the need for conservation of traditional practices and customs.

References

Ahearn, E A., Oelz, M., & Dhir, R K. (2019, April 16). Indigenous Peoples and Climate Change: Emerging Research on Traditional Knowledge and Livelihoods. https://papyrus.ilo.org/global/topics/indigenous-tribal/publications/WCMS_686780/lang--en/index.htm

Makondo, C C., & Thomas, D S. (2018, October 1). Climate change adaptation: Linking indigenous knowledge with western science for effective adaptation. Elsevier BV, 88, 83–91. https://doi.org/10.1016/j.envsci.2018.06.014.

4.2 Machine Generated Summaries

Disclaimer: The summaries in this chapter were generated from Springer Nature publications using extractive AI auto-summarization: An extraction-based summarizer aims to identify the most important sentences of a text using an algorithm and uses those original sentences to create the auto-summary (unlike generative AI). As the constituted sentences are machine selected, they may not fully reflect the body of the work, so we strongly advise that the original content is read and cited. The auto generated summaries were curated by the editor to meet Springer Nature publication standards. To cite this content, please refer to the original papers.

Machine generated keywords: traditional, knowledge, traditional knowledge, indigenous, farmer, agriculture, specie, food, fertility, farming, soil, practice, region, plant, livelihood.

4.2.1 Traditional Agriculture: A Climate-Smart Approach for Sustainable Food Production [115]

This is a machine-generated summary of:

Singh, Rinku; Singh, G. S.: Traditional agriculture: a climate-smart approach for sustainable food production [115].

Published in: Energy, Ecology and Environment (2017).

Link to original: https://doi.org/10.1007/s40974-017-0074-7.

Copyright of the summarized publication:

If you want to cite the papers, please refer to the original.

For technical reasons we could not place the page where the original quote is coming from.

Abstract-Summary

"Sustainable food production is one of the major challenges of the twenty-first century in the era of global environmental problems such as climate change, increasing population and natural resource degradation including soil degradation and biodiversity loss."

"Green Revolution though multiplied agricultural production several folds but at the huge environmental cost including climate change."

"Agriculture is one of the largest sectors that sustain livelihood to maximum number of people and contribute to climate change."

"Traditional agriculture is getting increased attention worldwide in context of sustainable food production in changing climate."

"The present article advocates traditional agriculture as a climate-smart approach for the sustainable food production and also deliberates the correlation between climate change and agriculture."

Introduction

"Green Revolution though boosted agriculture production but at the huge socio-ecological cost such as environmental pollution, biodiversity loss, climate change, land degradation, erosion of traditional agricultural knowledge and decline in human health and livelihood (Redclift [1]; Alteri [2]; Eakin and others [3]; Phungpracha and others [4]; Srivastava and others [5])."

"High productivity, biodiversity conservation, low energy inputs and climate change mitigation are some of the salient features of the traditional agriculture systems (Srivastava and others [5])."

"Traditional practices like agroforestry, intercropping, crop rotation, cover cropping, traditional organic composting and integrated crop-animal farming have potentials for enhancing crop productivity and mitigating climate change."

"Modern agriculture, however, enhanced food productivity but with the acceleration of several environmental problems such as climate change, food unsafety, biodiversity loss, soil degradation and environmental pollution (Zhang and others [7])."

"In the context of sustainable food production in changing climate, adoption of climate-smart traditional practices is an urgent need."

"The aim of this article is to advocate traditional agriculture as a climate-smart approach for sustainable food production."

Agriculture and Climate Change: A Two-Way Relationship

"Economic activities prompted by industrial revolution have been contributed to climate change through increasing GHGs emission (IPCC [8])."

"Agricultural activities from cropping to harvesting emit GHGs that cause climate change which in turn disturbs agriculture (Paustian and others [9])."

"Climate change is a severe threat to both food production and human health (McMichael and others [10])."

"Climate change affects agriculture directly through altering the agroecological conditions and indirectly by increasing demand of agricultural production (Schmidhuber and Tubiello [11])."

"Climate change is a serious threat to all aspects of agriculture including production, distribution, food accessibility and food prices (Tai and others [12]; Islam and Nursey-Bray [13])."

"Tropical and developing countries are at the greater risk to climate change as compared to temperate and developed countries and this scenario encounters current and future food production (Gornall and others [14]; Hillel and Rosenzweig [15]; Deryng and others [16]; Porter and others [17]; Challinor and others [18])."

Climate-Smart Agriculture: Principles and Objectives

"The CSA approach has three objectives: (1) sustainably enhancing agricultural productivity to support equitable increase in income, food security and development (2) increasing adaptive capacity and resilience to shocks at multiple levels, from farm to national and (3) reducing GHG emissions and increasing carbon sequestration where possible."

"Sustainable food production while reducing GHG emissions and increasing climate resilience of agricultural system is the foremost objective of CSA (Harvey and others [19]; Brandt and others [20])."

"According to Olayide and others [21], CSA is an emerging approach to enhance food production, biodiversity, environmental quality, agroecosystem resilience, livelihoods and economic development while addressing the climate change impacts."

"CSA has been getting a mounting attention particularly in developing world due to its capabilities to enhance agricultural productivity and agroecosystem resilience while reducing GHG emission (Grainger-Jones [22]; Long and others [23]; Mwongera and others [24])."

Traditional Agriculture: Concept and Agroecological Features

"Traditional agriculture is the result of the experiences delivered by the local farming practices through thousands of years (Pulido and Bocco [6])."

"Traditional farming practices contributed a significant role to the building of scientific knowledge in agriculture (Sandor and Furbee [25]; Singh and others [26])."

"Farmers throughout the world particularly in developing regions use local, traditional or landraces of both minor and major crops (Jackson and others [27])."

"Small farmers are stewards of the traditional agricultural practices, and globally about 84% of farms have area less than 2 ha that operate 12% of farmland (Altieri [28]; Lowder and others [29])."

"Smallholder farmers adjust to environmental changes through their indigenous knowledge and experience such as changing farming practices and cultivating adapted crops (Lasco and others [30])."

"High vegetational diversity and a multifaceted system of indigenous knowledge are the salient features of traditional farming systems in developing countries (Altieri [31]; Gliessman [32]; Altieri [33])."

Traditional Agriculture: Alternative Practices for Climate Change Mitigation

"Agroforestry, intercropping, crop rotation, cover cropping, traditional organic composting and integrated crop-animal farming are prominent traditional agricultural practices."

"Wheat–maize intercropping combined with conservation agricultural practices can be used for reducing CO_2 emission and increasing crop productivity (Hu and others [34])."

"Crop rotation is a sustainable approach that increases yield and water use efficiency while reducing soil erosion (Huang and others [35])."

"In crop rotation practice, species that enhance N in the soil can increase the phytomass production of the subsequent crops and consequently increasing SOM (Peoples and Baldock [36]; Raphael and others [37])."

"Cover cropping is a sustainable approach for enhancing soil health, soil microbial biomass and agroecosytem services such as nutrient cycling, water storage, weed and pest control and carbon sequestration (Schipanski and others [38]; Frasier and others [39]; Pinto and others [40])."

"Cover crops are sustainable farming tool that increase SOM and improve soil water dynamics (Steele and others [41]; Poeplau and Don [42]; Basche and others [43]; Duval and others [44])."

Traditional Agriculture: A Sustainable Approach for Climate Change Adaptation

"Farmers in different regions have different perceptions and adaptation strategies to adjust with change in climate."

"Farmers' perception to climate change play a great role not only to adaptation strategies but also in designing policy and integration of scientific and indigenous knowledge for climate change adaptation (Juana and others [45]; Ayal and Filho [46]; Woods and others [47])."

"Indigenous farmers have a long history of climate change adaptation through making changes in agriculture practices."

"Shifting planting and harvesting dates, crop diversification, integrated crop-livestock farming, cropping drought-resistant varieties and high yielding water sensitive crops are some of the sustainable approaches to climate change adaptation (Bradshaw and others [48]; Di Falco and others [49]; Di Falco and Veronesi [50]; Moniruzzaman [51])."

"Small farmers use several techniques to reduce climate-driven crop failure such as use of drought-tolerant local varieties, polyculture, agroforestry, water harvesting and conserving soil (Browder [52]; Altieri and Nicholls [53]; Chhetri and others [54])."

"In Ethiopia and South Africa, farmers use several drought adaptation strategies such as changing planting dates, adopting new crop varieties and migrating seasonally (Deressa and others [55]; Bryan and others [56]; Lei and others [57])."

Conclusion and Recommendations

"This review article advocates traditional agriculture as a climate-smart agriculture approach to sustainable food production."

"Traditional agricultural practices have potentials to adapt and mitigate climate change through their agroecological features."

"Traditional agriculture can be adopted as an alternative method for sustainable food production in changing climate."

"Besides mitigating climate change, traditional agriculture is also helpful for human health safety, natural resource management, energy conservation and socio-ecological integrity."

"These practices not only mitigate climate change but also enhance agricultural sustainability."

"Traditional practices coupled with modern sustainable farming practices would be a noble choice for climate change mitigation and adaptation."

"Increasing food productivity while addressing climate change impacts needs the integration of traditional and modern agricultural practices."

"Climate-smart agriculture practices both modern as well as traditional should be encouraged and used widely."

4.2.2 Beyond the Material: Knowledge Aspects in Seed Commoning [116]

This is a machine-generated summary of:

Sievers-Glotzbach, Stefanie; Euler, Johannes; Frison, Christine; Gmeiner, Nina; Kliem, Lea; Mazé, Armelle; Tschersich, Julia: Beyond the material: knowledge aspects in seed commoning [116].

Published in: Agriculture and Human Values (2020).

Link to original: https://doi.org/10.1007/s10460-020-10167-w.

If you want to cite the papers, please refer to the original.

For technical reasons we could not place the page where the original quote is coming from.

Abstract-Summary

"Core sustainability issues concerning the governance of seeds revolve around knowledge aspects, such as intellectual property rights over genetic information or the role of traditional knowledge in plant breeding, seed production and seed use."

"While the importance of knowledge management for efficient and equitable seed governance has been emphasized in the scientific discourse on Seed Commons, knowledge aspects have not yet been comprehensively studied."

"With this paper, we aim to (i) to analyze the governance of knowledge aspects in both global and local/regional Seed Commons, (ii) to highlight disconnections in knowledge governance between the local Seed Commons and global governance of plant genetic resources, and (iii) to investigate the contribution of knowledge commoning to environmentally sustainable and culturally adapted food systems."

Introduction

"The goals of the paper are thus (i) to analyze the governance of knowledge aspects in both global and local/regional Seed Commons, (ii) to highlight disconnections in knowledge governance between the local Seed Commons and global governance of plant genetic resources, and (iii) to investigate the contribution of knowledge commoning to environmentally sustainable and culturally adapted food systems."

"We will then analyze knowledge governance aspects of two local Seed Commons, which reflect the diversity of Seed Commons arrangements in practice."

"The main challenges for effective knowledge governance in both local and global Seed Commons are finding institutional arrangements, which fruitfully integrate aspects from both traditional and scientific knowledge systems, taking into account the complex interrelation between the knowledge-related, material and cultural aspects of seeds."

Theoretical Background

"We examine the conceptual distinction made in commons studies to investigate material resources (mainly natural resources) and immaterial resources (mainly knowledge)."

"A distinct research field with specific conceptual frameworks on the governance of Knowledge Commons has emerged alongside the study of Natural Resource Commons (Frischmann and others [59]; Hess and Ostrom [60]; Hess [61])."

"Hess and Ostrom ([62], 12) cite historical reasons for the fact that "most of the interdisciplinary work on the Knowledge Commons to date is not an outgrowth of the Natural Resource Commons literature."

"Knowledge Commons have been embedded in the narrative of the "second enclosure movement" (Boyle [63]): the enclosure of the 'intangible commons of the mind' that followed the historical enclosure of collectively managed natural resource systems."

"Knowledge Commons have featured in discussions on openness, inclusiveness and democratic processes, such as the town commons in the United States, which are shared spaces allowing free speech (Hess and Ostrom [62], 12)."

Knowledge Governance in Seed Commons: The International Seed Treaty

"A key objective of the Seed Treaty is the facilitated access to PGRFA (Treaty Article 13.1), which is an essential element in the development and exchange of related knowledge."

"While the Seed Treaty formally recognizes small-scale farmers' relevance in the generation and conservation of genetic resources, and in the management of knowledge relating to seeds, the Treaty does not include sufficient instruments to ensure the access and active participation of farmers for instance in the MLS, or compliance in implementation of Farmers' Rights on the national level."

"Farmer communities are supported through projects aimed at conservation and participatory breeding funded by the BSF, a more active role of farmers as active innovators as part of the MLS, an enhanced (direct) integration in deliberation processes of the Treaty, and a better integration of TAeK could enhance the contribution of the Seed Treaty to food sovereignty."

Knowledge Governance by Two Local Seed Commons

"The first is MASIPAG, a Philippine farmer-led network, which includes community seed banks, seed exchange practices and farmer-led breeding of locally adapted rice cultivars."

"The second case study describes Kultursaat, a German association of farmer-breeders, which breed organic vegetable varieties on farm and makes them accessible to farmers and breeders without claiming variety protection."

"Kultursaat's understanding of the link between knowledge, intellectual property rights and varieties and seeds as material goods, leads the organization to adopt a practice of seed governance where newly bred open-pollinated varieties fall under collective responsibility and knowledge on both the breeding process and variety characteristics is openly shared."

"The polycentric structure of Kultursaat, including decentralized breeding locations and decision-making capacities, contribute to knowledge commoning: site-specific agroecological knowledge is drawn upon to breed locally adapted varieties and autonomous decision-making structures allow to attend to local environmental and social circumstances flexibly."

Discussion

"While in both cases farmers and their knowledge play an important role, the emphasis with regard to knowledge creation, breeding and decision-making is on small-scale farmers within MASIPAG and on farmer-breeders for Kultursaat."

"For Kultursaat, the breeding community and the user community are distinct, since the Kultursaat breeders develop new organic varieties mainly for commercial farmers who neither breed themselves, nor are members of the association."

"The understanding that knowledge and culture are deeply embedded in seeds and the collective practices of seed sharing (MASIPAG) and breeding is fundamental for both local Seed Commons."

"Seeds (the material) are only shared with farmers outside of the community in conjunction with training that conveys the corresponding knowledge and MASIPAG's values."

"While for Kultursaat, the focus is on organic breeding and developing varieties that are suitable for organic agriculture, the core purpose of MASIPAG's work is to improve the livelihood of small-scale farmers."

Conclusion

"This investigation has revealed the importance of knowledge aspects for the functioning of Seed Commons at both at global and local levels, and the manifold and complex interrelations between informational, material and cultural aspects in their governance."

"The two cases of local Seed Commons investigated here differ in their main governance objectives: MASIPAG's primary goal is the empowerment of farmers, whereas Kultursaat primarily aims at developing organically-bred varieties for organic agriculture."

"Regarding the integration of local Seed Commons into the global governance of PGRFA, we found that the knowledge governance structures by the MLS pose obstacles to farming communities from directly accessing PGRFA for on farm breeding and also from actively participating in the MLS."

"Our investigation of Seed Commons points to the relevance of knowledge aspects across diverse Natural Resource Commons."

4.2.3 *Traditional Knowledge Affects Soil Management Ability of Smallholder Farmers in Marginal Areas [117]*

This is a machine-generated summary of:

Occelli, Martina; Mantino, Alberto; Ragaglini, Giorgio; Dell'Acqua, Matteo; Fadda, Carlo; Pè, Mario Enrico; Nuvolari, Alessandro: Traditional knowledge affects soil management ability of smallholder farmers in marginal areas [117].

Published in: Agronomy for Sustainable Development (2021).

Link to original: https://doi.org/10.1007/s13593-020-00664-x.

If you want to cite the papers, please refer to the original.

For technical reasons we could not place the page where the original quote is coming from.

Abstract-Summary

"We integrate agronomic and social sciences approaches to explore links between smallholder households' farming knowledge and soil fertility in an ethnopedological perspective."

"By analyzing soil compositions at tested households, we identify a novel measure of soil management ability, which provides an effective empirical characterization of the soil managing capacity of a household."

"Regression analysis is used to evaluate the effects of household knowledge on the soil management ability derived from laboratory analysis."

"We are able to disentangle a home learning and a social learning dimension of the household knowledge and appraise how they can result in virtuous and vicious cycles of soil management ability."

"We show that higher soil management ability is associated with farmers relying to a great extent on farming knowledge acquired within the household, as a result of practices slowly elaborated over the years."

"Lower soil management ability is linked to households valuing substantially farming knowledge acquired through neighbors and social gatherings."

"The present study is the first to formulate the concept of soil management ability and to investigate the effects of the presence and the types of farming knowledge on the soil management ability of smallholder farmers in remote areas."

Introduction

"We hypothesize that in these marginal areas, smallholder farmers able to preserve soil fertility demonstrate a higher management ability, possibly driven by access to a specific set of knowledge."

"The soil management ability is calculated stemming from the content of total nitrogen (N), phosphorus (P), and potassium (K) in the soil, which suggest among the smallholder farmers in the villages those capable of managing the soil fertility more effectively (Öborn and others [64]); this is especially true in subsistence, low-input rainfed farming systems as those analyzed in this work."

"Household's farming knowledge refers to the process of knowledge accumulation which leads to a certain degree of soil management ability for each smallholder farmer; it encompasses soil managerial notions as well as acquired information on farm resources maintenance and diversification (Laekemariam and others [65])."

Material and Methods

"The full logit model specification at farmer level was the following, for high and low soil farmers, respectively: With: β_{ik} = coefficient for home leaning, in each region k, for each farmer i X1, X2, $X3_{ik}$ = categorical variable for home learning, social learning, and education respectively, for each farmer i in region k, μ_{ijk} = coefficient attached to each j controls, in each region k, for each farmer I, Z_{ijk} = common set of resource and farmer's level controls j for each farmer i in each region k, k = study regions (Amhara-NW and Tigray-C) The econometric exercise was performed in R software: the package utilized is {miceadds}."

Results and Discussion

"The direction of the effect depends on the dimension of household's knowledge considered: the relation between home learning and soil management ability is significantly positive ($p < 0.10$), whenever resource level and farmer's level controls are inserted."

"The odds of being a low soil management ability farmer increase by 1.91 (exp(0.65)) ($p < 0.10$) times if the farmer values social learning knowledge and the effect is significant and robust to controls."

"The odds of being a low soil management ability farmer instead do not show a significant relation with the fact that the farmer values home learning knowledge (indeed, the odds ratio decrease by 0.74 (exp(− 0.30)) times but it is not statistically significant)."

"When the three dimensions of the household's farming knowledge are disentangled, surprising drifts emerge: home learning is persistently significant and positive across results, showing a virtuous cycle connecting local agricultural knowledge accumulated within the household and the soil management ability of that household."

Conclusions

"This study, combining a unique dataset composed by household surveys and topsoil sample data collected from households' main fields, identifies three dimensions of knowledge and formalizes the concept of soil management ability indicator, which iconizes the soil management capacity of a farmer relatively to neighbors."

"The work investigates the effect of household's farming knowledge on soil management ability gradients among smallholder farmers."

"The results of this work advocate for a holistic inclusion of the farmers' knowledge in all the agricultural production steps, not last during the soil and resource management phases."

"Further works should aim at codifying the exchange of farmers' knowledge among marginal communities at a finer degree of specificity, in the attempt to encompass further attitudinal aspects of the household's soil management ability, such as the degree of autonomy in experimentation as well as the attitude to soil managerial changes."

4.2.4 Indigenous Practices of Soil and Water Conservation for Sustainable Hill Agriculture and Improving Livelihood Security [118]

This is a machine-generated summary of:

Arora, Sanjay; Bhatt, Rajan; Sharma, Vikas; Hadda, Manjeet S.: Indigenous Practices of Soil and Water Conservation for Sustainable Hill Agriculture and Improving Livelihood Security [118].

Published in: Environmental Management (2022).

Link to original: https://doi.org/10.1007/s00267-022-01602-1.

If you want to cite the papers, please refer to the original.

For technical reasons we could not place the page where the original quote is coming from.

Abstract-Summary

"The soil and water resources are becoming the major constraints in agricultural production in the erosion-prone fragile ecosystem of the region."

"Due to intensive rains and sloping lands of the region, erosion of the topsoil becoming the major problem for practicing sustainable agriculture in the region, which further dissects the lands, reduces the fertility potentials and land productivities of the region."

"The root cause of this huge erosion in the region is the intensive rains within a short interval of time on the bare sloping hillsides and handling this problem in both ways could reduce their erosion damage."

"Farmers in the region do have the skills to manage the problem of soil erosion which they learned from their forefathers and have faith in them."

"Among different ITKs bunding of field, plowing before monsoon, filter strips, earthing-up in maize, mulching, compression of soil in sugarcane are the important ones for restoring the fertility of soils, reducing erosion losses, improving land productivity, and ultimately livelihoods in the region."

Introduction

"Further, intensive short duration rains on the sloping lands of the region, responsible for soil erosion, removing the fertile topsoil and reduction in the crop yield potential (Arora [66]; Arora and others [67]; Kukal and Bhatt [68]; Bhatt [69])."

"Resource poor farmers are not willing to adopt the latest and advanced techniques of soil and water conservation due to their lack of risk-bearing ability (Vashisht and others [70]; Arora and others [67])."

"Sustainable control of soil erosion is the only solution for enhancing the yield potentials as well as the livelihoods of the farmers."

"Attempts have been made in the region for refining these ITKs and understanding their role under diverse land uses which they adopted for controlling the losses caused by soil erosion (Vashisht and others [70])."

"The refinement of ITKs based on the current scientific interventions is a must for having better control of the soil erosion losses, enhancing water and land productivity, and livelihoods in the foothills of Shivalik in the North-West Himalayan belt."

Problems and Constraints

"Animals graze whatsoever fodder is present on the sloping lands followed by tufts burning by hot winds which enhance soil susceptibility to erosion both by water and wind."

"All this set a scene for soil erosion to take place under the impact of high-intensity summer monsoon rain showers over the sloping lands."

"Variation in the land, soil, rainfall, depth of underground water from place to place demands the development of need-based and location-specific technologies so that the production potential of the area can be realized on a sustainable basis (Arora and others [67])."

"Among different types of soil erosion, gully erosion is the most serious one in the region."

"The problem of soil erosion is thus quite serious in the region and farmers mostly depend on their indigenous technical knowledge (ITK) to raise and grow their crops sustainably and many of these techniques are scientifically sound also (Kukal and Bhatt [71]; Bhatt [72]).”

Indigenous Technical Knowledge for Conserving Traditional Soil and Water Resources

“In this, the kharif crop in the area is missed and only a rabi crop is taken, utilizing the water conserved in the soil profile during the summer monsoon season.”

“By doing so, the farmer believes that he can conserve water during the monsoon period into the soil which may ultimately be utilized by the succeeding rabi crop.”

“Gullies (the main culprit behind the soil erosion losses) are the main cause of concern for the farmers of the foothill region as it dissects the agricultural fields and hinder the tillage operations, thereby decreasing the production potential.”

“There are a large number of such indigenous techniques used by the farmers of the area, however, the most important and commonly used techniques are as follows: By simply plugging in the lower ordered gullies using a simple spade, then it will not be in a position to collect runoff water and divert to higher order ones and hence control soil and water loss.”

Farming Practices in Doon Valley

“The farmers usually construct terraces for cultivation in the Doon Valley which further helps them in reducing the erosion losses, conserving better moisture, etc. In the region, around 125 nalas (drains) exists, where rainwater is managed.”

“In the region, farmers tried to use the land to the maximum extent possible by preparing terraces (inwards) from top to bottom after constructing loose boulder retention walls (risers) and putting grass on them for proper land-stone interactions (Arora and Bhatt [73]).”

“On mild slopes, farmers construct shoulder bunds to protect their lands from soil erosion and grow vegetation over the bunds, particularly grasses for binding the soil.”

“In the foothill region of lower Shivalik’s of Punjab and Himachal Pradesh states, farmers have village ponds on their lands or community lands for storing runoff water which is further used for domestic and other purposes.”

Traditional Water Management in North-East Hills

“Scientists are exploring new ways for assuring water availability in the region for most of the year.”

“Research has revealed that by adopting new techniques, land productivity of the region could be enhanced to the tune of two to three times if the available water is utilized properly (Sharma [74]).”

“In the north-eastern hilly region, where bamboo plantation is common, the farmers of the Jaintia and Khasi hills have developed a unique bamboo drip irrigation system.”

"Water from an uphill source is trapped and brought to the plantation by the main bamboo channel."

"These water sources are far off from the plantations and the main bamboo channel runs hundreds of meters - in some cases even a few kilometers."

"The water is then regulated through a complex bamboo network of secondary and tertiary channels to all the parts and corners of a plantation, right up to the bottom of the hill."

Policy Implication, Suggestions, and Future Approaches

"Government is promoting capacity building and knowledge sharing programmes for farmers to share ITKs or Good practices that may be scientifically refined for widespread adoption to conserve natural resources and boost productivity in similar rainfed regions across the country."

"It is strongly believed that the ITKs in the foothill regions need to be replicated in other rainfed areas where soil and water conservation is of paramount concern with some modifications as per site-specific needs."

"The authors believe that most of the ITKs shall be easily adopted by small and marginal farmers in 65% of the rainfed farming area of the country having the substantial potential to meet the requirements of food, fodder, timber and livelihood demands of the nation."

Conclusions

"Indigenous Technical Knowledge system (ITKs) of the region for successfully growing crops by minimizing the erosion losses and conserving a higher fraction of the moisture is very important."

"Field bunding, pre-monsoon plowing, filter strips, earthing-up in maize, mulching, compression of soil in sugarcane have proved effective in preventing soil loss, and conserving in-situ moisture and soil fertility restoration are the chief ITKs."

"All these traditional practices adopted for ages and transmitted from one generation to other, are easy, helpful in conserving soil and water and optimizing crop yields apart from benefiting soil health."

"There is a need to refine these traditional practices as per site-specific needs on a scientific basis to realize their full potential and ultimately improve the soil fertility and socio-economic status of the farmers in the region."

4.2.5 Identifying Indigenous Practices for Cultivation of Wild Saprophytic Mushrooms: Responding to the Need for Sustainable Utilization of Natural Resources [119]

This is a machine-generated summary of:

Wendiro, Deborah; Wacoo, Alex Paul; Wise, Graham: Identifying indigenous practices for cultivation of wild saprophytic mushrooms: responding to the need for sustainable utilization of natural resources [119].

Published in: Journal of Ethnobiology and Ethnomedicine (2019).

Link to original: https://doi.org/10.1186/s13002-019-0342-z.

If you want to cite the papers, please refer to the original.

For technical reasons we could not place the page where the original quote is coming from.

Abstract-Summary

"Satisfying the increasing consumer demand for grassland or forestry products like wild mushrooms as food or medicine, requires innovative ethno-biological and industry development strategies to improve production capacity, while easing the pressure on diminishing natural resources and averting ecosystems degradation."

"This case study addresses traditional knowledge systems for artisanal mycoculture to identify cultivation practices that enhance sustainable utilization of natural resources."

"Focus groups and semi-structured interviews characterized artisanal production methods and identified locally used substrates for cultivation of different mushroom species."

"Artisanal practices were characterized for the cultivation of six wild saprophytic mushroom species including Volvariella speciosa (akasukusuku), two Termitomyces sp. (obunegyere and another locally unnamed species), Agaricus sp. (ensyabire) and Agrocybe sp. (emponzira), and one exotic Pleurotus sp. (oyster) that are used as food or medicine."

"on heaped cobs placed near homesteads; and primary decomposers growing on waste sorghum from brewing the traditional alcoholic drink, muramba for Pleurotus sp. (oyster), and banana and spear grass residue from banana juice processing like the Volvariella speciosa known as akasukusuku because it is associated with the banana plantation locally known in the Luganda language as olusuku and is usually heaped under ficus trees."

"Our description of artisanal mycoculture methods that respond to conservation and utilization pressures, demonstrates the value of addressing traditional knowledge to improve ethno-biology and mycoculture industry practice."

"Future research and industry development can utilize the result of this study to boost artisanal production of wild saprophytic mushrooms in Sub-Saharan countries, for food or medicinal consumption, and environment conservation."

Background

"We identify and describe the artisanal cultivation of wild saprophytic mushroom species as examples of innovative artisanal farming systems in Uganda."

"We note that access to natural resources to forage for mushrooms is increasingly being curtailed by natural resource management policies, conservation regulations, and agricultural development programs that are not in accord with ecosystem health, traditional systems of property ownership, or natural resource utilization and conservation."

"It is shown that as traditional farming communities are assimilated into the wider global economy, their farming systems for wild saprophytic mushroom are simply not sufficient to ensure local food security and community economic resilience."

"This study identifies artisanal methods for the cultivation of wild saprophytic mushrooms, which are based on indigenous knowledge used in traditional farming systems."

"Important for the development of community economic resilience, local varieties of wild saprophytic mushrooms earn more income for producers than commercial oyster mushrooms."

Methods

"The participative, innovation system–sensitive, and community-integrated methodology described above was used to identify innovative artisanal mycoculture practices and select participants for this study based on knowledge that is held across many complex government, cultural, language, and social divisions."

"Eight focus groups having between 8 and 20 participants were held across all five districts, at village locations that were identified by government innovation system actors."

"In accordance with Gill and others [75], focus groups were used to generate information on collective views, group norms, and community narratives on common innovation challenges, and to identify innovation practices used in the local region."

"Individual semi-structured interviews with key informants were important, in particular for discussions on mushroom growing practices and the preparation of herbal medicines as follows: Within the existing innovation systems framework, it was found out that females are located at the fringes of all production processes."

Results

"Obunegyere mushrooms bloom during the rainy season, and during this reproductive growth phase, growers lightly place piles of grass or banana leaves over sites where pinheads are seen; then, they water them to control moisture content until they grow to an appropriate size and are harvested."

"It was locally reported that despite the use of this method to manipulate the microclimate of sites within forests, production of wild mushrooms has reduced due to conversion of forests and grassland for agriculture, and deforestation for charcoal burning."

"Local growers reported that the environmental control method cannot overcome the seasonality of obunegyere mushrooms; therefore, current production cannot satisfy local needs for sustenance, and a growing market."

"In Kabale District, local growers reported that cattle manure is an easily accessible and low-cost local substrate for mushroom cultivation."

"After the composting period, the mushrooms locally known as ensyabire (Agaricus sp.) were reported to grow on the waste by natural spawning."

Role of the Traditional Techniques and Products in Local People's Subsistence

"In the grass waste or forest litter mushroom production systems zero tillage or minimum interference ensures access to resources such as rare mushroom species, medicinal herbs, fire wood, income, and conservation of endangered species such as termites and intricate biomes."

"In the banana residue mushroom production system enhanced composting increases recycling of nutrients such as nitrogen and microelements, conservation of moisture by providing mulch, and control of pests and diseases."

"The knowledge on mushroom production systems using different substrates that was gathered from interviews and focus groups has been analyzed in context with environmental and agricultural inputs for mushroom production."

"Control of the nutrition of the substrate was strong for all production methods and mushroom species, perhaps presenting a strong basis from which to plan future industry development activities."

Discussion

"For all artisanal mushroom production systems studied, communities have identified substrates that can be sourced at a sufficiently low price within each of their local regions to support the development of economically viable production systems in low or very low–income communities."

"Clear from existing artisanal production systems is that spawning method remains uncontrolled for each of the five wild mushroom species."

"Developing cost-effective spawning methods for each of the species is an essential next step for industry development, one that has been successfully achieved for community-based artisanal production of commercial oyster mushrooms."

"The level of control achieved for the artisanal production of oyster mushrooms, through the use of simple building structures, indicates the economic feasibility of using covered structures to improve the degree of control over environmental and agricultural inputs for each of the production systems of the five studied wild mushroom species."

Conclusions

"To economic development, such integration of traditional knowledge into a mushroom industry development program would improve resource utilization and enhance conservation."

"Future industry development programs can now extend on existing knowledge of the required substrates for mushroom production."

"The artisanal production of high value wild mushroom species can be developed in the same way that oyster mushrooms have even in the backyard."

"Integrated growing of currently wild mushrooms by recycling the byproducts of other agricultural systems as substrates, improves livelihoods, provides nutritious products, and recycles biomass thus enhancing environment conservation."

"This study posits that new contexts for cultivation of wild mushroom species should be rooted in existing traditional innovation frameworks and organic ecosystems which mimic the natural ecosystems, and that multifaceted technological and organizational innovations are of value for subsistence agriculture intensification, agroecology conservation, and agro-industry development activity."

4.2.6 Traditional Forest-Related Knowledge and Agrobiodiversity Preservation: The Case of the Chagras in the Indigenous Reserve of Monochoa (Colombia) [120]

This is a machine-generated summary of:

Hernandez Marentes, Maria Alejandra; Venturi, Martina; Scaramuzzi, Silvia; Focacci, Marco; Santoro, Antonio: Traditional forest-related knowledge and agrobiodiversity preservation: the case of the chagras in the Indigenous Reserve of Monochoa (Colombia) [120].

Published in: Biodiversity and Conservation (2021).

Link to original: https://doi.org/10.1007/s10531-021-02263-y.

If you want to cite the papers, please refer to the original.

For technical reasons we could not place the page where the original quote is coming from.

Abstract-Summary

"During the centuries, thanks to their traditional knowledge, indigenous populations have developed a deep relationship with the surrounding environment, as, living in remote places, they must be self-sufficient."

"The paper takes into consideration the Indigenous Reserve of Monochoa in Colombia as an example of how traditional knowledge can support a rich biodiversity conservation."

"Results highlighted the crucial role of the indigenous communities for biodiversity conservation."

"The preservation and adaptation of traditional knowledge and practices, a decentralized autonomous governance system demonstrates that local communities not only can be part of ecosystems with unique biodiversity, but that they can represent the main actors for an active conservation of biodiversity."

"Agroforestry systems based on traditional forest-related knowledge can therefore be an effective alternative to biodiversity and ecosystem services conservation based on strict nature protection where humans are perceived as a negative factor."

Introduction

"One of the most interesting practices is related to shifting cultivations, the most common traditional productive practice in the Amazon region (Coomes and others [76]; Padoch and Pinedo-Vásquez [77]; Junqueira and others [78]), and in particular to the agroforestry systems called chagras."

"Chagras are a complex of different activities related to agroforestry practices, carried out by local indigenous communities in the Amazon area, shared by different South American Countries, and have been defined as "a polyculture mode, in appearance simple, around which the other productive activities of the indigenous population revolve, but that implies a management of time, space and cultural identity that increases the system complexity" (Triana-Moreno and others [79])."

"The paper aims to highlight the connections between traditional agroforestry practices and local biodiversity conservation, providing a deep analysis of the local traditional knowledge related to Chagras and of the resulting agrobiodiversity."

"The agrobiodiversity linked to the presence of indigenous communities, is an important part of biocultural diversity and can be crucial for the conservation of biodiversity of natural rainforests."

Study Area, Materials and Methods

"72% of the area is classified as dense forest, corresponding to areas with vegetation characterized by a more or less continuous stratum, with tree cover more than 70%, canopy height greater than 15 m and which is located in areas that do not have periodic flooding processes (Gobernación del Caqueta [80])."

"The remaining 3% of the land corresponds to dense heterogeneous high flood forest, which corresponds to areas with vegetation characterized by a more or less continuous coverage (more than 70%), canopy height > 15 m and located in the fringes adjacent to the bodies of water, which correspond mainly to the plains subjected to periodical floods lasting more than two months (Gobernación del Caqueta [80])."

"This analysis was conducted in two steps, first of all by an assessment of the agrobiodiversity of the chagras through a deep bibliographic survey, and second one deepening the different species and varieties cultivated by the two main ethnic groups of the Indigenous Reserve of Monochoa."

Results

"According to Schroder and others [81], the choice of the plants depends on the following seven criteria: (1) soil characteristics; (2) species of sorrounding forest, as these will be replaced by their counterparts of chagra, for example wild guamo is replaced by guamo of chagra; (3) aim of the chagra (self-consumption, commercialization, urgent need for some crops); (4) dietary habits of the families; (5) historical-cultural factors (different ethnicities sow different species and varieties); (6) short-term personal and/or family circumstances; (7) availability of seeds."

"Even if indigenous people cut down some wild species during the socola, they recover them sowing counterparts of varieties that have been adapted in order to produce in the chagra, replicating the rainforest system due to the synergic relations built between the species."

Discussion

"This recognition represents a great advance at global level regarding the legislative matter, also because it is a recognition of the role of indigenous ethnic groups living in the Amazon for the preservation of these crucial ecosystems."

"The communities of the sector of the middle Caquetá river, on 23 August 1988, formed what is currently known as Association Council of the Regional Indigenous of the Middle Amazon-CRIMA with the aim of generate development from the perspective of indigenous cultures; this association is composed of 16 communities of four indigenous ethnic groups (uitoto, muinane, andoque and nonuya), under a structure that includes a Congress, a Board of Directors, an Executive Committee and five areas of coordination: Health, Own Territory-Government, Education, Women and Natural resources" (Instituto Amazónico de Investigaciones Científicas—SINCHI [82])."

"Results, in fact, highlight that the Reserve of Monochoa is characterized by high levels of agrobiodiversity related to the maintainance of traditional agroforestry practices, especially of the system called chagra."

Conclusions

"This includes obtaining resources for their families without the use of additional chemical fertilizers or pesticides, only relying on their traditional knowledge which represents the key factor for the Amazonian conservation, in order to give the same opportunities that they had, to the future generations."

"The research highlighted that indigenous traditional knowledge developed through the centuries, based on the respect and equilibrium with the surrounding environment is the key for the sustainable development and for the wellbeing of local communities as well as for the preservation of biodiversity and ecosystem services."

"Agroforestry systems based on traditional forest-related knowledge all over the world (Santoro and others [83]) can therefore represent an effective alternative to biodiversity and ecosystem services conservation based on strict nature protection where humans are perceived as a negative factor."

4.2.7 Socio-ecological and Religious Perspective of Agrobiodiversity Conservation: Issues, Concern and Priority for Sustainable Agriculture, Central Himalaya [121]

This is a machine-generated summary of:

Negi, Vikram S.; Maikhuri, R. K.: Socio-Ecological and Religious Perspective of Agrobiodiversity Conservation: Issues, Concern and Priority for Sustainable Agriculture, Central Himalaya [121].

Published in: Journal of Agricultural and Environmental Ethics (2012).

Link to original: https://doi.org/10.1007/s10806-012-9386-y.

If you want to cite the papers, please refer to the original.

For technical reasons we could not place the page where the original quote is coming from.

Abstract-Summary

"Rural community of the mountains has developed several indigenous and traditional methods of farming to conserve the crop diversity and rejoice agrodiversity with religious and cultural vehemence."

"Traditional food items are prepared during occasion, festivals, weddings, and other religious rituals from diversified agrodiversity are a mean to maintain agrodiversity in the agriculture system."

"The different traditional system of agriculture and indigenous method of maintaining soil fertility, socio-cultural and religious rituals has saved many crops that are under threatened category."

"All these system and practices are ignored nauseatingly in hill agriculture policy, where more emphasis was given for plain areas."

"There is a strong need to bring desirable changes in the agricultural policy, research, and development in reference to mountainous regions."

"The present paper describe present scenario of agriculture, traditional, and socio-cultural practices of retaining soil fertility and agrodiversity, policy dimensions, and strategies for management of the Himalayan agroecosystems."

Introduction

"Environmental, biological, socio-cultural, and economic factors prevailing in the Central Himalayan region have resulted in the evolution of diverse agroecosystems along with crop diversity."

"During recent past a decline in interest of local farming communities towards traditional crop cultivation has been observed as a result of climatic, cultural, and socio-economic transforms."

"Due to a variety of reasons the crops diversity is declining very fast in the region but undoubtedly many crop varieties are still conserved in the region because of their socio-cultural, medicinal, and religious values."

"At the same time, the traditional agriculture with diverse crops and cropping patterns are under great threat all over the country including Uttarakhand (Maikhuri and others [84])."

"The huge diversity in Mountain region has been maintained through a variety of crop compositions, indigenous method of maintaining soil fertility, socio-cultural, and religious rituals."

Methodology and Study Area

"The Central Himalayan region of India is spread over an area of 53,483 sq."

"Information about the number of landraces cultivated in the past (1980) and at present (2010), shift in landrace diversity of crops was obtained by interviewing the head of each selected household (Maikhuri and others [85, 86]; Negi and others [87])."

"To document ethno-medicinal, socio-cultural, and religious knowledge pertaining to crops, the farmers of different age groups (18–30, 31–60 and above 60) were interviewed and several village level meetings were organized for detailed information."

"Respondents were also asked to fill out a questionnaire to acquire information concerning cultural dishes and recipes prepared from traditional crops and their medicinal properties and uses in traditional healthcare system."

"We have surveyed the literature where empirical evidences showed that agrobiodiversity conservation goal would be achieved efficiently with proper documentation and giving values to socio-cultural activities of the local people through which they have been maintaining and conserving the valuable germplasm in the Himalayan agroecosystems."

Results and Discussion

"In this endeavor empirical evidence (Maikhuri and others [86, 88]; Palni and others [89]; Nautiyal and others [90]; Bisht and others [91]) supports the claim that the diversity of traditional livestock based farming system in fragile Himalayan landscape would be difficult to conserve without conserving the socio-cultural values of the people living in the region along with proper documentation of traditional ecological knowledge pertaining to local available biological resources."

"Under marginal production conditions in low-intensity agro-ecosystems, local breeds, and crop varieties tend to be adapted to a range of environments through a process of human selection based on farmers' preferences and traditional knowledge (Ceccarelli and others [92]; Cavatassi and others [93])."

"On-farm conservation is considered to be "the sustainable management of genetic diversity of locally developed traditional crop varieties with associated wild and weedy species or forms by farmers within traditional agriculture systems."

Conclusion and Recommendations

"Traditional crop diversity needs institutional and policy support for their conservation and sustainable land use development throughout the world, particularly in the Himalayan region."

"Appropriate policy incentives are needed to enhance livelihood options through utilization of agro-biodiversity resources for the benefit of marginal farmers."

"In the Himalayan region, indigenous knowledge, traditional cropping system, socio-cultural rituals, belief and practices of soil fertility as mention in this paper has enormous potential to facilitate agrodiversity conservation need to be promoted while formulating the policy implementation."

"Bioresources based simple technologies and institutional innovations developed by various well recognised institutions of national and international level need to incorporate in hill agriculture management plan by policy maker for better livelihood options."

4.2.8 Social Practices of Knowledge Mobilization for Sustainable Food Production: Nutrition Gardening and Fish Farming in the Kolli Hills of India [122]

This is a machine-generated summary of:

Hudson, Suraya; Krogman, Naomi; Beckie, Mary: Social practices of knowledge mobilization for sustainable food production: nutrition gardening and fish farming in the kolli hills of India [122].

Published in: Food Security (2016).

Link to original: https://doi.org/10.1007/s12571-016-0580-z.

If you want to cite the papers, please refer to the original.

For technical reasons we could not place the page where the original quote is coming from.

Abstract-Summary

"We examine how knowledge about sustainable food production is acquired and mobilized among small-scale farmers belonging to the Malayalis Tribe, who reside in the Kolli Hills region of Tamil Nadu, India."

"Using qualitative participatory research methods, we explored ways in which farmers are accessing, using and sharing knowledge about these practices, and barriers to equitable access."

"This research revealed that participants in both nutrition gardening and fish farming rely primarily on external experts (non-governmental agricultural research and extension) for formal training (e.g. workshops) and advice, despite a long history and traditional knowledge of gardening and river fishing."

"Agricultural knowledge is also mobilized within the region less formally between individuals and households through face-to-face contact or verbal communication via mobile phones."

"The use of additional communication technologies (e.g. computers) to enhance access and mobilization of knowledge was perceived to be most relevant and beneficial for young, educated people."

Introduction

"Fish farming and nutrition gardening can be characterized as sustainable food production practices as they use and enhance local knowledge and ecological resources to develop more self-reliant and resilient agri-food systems that produce nutritious food, and enable improved livelihoods for local communities (Altieri and others [94])."

"Knowledge plays a central role in the development of these systems, we set out to explore how small-scale farmers in the Kolli Hills access, use and share agricultural information, particularly related to nutrition gardening and fish farming."

"Following a description of data collection methods, the results are presented in five sections: an overview of knowledge mobilization over time in the Kolli Hills; current agricultural knowledge systems; traditional and new knowledge for nutrition gardening and fish farming; an exploration of the use of ICTs as tools for knowledge mobilization and finally; a discussion on barriers to equitable access to information."

Literature Review

"The knowledge that is generated in traditional farming systems is continually produced and reproduced by small-scale farmers through an iterative process of observation and experiential learning (theory and practice), and is strongly shaped by the socio-cultural and bio-physical contexts (Altieri [95]; Yano and Lanusosang [96])."

"Knowledge by itself is not enough to change practice, since practices are social and therefore reinforced by many elements such as norms, cultures and habits (8) Although knowledge is not the only factor influencing the development of new social practices, it provides a starting point to understand how they form and become routine, and for the purposes of this study, allows us to see how sustainable food production practices may become normalized through learning and sharing of traditional and new knowledge."

"It is the social learning that comes through changing social practices, often initiated or evolving with knowledge mobilization, that has the potential to support more resilient foods systems for marginalized, small-scale farmers."

Methods

"Qualitative methodology is well established in social practices and knowledge mobilization research as a useful approach to understanding learning processes and the nature of learning outcomes."

"One of the core initiatives of the APM project was to establish a Village Resource Center (VRC) in the main town of Semmedu and two Village Knowledge Centers (VKC) in the small villages of Asakadupatti and Alawadipatti as resource hubs for community members to access and share agricultural information, gain skills training (e.g., computer classes) and serve as venues for community meetings."

"PRA was used so that community members could be actively involved in the research process."

"These interviews were done with community leaders to understand more detailed accounts of the social practices that exist, with MSSRF staff to understand the goals, motivations and inner-workings of current and future projects and with the staff of other local NGOs to understand what other programs are being run in the area."

Findings and Discussion

"In order to further explore the formal and informal knowledge networks that exist in the Kolli Hills, a PRA was conducted to understand the role of traditional and new knowledge (Singh and Hensel [97]) for nutrition gardeners and fish farmers."

"Two groups of nutrition gardeners and two groups of fish farmers were asked to create a diagram that outlines where they learned the types of skills needed in order to maintain their respective practices."

"Nutrition gardeners also expressed the need to learn more about food preparation, when to harvest various vegetables and about marketing."

"All farmers, although satisfied overall with the training provided, would have liked to learn more about food preparation, when to harvest and how to expand the enterprise in order to sell excess fish in the market."

Use of ICT Tools for Knowledge Mobilization

"Despite the fact that mobile phone use in the Kolli Hills is increasingly prevalent, farmers said that they would prefer face-to-face communication to learn about agricultural information."

"Twenty households were asked about the ownership and use of mobile phones within their homes, only one of which said that they had none (five percent)."

"Another PRA showed the frequency of use of various technologies based on age, where farmers self-identified into two groups: younger and older generations."

"Farmers were asked to indicate how often they used each of the technologies that were already drawn into a diagram such as television, radio, mobile phones and fax machines."

"While farmers can gain access to agricultural information through the government-funded farm television program, mobile phones are used informally to speak to immediate family, relatives and friends when the distance is too far to travel by foot."

Barriers to Equitable Access to Information

"During one-on-one interviews, men and women of all ages were asked about their perceptions of access to information based on age, gender, income, geography and education level."

"Farmers in the Kolli Hills face limitations in their access to information given the remoteness of the area and the unreliability or lack of public transportation."

"Most farmers indicated that the use of these centers and particularly the use of computers were most suited to those who have higher levels of education and mostly young people who are most comfortable with and proficient in technology use and therefore most able to access information in this way."

"Farmers in the Kolli Hills access information through formalized systems such as MSSRF's knowledge centers, automated message programs or hotlines but barriers do exist due to distance to these centres, local perceptions about education, age and technology use, as well as gendered differences in access to mobile phones and travel to surrounding areas."

Conclusion

"This paper identifies ways that small-scale farmers in Kolli Hills communicated in the past, how they currently gain access to information about sustainable food production practices and what kinds of aspirations and opportunities exist for the future."

"Understanding knowledge mobilization of the Malayali Tribe can contribute to ensuring that all farmers, even those in remote areas, have access to information about sustainable food production practices so that they may take control of their own food production in ways that are environmentally sound and sustainable in the long run."

"By taking a social practices of knowledge mobilization approach, this study was able to investigate how agricultural information has been accessed over time, particularly how new communication channels enable the development and sharing of information about new practices that promote sustainable food production."

4.2.9 Assessment of Biodiversity Goods for the Sustainable Development of the Chagra in an Indigenous Community of the Colombian Amazon: Local Values of Crops [123]

This is a machine-generated summary of:

Garavito, Giovanny; Clavijo, Rafael; Luengas, Pilar; Palacios, Pablo; Arias, María Helena: Assessment of biodiversity goods for the sustainable development of the chagra in an indigenous community of the Colombian Amazon: local values of crops [123].

Published in: Journal of Ethnobiology and Ethnomedicine (2021).

Link to original: https://doi.org/10.1186/s13002-021-00453-0.

If you want to cite the papers, please refer to the original.

For technical reasons we could not place the page where the original quote is coming from.

Abstract-Summary

"There has been a reported loss of biodiversity and traditional knowledge associated with the chagras."

"This paper characterizes the cultivators, exploring knowledge and expressed value perception in the context of the Amazonian chagra of an indigenous community; also, this prioritizes species, under the optics of commercial opportunity."

"A semi-structured instrument was applied to 14 volunteers, asking about marketing preferences and use values of the species; later, a floristic inventory and prioritization workshop was developed."

"Open conversations showed that traditional knowledge is a matter of practice; and is maintained mainly by the older "grandfathers"."

"The Ziora-Amena community centralizes the handling of chagras in the community's older adults, who transmit their traditional knowledge to new generations through oral tradition."

"The perception of the trade stakeholder encourages research and development of endemic species, with health properties or ingredients for industry, which represent an opportunity of high added value for the region."

Introduction

"The chagra is also a space for transmission of traditional knowledge (TK) about management, sustainable use, and conservation of biodiversity, and where education about family and social relations of the community is provided [98, 99]."

"There is a crisis in the traditional and successful farming system of the indigenous communities, due to the lack of interest and preference for foreign foods, the decrease in varieties and species present thus the inability to satisfy the food requirements of the families (generating additional pressure on the extractive methods fishing, hunting, and gathering in the forest), and the absence of policies and proposals to revitalize the chagra [100]."

Methods

"The presence and abundance of the reported plant species in the active chagras was evaluated following previously described procedures [100, 101], with some modifications; briefly, in the chagra, two fixed-width linear transects of 40×2 m (80 m^2), arranged in the form of an "L" were carried out; in each one, information was collected on the population density of the 10 plant species previously reported by the participant; additionally, an ethnobotanical walk was carried out, tending to locate those species that, according to traditional knowledge, are located in particular areas of the chagra or in the stubble."

"As desirable characteristics of selection of prioritized plants, the equipment, together with the business stakeholder in the region, considered the perception of commercial factors (profitable marketing prices, affordable production costs, and ease of transport), social factors (traditional use in the chagra, absence of cultural restrictions for commercial exploitation), ecological factors (frequent harvesting, easy growth, possibility of multiple uses, sustainability in use, presence and abundance of the species in the active chagras, resistance to pests), and market differentiation factors (endemic species of the region and species without competition from large producers in the area) [101]."

Results

"TK in the management of the chagra related to the need to crop it with another vegetable species (as a strategy of plant protection); the need to seed the vegetable species at a certain time, and the distribution of the vegetable species in the area of the chagra are typical of the management of the chagra and evident in the open conversations held during the visit to the chagras, more than in the structured instrument applied."

"Regarding the perception of to whom the chagra plants can be sold, 79% of participants consider the Leticia market and tourists as the main trade niches."

"Our study prioritizes the plants: bejuco (sacha inchi), present in two of the chagras as a grass that grows robustly and abundantly; macambo and copoazú, two theobromas of important UVI in the community; mucuracaá (mucura), sacha ajo, uva caimarona, canangucho (aguaje), umari, and guama (native species of the Amazon); and jidoro, a plant reported by community as traditionally used in dyes for non-permanent tattoos."

Discussion

"In this community, it was evidenced that elders (traditional grandparents) maintain the knowledge about the management of chagra and medicinal plants (finding consistent with previous reports) [98]; taxa registered in this last category were provided by this stratum of the sample; the other age strata exercise the handling of chagra through practices mixed with contributions from other cultural origins."

"Specific aspects of the organization and traditional management of the chagra, such as the sowing of crops, planting at specific times and the distribution of plant species in the chagra area, were more evident in the open interviews, in situ, with the participants than in the specific questions of the instrument applied, making it clear that this knowledge is intrinsic to the work of the chagra and that it is evident in the accompaniment of the different activities of the chagra."

Conclusions

"The chagras of community have great biodiversity, the perception of participants allowed to highlight 38 taxa of interest for their commercialization, which belong to 28 different botanical families, predominating the neotropical species."

"Regarding the perception of marketing scenarios for the chagra plants, 79% of the participants consider the Leticia market and sales to tourists as their main marketing opportunities."

"Alternatives are needed that, in keeping with the social context and opportunities of the area, allow the recovery of interest in the chagra and thus preserve the traditional knowledge at risk today."

4.2.10 Socio-economic Factors Influencing Afzelia Africana Sm. Use Value and Traditional Knowledge in Uganda: Implications for Sustainable Management [124]

This is a machine-generated summary of:

Biara, Emmanuel; Egeru, Anthony; Mensah, Sylvanus; Salamula, Jenipher Biira; Kadigo, Mark Marvin: Socio-economic factors influencing Afzelia africana Sm. use value and traditional knowledge in Uganda: implications for sustainable management [124].

Published in: Environment, Development and Sustainability (2020).

Link to original: https://doi.org/10.1007/s10668-020-00673-6.

If you want to cite the papers, please refer to the original.

For technical reasons we could not place the page where the original quote is coming from.

Abstract-Summary

"We assessed use values of A. africana and users' traditional knowledge, and how they relate to plant parts and socio-economic factors including ethnicity, gender, education, age, marital status, profession, household size, income, land size and livestock ownership."

"Use values were assessed based on plant part value (PPV) and use value per use category (UV_k), while users' traditional knowledge was compared using overall use value (OUV) and reported use value (RUV)."

"Nine plant use categories were enumerated, with most dominant being material ($UV_k = 0.63$), followed by social (0.49) and fuel wood (0.41)."

"Men frequently mentioned the use in agriculture, for fuelwood, environment and medicine, while women reported social use."

"Although socio-cultural group did not influence significantly OUV and RUV, multivariate analyses revealed differentiation in use category according to socio-cultural group."

"Land size also predisposed informants to report more uses for the species."

"Taking these significant socio-economic factors into account in participative forest management will facilitate A. africana sustainable use."

Introduction

"The patterns of actual use of a target species provide insights into the most important and valued plant parts as well as user social categories or socio-economic factors to be considered in decision-making for biocultural conservation and strengthening traditional knowledge systems and for effective forest management (Beltrán-Rodríguez and others [102])."

"Several studies on human–environment interactions have shed light into the influence of various individual socio-cultural and demographic traits such as age, gender, formal educational level and ethnicity on plant traditional knowledge and use."

"In Uganda, there have been several studies on natural resources conservation and management, but very few addressed traditional ecological knowledge and the implications for sustainable use and management of woody plant species (Tabuti [103[; Okullo and others [104]; Nyamukuru and others [105]; Kyarikunda and others [106])."

"We assessed the use values and traditional knowledge of A. africana, and how they relate to the species plant parts and users' socio-economic factors including ethnicity, gender, education, age, marital status, profession, household size, income, land size and livestock ownership."

Materials and Methods

"The questionnaire was designed to record information on respondents' demographic and socio-economic factors including gender, age, socio-cultural group, main occupation, education background, household size, income, land size and livestock ownership."

"Each respondent was asked to disclose ethnobotanical information regarding the use of A. africana, including all the plant parts used as well as specific use for each part harvested."

"Data of OUV were normally distributed after logarithmic transformation, and one-way analyses of variance (ANOVA) were used to test significance of differences among socio-cultural groups, gender, education levels and age categories."

"We performed two principal component analyses (PCAs) to assess the patterns of variability of use category and association with (1) socio-cultural groups and (2) socio-cultural groups across gender."

"At individual informant level, we used RUV to test for significant effects of additional individual socio-economic factors on users' traditional knowledge."

Results

"Twenty-eight specific uses were reported across these use categories."

"Shade, games and clothing were cited as specific uses within the social use category, whereas fuel wood category included firewood and charcoal."

"The results of the correspondence analysis showed significant association between plant parts and use category ($\chi^2 = 733$; $p < 0.001$)."

"The effect of age was shown by a significant decrease in the reported use value as age increases."

"Land size had significant and positive effect on reported use value, suggesting that the bigger the size of informants' lands, the more knowledgeable they are in using A. africana."

Discussion

"The results from ANOVA, Poisson regression and PCA collectively showed that age, gender and socio-cultural groups are determinant variables of traditional knowledge and use value of plant species."

"Further, difference in the use of A. africana among the men and women of the same socio-cultural group could perhaps be linked to the predetermined social construction where men are perceived as those involved in "productive" activities and women only in the "reproductive" activities."

"Across socio-cultural groups, uses in agriculture, for fuelwood, environment and medicine were frequently mentioned by men, whereas social use was mentioned by women."

"The similarities in use of A. africana among men of different socio-cultural groups could be determined by the prevailing economic situation in the study area."

"Across socio-cultural groups, uses in agriculture, for fuelwood, environment and medicine were frequently mentioned by men, whereas social use was mentioned by women, suggesting a dominance of men over women in the utilization of A. africana."

Conclusion and Implications for Sustainability

"It points to a great deal of local knowledge of the plant and plant parts and its uses and applications; it also reveals how the species meets various demands, being exploited for material, social purposes, fuel wood and medicinal purposes; further, it reveals the important role that A. africana plays in the social–cultural life of these rural communities."

"The findings collectively support the ideas that age, gender and socio-cultural groups are determinant variables of traditional knowledge and use value of plant species."

"These findings are important for conservation of the species by establishing management strategies based on local demand and use."

"Participative actions should also involve (young people, men and landowners), local decision-makers to increase the awareness on importance of the species and develop win–win scenarios for its sustainable use and for the benefits of future generations."

4.2.11 Joining Smallholder Farmers' Traditional Knowledge with Metric Traits to Select Better Varieties of Ethiopian Wheat [125]

This is a machine-generated summary of:

Mancini, Chiara; Kidane, Yosef G.; Mengistu, Dejene K.; Melfa and Workaye Farmer Community, None; Pè, Mario Enrico; Fadda, Carlo; Dell'Acqua, Matteo: Joining smallholder farmers' traditional knowledge with metric traits to select better varieties of Ethiopian wheat [125].

Published in: Scientific Reports (2017).

Link to original: https://doi.org/10.1038/s41598-017-07628-4.

If you want to cite the papers, please refer to the original.

For technical reasons we could not place the page where the original quote is coming from.

Abstract-Summary

"Participatory methods only scratched the surface of the exploitability of farmers' traditional knowledge in breeding."

"30 smallholder farmers in each of two locations in Ethiopia provided quantitative evaluations of earliness, spike morphology, tillering capacity and overall quality on 400 wheat genotypes, mostly traditional varieties, yielding altogether 192,000 data points."

"Metric measurements of ten agronomic traits were simultaneously collected, allowing to systematically break down farmers' preferences on quantitative phenotypes."

"Farmer trait values were used to produce a ranking of the 400 wheat varieties identifying the trait combinations most desired by farmers."

"The study scale and methods lead to a better understanding of the quantitative basis of Ethiopian smallholder farmer preference in wheat, broadening the discussion for the future of local, sustainable breeding efforts accommodating farmers' knowledge."

Introduction

"In participatory approaches, farmers identify their preferred germplasm among collections of genotypes that are either landraces, MVs or nearly finished varieties."

"The exploration of farmer knowledge and preferences promoted by these methods allows to understand local priorities, and by this to select or produce genetic materials with increased chances of adoption and dissemination [107, 108]."

"Eight out of ten Ethiopian farmers are smallholders [109], often conducting subsistence agriculture based on landraces selected for their adaptive traits to local conditions."

"A better understanding of the traditional knowledge guiding smallholder farmers' choice of wheat genotypes may help in identifying and disseminating genetic materials aimed at addressing the requirements of local agriculture [110, 111]."

"60 smallholder farmers from two highland communities evaluated 400 Ethiopian wheat accessions for traits of their interest."

"We discuss the role of farmers' traditional knowledge to identify farmer varieties to be prioritized for local breeding and distribution."

Materials and Methods

"Thirty smallholder farmers growing wheat, 15 men and 15 women, were selected in each location and were involved in a focus group discussion (FGD) followed by a participatory evaluation (PE) in open field, held concurrently at the beginning of November."

"Farmers listed the traits they used to evaluate a wheat variety, which were ranked by importance."

"Three traits among those listed were chosen to evaluate the wheat varieties brought by the farmers and to conduct the PE: (i) earliness, as the maturation stage of the variety at the time of the field evaluation, (ii) tillering, as the capacity to produce secondary straws and spikes, and (iii) spike, as the appearance and morphology of the spike."

"In front of each plot, the rapporteur referred to each trait under evaluation, one at a time, and asked the farmers to express their individual scores asking the question: "What is your evaluation of [the trait] of this variety?"

Results and Discussion

"Farmers in the FGD were asked to prioritize traits in evaluating a wheat variety."

"The overall evaluation of varieties can be considered as the farmers' perception of the value of a specific wheat genotype, including hidden quality traits, as farmers naturally prefer material that they would grow in their own fields."

"Farmers conducted their overall evaluation considering all the metric traits at once, and this combination resulted in a preferred wheat genotype that would have been difficult to derive form agronomic phenotypes individually considered."

"Not only did farmers in the PE evaluate the agronomic traits that could be measured, but also and most importantly, they were driven by their own perception of the value of a wheat genotype."

"Having the farmers evaluate an untapped collection of wheat diversity allowed us to survey their preference towards specific trait combinations."

4.2.12 Traditional Agroforestry Systems: A Methodological Proposal for Its Analysis, Intervention, and Development [126]

This is a machine-generated summary of:

Gómez, Miguel Uribe; Bueno, Alejandro Lara; León, Artemio Cruz; Uribe Bernal, Johena I.; Hernández Aguirre, Sergio A.: Traditional agroforestry systems: a methodological proposal for its analysis, intervention, and development [126].

Published in: Agroforestry Systems (2021).

Link to original: https://doi.org/10.1007/s10457-021-00692-w.

If you want to cite the papers, please refer to the original.

For technical reasons we could not place the page where the original quote is coming from.

Abstract-Summary

"Agroforestry is an interdisciplinary science that allows the sustainable use of the land and promotes a holistic management of its components, such as trees, crops, and animals."

"Traditional agroforestry systems (TAS) are contemporary to the discovery of agriculture while agroforestry was established as a science in the second half of the last century."

"As in any other science, agroforestry research applies the scientific method and for the process of intervention and development of agroforestry systems there are several proposals, for example, one is the Design and Diagnosis (D&D) method."

"The main objective of this research is to find the method to diagnose of the TAS which allows the necessary intervention for sustainable development."

Introduction

"In this new research approach, the family production unit (FPU) is used as the experimental unit, as established by the Inter-American Institute for Agricultural Cooperation (IICA [112])."

"The study of FPUs is complex, since they are diverse and present a dynamic reality, based on the interest of each FPU according to the different levels of access to the factors of production (land, labor, and capital)."

"According to Dufumier [113], the fundamental element for understanding the type of land exploitation at the FPU level is the concept of production system, defined as the combination (in time and space) of available resources, labor force and means of production (land and capital) for agricultural, livestock and forestry production, since, in general, the FPU operates under different environmental and socioeconomic conditions."

"The referent of this research is traditional knowledge, the object of study is family farming, and the approach is the Agroforestry Production System."

Methodology

"This methodology is based on several information gathering tools: (1) conducting transects and landscape readings with the support of producers to inquire about the biophysical conditions of the study area; (2) recording observations made by participants in the study area for a given time; (3) conducting historical interviews with former producers in the area; (4) conducting technical–economic interviews with those responsible for the FPU; (6) developing typologies of producers; and (5) sharing preliminary results with those responsible for the FPU."

"Fieldwork was carried out using the following tools: (1) direct observation through field visits; (2) opinions with producers; (3) semi-structured interviews and surveys; and (4) community workshops to socialize information."

"The overall annual cost to cover the family's basic needs (food products, health, clothing, and education) was estimated as proposed by Devienne and Wybrecht [114]."

Results

"This agroforestry system is in areas with gentle or flat slopes where the tree component is not the most important because the soil types have an average depth of 80 to 100 cm, which allows the planting and development of corn, sorghum and vegetable crops and the use of agricultural machinery."

"The third agroecological zone is located between 1000 and 1200 masl, where the agricultural-livestock production system is predominant with seasonal corn, sorghum, and beans (Phaseolus vulgaris) crops used to feed livestock during the dry season."

"The analysis and systematization of the information made it possible to identify in the Sierra de Huautla the agroforestry livestock production systems whose main product is the sale of recently weaned calves with a live weight between 180 and 220 kg, and cull cows, with which the producer obtains cash income."

"The UPFs have land areas between 15 and 40 ha where they combine agricultural production with cattle and forestry with the sale of firewood and fence posts."

Discussion

"The differences and similarities of the family production units captured in the typology of producers are a function of the availability of natural resources, particularly land area, the intensification of the labor force and the use of technology."

"Farming families located in categories I and II do not have sufficient income to reach the desired minimum welfare line, so their level of social reproduction is at risk and, despite this, they continue to exist by selling their labor force outside their production unit, which indicates migration."

"The diagnosis of family production units and their typology are essential prerequisites for generating development and agroforestry management proposals, based on the availability of productive resources, local knowledge of traditional knowledge and their cultural worldview."

"The diagnosis of family production units and their typology are essential prerequisites for generating development and agroforestry management proposals, based on the availability of productive resources, local knowledge of traditional knowledge and their cultural worldview."

Bibliography

1. Redclift M (1989) The environmental consequences of Latin America's agricultural development: some thoughts on the Brundtland Commission report. World Dev 17:365–377
2. Altieri MA (2000) Multifunctional dimensions of ecologically-based agriculture in Latin America. Int J Sustain Dev World Ecol 7:62–75
3. Eakin H, Webhe M, Avila C, Sanchez Torres G, Bojorquez-Tapia LA (2007) Climate change and vulnerability. Earthscan, London
4. Phungpracha E, Kansuntisukmongkon K, Panya O (2016) Traditional ecological knowledge in Thailand: mechanisms and contributions to food security. Kasetsart Journal of Social Sciences 37(2):82–87
5. Srivastava P, Singh R, Tripathi S, Raghubanshi AS (2016) An urgent need for sustainable thinking in agriculture: an Indian scenario. Ecol Ind 67:611–622
6. Pulido JS, Bocco G (2003) The traditional farming system of a Mexican indigenous community: the case of Nuevo San Juan Parangaricutiro, Michoaca´n, Mexico. Geoderma 111:249–265
7. Zhang Y, Min Q, Li H, He L, Zhang C, Yang L (2017a) A conservation approach of globally important agricultural heritage systems (GIAHS): improving traditional agricultural patterns and promoting scale-production. Sustainability 9(2):295
8. IPCC, Intergovernmental Panel on Climate Change (2007a) Synthesis report. Contribution of working groups I, Ii and Iii to the fourth assessment report of the intergovernmental panel on climate change. IPCC, Geneva, Switzerland
9. Paustian KA, Andrén O, Janzen HH, Lal R, Smith P, Tian G, Woomer PL (1997) Agricultural soils as a sink to mitigate CO2 emissions. Soil Use Manag 13:230–244
10. McMichael AJ, Powles JW, Butler CD, Uauy R (2007) Food, livestock production, energy, climate change, and health. Lancet 370(9594):1253–1263
11. Schmidhuber J, Tubiello FN (2007) Global food security under climate change. Proc Natl Acad Sci 104(50):19703–19708

12. Tai AP, Martin MV, Heald CL (2014) Threat to future global food security from climate change and ozone air pollution. Nat Clim Change 4(9):817–821
13. Islam MT, Nursey-Bray M (2017) Adaptation to climate change in agriculture in Bangladesh: the role of formal institutions. J Environ Manage 200:347–358
14. Gornall J, Betts R, Burke E, Clark R, Camp J, Willett K, Wiltshire A (2010) Implications of climate change for agricultural productivity in the early twenty-first century. Philos Trans R Soc London B Biol Sci 365(1554):2973–2989
15. Hillel D, Rosenzweig C (2010) Climate change and agroecosystems: main findings and future research directions ICP series on climate change impacts, adaptation, and mitigation, vol 1. In: Hillel D, Rosenzweig C (eds) Handbook of climate change and agroecosystems impacts, adaptation, and mitigation. Imperial College Press, London
16. Deryng D, Conway D, Ramankutty N, Price J, Warren R (2014) Global crop yield response to extreme heat stress under multiple climate change futures. Environ Res Lett 9(3):034011
17. Porter JR, Xie L, Challinor AJ, Cochrane K, Howden SM, Iqbal MM, Lobell DB, Travasso MI (2014) Food security and food production systems. In: Field CB, Barros VR, Dokken DJ, Mach KJ, Mastrandrea MD, Bilir TE, Chatterjee M, Ebi KL, Estrada YO, Genova RC, Girma B, Kissel ES, Levy AN, MacCracken S, Mastrandrea PR, White LL (eds) Climate change 2014: impacts, adaptation, and vulnerability. Part A: global and sectoral aspects, pp. 485–533. Contribution of Working Group II to the Fifth Assessment Report of the Intergovernmental Panel on Climate Change. Cambridge University Press, Cambridge
18. Challinor AJ, Watson J, Lobell DB, Howden SM, Smith DR, Chhetri N (2014) A meta-analysis of crop yield under climate change and adaptation. Nat Clim Change 4:287–291
19. Harvey CA, Chacón M, Donatti CI, Garen E, Hannah L, Andrade A, Bede L, Brown D, Calle A, Chará J, Clement C, Gray E, Hoang MH, Minang P, Rodríguez AM, Seeberg-Elverfeldt C, Semroc B, Shames S, Smukler S, Somarriba E, Torquebiau E, van Etten J, Wollenberg E (2014) Climate-smart landscapes: opportunities and challenges for integrating adaptation and mitigation in tropical agriculture. Conserv Lett 7:77–90. https://doi.org/10.1111/conl.12066
20. Brandt P, Kvakić M, Butterbach-Bahl K, Rufino MC (2015) How to target climate-smart agriculture? Concept and application of the consensus-driven decision support framework "target CSA". Agric Syst 151:234
21. Olayide OE, Tetteh IK, Popoola L (2016) Differential impacts of rainfall and irrigation on agricultural production in Nigeria: Any lessons for climate-smart agriculture? Agric Water Manag 178:30–36
22. Grainger-Jones E (2011) Climate-smart smallholder agriculture: What's different? No. 3. IFAD Occasional Paper, Rome
23. Long TB, Blok V, Coninx I (2016) Barriers to the adoption and diffusion of technological innovations for climate-smart agriculture in Europe: evidence from the Netherlands, France, Switzerland and Italy. J Clean Prod 112:9–21
24. Mwongera C, Shikuku KM, Twyman J, Läderach P, Ampaire E, Van Asten P, Twomlowd S, Winowiecki LA (2017) Climate smart agriculture rapid appraisal (CSA-RA): a tool for prioritizing context-specific climate smart agriculture technologies. Agric Syst 151:192–203
25. Sandor JA, Furbee L (1996) Indigenous knowledge and classification of soils in the Andes of southern Peru. Soil Sci Soc Am J 60:1502–1512
26. Singh GS, Ram SC, Kuniyal JC (1997a) Changing traditional land use patterns in the Great Himalayas: a case study of Lahaul Valley. J Environ Syst 25:195–211
27. Jackson LE, Pascual U, Hodgkin T (2007) Utilizing and conserving agrobiodiversity in agricultural landscapes. Agr Ecosyst Environ 121(3):196–210
28. Altieri MA (2004) Linking ecologists and traditional farmers in the search for sustainable agriculture. Front Ecol Environ 2(1):35–42
29. Lowder SK, Skoet J, Raney T (2016) The number, size, and distribution of farms, smallholder farms, and family farms worldwide. World Dev 87:16–29
30. Lasco RD, Delfino RJP, Catacutan DC, Simelton ES, Wilson DM (2014) Climate risk adaptation by smallholder farmers: the roles of trees and agroforestry. Curr Opin Environ Sustain 6:83–88

31. Altieri MA (1993) Ethnoscience and biodiversity: key elements in the design of sustainable pest management systems for small farmers in developing countries. Agr Ecosyst Environ 46(1–4):257–272
32. Gliessman SR (1998) Agroecology: ecological process in sustainable agriculture. Ann Arbor Press, Ann Arbor
33. Altieri MA (2002) Agroecology: the science of natural resource management for poor farmers in marginal environments. Agr Ecosyst Environ 93(1):1–24
34. Hu F, Feng F, Zhao C, Chai Q, Yu A, Yin W, Gan Y (2017) Integration of wheat-maize intercropping with conservation practices reduces CO2 emissions and enhances water use in dry areas. Soil Till Res 169:44–53
35. Huang M, Shao M, Zhang L, Li Y (2003) Water use efficiency and sustainability of different long-term crop rotation systems in the Loess Plateau of China. Soil Till Res 72(1):95–104
36. Peoples MB, Baldock JA (2001) Nitrogen dynamics of pastures: nitrogen fixation inputs, the impact of legumes on soil nitrogen fertility, and the contributions of fixed nitrogen to Australian farming systems. Aust J Exp Agric 41(3):327–346
37. Raphael JP, Calonego JC, Milori DMB, Rosolem CA (2016) Soil organic matter in crop rotations under no-till. Soil Till Res 155:45–53
38. Schipanski ME, Barbercheck M, Douglas MR, Finney DM, Haider K, Kaye JP, White C (2014) A framework for evaluating ecosystem services provided by cover crops in agroecosystems. Agric Syst 125:12–22
39. Frasier I, Quiroga A, Noellemeyer E (2016) Effect of different cover crops on C and N cycling in sorghum NT systems. Sci Total Environ 562:628–639
40. Pinto P, Long MEF, Piñeiro G (2017) Including cover crops during fallow periods for increasing ecosystem services: Is it possible in croplands of Southern South America? Agr Ecosyst Environ 248:48–57
41. Steele MK, Coale FJ, Hill RL (2012) Winter annual cover crop impacts on no-till soil physical properties and organic matter. Soil Sci Soc Am J 76:2164–2173
42. Poeplau C, Don A (2015) Carbon sequestration in agricultural soils via cultivation of cover crops-a meta-analysis. Agric Ecosyst Environ 200:33–41
43. Basche AD, Kaspar TC, Archontoulis SV, Jaynes DB, Sauer TJ, Parkin TB, Miguez FE (2016) Soil water improvements with the long-term use of a winter rye cover crop. Agric Water Manag 172:40–50
44. Duval ME, Galantini JA, Capurro JE, Martinez JM (2016) Winter cover crops in soybean monoculture: effects on soil organic carbon and its fractions. Soil Till Res 161:95–105
45. Juana J, Kahaka Z, Okurut F (2013) Farmers' perceptions and adaptations to climate change in sub-Sahara Africa: a synthesis of empirical studies and implications for public policy in African agriculture. J Agric Sci 5(4):2013
46. Ayal DY, Leal Filho W (2017) Farmers' perceptions of climate variability and its adverse impacts on crop and livestock production in Ethiopia. J Arid Environ 140:20–28
47. Woods BA, Nielsen HØ, Pedersen AB, Kristofersson D (2017) Farmers' perceptions of climate change and their likely responses in Danish agriculture. Land Use Policy 65:109–120
48. Bradshaw B, Dollan H, Smit B (2004) Farm-level adaptation to climate variability and change: crop diversification in the Canadian prairies. Clim Chang 67:119–141
49. Di Falco S, Veronesi M, Yesuf M (2011) Does adaptation to climate change provide food security? Microevidence from Ethiopia. Am J Agric Econ 93(3):829–846
50. Di Falco S, Veronesi M (2011) On adaptation to climate change and risk exposure in the Nile basin of Ethiopia. IED Working Paper 15
51. Moniruzzaman S (2015) Crop choice as climate change adaptation: evidence from Bangladesh. Ecol Econ 118:90–98
52. Browder JO (1989) Fragile lands in Latin America: strategies for sustainable development. Westview Press, Boulder
53. Altieri MA, Nicholls CI (2008) Scaling up agroecological approaches for food sovereignty in Latin America. Development 51(4):472–480

54. Chhetri N, Chaudhary P, Tiwari PR, Yadaw RB (2012) Institutional and technological innovation: Understanding agricultural adaptation to climate change in Nepal. Appl Geogr 33:142–150
55. Deressa TT, Hassan RM, Ringler C, Alemu T, Yesuf M (2009) Determinants of farmers' choice of adaptation methods to climate change in the Nile Basin of Ethiopia. Glob Environ Chang 19:248–255. https://doi.org/10.1016/j.gloenvcha.2009.01.002
56. Bryan E, Deressa TT, Gbetibouo GA, Ringler C (2009) Adaptation to climate change in Ethiopia and South Africa: options and constraints. Environ Sci Policy 12:413–426
57. Lei Y, Liu C, Zhang L, Luo S (2016) How smallholder farmers adapt to agricultural drought in a changing climate: A case study in southern China. Land Use Policy 55:300–308
58. Halewood et al. 2020. Enhancing farmers' agency in the global crop commons through use of biocultural community protocols. Symposium: Seed as a commons—exploring innovative concepts and practices of governing seed and varieties. Agriculture and Human Values.
59. Frischmann, Brett M., Michael J. Madison, and Katherine Jo Strandburg, ed. 2014. Governing Knowledge Commons. Oxford; New York: Oxford University Press.
60. Hess, Charlotte, and Elinor Ostrom. 2003. Ideas, artefacts and facilities: information as a common-pool-resource. Law and Contemporary Problems 66: 111–145.
61. Hess, Charlotte. 2012. The Unfolding of the Knowledge Commons. St Antony's International Review 8 (1): 13–24.
62. Hess, Charlotte, and Elinor Ostrom (eds.). 2007. Understanding knowledge as a commons: from theory to practice. Cambridge, Mass.: MIT Press.
63. Boyle, James. 2003. The Second Enclosure Movement and the Construction of the Public Domain. Law and Contemporary Problems 66: 33–74. https://doi.org/10.2139/ssrn.470983 .
64. Öborn I et al (2005) Critical aspects of potassium management in agricultural systems. Soil Use Manag 21(1):102–112. https://doi.org/10.1111/j.1475-2743.2005.tb00414.x
65. Laekemariam F, Kibret K, Shiferaw H (2018) Potassium (K)-to-magnesium (Mg) ratio, its spatial variability and implications to potential Mg-induced K deficiency in nitisols of southern Ethiopia. Agric Food Security 7(1):13. https://doi.org/10.1186/s40066-018-0165-5
66. Arora S (2006) Preliminary assessment of soil and water conservation status in drought prone foothill region of North-West India. J World Asso Soil Water Conser 5:55–63
67. Arora S, Hadda MS, Kohli A (2009) Soil moisture conservation for sustainable agriculture in rainfed foothill region. In: Bhatt GM, et al. (Eds.) Geoenvironment-Challenges Ahead. McMillan Publishers India Ltd., New Delhi, p 194–206.
68. Kukal SS, Bhatt R (2015) Extent, distribution and patterns of gully erosion in lower Shiwaliks of Punjab, India. Environ Ecolo 32(2):401–404
69. Bhatt R (2018) Tackling gully erosion in foothills of lower Shivaliks. Soil Water Conser Today 13(3):4
70. Vashisht BB, Maharjan B, Sharma S, Kaur S (2020) Soil quality and its potential indicators under different land use systems in the Shivaliks of Indian Punjab. Sustainability 12(8):3490. https://doi.org/10.3390/su12083490
71. Kukal SS, Bhatt R (2006) An appraisal of the indigenous technology for soil and water conservation in Shivaliks of Punjab. Department of Soils, Punjab Agricultural University, Ludhiana, India
72. Bhatt R (2013) Gully erosion management in foothills of lower Shivaliks. Soil Water Conser Today 8(4):5–6
73. Arora S, Bhatt R (2006) Effect of in-situ soil moisture conservation practices on soil moisture storage and yield of maize (Zea mays L.) in a rainfed foothill region of North-west India. Trop Agric 82(3):37–42
74. Sharma UC (2006) Water management under scarcity situations for food security. In: Jalali et al. (Eds.), Sustainable and Integrated nutrient management for north-western Himalayas, SKUAST-J, Jammu, pp. 34–49.
75. Gill P, Stewart K, Treasure E, Chadwick B. Methods of data collection in qualitative research: Interviews and focus groups. Brit Dental J. 2008;204(6):291–5.

76. Coomes OT, Takasaki Y, Rhemtulla JM (2017) What fate for swidden agriculture under land constraint in tropical forests? Lessons from a long-term study in an Amazonian peasant community. J Rural Stud 54:39–51. https://doi.org/10.1016/j.jrurstud.2017.06.002
77. Padoch C, Pinedo-Vásquez M (2010) Saving slash-and-burn to save biodiversity. Biotropica 42(5):550–552. https://doi.org/10.1111/j.1744-7429.2010.00681.x
78. Junqueira AB, Almekinders CJM, Stomph T-H, Clement CR, Struik PC (2016) The role of Amazonian anthropogenic soils in shifting cultivation: learning from farmers' rationales. Ecol Soc 21(1):12. https://doi.org/10.5751/ES-08140-210112
79. Triana-Moreno LA, Rodríguez NC, García J (2016) Dinámica del sistema agroforestal de chagras como eje de la producción indígena en el Trapecio Amazónico (Colombia). (Vol. Vol.24 No. 1). Bogotá: Agronomía Colombiana.
80. Gobernación del Caqueta (2012) Plan Municipal para la gestión del riesgo de desastres municipio de solano. Obtenido de Decreto No. 055 del 7 de Mayo de 2012: https://caqueta.micolombiadigital.gov.co/sites/caqueta/content/files/000623/31145_pmgrd-solano-final.pdf
81. Schroder T, Castro B, Román O, Jekone M (1987) Estudio de un sistema agrícola tradicional en Araracuara. Colombia Amazónica 2(2):31–50
82. Instituto Amazónico de Investigaciones Científicas—SINCHI (2017) Moniya ringo. Mujer de abundancia y reproducción: estudio de caso de la chagra. Editorial Kimpres, Bogotá
83. Santoro A, Venturi M, Bertani R, Agnoletti M (2020) A review of the role of forests and agroforestry systems in the FAO Globally Important Agricultural Heritage Systems (GIAHS) programme. Forests 11:860. https://doi.org/10.3390/f11080860
84. Maikhuri, R. K., Rana, U., Semwal, R. L., & Rao, K. S. (2000). Agriculture of Uttarakhand: Issues and management prospects for economic development. In M. C. Sati & S. P. Sati (Eds.), Uttaranchal statehood: Dimensions for development (pp. 151–167). New Delhi: Indus Publishing Co.
85. Maikhuri, R. K., Semwal, R. L., Rao, K. S., & Saxena, K. G. (1997). Eroding traditional crop diversity imperils the sustainability of agriculture systems in Central Himalaya. Current Science, 73(9), 777–782.
86. Maikhuri, R. K., Rao, K. S., & Semwal, R. L. (2001). Changing scenario of Himalayan agroecosystems: Loss of agrobiodiversity, an indicator of environmental change in Central Himalaya, India. The Environmentalist, 21, 23–39.
87. Negi, V. S., Maikhuri, R. K., Rawat, L. S., & Bahuguna, A. (2009). Traditional agriculture in transition: A case of Har-ki Doon Valley (Govind pashu vihar sanctuary and national park) in Central Himalaya. International Journal of Sustainable Development and World Ecology, 16(5), 313–321.
88. Maikhuri, R. K., Rao, K. S., & Saxena, K. G. (1996). Traditional crop diversity for sustainable development of Central Himalayan agroecosystems. International Journal of Sustainable Development and World Ecology, 2, 1–24.
89. Palni, L. M. S., Maikhuri, R. K., & Rao, K. S. (1998). Conservation of the Himalayan agroecosytstems: Issues and priorities. Technical Paper V. In: Eco-regional co-operation for biodiversity conservation in the Himalaya (253–290). UNDP.
90. Nautiyal, S., Rajan, K. S., & Shibasaki, R. (2005). Interaction of biodiversity and economic welfare-a case study from Himalayas of India. Journal of Environmental Informatics, 6(2), 111–119.
91. Bisht, I. S., Rao, K. S., Bhandari, D. C., Nautiyal, S., Maikhuri, R. K., & Dhillon, B. S. (2006). A suitable site for in situ (on-farm) management of plant diversity in traditional agroecosystems of western Himalaya in Uttaranchal state: A case study. Genetic Resources and Crop Evolution, 53, 1333–1350.
92. Ceccarelli, S., Grando, S., & Baum, M. (2007). Participatory plant breeding in water-limited environments. Experimental Agriculture, 43(4), 1–25.
93. Cavatassi, R., Lipper, L., & Narloch, U. (2011). Modern variety adoption and risk management in drought prone areas: Insights from the sorghum farmers of eastern Ethiopia. Agricultural Economics, 42(3), 279–292.

94. Altieri, M., et al. (2012). Agroecologically Efficient agricultural systems for smallholder farmers: contributions to food sovereignty. Agronomy for Sustainable Development, 32(1), 1–13.
95. Altieri, M. (2004). Linking ecologists and traditional farmers in the search for sustainable agriculture. The Ecological Society of America, 2(1), 35–42.
96. Yano, K., & Lanusosang, T. (2013). Globalisation and its impact on agriculture: an overview of Kohima District, Nagaland, India. International Journal of Bio- Resource and Stress Management, 4(4), 651–654.
97. Singh, S., & Hensel, O. (2014). Impact of extension education on improving knowledge of sustainable technical agricultural practices. CIGR Journal, 16(1), 198–206.
98. Acosta LE, Pérez MN, Juragaro LA, Nonokudo H, Sánchez G, Zafiama ÁM, et al. La chagra en la chorrera: más que una producción de subsistencia, es una fuente de comunicación y alimento físico y espiritual, de los Hijos del tabaco, la coca y la yuca dulce: los retos de la nuevas generaciones para las prácticas culturales y los saberes tradicionales asociados a la biodiversidad. Instituto Amazónico de Investigaciones Científicas, Sinchi. Asociación Zonal Indígena de Cabildos y Autoridades Tradicionales de La Chorrera – AZICATCH; 2011.
99. Arbelaes JJ, Martinez LB. Diagnostico socio agroforestal para la diversificación de chagras y fortalecimiento de la seguridad alimentaria en el resguardo indígena km 6 y 11, parcialidad comunidad km 11 N+maira Na+mek+ Ib+r+ (patio de ciencia dulce), municipio de Leticia, departamento de Amazonas. In: Universidad Nacional Abierta y a Distancia - UNAD. 2012. https://repository.unad.edu.co/handle/10596/1031 . Accessed Jan 2020.
100. Sierra SM, Raz L. Uso y Manejo de las Especies Sembradas en las Chagras de dos Comunidades Murui-Muinane de la Amazonia Colombia. Ethnobot Res Appl. 2014;12:473. https://doi.org/10.17348/era.12.0.473-495 .
101. Arias JC, Cárdenas D. Manual de identificación, selección y evaluación oferta de productos forestales no maderables: Proyecto Manejo integral y Sostenible de los Bosques de Tarapacá (Amazonas). 1st ed: Inst. Amazónico de Investigaciones Científicas, SINCHI; 2007.
102. Beltrán-Rodríguez, L., Ortiz-Sánchez, A., Mariano, N. A., Maldonado-Almanza, B., & Reyes-García, V. (2014). Factors affecting ethnobotanical knowledge in a mestizo community of the Sierra de Huautla Biosphere Reserve, Mexico. Journal of Ethnobiology and Ethnomedicine, 10, 14. https://doi.org/10.1186/1746-4269-10-14 .
103. Tabuti, J. R. (2007). The uses, local perceptions and ecological status of 16 woody species of Gadumire Sub-county, Uganda. Biodiversity and Conservation, 16(6), 1901–1915.
104. Okullo, J. B. L., Obua, J., & Okello, G. (2004). Use of indigenous knowledge in predicting fruit production of shea butter tree in agroforestry parklands of north-eastern Uganda. Uganda Journal of Agricultural Sciences, 9(1), 360–366.
105. Nyamukuru, A., Nabatanzi, A., Mpiira, S., & Tabuti, J. R. (2015). Locally preferred woody species and their management in Kiruhura and Arua districts, Uganda. Ethnobotany Research and Applications, 14, 49–61.
106. Kyarikunda, M., Nyamukuru, A., Mulindwa, D., & Tabuti, J. R. (2017). Agroforestry and management of trees in Bunya County, Mayuge District, Uganda. International Journal of Forestry Research. https://doi.org/10.1155/2017/3046924 .
107. Assefa, T. et al. Participatory Plant Breeding with Traders and Farmers for White Pea Bean in Ethiopia. J. Agric. Educ. Ext. 20, 497–512 (2014).
108. Singh, Y. P. et al. Farmers' participatory varietal selection: A sustainable crop improvement approach for the 21st century. Agroecol. Sustain. Food Syst. 38, 427–444 (2014).
109. Salami, A., Kamara, A. B. & Brixiova, Z. Smallholder agriculture in East Africa: Trends, constraints and opportunities (African Development Bank Tunis, 2010).
110. Ceccarelli, S. et al. A methodological study on participatory barley breeding I. Selection phase. Euphytica 111, 91–104 (2000).
111. Ceccarelli, S. et al. A methodological study on participatory barley breeding II. Response to selection. Euphytica 133, 185–200 (2003).
112. IICA [Instituto Interamericano de Cooperación para la Agricultura] (2016) La agricultura familiar en las Américas: Principios y conceptos que guían la cooperación técnica del

IICA. San José, Costa Rica. 36 p. http://repositorio.bibliotecaorton.catie.ac.cr/bitstream/handle/11554/11028/La_agricultura_familiar_en_las_Americas_Principios_y_conceptos_que_guian_la_cooperacion_tecnica_del_IICA.pdf?sequence=1&isAllowed=y
113. Dufumier M (1996) Les projets de développement agricole: manuel d'expertise. Karthala. http://books.google.com.mx/books?hl=es&lr=&id=J6cd3KsuOpgC&oi=fnd&pg=PA183&dq
114. Devienne S, Wybrecht B (2003) Analyser le fonctionnement d'une exploitation. In: Mémento de l'Agronome, chapitre 3–2; Paris: CIRAD-GRET
115. Singh, Rinku; Singh, G. S. Traditional agriculture: a climate-smart approach for sustainable food production. Energy, Ecology and Environment (2017). https://doi.org/10.1007/s40974-017-0074-7
116. Sievers-Glotzbach, Stefanie; Euler, Johannes; Frison, Christine; Gmeiner, Nina; Kliem, Lea; Mazé, Armelle; Tschersich, Julia Beyond the material: knowledge aspects in seed commoning. Agriculture and Human Values (2020). https://doi.org/10.1007/s10460-020-10167-w
117. Occelli, Martina; Mantino, Alberto; Ragaglini, Giorgio; Dell'Acqua, Matteo; Fadda, Carlo; Pè, Mario Enrico; Nuvolari, Alessandro Traditional knowledge affects soil management ability of smallholder farmers in marginal areas. Agronomy for Sustainable Development (2021). https://doi.org/10.1007/s13593-020-00664-x
118. Arora, Sanjay; Bhatt, Rajan; Sharma, Vikas; Hadda, Manjeet S. Indigenous Practices of Soil and Water Conservation for Sustainable Hill Agriculture and Improving Livelihood Security. Environmental Management (2022). https://doi.org/10.1007/s00267-022-01602-1
119. Wendiro, Deborah; Wacoo, Alex Paul; Wise, Graham Identifying indigenous practices for cultivation of wild saprophytic mushrooms: responding to the need for sustainable utilization of natural resources. Journal of Ethnobiology and Ethnomedicine (2019). https://doi.org/10.1186/s13002-019-0342-z
120. Hernandez Marentes, Maria Alejandra; Venturi, Martina; Scaramuzzi, Silvia; Focacci, Marco; Santoro, Antonio Traditional forest-related knowledge and agrobiodiversity preservation: the case of the chagras in the Indigenous Reserve of Monochoa (Colombia). Biodiversity and Conservation (2021). https://doi.org/10.1007/s10531-021-02263-y
121. Negi, Vikram S.; Maikhuri, R. K. Socio-Ecological and Religious Perspective of Agrobiodiversity Conservation: Issues, Concern and Priority for Sustainable Agriculture, Central Himalaya. Journal of Agricultural and Environmental Ethics (2012). https://doi.org/10.1007/s10806-012-9386-y
122. Hudson, Suraya; Krogman, Naomi; Beckie, Mary Social practices of knowledge mobilization for sustainable food production: nutrition gardening and fish farming in the kolli hills of India. Food Security (2016). https://doi.org/10.1007/s12571-016-0580-z
123. Garavito, Giovanny; Clavijo, Rafael; Luengas, Pilar; Palacios, Pablo; Arias, María Helena Assessment of biodiversity goods for the sustainable development of the chagra in an indigenous community of the Colombian Amazon: local values of crops. Journal of Ethnobiology and Ethnomedicine (2021). https://doi.org/10.1186/s13002-021-00453-0
124. Biara, Emmanuel; Egeru, Anthony; Mensah, Sylvanus; Salamula, Jenipher Biira; Kadigo, Mark Marvin Socio-economic factors influencing Afzelia africana Sm. use value and traditional knowledge in Uganda: implications for sustainable management. Environment, Development and Sustainability (2020). https://doi.org/10.1007/s10668-020-00673-6
125. Mancini, Chiara; Kidane, Yosef G.; Mengistu, Dejene K.; Melfa and Workaye Farmer Community, None; Pè, Mario Enrico; Fadda, Carlo; Dell'Acqua, Matteo Joining smallholder farmers' traditional knowledge with metric traits to select better varieties of Ethiopian wheat. Scientific Reports (2017). https://doi.org/10.1038/s41598-017-07628-4
126. Gómez, Miguel Uribe; Bueno, Alejandro Lara; León, Artemio Cruz; Uribe Bernal, Johena I.; Hernández Aguirre, Sergio A. Traditional agroforestry systems: a methodological proposal for its analysis, intervention, and development. Agroforestry Systems (2021). https://doi.org/10.1007/s10457-021-00692-w

Chapter 5
Food Security, Nutrition, and Hunger

5.1 Introduction by Editor

Food security is one of the greatest concerns among developing countries. A long history of crop failure and mismanagement of the food system has led to multiple famines in several nations. No government wants to be in a situation where even a part of their country faces food shortages. However, according to the recent estimates, from 2014 onwards, the number of people who are undernourished started to gradually rise until, in 2020, the world saw an unprecedented setback in its efforts to end hunger (FAO, 2021).

Food security, nutrition, and hunger are interconnected issues that have significant implications for the well-being of individuals, communities, and nations worldwide. Despite the rapid progress in global food production over the past few decades, the world still faces the persistent challenge of ensuring that all people have access to sufficient, safe, and nutritious food. Food security encompasses the availability, access, utilization, and stability of food supplies. While global food production has kept pace with demand, the uneven distribution of resources, economic disparities, and various socio-economic factors continue to hinder equitable access to food, particularly in developing countries. Environmental pressures, such as climate change, population growth, and urbanization, further exacerbate the challenges of food security, especially in regions like Africa, India, and South America.

Hunger, lack of nutrition and food insecurity are formidable challenges in human development. They are linked to poor health outcomes, mortality and stunted economic growth (Misselhorn et al., 2012). Grain loss during storage and food waste in developed nations are significant contributors to this problem, highlighting the need for more efficient and sustainable food systems. Moreover, the intensification of agricultural practices has led to concerns about the nutritional quality of food. Studies have shown that processing and distribution methods can diminish the nutrient content of produce (Santeramo & Lamonaca, 2021; Session, 2014; Misselhorn et al., 2012; Pinstrup-Andersen & Pandya-Lorch, 1998).

G. Nulkar (ed.), *The Economics of Sustainable Development*,
https://doi.org/10.1007/978-981-99-7379-8_5

Addressing the complex challenge of food security, nutrition, and hunger requires a multifaceted approach that combines policy interventions, technological innovations, and collaborative efforts across multiple stakeholders (Santeramo & Lamonaca, 2021; Pinstrup-Andersen & Pandya-Lorch, 1998; Misselhorn et al., 2012; Session, 2014). Investing in agriculture, strengthening farmer organizations, and promoting adaptive research to solve farmer-specific problems are some of the strategies that have been identified as crucial in achieving food security in regions like East Africa (Lokuruka, 2021).

Insufficient nutrition afflicts over 2 billion people across the world (Fanzo, 2018). The Global Nutrition Report (2018) observes that malnutrition is a widespread issue that manifests in various forms. From childhood through old age, it has an impact on the majority of people on the planet. No nation is unaffected. It affects everyone, regardless of where they live, their age, their socioeconomic status, or their gender. In recent years, concerns over overweight and obesity, the other facets of malnutrition, are rising. This has multiple causes, including excessive eating of energy-dense meals, insufficient exercise, and heredity. Conflict, climate variability and extremes, and economic slowdowns and downturns are three of the main factors that have recently affected food security and nutrition. These factors are exacerbated by the root causes of poverty and very high and persistent levels of inequality (FAO, 2021). The FAO report observes that over 2.37 billion people worldwide are exposed to moderate or severe food insecurity. Of these, half are located in Asia, one-third in Africa, and 11 percent in Latin America and the Caribbean.

The UN has noted various issues with India's food security. First off, agricultural growth rates have fluctuated as farming turns more susceptible to climate change. Second, rapidly declining biological diversity and land degradation pose an imminent threat to India's food. Overuse of chemical fertilisers has turned large tracts of farmland barren. Third, in nearly 30% of the administrative blocks of India, groundwater extraction is turning unsustainable. Many farmers have been compelled to seek alternative sources of income due to the gradual fall in farm holding size and productivity (United Nations, 2022).

In the holy grail of sustainable development, food security and nutrition are critical challenges for developing countries. The second of the United Nations Sustainable Development Goals (SDG) is about Zero Hunger. The SDGs raise this concern and sets 8 targets to overcome the risks. The SDG mentions that food security necessitates a multifaceted strategy, ranging from social protection to ensure healthy food, to the transformation of food systems to create a more equitable and sustainable society. This requires investments in improvement of living standards, investments in rural and urban regions and social protection.

In this chapter we present research from African countries, India's fight against hunger, the multiple dimensions of food security and studies on the role of agroecosystem resilience in the changing climate.

References

Fanzo, J. e. (2018). Global Nutrition Report: Shining a light to spur action on nutrition.

FAO, I. U. (2021). The State of Food Security and Nutrition in the World. Rome: FAO. https://doi.org/10.4060/cb4474en

United Nations. (2022, December 31). Nutrition and Food Security. Retrieved from United Nations India: https://india.un.org/en/171969-nutrition-and-food-security

Lokuruka, M. (2021, January 20). Food and Nutrition Security in East Africa (Kenya, Uganda and Tanzania): Status, Challenges and Prospects. IntechOpen. https://doi.org/10.5772/intechopen.95036

Misselhorn, A., Aggarwal, P., Ericksen, P J., Gregory, P., Horn-Phathanothai, L., Ingram, J., & Wiebe, K. (2012, February 1). A vision for attaining food security. Elsevier BV, 4(1), 7-17. https://doi.org/10.1016/j.cosust.2012.01.008

Pinstrup-Andersen, P., & Pandya-Lorch, R. (1998, July 1). Food security and sustainable use of natural resources: a 2020 Vision. Elsevier BV, 26(1), 1-10. https://doi.org/10.1016/s0921-8009(97)00067-0

Santeramo, F G., & Lamonaca, E. (2021, April 22). Food Loss–Food Waste–Food Security: A New Research Agenda. Multidisciplinary Digital Publishing Institute, 13(9), 4642-4642. https://doi.org/10.3390/su13094642

5.2 Machine Generated Summaries

Disclaimer: The summaries in this chapter were generated from Springer Nature publications using extractive AI auto-summarization: An extraction-based summarizer aims to identify the most important sentences of a text using an algorithm and uses those original sentences to create the auto-summary (unlike generative AI). As the constituted sentences are machine selected, they may not fully reflect the body of the work, so we strongly advise that the original content is read and cited. The auto generated summaries were curated by the editor to meet Springer Nature publication standards. To cite this content, please refer to the original papers.

Machine generated keywords: food security, food, security, hunger, nutrition, household, rural, insecurity, food insecurity, status, fao, adequate, child, access, Africa.

5.2.1 The Impact of Logistics on Four Dimensions of Food Security in Developing Countries [58]

This is a machine-generated summary of:

Subramaniam, Yogeeswari; Masron, Tajul Ariffin; Naseem, Niaz Ahmad Mohd: The Impact of Logistics on Four Dimensions of Food Security in Developing Countries [58]

Published in: Journal of the Knowledge Economy (2022)

Link to original: https://doi.org/10.1007/s13132-022-01037-3

If you want to cite the papers, please refer to the original.

For technical reasons we could not place the page where the original quote is coming from.

Abstract-Summary

"Considering the complicated nature of food security, this study aims to investigate the impact of logistic performance on food security in 51 developing countries covering the period 2010–2016, under 4 unique dimensions of food security."

"Applying the generalized method of moments (GMM), the results provide supportive evidence that the level of food security tends to be improving in countries with a higher level of logistic performance."

"The overall result suggests that policymakers should improve the level of logistic performance, which is generally far below that in developed countries, so that it can form the fundamental ground toward alleviating hunger and improving the food supply."

Introduction

"To North America, the South American region shows consistent improvement in the hunger index in all 6 countries."

"Food security continues to be a major global challenge even though 72 developing countries out of 129 countries have reached the hunger target of the Millennium Development Goals (MDGs) of the United Nations (United Nations [1])."

"If a country takes this challenge seriously, the government may need to pay more attention to the role of logistics in improving food security."

"To confirm this hypothesis, this study examines the implication of logistic improvement on food security in developing countries, with a special interest in its implication on each dimension of food security; 51 countries have been utilized as a sample for a period between 2010 and 2016 using the regression analysis of GMM."

Literature Review

"Empirical studies by Ehrlich and others [2], Gilland [3], Faisal and Parveen [4], Godfray and others [5], Schneider and others [6], Tian and others [7], and Hall and others [8] reveal that population growth hurts food security."

"The result shows that more population significantly decreases the food supply and thus can play a role in addressing food security problems."

"This study can synthesize from past studies that population growth negatively affects food security."

"This implies that more land significantly increases the food crops and thus can play a role in addressing food security problems."

"Land as a static resource plays an essential role and foundation in the production of crops as well as making more food available for a growing population."

"It suggests that environmental degradation poses significant threats to food security due to changes in crop productivity and food supply."

"Empirical studies such as Hanjra and Qureshi [9], Otsuka [10], Masters and others [11], Koizumi [12], Campbell and others [13], Grzelak [14], and Elmes [15] shift the focus on the role of income inequality and its implication on food security."

Methodology

"The final estimating model will then be Given the panel nature of our data, we adopted the panel data analysis which consisted of pooled, fixed-effect, and random-effect models."

"In the fixed-effect model, the intercept may differ across the subjects, and the intercept does or does not vary over time, which is known as time-variant or time-invariant (Gujarati & Porter [16])."

"The fixed-effect model is expensive in terms of the degree of freedom if there are several cross-sectional units."

"In the random-effect model, the individual error components are not correlated with each other and are not correlated with the cross-section and time-series units."

"The difference between the fixed-effect and random-effect is that each cross-sectional unit has a fixed intercept value in all cross-sectional units."

"The difference GMM estimator led to incorrect inferences if the explanatory variables are persistent, and the lagged levels of the variables become weak instruments (Arellano & Bover [17])."

Result

"This highlights that better logistic performance promotes a sizeable increase in food security in developing countries."

"Concerning the interaction term of (lnLPI*lnGII), we get a positive sign suggesting that the impact of logistic performance on food security can further increase with the improvement in the country's level of innovation."

"This indicates that better logistic performance in countries will lead to increased food availability in terms of quantities and qualities of food that are either locally produced or supplied by imports, accessibility by improving food distribution, utilization in terms of food quality and safety throughout the supply chain activities of transport and stability by stabilizing the other three dimensions."

"Higher logistics level performance tends to increase the food supply chains' sustainability, leading to an increase in food production (availability), food distribution (accessibility), and consumption levels (utilization)."

Conclusion

"This paper focuses on providing insights into the neglected role of logistic performance on food security in developing countries over the period 2010–2016."

"The relationship between logistic performance and food security has been proven by applying the static panel methodology."

"The findings of this study display that the significant improvement in logistic performance could contribute to the increment in food security and may allow households and individuals to be food secure at all times."

"Future studies may expand his analysis on a country-specific level to identify more precise implications at the country-specific level once published long-term time series data on measures of the logistic variables becomes available."

5.2.2 Agriculture and Food Security in North Africa: A Theoretical and Empirical Approach [59]

This is a machine-generated summary of:

Abdelhedi, Imen Turki; Zouari, Sonia Zouari: Agriculture and Food Security in North Africa: a Theoretical and Empirical Approach [59]

Published in: Journal of the Knowledge Economy (2018).

Link to original: https://doi.org/10.1007/s13132-018-0528-y.

Copyright of the summarized publication:

If you want to cite the papers, please refer to the original.

For technical reasons we could not place the page where the original quote is coming from.

Abstract-Summary

"This paper aims at examining the impact of agriculture on food security in North Africa (Tunisia, Algeria, Morocco, and Egypt) over the 2000/2010 period."

"The results that were achieved by using a fixed effect model show that there is a positive and significant impact of the agricultural added value on food security."

"Since agriculture is a crucial instrument for food security, it is worth being considered now as a fully-fledged sector."

Introduction

"They applied the micro-simulation and an agent-based model to show that there is a negative relationship between price instability and household food security."

"Using the VAR approach, Applanaidu and others [18] focused on the dynamic relationship between food security in Malaysia and macroeconomic variables, such as the biodiesel production, the exchange rate, the public expenditure, the GDP, the index of food prices, and population growth."

"Unlike most authors, Negash and Johan [19] showed a positive impact of bio-fuel production, notably those based on castor oil, on food security in Ethiopia."

"The purpose of this paper is to set up a link between food security and agricultural development in North Africa in order to check the correlation between agriculture and food security in this region."

"Empirically, we will then limit ourselves to the panel data econometrics over a period running from 2000 to 2010 to check the correlation between agriculture and food security in this region (North Africa)."

Literature Review

"Investments in the agricultural sector and in the rural areas of the developing countries are certainly necessary to increase agricultural production and create more jobs."

"Several authors state that the increase of farm income would reduce poverty more efficiently than that of non-farm income (Cervantes-Godoy and Dewbre [20]; BM [21]; Bresciani and Valdes [22]; Amsterdam and Audrey [23]; Ravillion and Chen [24]), because most of the poor live in rural areas and depend on the agricultural sector for food."

"Several contemporary writers argue that agriculture dominated by small farms, but not agriculture on a large scale, currently represents the backbone of food security in the developing countries (Chappell and Lavalle [25]; Horlings and Marsden [26]; Tscharnke and others [27])."

"Countries, such as Morocco, Algeria, and Egypt, are, indeed, zones of "high food seismicity," reflecting the chronic imbalances between agricultural production and the population needs (Hervieu and Abis [28])."

Agricultural Situation in North Africa

"Irrigation is a key factor in agricultural productivity."

"The climate change severely threatens the rain-fed agriculture which predominates in most of the countries of the region, while Egypt has 100% irrigated agriculture thanks to the Nile River (Rastoin and others [29])."

"Arable land represents only a small proportion of the entire area, mainly in Tunisia, Algeria, and Egypt where this rate is less than 3.5% (White [30])."

"Due to the water and on land constraints (population pressure), intensive agriculture is concentrated throughout the Nile in Egypt and the coastal areas in the Maghreb."

Empirical Validation

"In our econometric analysis, we used panel data to identify the factors of food security and answer the main question of this manuscript, namely, if the impact of agriculture and the priority of this sector stimulate food security in North Africa, by examining the determinants of malnutrition in this region."

"There are four the independent variables, namely (1) the agricultural added (lAGR_PIB), which is measured by the value added of agriculture in percentage of GDP; (2) the purchasing power (PP) which is measured by gross national income at purchasing power parity; (3) population growth (CROISS_DEMOG) measured by the rate of population growth; and (4) the food trade balance (XM), of which the export and import food are measured by the percentage of the exports and imports of goods."

"The latter points out that the most influential variables on food security are the level of agricultural and food production, the quality of production; the related poverty reports inequality and food cultural identity."

Discussion

"The basic principle of this proclamation is to give the family agriculture the importance it deserves because it has a crucial role not only in improving food security but also in reducing poverty and rural exodus, decreasing unemployment and managing natural resources... At the same time, the International Fund for Agricultural Development (IFAD) considers that family farmers have the opportunity and the obligation to be the precursors of change of global agriculture."

"They will have to focus on the factors that explain the family farming: facilitating access to markets, and to land (exemption of the agricultural vocation lands leased from registration fees and taxes on income), rescheduling farmers' debts, promoting research in issues of food security and sustainable agricultural development, creating an interaction between the researchers and the important players in the practical field, and making them aware of the major role of small family farmers in agricultural development, particularly women."

Conclusions

"The econometric study carried out on the North African countries over the 2000/2010 period showed that some determinants have a major impact on the food security in this region."

"The estimates in this article, through a fixed effect model, showed a positive and significant relationship between agricultural development and food security in North Africa."

"This research represents a real empirical contribution in an attempt to identify the factors that explain food security in North Africa, because most of the studies did not empirically tackle the food quantitative aspect security in this region."

"And to our knowledge, there are no empirical studies that modeled the impact of agriculture on food security, at least at the level of this sub-region."

"Our study on food security in North Africa covers only the 2000/2010 period, which means that it does not cover the post-revolutionary period."

5.2.3 Agroecosystem Health, Agroecosystem Resilience, and Food Security [60]

This is a machine-generated summary of:

Hoy, Casey W.: Agroecosystem health, agroecosystem resilience, and food security [60]

Published in: Journal of Environmental Studies and Sciences (2015).

Link to original: 10.1007/s13412-015-0322-0.

If you want to cite the papers, please refer to the original.

For technical reasons we could not place the page where the original quote is coming from.

Abstract-Summary

"Indicators that characterize biophysical and social conditions including soil health, biodiversity, topography, farm economics, land economics, and social organization can be combined using analytical hierarchy process to map agroecosystem health across a landscape."

"Resilience and agroecosystem health overlap largely because both rely on diversity, in biological and physical as well as human cultural, social, and economic terms."

"The Agroecosystems Management Program at The Ohio State University has approached research and outreach to improve agroecosystem health, resilience, and food security by encouraging self-organizing social networks for economic development around local and regional agricultural supply chains, encouraging farm enterprise diversity at a wider range of farming scales, and conducting research to monitor and estimate the benefits of such diversification."

"Although challenging, such adaptive management experiments may create and encourage new opportunities for managing agroecosystem health, and with it, resilient food production and security."

Introduction

"Six key variables were proposed to describe a minimum set of conditions required to quantify agroecosystem health as a combination of the properties of agroecosystems described above (productivity, stability, sustainability, equitability): soil health, biodiversity, topography, farm economics, land economics, and social organization."

"Based on the conceptual model of agroecosystem health described above, and how food security is expected to emerge from healthy agroecosystems, our approach to managing agroecosystems for both health and resilience has focused on promoting self-organizing social network behavior to build greater farm enterprise diversity into agroecosystems."

"Associated benefits to agroecosystem health should include biological diversity, accruing from greater landscape heterogeneity in crops and production systems at a range of scales from very local to regional."

"Greater and more equitable access to tools that assist entry to the economy may also contribute to the self-organizing behavior of economic systems that enhance diversity in agriculture, ultimately improving agroecosystem health and resilience, and improving food security via more consistent access to fresh food at local and regional scales."

Conclusions

"Social and cultural diversity could add to agroecosystem health if accompanied by sufficient self-organizing capacity to spur innovation."

"Entrepreneurial training and support for collaborative business plan development, with greater efficiency and lower risk in a diversified business ecosystem would contribute needed self-organizing capacity at local scales."

"Cultural and social diversity that is sufficiently harmonious to self-organize in support of the biological diversity needed to achieve agroecosystem health and resilience will be key to resilient food security as a property that emerges from healthy agroecosystems."

"Valuable would be a focus on the mechanistic relationships between social and biological diversity, including self-organizing processes and their contributions to the properties of agroecosystem health."

"Our economic opportunity is to meet basic needs that the global economy does not meet, such as food security, at local scales and via an alternative business ecosystem that supports agroecosystem health."

5.2.4 Pathways Leading Women in Farming Households to Food Security: An Empirical Study in Central Tunisia [61]

This is a machine-generated summary of:

Gaillard, Cédric; Bosc, Pierre-Marie; El-Ati, Jalila; Dury, Sandrine: Pathways leading women in farming households to food security: an empirical study in Central Tunisia [61]

Published in: Review of Agricultural, Food and Environmental Studies (2022).

Link to original: https://doi.org/10.1007/s41130-022-00167-1.

If you want to cite the papers, please refer to the original.

For technical reasons we could not place the page where the original quote is coming from.

Abstract-Summary

"This paper examines the complex relations between farm, household and individual characteristics with the food insecurity of women living in the rural area of Sidi Bouzid, cradle of the 2010 Jasmine Revolution in Tunisia."

"According to the Individual Food Insecurity Index and based on a survey of 290 women living in farming households, more than 40% of women in the area reported experiencing some form of food insecurity."

"Women working and earning an income from agriculture, on and off the family farm, were more often food insecure than other women."

"Moderate and severe insecurity risks were also reduced by household-level factors, such as access to off-farm income by family members other than women."

Introduction

"Studies examining the relationship between agricultural production and nutrition outcomes across countries in sub-Saharan Africa and South Asia (Carletto and others, [31]; Omotesho and others, [32]; Mitiku and others, [33]; Sidhu and others, [34]) support the hypothesis that agricultural production in a farming household has direct and important links with the dietary pattern and food security of each of its members."

"The role of farm production factors (capital and labour) in the improvement of food security has been highlighted in different African contexts where farming households with access to good quality agricultural land or livestock (Muhoyi and others [35]), to an irrigation system (Nkhata [36]) or motorised equipment (Sohoulande Djebou and others [37]) or a greater labour force (Agidew & Singh [38]), are more likely to be food secure than their counterparts."

"Other non-agricultural factors describing household resources play a role in the pathway to food and nutrition security."

"This paper, based on a survey of 290 farm households with at least one adult woman, aims to contribute to filling this gap and to explore the determinants of women's food insecurity, measured by an individual indicator derived from the Household Food Insecurity Access Scale (HFIAS) by considering factors of the family farm, other household livelihoods and women's individual factors."

Background Information

"The most vulnerable populations continue to face difficulties in meeting food needs and under-nutrition continues to exist in rural areas with economic difficulties."

"The same report underlined the difficulties of family farms in central regions of Tunisia, which produce olives, fruit and vegetables, in gaining access to finance, inputs and information, and in marketing their products."

"In Sidi Bouzid, women, who account for 60% of the total workforce in the agricultural sector and have a higher unemployment rate than men (INS [39]), have been particularly affected by the area's economic difficulties."

"According to a report by the Overseas Development Institute [40], women farm workers face informal and precarious working conditions."

"Women in rural central Tunisia are obliged to work on the farm without obtaining their own remuneration and combine their agricultural tasks with domestic tasks."

Data and Methods

"We estimated a model to explain women's food insecurity (as a dichotomous variable) by the value of farm production and a set of other variables."

"In one case (3), the variable was equal to 0 when women were food secure by IFIAS score, 1 when women were mildly to severely food insecure."

"The model can be written as:where the explanatory variables on the B vector contain the two asset scores (human assets and physical and natural assets scores) described in the 'Scoring of farm assets' section and the categorial variable of the (market) orientation of production (self-consumption, mixed, sold)."

"The effects of the scores of farm assets (human and physical assets) were deemed important to observe to consider their potential indirect effect, via the value of production, on women's food security."

Results

"In the IV estimation, an increase in the value of agricultural production was associated with a decrease in women's risk of food insecurity."

"Having income from the family farm, having income from another farm and working time on the farm were positively and significantly associated with an increased risk of women's food insecurity."

"An increase in household size was associated with increased severe food insecurity, while off-farm income from household members other than women reduces this risk."

"As in the previous model, while the effects of women's income affected their food insecurity in different ways, income from other farms was not associated with an increased risk of food insecurity."

"Married women with the main responsibility for the farm were associated with an over-risk of food insecurity compared to woman without farm responsibilities."

"Women's labour on the family farm was also positively associated with farm production."

Discussion and Policy Implications

"We have examined the relationship between farm structure, agricultural production, household and individual incomes and perceived food insecurity among women living in farming households in Sidi Bouzid, Tunisia."

"When moderate and severe food insecurity were observed, household level characteristics also played a protective role (household non-farm income) or aggravating role (family size)."

"As observed in empirical studies dealing with the linkages between agricultural production and food security at the household level, our results showed that an increase in agricultural production on the family farm, with equivalent wealth and farm assets, also reduces the risk of food insecurity for women living in these households."

"The time women spend working on the family farm, although positively correlated with an improvement in agricultural production (and thereby an indirect positive effect on food insecurity), is associated with an increased risk of food insecurity, but this direct effect disappears when it comes to estimating more severe insecurity."

Conclusion

"Family farming faces many obstacles and struggles to ensure food security for women."

"Food insecurity is more severe among women living on less well endowed, less productive farms with larger families."

"Women who have an income from agriculture, often constrained by a lack of resources, are also the most food insecure."

"Non-agricultural income, received by 10% of women, is associated with a better level of women's education and better food security."

"Putting the emphasis on agriculture is insufficient to enhance the living conditions and food security of women living in agricultural households."

5.2.5 Assessing Food Security Among Young Farmers in Africa: Evidence from Kenya, Nigeria, and Uganda [62]

This is a machine-generated summary of:

Adeyanju, Dolapo; Mburu, John; Gituro, Wainaina; Chumo, Chepchumba; Mignouna, Djana; Ogunniyi, Adebayo; Akomolafe, John Kehinde; Ejima, Joseph: Assessing food security among young farmers in Africa: evidence from Kenya, Nigeria, and Uganda [62]

Published in: Agricultural and Food Economics (2023).

Link to original: https://doi.org/10.1186/s40100-023-00246-x.

If you want to cite the papers, please refer to the original.

For technical reasons we could not place the page where the original quote is coming from.

Abstract-Summary

""There is a dearth of empirical evidence on youth food security status in Africa."

"We assessed the level and determinants of food security among young farmers in Africa."

"The Food Consumption Score, which combines dietary diversity and consumption frequency was used to assess food security status while the determinants of food security were identified using a logistic regression model."

"The majority of the respondents had an unacceptable food consumption score, suggesting that despite being food producers, young farmers are still food insecure."

"The odds of being food secure was positively determined by access to extension services, participation in the ENABLE TAAT business incubation programme, and access to market information but, negatively by access to credit, number of employees, Covid-19 pandemic, and location."

"The food security status of young female farmers was positively influenced by age, suggesting that younger youths are less food secure compared to older ones."

"These results suggest that more efforts should be directed towards improving the food security of young African farmers and that policy- and programme-level interventions should support access to extension services, market information, and land."

Introduction

"Studies have been conducted on youth food security in Africa, the bulk of available pieces of literature focus mostly on children, university students, and other socially vulnerable groups (Davidson and Morrell [41]; Larson and others [42]; Leung and others [43]; Loftus and others [44]; McArthur and others [45]; Ragasa and others [46]; Tong and others [47]) while studies on the food security status of young farmers are quite limited."

"What are the determinants of food security among young African farmers?"

"For better insights, the result was disaggregated by gender to assess the common and different factors that influence food security among young male and female farmers."

"This implies that efforts to improve the food security status of young farmers must consider these factors for better outcomes."

"The current study aims to fill this research gap by assessing the food security status and identifying factors affecting food security among young African farmers."

"In addition to identifying factors that influence food security among young farmers, we assessed whether significant socioeconomic correlates of food security varied by gender."

Materials and Methods

"The sampling frame for each country consisted of a complete list of young farmers registered in the database of the ENABLE-TAAT programme in the reference year which included 744, 805, and 1119 young farmers in Kenya, Nigeria, and Uganda, respectively."

"Respondents were asked how many days they had consumed different food items in the week before the survey."

"The result also indicates that respondents across the three countries had an average farmland of 2 hectares, suggesting that young farmers engaged in food production operate on small farmlands which could be a result of limited access to land."

"Compared to Nigeria, where only one-third of the respondents reside in rural areas, most of the respondents in Kenya and Uganda (94% and 90, respectively) are domiciled in rural areas, indicating some demographic similarities between the eastern countries."

Results and Discussions

"Even though this is not expected since credit is expected to raise production and contradicts some literature where credit support is highlighted as a crucial aspect of promoting household food security (Aidoo and others [48]; Iftikhar and Mahmood [49]; Matchaya and Chilonda [50]), the results could be attributed to the lack of creditworthiness among young farmers, who would most likely explore informal credit sources with high-interest rates and ridiculous repayment conditions."

"These findings highlight the relevance of extension services, agribusiness empowerment programmes (ENABLE-TAAT), and access to market information in boosting the food security status of young farmers."

"The significance of the ENABLE TAAT programme in improving food consumption across the three groups suggests the need to promote and facilitate young farmers' participation in agribusiness incubation programmes."

"These results show that participation in agribusiness empowerment programmes is crucial to empowering young farmers to improve their skills and productivity, which is likely to impact their level of food consumption and, subsequently, food security status."

5.2.6 Fight Against Hunger and Poverty in Rural India [63]

This is a machine-generated summary of:

Kumar, Praduman: Fight against hunger and poverty in rural India [63]

Published in: Journal of Social and Economic Development (2022).

Link to original: https://doi.org/10.1007/s40847-022-00217-w2.

If you want to cite the papers, please refer to the original.

For technical reasons we could not place the page where the original quote is coming from.

Abstract-Summary

"Small farmers constitute more than 50 per cent of the country's farming population and account for more than half of the number of hungry and poor people."

"It looks at the contribution of small farmers to the national and household food security and their own nutritional and hunger status."

Introduction

"I have chosen the theme of empowering the small farmers towards a hunger-free rural India for this address in the hope that it would give us an opportunity to have a look on the contribution of small farmers to the national and household food security and their own nutritional and hunger status."

"How to empower these smallholders to fight against hunger and poverty is a challenge before policy planners and agricultural scientists?"

"This agricultural transformation and the associated broad-based economic growth have been instrumental in increasing per capita real income—almost halving the incidence level of poverty and making the nation virtually self-sufficient in food."

"What is the current status of small farmers in relation to his/her contribution to the household and national food security and poverty situations?"

"How can the role of small farmers be enhanced in fighting against hunger and poverty?"

Increasing Role of Small Farmers in Food Supply

"Small farmers, including marginal farmers (cultivating less than 2 hectares of land per farm household) as per the Agricultural Census 1990–91, constitute about 78% of the farmers in the country."

"This is borne out by the changes in area under food crops at various farm size holdings and changes in their share in total food production of the country over time."

"The participation of marginal and small farmers in the growth process has thus largely contributed to the green revolution as well as food-secure India."

"The distribution of agricultural production according to farm size categories shows that the contribution of small farmers to the country's total rice production has increased from 38% in 1970–71 to 49% in 1990–91."

"The share of farmers with 2–4 ha holdings, in the total foodgrains production, has increased only by 3% during 1970–71 to 1990–91 and is expected to increase only marginally by 2010."

Nutritional Intake and Dynamics of Undernourishment

"These results have revealed that the agricultural landless labourers and marginal farm size households constituted more than 82% of the total poor and 73% of the undernourished population in year 1999–00, and alarmingly, their number has been increasing with time."

"If India is to meet the target of the World Food Summit 1996 towards reducing by half the number of poor people in the world by 2015, and the Indian Prime Minister's call for doubling the food production and making India hunger-free in a decade, then the highest priority must be accorded to the landless agricultural labourers and marginal farm households for raising their income and employment levels."

"The custom-hire service may play a significant role in increasing agricultural production and meeting the nutritional needs of poor farm households in rural India."

"Addressal of the nutritional needs of this group means targeting about 87% of the undernourished farm households in the rural India."

Accent on Empowering Small Farmers

"Raising of the agricultural productivity requires regular investments in human resource development, agricultural research and development, improved information and extension, markets, roads and related infrastructural development, efficient small-scale, farmer-controlled irrigation technologies, and custom-hiring services."

"Improved agricultural technology, irrigation facilities, livestock development and literacy level will be most important instruments for improving the nutritional security of small farm households."

"Livestock sector should be accorded higher priority with multiple objectives of diversifying agriculture, raising income and meeting the nutritional security of the poor farm households in rural India."

"In the face of shrinking natural resources and ever increasing demand for more food and higher agricultural production, arising due to high population and income growths, agricultural intensification is the main course for future growth of agriculture."

"Besides developing technologies for promoting agricultural intensification, the country must accord higher attention to the development of technologies that will facilitate agricultural diversification, particularly towards intensive production of fruits, vegetables, flowers and other high-value crops having potential to increase income levels."

Conclusions and Policy Suggestions

"The contribution of small farmers to the national and household food-security has been increasing steadily."

"Improved agricultural technology, irrigation facilities, livestock development, aquaculture enhancement, and literacy level are the most important instruments for improving the nutritional security of poor farm households in the rural India."

"Education, skill development, livestock status, irrigation facilities, agricultural diversification, infrastructural development, aquaculture enhancement and micro-credit support have tremendous impact on rural employment opportunities and alleviation of hunger and poverty."

"Efficient resource management and science-led productivity growth are the must for rationally and effectively addressing the interrelated and complex problems of small landholders."

5.2.7 Beyond "Surpluses": Food Security in Changing Context of Indian Economy [64]

This is a machine-generated summary of:

Rao, V. M.: Beyond "surpluses": food security in changing context of Indian economy [64]

Published in: Journal of Social and Economic Development (2022).

Link to original: https://doi.org/10.1007/s40847-022-00229-6.

If you want to cite the papers, please refer to the original.

For technical reasons we could not place the page where the original quote is coming from.

Abstract-Summary

"The transition of the food security system in India from security to "surpluses" marks the end of an important phase, signifying the crossing by the country of the Malthusian hurdle of widening gap between the food supplies and the needs of a growing population."

"This lecture, therefore, focuses on the past performance, present status, and the priorities for the future directions of change in the food security system in India."

Introduction

"I have chosen the theme of food security for this address in the hope that it would give us an opportunity to have a look at the development issues which are of keen concern and interest to Professor Venkataramanan."

"The recent trends in agricultural production and the on-going policy reforms for liberalization of the economy and reduction in government deficits are the kind of changes which offer considerable scope for improvements if the food situation is comfortable."

"If our luck with monsoon holds for some more years, the situation is likely to be very congenial for major systemic changes in food security which until very recently were considered impractical and even risky, to implement."

"The present phase in our economy and agriculture would indeed be an appropriate time to give some serious thought to the past performance, present status and the priorities for the future directions of change in the food security system."

Food Security: A Retrospect

"It is not suggested that the policies of the latter phase suddenly emerged in the mid-sixties, but an economic historian is likely to discover that the traumatic food crisis of that period along with fears about the political repercussions of large and persistent dependence on imports did bring about a radical change in the perception at the highest level of policy making about the importance of having a country-wide system of food security based on domestic capabilities to increase production."

"Important to note is that the food security system evolved as an integral part of a development strategy which brought about a sticking technological change in selected food crops, provided effective price and market support to farmers and deployed a wide range of measures to generate employment and income for the rural poor with a view to improving their level of well-being including better physical and economic access to food grains."

Revamping Food Security: An Optimistic Scenario

"This addition this tier would implement the price policy formulated at the national level and hold operational stocks of food grains in dispersed locations to ensure intra-year stability of prices and to reach food grains to areas in which the normal trade channels are weak and sluggish."

"Price support operations, moderation of intra-year fluctuations in prices, reaching food to areas remaining out of the mainstream market economy and strengthening the economic access to food of genuinely needy groups by selective and tight targeting could all be located at the state and lower tiers."

"With the latest round of agreement among the GATT nations to progressively reduce the barriers to international trade, the food security system could benefit from increased access to international markets through imports at competitive prices and the promotion of exports in which the country has a comparative advantage."

Conclusion

"In the evolution of food security system in India, the transition from security to "surpluses" marks the end of an important phase signifying the crossing by the country of the Malthusian hurdle of widening the gap between the food supplies and the needs of a growing population."

"The strategy and policies which helped the country in overcoming the hurdle would need substantive changes for the next phase of moving forward from "surpluses" to genuine adequacy of food."

"The broader implication for the future is that an effective food security system would need not only increase in food production and its equitable distribution but also an achievement of these improvements through a development process in which all areas, groups and sectors participate."

5.2.8 Do Various Dimensions of Food Security Affect Household Welfare Differently? An Empirical Investigation of India's National Food Security Act [65]

This is a machine-generated summary of:

Satapathy, Jyotirmayee; Nayak, Narayan Chandra; Mahakud, Jitendra: Do various dimensions of food security affect household welfare differently? An empirical investigation of India's national food security act [65]

Published in: Letters in Spatial and Resource Sciences (2022).

Link to original: https://doi.org/10.1007/s12076-022-00312-w.

If you want to cite the papers, please refer to the original.

For technical reasons we could not place the page where the original quote is coming from.

Abstract-Summary

"This paper examines the impact of various dimensions of food security under India's National Food Security Act on household welfare based on a primary survey over three different states."

"The two-stage residual inclusion approach is employed to control for any possible endogeneity and reverse causality problems between food security and welfare dimensions."

"The dimensions of accessibility and utilization seem to improve several welfare dimensions."

"To availability, accessibility, and utilization, awareness building can lift more beneficiaries to the acceptable range of food consumption scores."

Introduction

"To eradicate abject poverty and hunger, India introduced the National Food Security Act (NFSA) in 2013, a flagship food subsidy program."

"While bridging the food deficiency, this food security program is expected to improve health and labor productivity to enhance human welfare."

"It may, however, be significant to note that the impact of food security on welfare does not necessarily depend upon the mere availability of food at subsidized rates."

"The notion of food security encompasses four different dimensions, namely, availability, accessibility, utilization and stability."

"Given the all-encompassing meaning of food security, one may like to know how India's food subsidy program performs on all these different dimensions and how these individual dimensions affect household welfare."

"While examining the role of various dimensions of food security on household welfare, it introduces other relevant household characteristics as control."

Literature Review

"Definition, one can identify four distinct dimensions of food security: physical availability of food, economic and physical access to food, food utilization, and stability over time (World Food Summit [54])."

"The utilization aspect is a critical dimension of food security towards enhancing the wellbeing of the households."

"Food availability and accessibility dimensions reflect food security's demand and supply sides, respectively."

"The utilization dimension includes the knowledge and facilities available to the households for proper food usage."

"The studies analyzing how various dimensions of food security affect various dimensions of household welfare are limited."

"This study thus attempts to analyze the impact of three different dimensions of food security: availability, accessibility, and utilization on the welfare of the beneficiary households, the latter covering dimensions like FCB, IT, SSE, EO, and FCS."

Data and Methodology

"The beneficiary households' food security index and the indices of its various dimensions are formed by employing PCA."

"Food production can be increased with the increase in the landholding of the households, leading to food security and welfare."

"To find out how different dimensions of food security affect household welfare, the four dimensions of welfare are considered dependent variables."

"While trying to find out how various dimensions of food security affect various dimensions of welfare, as mentioned earlier, we have taken several household characteristics as control."

"The two-stage residual inclusion (2SRI) approach is applied to determine the effect of food security on household welfare (Terza and others [51]; Tesfaye and Tirivayi [52]; Wooldridge [53])."

"The 2SRI approach deals with possible endogeneity and reverse causality problems between food security and household welfare dimensions."

"The food security dimensions are endogenous terms, and household characteristics are taken as control variables while keeping welfare dimensions dependent in the second stage regression."

Results and Discussion

"Nearly 3/5th of the sample households revealed that food grains available under the program are insufficient."

"About 3/4th of the sample households were satisfied with the quality of food grains available under the program."

"A large majority of the sample households (90.65%) utilize the subsidized food grains for self-consumption."

"This provides prima facie evidences, if not conclusive, that household welfare seems to be associated with the availability, accessibility, and utilization dimensions of food security."

"The ease of access to food grains from the FPSs enhances the FCS of the households."

"The majority of households utilize the food grains for their own consumption, thus ensuring improved FCB and reducing food shortage and dependence on others (Beatty and Tuttle [55])."

"The proper utilization of food seems to help households transfer incomes to alternative uses (Kim [56])."

Conclusion

"While most of the earlier studies have tried to find out the determinants of food security, this study explored the impacts of the different dimensions of food security on household welfare, specifically for the beneficiaries covered under the NFSA."

"While examining the impact of various dimensions of food security on various welfare indicators, all the relevant household characteristics were taken as control."

"The dimensions of accessibility and utilization seem to improve several welfare dimensions of the households, justifying the income effect of the food subsidy program."

"The majority of the households utilize the food grains available under the program for their own consumption purposes, thus ensuring better food consumption behavior, which is also reflected in the income transfer and educational outcome."

"To availability, accessibility, and utilization of food grains, awareness building regarding healthy practices can lift more beneficiaries from the poor and borderline category to the acceptable range of FCS and improve the overall food security of all the sections of people."

5.2.9 Mainstreaming Nutrition for Better Development Outcomes [66]

This is a machine-generated summary of:

Bustamante Izquierdo, Juana Paola; Bellucci, Daniela: Mainstreaming Nutrition for Better Development Outcomes [66]

Published in: Transition Studies Review (2009).

Link to original: https://doi.org/10.1007/s11300-009-0087-9.

Copyright of the summarized publication:

If you want to cite the papers, please refer to the original.

For technical reasons we could not place the page where the original quote is coming from.

Abstract-Summary

"Even though the millennium development goals (MDGs) tackle the main economic and social problems in developing countries, nutrition is not addressed adequately despite its important role in achieving the MDGs."

"It is therefore crucial to mainstream nutrition by strengthening public-private partnerships, increased political and technical capacity and by achieving food security."

"The paper aims to discuss how strong partnerships bringing public and private efforts together are important to the success of increasing nutritional programs by providing a redesign and marketing of products and services, other services, and funding."

"The study found that the relation between agriculture and nutrition is a key component for achieving food safety in order to mainstream nutrition into the MDGs."

Introduction

"We will discuss how strong public-private partnerships, achieving food security, and increasing political and technical capacity are important to mainstream nutrition."

"The private sector can contribute and has to take part in various food and nutrition issues (like food fortification programs) through the development of global partnerships (goal 8)."

"Given the growing demand of food and the lack of resources that we are facing nowadays because of the increase in population and the energy problem, the private sector should become aware and take part in the solution to this social problem by applying good business practices which consist in using private skills, technology and scientific research in order to address nutrition improvements."

"Many nutritional programs focus on food security and food supplementation where poor access to health services or poor child-care practices are the main causes of malnutrition ("food first mismatch" World Bank [57], p. 89)."

Conclusions

"Nutrition is a key topic for achieving the MDGs, we have discussed the importance to mainstream nutrition by strengthening public-private partnerships, increasing political and technical capacity and achieving food security."

"Strong partnerships bringing public and private efforts together are important elements to the success of increasing nutritional programs to channel the capacities of the private sector into the public good by providing a redesign and marketing of products and services, other services not related with the core business (promoting knowledge of better nutrition), and funding."

"In order to achieve food security, Biotechnology has a great potential for increasing food production and food processing, but it requires cooperation between developed and developing countries in order to transfer research and best practices in how agronomic and nutritional requirements can be better matched."

5.2.10 Agricultural Growth, Farmers' Income and Nutrition Security: Linkages and Challenges [67]

This is a machine-generated summary of:

Chand, Ramesh: Agricultural growth, farmers' income and nutrition security: linkages and challenges [67]

Published in: Journal of Social and Economic Development (2022).

Link to original: https://doi.org/10.1007/s40847-022-00200-5.

If you want to cite the papers, please refer to the original.

For technical reasons we could not place the page where the original quote is coming from.

Abstract-Summary

"It discusses, with evidence, how and when the turnaround in agriculture took place, for developing proper understanding of India's agriculture growth story."

Introduction

"Agriculture sector in India and in many other countries faced several tough challenges for couple of years after implementation of the WTO agreement on agriculture in year 1995, primarily due to sharp fall in prices of agricultural commodities."

"The fall in prices led to perceptible decline in growth of agricultural output after the mid-1990s which had several consequences including widespread agrarian distress."

"It was a great challenge and formidable task to arrest the decline and to reverse the slowing growth of agriculture sector."

"The last nine years (2004–05 to 2012–13) have witnessed impressive revival of agriculture growth rate even though growth rate of non-agriculture sector decelerated in recent years."

"We have tried to understand effect of varying growth rate in agriculture on producers by using farmer income as an indicator and on consumers by using nutrition level as the indicator."

Growth Trend and Composition

"The country achieved close to 5 per cent average annual rate of growth in agriculture during 8th Plan (1992–93 to 1996–97) and fixed a target of 4.5% growth for the 9th Plan (1997–2002)."

"Unlike the previous two Five Year Plans, the 11th Plan recorded an average growth rate of 4.06 per cent in the agricultural GDP."

"The growth rates reveal that after growing at 2.5 per cent for 10 years, during 9th and 10th Plan, agriculture growth in the subsequent period has accelerated to 3.5 per cent level."

"Both vegetable and fruit production increased by more than 60% in 8 years after 2003–05 which is much higher than the growth in the previous period."

"These growth rates have taken fruit production to 77 mt and vegetable production to 158 mt during 2011–13."

Agricultural Growth at State Level

"The states like Madhya Pradesh, Karnataka, Rajasthan, Jharkhand and Chhattisgarh achieved more than 5% annual growth rate in agriculture, and Gujarat, Tamil Nadu, Maharashtra and Andhra Pradesh exceeded the national target of agriculture growth."

"Haryana, recorded close to 4% annual growth in NSDP agriculture even with high level of productivity."

"The state of Punjab comes at the bottom in the list of states which recorded positive growth in agriculture, with only 1.5% annual growth."

"In the North West Himalayan region, agriculture growth rate in Jammu & Kashmir and Uttarakhand was around 2% whereas agriculture was stagnant in the state of Himachal Pradesh."

Initiatives and Factors Underlying the Achievements

"Higher use of productivity enhancing inputs like fertilizer and quality seed Expansion of irrigation and increase in agricultural investments supported by public sector capital formation Substantial increase in the supply of institutional credit to agriculture."

"As seed is the carrier of technology, the growth in supply of quality seed has been a major factor for increase in agriculture production in recent years."

"Electricity consumption in agriculture sector was 85.7 thousand GWh in year 1995–96."

"The consumption increased to 97.2 thousand GWh by the year 1998–99 and sharply declined thereafter."

"The consumption of electricity in agriculture picked up in year 2006–07 and the upward trend continued thereafter."

"The electricity consumption reached 129 thousand GWh in year 2010–11."

"Between 1998–99 and 2005–06 electricity consumption in agriculture declined by 7% and in next 5 years it increased by as much as 43%."

Farmers' Income

"In the absence of this crucial information, pieces of information like income from one crop, higher rise in input prices compared to output prices, slow growth in output and value added, rise in wages, rising indebtedness, rising gap between farm harvest and retail prices are used to prove decline in farm income."

"Growth in farm income has been higher than growth in output."

"After 2004–5 both output and farm income have seen historical growth rate."

"After 2004–5, growth rate in agriculture and farm income started accelerating and number of farmers suicides started falling."

Nutrition

"This raises serious question that why in India undernutrition is not falling when per capita food production has risen by more than 1 per cent for several years."

"Indian consumers show decline in per capita intake of cereals, during last two decades, which are the major source of dietary energy."

"Its implication for nutrition security is how to check decline in intake of basic nutrient like energy and protein resulting from consumers preference away from cereals."

"The only way to ensure adequate nutrition consistent with preference of Indian consumers is increase in intake of livestock products like dairy products, eggs and meat."

"Its implication for nutrition security is how to check decline in intake of basic nutrient like energy and protein resulting from consumers preference away from cereals."

"The only way to ensure adequate nutrition consistent with preference of Indian consumers is increase in intake of livestock products like dairy products, eggs, meat and fish."

Conclusions and Lessons

"The period from mid-1990s to mid-2000s witnessed slowdown of agriculture growth from above 3.5 per cent to below 2 per cent."

"This has been followed by a sharp turnaround in year 2005–06 which took agriculture back to above 3.5% growth."

"The most important factor for improved and impressive performance of agriculture, post-2004–05, has been the increase in the prices received by the farmers."

"Slowdown of agriculture growth and its recovery in response to changes in price and non-price factors clearly establish that Indian farmers respond rather strongly to various types of incentives."

"State level comparison of agriculture growth offers useful lessons."

"As no country can afford to keep real agricultural prices rising for a very long time, therefore, ways and means have to be devised to sustain profitability incentive."

"The agricultural development strategy in the last ten years has focused on production and MSPs."

Bibliography

1. United Nations. (2018). World Economic Situation and Prospects 2018. United Nations publication
2. Ehrlich, P. R., Ehrlich, A. H., & Daily, G. C. (1993). Food security, population and environment. Population and Development Review, 19, 1–32
3. Gilland, B. (2002). World population and food supply: Can food production keep pace with population growth in the next half-century? Food Policy, 27(1), 47–63
4. Faisal, I. M., & Parveen, S. (2004). Food security in the face of climate change, population growth, and resource constraints: Implications for Bangladesh. Environmental Management, 34(4), 487–498
5. Godfray, H. C. J., Beddington, J. R., Crute, I. R., Haddad, L., Lawrence, D., Muir, J. F., & Toulmin, C. (2010). Food security: The challenge of feeding 9 billion people. Science, 327(5967), 812–818

6. Schneider, U. A., et al. (2011). Impacts of population growth, economic development, and technical change on global food production and consumption. Agricultural Systems, 104(2), 204–215
7. Tian, H., Ren, W., Tao, B., Sun, G., Chappelka, A., Wang, X., ... & Melillo, M. J. (2016). Climate extremes and ozone pollution: a growing threat to China's food security. Ecosystem Health and Sustainability, 2(1)
8. Hall, C., Dawson, T. P., Macdiarmid, J. I., Matthews, R. B., & Smith, P. (2017). The impact of population growth and climate change on food security in Africa: Looking ahead to 2050. International Journal of Agricultural Sustainability, 15(2), 124–135
9. Hanjra, M. A., & Qureshi, M. E. (2010). Global water crisis and future food security in an era of climate change. Food Policy, 35(5), 365–377
10. Otsuka, K. (2013). Food insecurity, income inequality, and the changing comparative advantage in world agriculture. Agricultural Economics, 44(s1), 7–18
11. Masters, W. A., Djurfeldt, A. A., De Haan, C., Hazell, P., Jayne, T., Jirström, M., & Reardon, T. (2013). Urbanization and farm size in Asia and Africa: Implications for food security and agricultural research. Global Food Security, 2(3), 156–165
12. Koizumi, T. (2015). Biofuel and food security. Renewable and Sustainable Energy Reviews, 52(December), 829–841
13. Campbell, B. M., Vermeulen, S. J., Aggarwal, P. K., Corner-Dolloff, C., Girvetz, E., Loboguerrero, A. M., & Wollenberg, E. (2016). Reducing risks to food security from climate change. Global Food Security, 11(December), 34–43
14. Grzelak, A. (2017). Income inequality and food security in the light of the experience of the OECD countries. Contemporary issues in business, management and education. 5th International Scientific Conference. Vilnius Gediminas Technical University, 384–392
15. Elmes, M. B. (2018). Economic inequality, food insecurity, and the erosion of equality of capabilities in the United States. Business & Society, 57(6), 1045–1074
16. Gujarati, D. N., & Porter, D. C. (2009). Basic econometric, 5th Edition. The McGraw-Hill. New York
17. Arellano, M., & Bover, O. (1995). Another look at the instrumental variable estimation of error-components models. Journal of Econometrics, 68(1), 29–51
18. Applanaidu, S.D., Abu Bakar, N.A. and Baharudin, A.H. (2014). An econometric analysis of food security and related macroeconomic variables in Malaysia: a Vector Autoregressive Approach (VAR), UMK Procedia. Elsevier. ScienceDirect. In International Agribusiness Marketing Conference 2013, IAMC 2013, 22–23 October 2013, Kuala Lumpur, Selangor, Malaysia, 93–100
19. Negash, M. and Johan, F.M. (2013). Biofuels and food security: micro-evidence from Ethiopia. Energy Policy. Elsevier. ScienceDirect, 61, 963–973
20. Cervantes-Godoy, D. and Dewbre, J. (2010). Economic importance of agriculture for poverty reduction. OECD Food. Agriculture and Fisheries Working Papers, OECD Publishing, 23
21. BM, (2008). Rapport sur le développement dans le monde 2008 – L'agriculture au service du développement, Banque Mondiale, Washington D.C, 246
22. Bresciani, F.and Valdes, A. (2007). Beyond food production: the role of agriculture in poverty reduction. Northampton, MA : Edward Elgar and the FAO.232
23. Amsterdam, L., & et Audrey, L. (2007). La pauvreté absolue des enfants en Haiti au 21 ème siècle. Rapport préliminaire Haiti, 2–58
24. Ravillion, M., & Chen, S. (2007). China's (uneven) progress against poverty. J Dev Econ, 82(1), 1–42
25. Chappell, M. J., & Lavalle, L. A. (2011). Food security and biodiversity: can we have both ? Agric Hum Values, 28, 3–26
26. Horlings, L. G., & Marsden, T. K. (2011). Towards the real green revolution ? Exploring the conceptual dimensions of a new ecological modernization of agriculture that could 'feed the word'. Glob Environ Chang, 21, 441–452
27. Tscharnke, T., Clough, Y., Wanger, T. W., Jackson, L., Motzke, I., Perfecto, I., Vandermeer, J., & Whitbread, A. (2012). Global food security, biodiversity conservation and the future of agricultural intensification. Biol Conserv, Elsevier, 151, 53–59

28. Hervieu, B. and Abis, S., (2009). Euro-méditerranée. La sécurité alimentaire, une priorité politique. CIHEAM. Déméter, Économie et stratégies agricoles. A.Colin, Paris, 9–52
29. Rastoin, J.L., Bourgeois, L., Cheriet, F. and Movahedi, N. (2012). Pour une politique agricole et agroalimentaire euro-méditerranéenne. Rapport « Construire la Méditerranée », Institut de Prospective Economique du Monde Méditerranéen IPEMED, 24
30. White, C. (2009). Analyse comparative des systèmes agro-alimentaires et des politiques agricoles dans les pays méditerranéens. MEDIT. 2,2001, 13–23
31. Carletto, G., Ruel, M., Winters, P., & Zezza, A. (2015). Farm-level pathways to improved nutritional status: Introduction to the special issue. The Journal of Development Studies, 51(8), 945–957
32. Omotesho, O. A., Adewumi, M. O., Muhammad-Lawal, A., & Ayinde, O. E. (2016). Determinants of food security among the rural farming households in Kwara State, Nigeria. African Journal of General Agriculture, 2(1)
33. Mitiku, A., Fufa, B., & Tadese, B. (2012). Empirical analysis of the determinants of rural households food security in Southern Ethiopia: The case of Shashemene District. Basic Research Journal of Agricultural Science and Review, 1(6), 132–138
34. Sidhu, R. S., Kaur, I., & Vatta, K. (2008). Food and nutritional insecurity and its determinants in food surplus areas: The case study of Punjab state. Agricultural Economics Research Review, 21(1), 91–98
35. Muhoyi, E., Mukura, T. J., Ndedzu, D., Makova, T., & Munamati, O. (2014). Determinants of household food security in Murehwa District, Zimbabwe. Journal of Economics and Sustainable Development, 5(3), 2014
36. Nkhata, R. (2014). Does irrigation have an impact on food security and poverty: Evidence from Bwanje Valley Irrigation Scheme in Malawi. Working paper n°4, International Food Policy Resources Institute
37. Sohoulande Djebou, D., Price, E., Kibriya, S., & Ahn, J. (2017). Comparative analysis of agricultural assets, incomes and food security of rural households in Ghana. Senegal and Liberia. Agriculture, 7(5), 38
38. Agidew, A. M. A., & Singh, K. N. (2018). Determinants of food insecurity in the rural farm households in South Wollo Zone of Ethiopia: The case of the Teleyayen sub-watershed. Agricultural and Food Economics, 6(1), 10
39. Institut national de la statistique (INS). (2014). Recensement général de la population et de l'habitat. Available on http://census.ins.tn/recensement
40. Overseas Development Institute. (2014). Building Momentum: Women's empowerment in Tunisia. https://www.odi.org/sites/odi.org.uk/files/odi-assets/publications-opinion-files/10319.pdf
41. Davidson AR, Morrell JS (2020) Food insecurity prevalence among university students in New Hampshire. J Hunger Environ Nutr 15(1):118–127
42. Larson N, Laska MN, Neumark-Sztainer D (2020) Food insecurity, diet quality, home food availability, and health risk behaviors among emerging adults: findings from the EAT 2010–2018 study. Am J Public Health 110(9):1422–1428
43. Leung CW, Wolfson JA, Lahne J, Barry MR, Kasper N, Cohen AJ (2019) Associations between food security status and diet-related outcomes among students at a large, public midwestern university. J Acad Nutr Diet 119(10):1623–163
44. Loftus EI, Lachaud J, Hwang SW, Mejia-Lancheros C (2021) Food insecurity and mental health outcomes among homeless adults: a scoping review. Public Health Nutr 24(7):1766–1777
45. McArthur LH, Ball L, Danek AC, Holbert D (2018) A high prevalence of food insecurity among university students in Appalachia reflects a need for educational interventions and policy advocacy. J Nutr Educ Behav 50(6):564–572
46. Ragasa C, Aberman N-L, Mingote CA (2019) Does providing agricultural and nutrition information to both men and women improve household food security? Evidence from Malawi. Glob Food Sec 20:45–59
47. Tong M, Tieu L, Lee CT, Ponath C, Guzman D, Kushel M (2019) Factors associated with food insecurity among older homeless adults: results from the HOPE HOME study. J Public Health 41(2):240–249

48. Aidoo R, Mensah JO, Tuffour T (2013) Determinants of household food security in the Sekyere-Afram plains district of Ghana. Eur Sci J 9(21):514
49. Iftikhar S, Mahmood HZ (2017) Ranking and relationship of agricultural credit with food security: a district level analysis. Cogent Food Agric 3(1):1333242
50. Matchaya G, Chilonda P (2012) Estimating effects of constraints on food security in Malawi: policy lessons from regressions quantiles. Appl Econ Int Dev 12(2):165–191
51. Tesfaye, W., Tirivayi, N.: The impacts of postharvest storage innovations on food security and welfare in Ethiopia. Food Policy 75, 52–67 (2018)
52. Thomson, A., Metz, M.: Implications of Economic Policy for Food Security: A Training Manual 40. Food and Agriculture Organization of the United Nations, Rome (1998)
53. World Food Programme: Food consumption analysis: Calculation and use of the food consumption score in food security analysis. United Nations Vulnerability Analysis and Mapping Branch (2008)
54. Garces, E., Thomas, D., Currie, J.: Longer-term effects of head start. Am. Econ. Rev. 92(4), 999–1012 (2002)
55. Beatty, T.K.M., Tuttle, C.J.: Expenditure response to increases in in-kind transfers: Evidence from the supplemental nutrition assistance program. Am. J. Agr. Econ. 97(2), 390–404 (2015)
56. Kim, J.J., Rabbitt, M.P., Tuttle, C.: Changes in low-income households' spending and time use patterns in response to the 2013 sunset of the ARRA-SNAP Benefit. Appl. Econom. Perspect. Policy. 42(4), 777–795 (2020)
57. World Bank (2006) Repositioning nutrition as central to development: a strategy for large-scale action. Developing issues and engagements in malnutrition. World Bank
58. Subramaniam, Yogeeswari; Masron, Tajul Ariffin; Naseem, Niaz Ahmad Mohd The Impact of Logistics on Four Dimensions of Food Security in Developing Countries. Journal of the Knowledge Economy (2022). https://doi.org/10.1007/s13132-022-01037-3
59. Abdelhedi, Imen Turki; Zouari, Sonia Zouari Agriculture and Food Security in North Africa: a Theoretical and Empirical Approach. Journal of the Knowledge Economy (2018). https://doi.org/10.1007/s13132-018-0528-y
60. Hoy, Casey W. Agroecosystem health, agroecosystem resilience, and food security. Journal of Environmental Studies and Sciences (2015). https://doi.org/10.1007/s13412-015-0322-0
61. Gaillard, Cédric; Bosc, Pierre-Marie; El-Ati, Jalila; Dury, Sandrine Pathways leading women in farming households to food security: an empirical study in Central Tunisia. Review of Agricultural, Food and Environmental Studies (2022). https://doi.org/10.1007/s41130-022-00167-1
62. Adeyanju, Dolapo; Mburu, John; Gituro, Wainaina; Chumo, Chepchumba; Mignouna, Djana; Ogunniyi, Adebayo; Akomolafe, John Kehinde; Ejima, Joseph Assessing food security among young farmers in Africa: evidence from Kenya, Nigeria, and Uganda. Agricultural and Food Economics (2023). https://doi.org/10.1186/s40100-023-00246-x
63. Kumar, Praduman Fight against hunger and poverty in rural India. Journal of Social and Economic Development (2022). https://doi.org/10.1007/s40847-022-00217-w
64. Rao, V. M. Beyond "surpluses": food security in changing context of Indian economy. Journal of Social and Economic Development (2022). https://doi.org/10.1007/s40847-022-00229-6
65. Satapathy, Jyotirmayee; Nayak, Narayan Chandra; Mahakud, Jitendra Do various dimensions of food security affect household welfare differently? An empirical investigation of India's national food security act. Letters in Spatial and Resource Sciences (2022). https://doi.org/10.1007/s12076-022-00312-w
56. Bustamante Izquierdo, Juana Paola; Bellucci, Daniela Mainstreaming Nutrition for Better Development Outcomes. Transition Studies Review (2009). https://doi.org/10.1007/s11300-009-0087-9
57. Chand, Ramesh Agricultural growth, farmers' income and nutrition security: linkages and challenges. Journal of Social and Economic Development (2022). https://doi.org/10.1007/s40847-022-00200-5

Chapter 6
What is the Unsustainable Component in Economic Growth

6.1 Introduction by Editor

Economic growth is regarded as the primary driver of progress and prosperity, with policymakers and governments worldwide striving to achieve sustained growth. It is thus comprehensible that economic expansion is central to national economic policies. Especially for developing countries like India, economic growth is indispensable. However, the pursuit of continuous economic expansion has come at a significant cost to the environment and exacerbated social inequalities. Every Dollar produced in the economy imposes negative externalities that acts against the environment and some sections of societies. Rapid consumption of natural resources depletes their stocks and affects economic growth, while emissions and pollutions deteriorate natural ecosystems and weaken the life-sustaining capacity of the planet. Climate change, the global manifestation of industrial emissions has led to an increased frequency and intensity of natural disasters. This imposes an undesirable expenditure, but adds up in the GDP. Notwithstanding its limitations, a part of the GDP does contribute to positive social externalities, which constitutes 'good' growth. In the holy grail of sustainable development, this 'good growth' of which 'green growth' is one part, is what Indian economy must move towards.

Economic growth is inexorably connected to the supply of natural capital. According to government estimates, India's resource consumption increased six times, from 1.5 billion tonnes in 1970 to 7 billion tonnes in 2015.[1] The depletion of natural resources either pushes up prices, or imposes a dependence on imports. Today, India is highly dependent on imports for critical resources like Molybdenum (100%), Nickel (100%), Cobalt (100%), Copper (95%), and Oil (70%), among others[5]. As imports rise, so does the nation's fiscal deficit. Moreover, as natural resource prices are pushed up, the competitiveness of the Indian manufacturing sector is affected.

[1] Draft National Resource Policy 2019. https://moef.gov.in/wp-content/uploads/2019/07/Draft-National-Resourc.pdf last accessed 16th November 2021.

G. Nulkar (ed.), *The Economics of Sustainable Development*,
https://doi.org/10.1007/978-981-99-7379-8_6

On the one hand the non-renewable resources are vital for economic growth, on the other, the condition of renewable resources and the natural environment affect the health of society. India's economic growth has been severely debilitating to India's natural environment. Water, air, and soil, the backbones of natural ecosystems, are continuously polluted by human activities. Degraded ecosystems cannot sustain biodiversity, and the landscape's ecological productivity diminishes over time. Government data reveals that in the last 30 years, forests about two-thirds of the size of Haryana have been lost to encroachments (15,000 sq km) and 23,716 industrial projects (14,000 sq km). Recently officials confirmed in the Parliament that up to 25,000 hectares of forests – 250 sq. km, or more than twice the area of Chandigarh city – are handed over for non-forestry activities every year. These include defence projects, dams, mining, power plants, industries and roads. The loss of habitats has broken the food chain and driven wildlife into human settlements. Elephants, leopards, Sambar, Chinkara, Blackbucks, wild pigs, and other species are moving into farms and villages. Tribal communities and marginal farmers are most susceptible to crop damage and face a significant loss of income.

Furthermore, the economy depends profoundly on ecosystem services like the hydrological cycle, carbon cycle, renewal of soil, pollination of crops, decomposition of waste, and many others. However, the threat of climate change looms large over them. India has been experiencing intense rainfall, unprecedented floods, droughts, and cyclones, at an alarming frequency. Each one of these natural disasters has contributed to the loss of forest cover and the decline in ecosystem productivity. The Millennium Ecosystem Assessment[2] reported that over 60 percent of ecosystem services they studied were being degraded or used unsustainably.

This unsustainable part of the economic growth, is what nations must address in the path of sustainable development. In this chapter, we look at studies from Asian countries, China, and Iran. The chapter examines aspects of economic growth from income inequality, environmental degradation, growth-povery nexus, and energy consumption. The chapter offers readers multiple perspectives of the negative externalities of economic growth. Economic growth that comes at the cost of nature, exclusion of few communities, and without considering the needs of future generations, cannot be sustainable. The chapter shows several examples of 'good growth', that can be replicated.

6.2 Machine Generated Summaries

Disclaimer: The summaries in this chapter were generated from Springer Nature publications using extractive AI auto-summarization: An extraction-based summarizer aims to identify the most important sentences of a text using an algorithm and

[2] Millennium Ecosystem Assessment, 2005. Ecosystems and Human Well-being: Synthesis. Island Press, Washington, DC. https://www.millenniumassessment.org/documents/document.356.aspx.pdf Last accessed 19th November 2021.

uses those original sentences to create the auto-summary (unlike generative AI). As the constituted sentences are machine selected, they may not fully reflect the body of the work, so we strongly advise that the original content is read and cited. The auto generated summaries were curated by the editor to meet Springer Nature publication standards. To cite this content, please refer to the original papers.

Machine generated keywords: emission, energy consumption, gdp, energy, carbon emission, growth, poverty, economist, capita, per capita, economic growth, inequality, gdp per, income inequality, carbon.

6.2.1 Inequality and Unsustainable Growth: Two Sides of the Same Coin? [1]

This is a machine-generated summary of:

Berg, Andrew G.; Ostry, Jonathan D.: Inequality and Unsustainable Growth: Two Sides of the Same Coin? [1]

Published in: IMF Economic Review (2017).

Link to original: https://doi.org/10.1057/s41308-017-0030-8

If you want to cite the papers, please refer to the original.

For technical reasons we could not place the page where the original quote is coming from.

Abstract-Summary

"The relationship between income inequality and economic growth is complex and the evidence mixed."

"This paper focuses on the connections between income inequality and the fragility of economic growth."

"Over longer horizons, avoiding excessive inequality and sustaining economic growth may be two sides of the same coin."

Introduction

"This phenomenon—the ability of some countries to sustain high per capita growth over very long periods, in some cases many decades, and the inability of others to enjoy the same fate—is poorly understood by economists and policy makers."

"We document an empirical regularity between the level of inequality and the sustainability of economic growth."

"Some inequality is integral to the effective functioning of a market economy and the incentives needed for investment and growth, and there are important instances of sustained growth involving increasing income inequality (Chaudhuri and Ravallion [2])."

"Too much inequality might be destructive to growth."

"Analyses have recognized the complex linkages among income distribution, growth, and policies to counter inequality."

"We ask whether growth can in fact be sustained in the face of a highly uneven income distribution."

The Hills and Valleys of Growth

"Following the broader growth literature, the typical approach was to relate a country's income distribution at the beginning of a long sample (say, 1965–1985) to the growth rate during that period, controlling for a few key variables such as initial per capita income."

"If this were indeed the common pattern, the most interesting question, indeed the only important question about growth, would be how to explain why some countries grow faster than others over long periods."

"Looking at such pictures, Pritchett [3] and other authors have been struck that an understanding of growth must involve looking more closely at the turning points, meaning not the ups and downs of growth over business cycle horizons, but rather why some countries are able to keep growing for long periods of time, whereas others see growth downbreaks after just a few years, followed by stagnation or decay."

Income Distribution and Growth Sustainability

"Inequality: There is indeed a large and statistically significant association between income inequality and growth duration."

"Lower external debt is associated with longer spells; a decrease from 44 to 39 percent in the ratio of external debt to GDP suggests an increase in the duration of the growth spell of about 2 percent."

"The key result from the joint analysis is that income distribution survives as one of the most robust and important factors associated with growth duration."

"Inequality also preserves its significance more systematically across different samples and definitions of growth spells than the other variables."

"The multivariate estimates would suggest that the associated decline in Brazil's Gini would, other things equal, increase the expected length of a growth spell by some 40 percent."

"We find some evidence to support the idea that higher ethnic fractionalization is associated with shorter growth spells, but the effect varies substantially across samples and is often not statistically significant."

Interpretation and Tentative Policy Implications

"The main results in this paper are that (1) increasing the length of growth spells, rather than just getting growth going, is critical to achieving income gains over the long term and (2) countries with more equal income distributions tend to have significantly longer growth spells."

"Rather, growth and inequality-reducing policies are likely to reinforce one another and help to establish the foundations for a sustainable expansion."

"Of course policies affect growth directly as well as through income distribution."

"The positive direct effect of certain policies on growth may outweigh their negative effects on income distribution."

"Despite the uncertainties and empirical challenges, there is surely policy scope to improve income distribution without undermining incentives—perhaps even improving them—and therefore contribute to lengthening the duration of growth spells."

6.2.2 The Nexus Between Carbon Emissions, Poverty, Economic Growth, and Logistics Operations-Empirical Evidence from Southeast Asian Countries [4]

This is a machine-generated summary of:

Khan, Syed Abdul Rehman: The nexus between carbon emissions, poverty, economic growth, and logistics operations-empirical evidence from southeast Asian countries [4].

Published in: Environmental Science and Pollution Research (2019).

Link to original: https://doi.org/10.1007/s11356-019-04829-4.

If you want to cite the papers, please refer to the original.

For technical reasons we could not place the page where the original quote is coming from.

Abstract-Summary

"This research examines the role of poverty and logistical operations under the circumstance of environmental deterioration with panel data of ASEAN states from 2007 to 2017."

"The results indicate that poverty and logistical operations have significant and positive relationship with greater environmental degradation."

"Reduction in environmental degradation can be achieved through reduction in poverty and encouraging renewable energy and green practices in logistical operations."

"This study also provides detailed policy implications to regulatory bodies and corporate sector in order to improve environmental sustainability through adoption of green practices and reduction in poverty."

Introduction

"Many developing countries are in the middle stage or early age of their economic growth and tend to further destruct the environmental beauty through business and logistical operations to fasten their economic growth on the price of environmental degradation."

"In order to attain a rapid economic development, some ASEAN countries violate the environmental laws and destroy the environment beauty when overuse natural resources (Lau and others [5])."

"Due to scarcity of researches on poverty-environment and logistics-environment linkage, environmental degradation issues are likely to be unseen behind the poverty and cross border logistical movement, particularly in developing countries such as ASEAN member states, which was highlighted by the Kuznet curve."

"Main contribution of this research is to find out the linkage between poverty, logistical operations, and environmental degradation in ASEAN member state countries."

Literature Review

"Many studies found out the association between logistical operations and economic development, while there are very limited studies conducted to discover the impacts of logistic activities and poverty on environmental beauty (Aldakhil and others [6]; Zaman and Shamsuddin [7]; Zhu and Sarkis [8])."

"Many researchers conducted studies to find out association between logistical performance, environmental sustainability, and economic development (Khan and

Dong [9]; Zailani and others [10]; Benitez-Amado and others [11]; Büyüközkan and çifçi [12]; Zhu and others [13]; Benitez and Walczuch [14]; Chien and Shih [15]; Green Jr and others [16]; Zhu and Sarkis [8])."

"A research was conducted by Zhang and Yang [17] to evaluate the effects of logistical and supply chain operations on firms' environmental and economic performance."

"The findings provide evidence that green logistics operation uplift economic performance of firms but have minor effects on environmental sustainability."

"A research was conducted in a wood industry, and the results showed that whether firms use green practices in their logistical operations or not has no significant effect on the buyers (Christmann [18])."

"Based on above cited studies, the following hypotheses are developed: H2: Logistical operations are positively and significantly correlated with environmental degradations."

Research Methodology

"The connection between logistical operations, poverty, and environmental degradations in a panel of ASEAN countries is drawn in this study."

"Due to the nature of panel data, the standard models such as pooled OLS (ordinary least square), RE (random effect) model, and FE (fixed effect) models are not suitable, because of presence of country-specific effects, endogenity of explanatory variables, and/or lagged dependent variable (Ibrahim and Law [19])."

"DIF-GMM (first difference) needs to remove the country-specific effects provided by First difference into consideration, the lagged endogenous variable $LED_{i,t-1} - LED_{i,t-2}$ is associated with the error $\mu_{i,t} - \mu_{i,t-1}$, imposing a bias on the model estimation (Masron and Subramaniam [20]; Khan and others [21])."

"To handle this issue, the generalized method of moment's difference estimator uses the lagged level of the independent variables as instruments under the situation that the term of disturbance is not auto correlated/serially correlated and that the explanatory variables are independent (Arellano and Bond [22])."

Analysis and Results with Discussion

"Khan and others [21] empirical study examined the role of GDP growth in environmental degradations, and they found that most developed countries' economic activities have no significant effect on environmental degradation, but developing countries' economic growth especially China, India, Malaysia, Viet Nam, and Philippine GDP per capita growth are positively correlated with poor environmental performance."

"Borin and others [23] considering environmental sustainability, maintaining the natural resources, implementing waste management and carbon emission reduction, and adopting renewable energy and green practices in logistical operations would improve environmental beauty without compromising economic growth."

"The findings revealed that logistics and GDP are strongly correlated with economic growth, but logistical operations less affect environmental beauty in the context of BRICS countries due to green practices and strict environmental policies implemented by governmental bodies."

Conclusion and Policy Analysis

"The findings show that logistical operations and poverty have significant and positive relationship with environmental degradation, while it has been confirmed that logistical operations and poverty are the primary contributors to poor environmental performance in ASEAN countries."

"Mitigating environmental degradation can be realized through reduction in poverty and improvement in logistical operations in a way that relates to minifying the environment degradation."

"Improving environmental sustainability is not only the responsibility of regulatory authority, but corporate sector should also come forward to take their responsibility in terms to implement green practices in their logistical and business operations so as to reduce harmful effects on environmental sustainability."

"Firms and governmental bodies need to conduct environmental friendly awareness programs and training in poor areas to help people to understand the importance of environmental sustainability under the initiative of CSR (Corporate Social Responsibilities) activities."

6.2.3 Does Economic Prosperity Lead to Environmental Sustainability in Developing Economies? Environmental Kuznets Curve Theory [24]

This is a machine-generated summary of:

Ahmad, Munir; Muslija, Adnan; Satrovic, Elma: Does economic prosperity lead to environmental sustainability in developing economies? Environmental Kuznets curve theory [24].

Published in: Environmental Science and Pollution Research (2021).

Link to original: https://doi.org/10.1007/s11356-020-12276-9.

If you want to cite the papers, please refer to the original.

For technical reasons we could not place the page where the original quote is coming from.

Abstract-Summary

"Since developing countries experience economic and environmental sustainability challenges, it is desirable digging into the linkages between economic and environmental parameters."

"The purpose of this work is to evaluate the existence of the environmental Kuznets curve (EKC) theory (i.e., the inverse U-shape connection between real GDP per capita and per capita carbon dioxide emissions) in the sample of 11 developing countries."

"By using balanced annual panel data in the period between 1992 and 2014 and two alternative estimation techniques, we explored the potential inverted U-shaped linkage between carbon dioxide emissions and real GDP per capita in the sample of interest."

"This work's findings provide the supportive evidence to the inverse U-shaped linkage in the long-run, indicating that an increase in real GDP per capita and electricity consumption tends to mitigate long-run carbon dioxide emissions in the developing countries, for the whole sample."

"It implicated that these countries are on the path of attaining environmental sustainability in the long-run."

"It manifested that these countries need to design strategies directed to reduce carbon dioxide emissions from economic activity and electricity generation through efficiency improvement or promotion of renewables."

"The findings suggest that country-specific targeted action plans should be implemented to ensure the environmental sustainability in the developing world."

Introduction

"Apart from that, the empirical evidence on the validity of the environmental Kuznets curve (EKC) theory that includes electricity consumption remains scarce in the sample of interest."

"After some point, pollution records a decline due to the following reasons (Lieb [25]): stricter environmental policies are induced due to the fear of irrevocable catastrophe increasing the demand for a better quality of the environment; one emitter of greenhouse gases is replaced by another and consequently reduced; technological progress enables the production increase with the less emission of greenhouse gases."

"None of the available studies investigated the validity of the EKC theory in the sample of selected developing countries in a simultaneous manner, although these countries represent one of the top CO_2 emitters in the world."

"In view of the above-stated deficiencies of previous studies, this work pursues the investigation of EKC theory's relevance in the developing countries."

Literature Review

"In a different research, over the period from 1970 to 2017 for six West African countries, the EKC theory's validity was explored by using human resources and bio-capacity as additional carbon emission determinants."

"The empirical findings found that, a U-shaped association among the development of economy and CO_2 discharge contrary to an inverted U-shaped EKC theory, present in the high-, medium-, and low-emission economies (Halliru and others [26])."

"The research seeks to study the nexus of CO_2 pollution (per capita), GDP (per capita), and per capita natural gas use, testing the authenticity of the EKC theory and evaluating natural gas efficiency in a group of 14 Asia Pacific nations from 1970 to 2016 (Dong and others [27])."

"The EKC theory hypothesized an inverse-U-shaped connection between ecological destruction and economic development."

"Within the context of the EKC theory for the top 10 creative nations, Gormus and Aydin [28] explored the relationship between ecological footprint, economic development, green innovation, and energy use."

Methods

"We use a comprehensive dataset for the sample of 11 developing countries in the period between 1992 and 2014 to explore the linkages between carbon dioxide emission, income, and electricity consumption."

"This paper also employs the electric power consumption (kWh per capita), which proxied for the electricity consumption (ELC)."

"This paper employs only one indicator of environmental depletion."

"As the cross-sectional dependence is likely to occur, we have employed the first and second generation of unit root test."

"As of the second generation, this paper suggests a simple panel unit root test in the presence of cross-section dependence—the CIPS test."

"If the co-integration link is present, this paper will further explore the long-run elasticities by employing the FMOLS."

"This work employs the Dumitrescu-Hurlin causality (DHC), postulated by Dumitrescu and Hurlin [29], to inspect the causal links' direction among the indicators of interest."

Results and Discussion

"Brazil, Mexico, Malaysia, Russia, Turkey, and South Africa record the values above the group average, whereas the other countries have a real GDPPC far below the group average."

"First-generation unit root tests suggest that null on the unit root cannot be rejected for the variables in log levels, whereas variables revealed stationarity (alternative hypothesis) at first difference."

"As of Turkey, it has recorded a significant increase in GDPPC in the period of interest, catching up with the advanced world economies."

"Even though the Turkish economy was seriously influenced by adverse shocks, GDPPC has recorded a significant growth of almost 7% in the period between 2010 and 2017 (OECD [30])."

"With the growth performance, the country records a significant increase in carbon dioxide emissions (Jalandoni [31])."

"Panel causality test reveals a bidirectional causal link between real GDPPC and $PCCO_2$ emissions; electricity consumption and $PCCO_2$ emissions; and real GDPPC and electricity consumption."

Conclusion and Recommendations

"Since the under-analysis developing countries recorded a significant growth performance in the last few decades, the substantial economic and population growth were accompanied by escalated electricity use and greenhouse gas emissions."

"The core findings are as follows: (1) both Pedroni and the Westerlund's approach established a long-run co-integration association among the variables of interest. (2) For the whole sample of 11 countries, the inverse U-shaped link was existent between economic development (real GDPPC) and carbon dioxide emissions."

"Energy efficiency is recognized as an important target not only in the case of China but also for the other countries of interest."

"Energy efficiency is recognized to play an important role in reducing carbon dioxide and other greenhouse gas emissions (Jabeen and others [32])."

"It would be useful to compare these countries with the other developed regions to allocate and use best practices that achieve environmental-friendly economic growth supporting the energy efficiency targets."

6.2.4 More Growth? An Unfeasible Option to Overcome Critical Energy Constraints and Climate Change [33]

This is a machine-generated summary of:

Capellán-Pérez, Iñigo; Mediavilla, Margarita; de Castro, Carlos; Carpintero, Óscar; Miguel, Luis Javier: More growth? An unfeasible option to overcome critical energy constraints and climate change [33].

Published in: Sustainability Science (2015).

Link to original: https://doi.org/10.1007/s11625-015-0299-3.

If you want to cite the papers, please refer to the original.

For technical reasons we could not place the page where the original quote is coming from.

Abstract-Summary

"The end of the era of cheap and abundant energy flows brings the issue of economic growth into question, stimulating research for alternatives as the de-growth proposal."

"The present paper applies the system dynamic global model WoLiM that allows economic, energy and climate dynamics to be analyzed in an integrated way."

"The results show that, if the growth paradigm is maintained, the decrease in fossil fuel extraction can only be partially compensated by renewable energies, alternative policies and efficiency improvements, very likely causing systemic energy shortage in the next decades."

"If a massive transition to coal would be promoted to try to compensate the decline of oil and gas and maintain economic growth, the climate would be then very deeply disturbed."

"The results suggest that growth and globalization scenarios are, not only undesirable from the environmental point of view, but also not feasible."

"Regionalization scenarios without abandoning the current growth GDP focus would set the grounds for a pessimistic panorama from the point of view of peace, democracy and equity."

Introduction

"Capellán-Pérez and others [34] showed that GDP growth scenarios have the potential to lead the socioeconomic system to structural energy scarcity in the next two decades in a context of climate deterioration, suggesting that an energy demand-driven transition as performed in the past might be unfeasible without an authentic paradigm shift."

"no-GDP-growth" socioeconomic paradigm scenarios are modeled in this paper to face the energy crisis, with special attention being paid to reach "acceptable" climate outputs by 2050."

"A second group has explored the long-term economic implications of the depletion of high-quality energy resources at national (Ayres [35]; D'Alessandro and others [36]; García-Olivares and Ballabrera-Poy [37]) and global level (Meadows and others [38, 39]; Nel and Cooper [40]; de Castro and others [41]; Yücel and Barlas [42]; Dale and others [43]; Capellán-Pérez and others [34]), concluding that growth of developed countries could cease in the next few decades."

"This paper contributes to this gap in the literature proposing an explorative "no-GDP-growth focus" modeling exercise from the global energy and climate perspectives."

Methodology

"We have developed five scenarios: three of them are "growth and coal" (BAU, A and B), focused on economic growth with different degrees of technological development and environmental concern (implementing GEA storylines, but assuming that the coal supply is not constrained and strongly promoted)."

"A combination of GEA's storylines "Economic optimism" and "Reforming Markets" with a strong focus on the mechanism of competitive, efficient market, free trade and associated rapid economic growth and technological development and diffusion (assuming "Environmental Kuznets Curve" for environmental impacts and higher extraction rates for non-renewable sources), but including some additional policies aimed at correcting some market failures."

"Scenario C: "Regional competition + no-GDP-growth focus" similar to scenario "Regional Competition" from GEA storylines in most aspects (smaller technological development, trade barriers, reactive environmental policies), but its GDPpc growth is set to the maximum level that avoids any energy shortage in any sector for the period studied."

Results and Feasibility of Scenarios

"In Scenario C, where the economy adjusts to the energy development constraints, the GDPpc is reduced 0.4% per year, dropping below 9000$ by 2050."

"The "no-GDP-growth focus" scenarios manage to deal with the energy development constraints by reducing between 20% (C) and 10% (D) the TPES demand for Transportation in relation to the 2010 level."

"Its massive use in the scenarios where no coal extraction constraints are considered (BAU, A and B), including the high CO_2 emitter coal-to-liquids process (Höök and others [44]), would lead to a stunning emissions level along the first half of the century."

"The only scenarios that seem able to escape from both critical energy constraints and dangerous climate change are the "no-GDP-growth focus" scenarios C and D."

Exploring the Implications of a Feasible, Sustainable and Desirable Global Future

"A stabilization of the average world GDPpc in an equitable context would imply that the current industrialized countries would have to decrease their income to allow the South to increase their material needs (Ayres [45]; Kerschner [46]; Schneider and others [47]; Kallis [48])."

"This section focuses on the implications of scenario D: "The de-growth movement calls for a decrease in material and energy consumption in countries that exceed their 'allowable ecological footprint' and acknowledges the allowance for Southern countries or societies, where ecological impacts are low relative to their biocapacity, to increase their material consumption and thus their ecological footprint" (Martínez-Alier and others [49])."

"The Scenario D presented in this paper is, in fact, very close to the scenarios 8 and 9 from Meadows and others [39], where more than 30 years of prosperity were obtained by implementing similar intensive transition policies and aiming for a modest increase in the global average Industrial output level per capita."

Conclusion

"We perform an exploratory scenario exercise that sheds light on the potential dangers of not addressing these dynamics properly in the context of the current energy and climate change crises."

"Deglobalization scenarios that do not include an anticipated and conscious abandoning of the growth paradigm would very likely face significant threats from the interplay of critical energy constraints with a troubling context of increasing regionalization, countries' self-reliance and climate deterioration."

"In an ideal sustainable de-growth scenario, assuming that global and equitable convergence is reached by 2050, the most energy intensive consuming countries (i.e., the most industrialized) should reduce, on average, their per capita energy use rate by at least 4 times and decrease their GDPpc roughly to the current world average."

6.2.5 The Nexus Between Income Inequality, Economic Growth and Environmental Degradation in Pakistan [50]

This is a machine-generated summary of:

Baloch, Amdadullah; Shah, Said Zamin; Noor, Zaleha Mohd; Magsi, Hussian Bakhsh: The nexus between income inequality, economic growth and environmental degradation in Pakistan [50].

Published in: GeoJournal (2017).

Link to original: https://doi.org/10.1007/s10708-016-9766-3.

If you want to cite the papers, please refer to the original.

For technical reasons we could not place the page where the original quote is coming from.

Abstract-Summary

"This study examines the impact of income inequality and economic growth on environmental degradation in Pakistan using ARDL bounds testing approach for the period 1966–2011."

"Empirical results for the aggregate CO_2 emissions and its four sources such as CO_2 emissions from solid fuel, liquid fuel and gaseous fuel consumption as well as electricity and heat production confirm the existence of long run co-integrated relationship between income inequality, economic growth and environment degradation."

"Besides the negative impact of industrial share and population density on CO_2 emissions, we also confirm that economic growth in Pakistan comes up with higher emissions."

"This study suggests that the distribution of income matters to aggregate carbon emissions and focus should be made on sustained economic growth to reduce pollutants and hence CO_2 emission in the study area."

Introduction

"The relationship between income inequality and the environmental degradation has been ignored in majority of developing countries including Pakistan."

"Pakistan's rising income inequality and environmental degradation are concerning issues."

"This study aims to analyse the impact of income inequality and GDP per capita on environmental degradation in an ideal country like Pakistan."

"The urgency to address the twins' problems of higher income inequality and environmental degradation in Pakistan is due to the fact that both are not only cutting the national GDP growth but also as a serious development concern, the increasing urbanization present additional environmental challenges (World Bank [51]).”

“This study contributes by exploring the effects of income inequality and economic growth on environmental degradation in Pakistan through the autoregressive distributed lag (ARDL) framework for the period 1966–2011.”

Literature Review

“The main focus of this group of studies is to test the existence of EKC and to find out the relationship between different environmental indicators, economic growth and income inequality.”

“As far as literature on the relationship between income inequality, economic growth and environmental degradation is concerned in Pakistan, there are very limited studies even on the dynamic impact of economic growth and related variables on CO_2 emission with inconclusive outcomes.”

“There is no relevant study on the impact of income inequality on environmental degradation in the country and also the existing studies on testing EKC hypothesis are not able to conclude about the nature of growth-emissions nexus due to their analytical techniques and others limitations.”

Empirical Specification, Estimation Techniques and Data and Data Sources

“While considering the nexus between income inequality, economic growth and environmental degradation, we adopt our empirical framework similar to the empirical studies by Baek and Gweisah [52] and Hao and others [53] to represent the long run relationship between income inequality and environmental degradation and its major determinants.”

“ARDL approach to co-integration redresses the endogeneity problem and it generally provides unbiased and consistent estimates of the long run coefficients and valid t-statistics even when some of the explanatory variables are endogenous.”

“If the computed F-statistic is higher than the upper bound critical value, this suggests that a long run relationship exists between the dependent variable and its determinants.”

“Gini coefficient is used as proxy for income inequality, for economic growth, GDP per capita and its square (constant 2005 US$) to explore the EKC hypothesis, industrial share measured as proportion of industry value added to GDP by following Hao and others [53] and Zhang and others [54] and population density measured as people per sq.”

Empirical Analysis

“The estimated results show that income inequality has a positive and statistically significant effect on environmental degradation in all models; [3, 5–8] at the 5% level in the long run.”

"The long run relationship between income inequality and environmental degradation is supported by the coefficient of the lagged error-correction term in all models, which is negative and statistically significant, as expected."

"The both, long run and short run results suggest that higher level of environmental degradation is associated with a higher Gini coefficient (greater income inequality) in Pakistan."

"The estimated coefficients of per capita income in model [2] are found to be positive and statistically significant at the 1% level in both the long run and short run."

"Apart from model [6], the short run coefficients of GDP per capita are statistically significant at the 5% level in all models, suggesting the short run relationship exists between income and CO_2 emissions in Pakistan."

Conclusion

"This study evaluates the nexus between income inequality, economic growth and CO_2 emissions in Pakistan by using annual data for the period 1966–2011 through employing the Auto Regressive Distributed Lag (ARDL) bounds testing approach to co-integration to investigate the existence of long run relationships and EKC."

"The estimated results confirm a long run relationship between income inequality and environment degradation."

"The empirical results illustrate that there is a co-integrated and stable long run relationship among carbon emissions per capita (and its four sources), income inequality, GDP per capita, industry share and population density."

"The study recommends the need for sustainability of economic growth for policy makers to reduce the ongoing environmental degradation on one hand while on other hand, it suggest to encounter measures to exploit the rising income inequality in the country."

6.2.6 Beyond Economics! The (Evolving) Role of Law in the Eradication of Extreme Poverty [55]

This is a machine-generated summary of:

Arimoro, Augustine Edobor: Beyond Economics! The (Evolving) Role of Law in the Eradication of Extreme Poverty [55].

Published in: Journal of Human Rights and Social Work (2023).

Link to original: https://doi.org/10.1007/s41134-023-00247-2.

Copyright of the summarized publication:

The Author(s) 2023.

If you want to cite the papers, please refer to the original.

For technical reasons we could not place the page where the original quote is coming from.

Abstract-Summary

"Extreme poverty is a complex and multifaceted challenge that cannot be solely addressed through economic interventions."

"This has legal and human rights implications, particularly in regions such as Sub-Saharan Africa where extreme poverty is concentrated."

"In light of these concerns, this article critically examines the existing literature on poverty economics and law and presents an analysis of key data."

Introduction

"Ironically, however, linking prosperity with per capita gross domestic product (GDP) does not necessarily mean that it reflects most of the population's wealth or poverty level."

"While a high GDP may reflect that a country is prosperous, it may mispresent the level of poverty experienced by the ordinary citizens of that country."

"This article aims to discuss the (evolving role) of the law and legal systems around the world in the quest to eradicate extreme poverty."

"Although economists have and do play a core role in poverty reduction and eradication (Si and others, [56]), there is a key role for law and legal systems that should be exploited in countries with high poverty rates."

"The overarching question that this paper seeks to deal with is this: how might the law and legal systems promote prosperity in regions with high levels of poverty around the globe?".

Conceptual Analysis

"Those living in extreme poverty are estimated to be around 648 million as of 2019 (The World Bank, [57])."

"The number of people living in extreme poverty increased by an estimated 50 million because of the pandemic and global economic meltdown (Suckling and others, [58])."

"Sub-Saharan Africa (SSA) has the largest share of people living in extreme poverty estimated to be around 458 million compared to the rest of the world with 240 million people."

"The data reviewed for this study shows that 26 countries in SSA have seen the number of people living in extreme poverty increase significantly between the years 2010 and 2020."

"It remains to be noted that the determinant of extreme poverty increase/reduction above is based on economic measurements which may not necessarily reflect the true poverty situation of the people."

Extreme Poverty in Sub-Saharan Africa

"To deal with the problem of extreme poverty in SSA, it is pertinent to understand poverty in the context of the region."

"Of the paper, World Bank poverty data on SSA is analysed."

"The data in focus in this section are the percentage of the population in the respective SSA countries that live below the international poverty line (monetary poverty) and where they can be found."

"It, therefore, means that to deal with the challenge of extreme poverty in SSA, policies that focus on rural development should be key in addressing the challenge of extreme poverty."

"It is worthy of note that although there is a large concentration of people living in extreme poverty in SSA, a few countries in the region are not as burdened by the high population of persons living in extreme poverty as do the majority."

A Human Rights Approach

"The position of the UN, among others, is that the adoption of an HRBA in tackling extreme poverty should be anchored on the norms and values enunciated in international human rights law (United Nations [59, 60])."

"The norms set out in international human rights law require that States take their international human rights obligations into account when formulating and implementing policies affecting the lives of persons living in poverty."

"The United Nations Committee on Economic, Social and Cultural Rights (UNCESCR) states as follows: Anti-poverty policies are more likely to be effective, sustainable, inclusive, equitable and meaningful to those living in poverty if they are based on international human rights."

"Other international human rights instruments provide a clear framework for States to apply in their national strategies to eliminate extreme poverty."

A Development and Justice Approach

"Making information about access to justice available to people living in poverty is a key part of the access to justice approach (Carmona & Donald [61])."

"All countries that have agreed to the SDGs have pledged to "promote the rule of law at the national and international levels, and ensure equal access to justice for all," as part of SDG 16 (OECD [62])."

"Following the above, SSA countries can adopt a social welfare approach in dealing with extreme poverty."

"The argument here is that unless realistic standards align with the needs of the citizens of the Global South and especially SSA (where there is the largest concentration of people living in extreme poverty are found), more and more pyrrhic victories would be declared by the governments using questionable standards of measurements which suits the fancies of international organisations."

Conclusion and Recommendations

"The overarching discussion in this article is the role that law and the legal system can play in eliminating extreme poverty around the world and especially in the SSA region which has the highest concentration of people living in extreme poverty in the world."

"The article argues that the parameters used for assessing extreme poverty and extreme poverty based on economic patterns may not show the true picture of whether governments are doing enough or the right things to end extreme poverty."

"This article resonates with the view that extreme poverty is indeed a human rights violation."

"The article evaluates three approaches to dealing with poverty."

"The article recognises that the elimination of extreme poverty in the world is a task that must be done before the year 2030."

"Be that as it may, the article analyses data that shows that SSA accounts for the largest concentration of people living in extreme poverty compared to other regions of the world."

Endnotes

"[1] 'Poverty and the International Covenant on Economic, Social and Cultural Rights' Statement adopted on 4 May 2001 by the Committee on Economic, Social and Cultural Rights (E/C 12/2001/10) para 3."

"[9]Resolution adopted by the General Assembly on 17 December 2018."

6.2.7 Industrialization, Urbanization, and Carbon Emission Efficiency of Yangtze River Economic Belt—Empirical Analysis Based on Stochastic Frontier Model [63]

This is a machine-generated summary of:

Zhang, Caiqing; Chen, Panyu: Industrialization, urbanization, and carbon emission efficiency of Yangtze River Economic Belt—empirical analysis based on stochastic frontier model [63].

Published in: Environmental Science and Pollution Research (2021).

Link to original: https://doi.org/10.1007/s11356-021-15309-z1.

If you want to cite the papers, please refer to the original.

For technical reasons we could not place the page where the original quote is coming from.

Abstract-Summary

"Carbon emission efficiency directly determines the level of green economic development."

"Based on the panel data of China's Yangtze River Economic Belt (YEB) from 2008 to 2017, this paper uses the stochastic frontier analysis (SFA) model to analyze the overall carbon emission efficiency level, influencing factors, and changing trends, with a view to discussing the relationship between economic development and carbon emission efficiency."

"The results suggest, first, the overall carbon emission efficiency of the YEB is on an upward trend, but there is still much room for improvement."

"As industrialization and urbanization progress, the impact on carbon emission efficiency shows a downward and then upward trend."

"Efforts to improve carbon emission efficiency in the YEB should focus on transforming the economic growth model, adjusting the industrial structure, improving the energy consumption structure, and innovating green technology."

Introduction

"Some studies have explored the impact of industrialization and urbanization on carbon emission efficiency."

"Li and others [64] analyzed the carbon emission data of 30 provinces in China from 1990 to 2010 and concluded that urbanization has the most significant impact on carbon emissions, while industrialization is not the main factor."

"Other scholars have conducted carbon emission efficiency studies on different urban agglomerations in China, such as Beijing-Tianjin-Hebei region (Wang and others [65]), Guangdong-Hong Kong-Macao Greater Bay area (Lin and Li [66]), etc. Considering the above background, based on the panel data from 2008 to 2017, this paper uses the SFA model to study the 10-year data of 11 provinces in the YEB."

"This paper selects industrialization level, urbanization level, technological progress, energy consumption structure, government intervention, and foreign trade as environmental variables, and quantifies the impact of each variable through the model [5]. To deeply study the impact of industrialization and urbanization on carbon emission efficiency, the square of industrialization and urbanization are introduced as new variables [6]. Most studies stick to national or provincial data."

Study Area

"It is necessary to know how industrialization and urbanization affect YEB's carbon emissions."

"The secondary industry accounts for a relatively high proportion, especially Anhui and Jiangxi, indicates that industry is an important way for the development of the YEB."

"The urbanization level of the YEB presents a ladder-like trend of high in the east and low in the west."

"Shanghai's urbanization rate is close to 90%, exceeding the average level of major developed countries in the world, Jiangsu and Zhejiang are also at a relatively high level."

"Other provinces have relatively low urbanization levels, especially Guizhou, where the urbanization rate is 15% lower than the national average, the scale of urbanization is smaller and development is backward."

Methods

"This paper selects Battese and Coelli [67] as the research model, which can not only study the carbon emission efficiency, but also analyze the influencing factors."

"Carbon emission efficiency is influenced by various factors and the degree of influence is different."

"This paper selects industrialization level, urbanization level, technological progress, energy consumption structure, government intervention, and foreign trade as environmental variables [2]. Industrialization level."

"The industrial structure reflects the level of economic development in a region and is closely related to carbon emission efficiency."

"The urbanization level affects carbon emission efficiency [5]. Technological progress."

"Foreign trade will promote technological efficiency, which will have a positive impact on carbon emission efficiency."

"Put the six influencing factors mentioned above into the model to build model 1, expressed as follows: The impact of each environmental variable on carbon emission efficiency is complex, which may be linear or double-sided."

Results and Discussion

"The coefficient Z_{11} is -0.792 in model 2, indicating that the industrialization level has a U-shaped effect on carbon emission efficiency."

"It is suggested in model 1 that industrialization level restrains the carbon emission efficiency of the YEB, indicating that the effect of industrialization on efficiency is in the left half of the U-shaped function, and the level of industrialization still needs to be improved."

"This paper uses the DEA-BCC model (carbon emission as input) and the SBM-DEA model (carbon emission as undesired output) to measure the carbon emission efficiency of 11 provinces and compared with the results of the SFA model."

"The test results show that $Z = -9.104$, $P < 0.001$ (SFA and BCC), $Z = -8.954$, $P < 0.001$ (SFA and SBM), rejecting the null hypothesis at the 1% significance level, indicating that the carbon emission efficiency calculation result of the DEA model is higher than that of the SFA model."

Conclusion and Policy Recommendations

"To study the impact of industrialization and urbanization on the carbon emission efficiency of the YEB, the data from 2008 to 2017 are selected, and first, the current status of carbon emissions in the YEB is analyzed."

"The SFA model is used to study and measure the carbon emission efficiency of 11 provinces, and the influence of industrialization, urbanization, and other factors is analyzed."

"The driving effect of industrialization and urbanization on economic development exceeds their carbon emission effect, and carbon emission efficiency is gradually increasing."

"The use of these clean energy sources can promote carbon emission efficiency [3]. For the urbanization Optimize population structure."

"The government should increase investment in low-carbon technologies through special funds, financial allocations, etc. Promoting technological innovation and development through key technology research, to improve energy efficiency [6]. In addition, reasonably develop foreign trade and scientifically introduce foreign capital can also improve carbon emission efficiency."

6.2.8 The Relationship Between FDI, CO_2 Emissions, and Energy Consumption in Asia–Pacific Economic Cooperation Countries [68]

This is a machine-generated summary of:

Balli, Esra; Sigeze, Ciler; Ugur, Mehmet Sedat; Çatık, Abdurrahman Nazif: The relationship between FDI, CO_2 emissions, and energy consumption in Asia–Pacific economic cooperation countries [68].

Published in: Environmental Science and Pollution Research (2021).

Link to original: https://doi.org/10.1007/s11356-021-17494-3.

If you want to cite the papers, please refer to the original.

For technical reasons we could not place the page where the original quote is coming from.

Abstract-Summary

"This paper investigates the relationship between CO_2 emissions, energy consumption, economic growth, and foreign direct investment for a sample of Asia–Pacific

Economic Cooperation Countries (APEC) countries from 1981:Q1 to 2021:Q1 employing panel data methodology."

"The cointegration test developed by Westerlund [69] reveals a long-run equilibrium between CO_2 emissions, energy consumption, economic growth, and foreign direct investment."

"The cointegration test results also show that the impact of Gross Domestic Product (GDP) on CO_2 emissions varies by country in the estimation sample."

"To the mixed evidence on the effects of other variables, the increase in energy consumption is positively and significantly affecting CO_2 emissions in all APEC countries."

"Emirmahmutoglu and Kose Econ Model 28:870–876, [70]'s panel causality test results show a bidirectional relationship between FDI and CO_2 emissions in Japan."

Introduction

"The question of whether APEC can boost economic growth by consuming more energy without harming the environment becomes critical, making those countries an interesting case study for examining the role of FDI in the nexus of economic growth, energy, and environment."

"Although FDI helps host countries' economic development, it can also lead to environmental degradation by spreading industrial activity and encouraging the use of polluting industrial goods."

"According to the pollution haven hypothesis, increasing FDI flows has a detrimental effect on the ecosystem of host countries where the regulatory regimes are less stringent or non-existent, as supported by several empirical studies (Gray [71]; Xing and Kolstad [72]; Mihci and others [73]; He [74]; Wagner and Timmins [75]; Tang [76]; Solarin and others [77])."

"Given the above background, the main objective of this article is to comprehend the relationship between economic growth, carbon dioxide (CO_2) emissions, energy consumption, and FDI for APEC countries."

Literature Review

"The relationship between environmental degradation, energy consumption, foreign direct investment, and economic growth has been investigated extensively in the literature in different countries utilizing different methodologies."

"Despite the fact that many studies have been conducted to investigate the relationship between environmental pollution and income, the results are still inconclusive due to differences between countries or the sample periods analyzed."

"The growth hypothesis is based on the assumption of unidirectional causality, implying that energy consumption positively impacts economic growth."

"Zhang and Cheng [78]'s study reveals a unidirectional causal relationship from growth to energy consumption for China."

"The empirical studies including Erdal and others [79] for Turkey, Apergis and Payne [80] for CIS countries, Belloumi [81] for Tunisia, Dagher and Yacoubian [82] for Lebanon, Zhixin and Xin [83] and Wang and others [84] for China, Magazzino [85] for Italy, Antonakakis and others [86] for 106 countries, and Rahman [87] for BRICS and ASEAN countries confirm the bidirectional causal relationship."

"The neutrality hypothesis indicates that economic growth and energy consumption do not have a causal relationship."

"Using the Toda-Yomamota causality test, Gokmenoglu and Taspinar [88] discover a bidirectional causal relationship between CO_2 emissions, FDI, and energy consumption."

Model and Data

"The main objective of this article is to examine the long-run relationship among CO_2 emissions, energy consumption, economic growth, and foreign direct investment in APEC countries, namely, Australia, Canada, China, Indonesia, Japan, Malaysia, Mexico, Philippines, Singapore, Thailand, and the USA."

"In contrast to previous studies that used annual time series, this paper utilizes quarterly data from the Refinitiv Eikon Datastream database on CO_2 emissions, primary energy consumption, economic growth, and foreign direct investment from 1981:Q1 to 2021:Q1 (Refinitiv Eikon Datastream [89])."

"The mean of CO_2 emissions is highest for the USA, followed by China and Japan."

"The mean value of the energy consumption variable is highest for the USA, followed by China and Japan."

"Singapore seems to have the lowest average energy consumption among the selected APEC countries."

Methodology

"As a solution to this problem, for (N, T) $\rightarrow \infty$, the existence of cross-sectional dependence among the variables of APEC countries is examined with cross-sectional dependence (CD) test proposed by Pesaran [90]."

"The CD statistic testing the null hypothesis of zero dependence across the panel unit is computed as follows: where N and T denote the number of cross-sections and the estimation period, respectively."

"CIPS statistic is computed by taking the average of individual CADF statistics based on the following formula: After determining the cross-sectional dependence and order of integration of variables, we used Westerlund [69] cointegration technique to determine the long-run relationship between variables under the cross-sectional dependence."

"To control heterogeneity and cross-sectional dependence in a panel model, Emirmahmutoglu and Kose [70] developed a panel Granger causality test."

Empirical Results and Discussion

"For all countries, including Indonesia and the Philippines, the elasticity of energy consumption to CO_2 emissions is positive, indicating that increasing energy consumption contributes to CO_2 emissions."

"The findings show that energy consumption has a positive and substantial effect on CO_2 emissions; a 1% increase in energy consumption leads to a 0.802% increase in the APEC nations' CO_2 emissions."

"A unidirectional relationship running from GDP to CO_2 emissions is evidenced for the Philippines, Singapore, Japan, and Thailand, indicating that a rise in GDP may increase CO_2 emissions in these countries."

"The findings of the study indicate a unidirectional causality from GDP to energy consumption for Indonesia, Mexico, the Philippines, and Thailand, supporting the conservation hypothesis."

"The findings of the causality between FDI and CO_2 emissions support the pollution haven hypothesis for overall panel countries."

Conclusions

"Using panel data methodology, this research explored the relationship between CO_2 emissions, energy consumption, economic growth, and foreign direct investment for a sample of APEC countries from 1981:Q1 to 2021:Q1."

"The panel cointegration test results indicate that the variables are cointegrated, and there exists a long-run relationship between the variables in APEC countries."

"PMG results also show that energy consumption was found to have a positive and significant effect on CO_2 emissions both in the long and short run."

"Due to the differences in the economic structures of APEC countries, the impacts of energy consumption, economic growth, and foreign direct investment on CO_2 emissions differ in a cross-country analysis."

"Our findings indicate that foreign direct investment and energy consumption contribute to CO_2 emissions in a subset of APEC nations."

6.2.9 Finance-Growth-Poverty Nexus: A Re-assessment of the Trickle-Down Hypothesis in China [91]

This is a machine-generated summary of:

Ho, Sin-Yu; Iyke, Bernard Njindan: Finance-growth-poverty nexus: a re-assessment of the trickle-down hypothesis in China [91].

Published in: Economic Change and Restructuring (2017).

Link to original: https://doi.org/10.1007/s10644-017-9203-8.

If you want to cite the papers, please refer to the original.

For technical reasons we could not place the page where the original quote is coming from.

Abstract-Summary

"Economic growth and financial development are two such crucial tools for overcoming poverty, as frequently pointed out by economists."

"These tools work through the so-called trickle-down hypothesis, which contends that a well-functioning financial system would enhance poverty reduction by promoting economic growth."

"The empirical test of the trickle-down hypothesis for China is scant."

"We utilized two standard proxies for financial development, namely: the domestic credit to private sector by banks as percentage of GDP, and money and quasi money as percentage of GDP; annual percentage change in real GDP per capita to proxy economic growth; and a standard proxy for poverty reduction namely: the household final consumption expenditure per capita growth."

"By accounting for structural breaks in our empirical specifications, we found overwhelming support for the trickle-down hypothesis at the national level."

"We found financial development to cause economic growth, which in turn causes poverty reduction in China at the national level."

Introduction

"These statistics show clearly that financial development and economic growth have been vital for poverty reduction in China."

"The few studies that attempted to examine the role of financial development and economic growth in poverty reduction in China are Ding and others [92], Wen and others [93], and Park and others [94]."

"Against this background, this paper seeks to empirically assess the causal linkages between financial development, economic growth, and poverty reduction in China by controlling for structural breaks."

"We utilized two standard proxies for financial development, namely: the domestic credit to private sector by banks as percentage of GDP, and money and quasi money (M2) as percentage of GDP (weighted average); annual percentage change in real GDP per capita to proxy economic growth; and a standard proxy for poverty reduction namely: the household final consumption expenditure per capita growth (annual percentage)."

"We found financial development to cause economic growth, which in turn causes poverty reduction in China at the national level."

"We present an overview of financial development, economic growth and poverty reduction in China."

Financial Development, Economic Growth, and Poverty Reduction in China

"This period also witnessed other impressive financial reforms such as further liberalization of the interest rate, and the establishment of China Banking Regulatory Commission in 2003 (Hasan and Zhou [95]; Turner and others [96]; Ye and others [97]; Zhang and others [98])."

"To the banking and financial sector reforms, other major economic reforms have been crucial in driving the sustained and extraordinarily high growth levels in China."

"The fundamental institutional change was the land reform, characterized by the implementation of the household responsibility system in rural China which greatly stimulated farmers' incentive for economic activity."

"Regarding the economic growth, the Chinese economy has grown at a record-setting rate of 8.86% annually during the period of 1985–2014, due to the financial sector reforms and other major economic reforms we have discussed in the previous section (WDI [99])."

Review of the Related Literature

"Given the theoretical implication of financial development on economic growth and poverty reduction, there have been an increasing number of studies examining the relationship among the three."

"Some of the country-specific studies that have attempted to examine the relationship between financial sector development and poverty reduction are Abosedra and others [100], Sehrawat and Giri [101], Uddin and others [102], Inoue and Hamori [103], Ding and others [92], Geda and others [104], Burgess and Pande [105], Quartey [106], Wen and others [93], Husain [107], and Park and others [94], among others."

"Sehrawat and Giri [101], while examining the relationship between financial sector development and poverty reduction in India during 1970–2012, find financial development and economic growth to reduce poverty in the short and the long run."

"In a very recent study, Sehrawat and Giri [108] examine the contribution of financial development to poverty reduction in 11 South Asian developing countries using panel data set over the period of 1990–2012."

"There are China-specific studies that do not examine the direct link between financial development and poverty reduction but offer insights into the linkages between financial development and growth."

Methodology and Data

"We utilize the Augmented Dickey-Fuller (ADF), the Dickey-Fuller generalized least squares (DF-GLS), and the Perron tests to examine the unit root or stationarity properties of financial development, economic growth and poverty."

"We use a modified unrestricted vector error correction model (UECM) to test for causal relations among financial development, economic growth and poverty."

"By previous studies (see, for example, Abosedra and others [100]), the conventional approaches for testing for cointegrating relationships assume that the relationship between variables are stable over time."

"We avoid this pitfall by modifying the ARDL bounds testing approach to account for structural breaks and using this approach to test for cointegrating relationships among the variables."

"One desirable property of the ARDL bounds testing approach to cointegration is that once we have established the existence of cointegrating relationships among the variables, it is straight forward to derive the associated error correction model without loss of information."

Empirical Results

"For the first model (i.e. the model which uses MC as the proxy for financial development), there is evidence of cointegrating relationships among poverty reduction, financial development and economic growth, if poverty reduction is the dependent variable."

"For the second model (i.e. the model which uses DC as the proxy for financial development), there is evidence of cointegrating relationships among poverty reduction, financial development and economic growth, if poverty reduction and financial development are the dependent variables."

"To test for joint short-long-run causal relationships, we must perform a joint test of statistical significance of both the lagged error correction term and the first differenced explanatory variables."

"The results show bidirectional causality between financial development and economic growth in the short run, regardless of the whether financial development is proxied by domestic credit to private sector by banks as percentage of GDP or money and quasi money (M2) as percentage of GDP."

Conclusion and Policy Implications

"We had one aim, which was to examine whether financial development led to economic growth, which in turn facilitated poverty reduction in China."

"A country that has seen a rapid fall in poverty alongside a rapid rise in financial development and economic growth is China, making the country more suitable for investigating the trickle-down hypothesis."

"At the national level, financial development and economic growth are pro-poor, in the case of China."

"A broad range of financial, institutional, and structural reforms has led to expansive financial sector development and economic growth in China during the past three decades."

"In the area of governance, financial development and economic growth would only continue to be pro-poor as it is if China adheres to transparent governance and accountability in order to effectively consolidate its recent gains and channel them to the appropriate sectors of the economy."

6.2.10 An Investigation of the Relationship Among Economic Growth, Agricultural Expansion and Chemical Pollution in Iran Through Decoupling Index Analysis [109]

This is a machine-generated summary of:

Najafabadi, Mostafa Mardani; Mirzaei, Abbas; Laskookalayeh, Somayeh Shirzadi; Azarm, Hassan: An investigation of the relationship among economic growth, agricultural expansion and chemical pollution in Iran through decoupling index analysis [109].

Published in: Environmental Science and Pollution Research (2022).

Link to original: https://doi.org/10.1007/s11356-022-21004-4.

If you want to cite the papers, please refer to the original.

For technical reasons we could not place the page where the original quote is coming from.

Abstract-Summary

"The use of decoupling index to decouple the link between agricultural production and pollution caused by the consumption of chemical inputs, such as fertilizers and pesticides, has been emphasized; Therefore, in the present study, the decoupling index first is calculated in relation to the emission of pollution caused by the use of chemical inputs in the process of agricultural production during the period of 1991–2016 in Iran."

"The results showed that in the recent years, pollution indicators in relation to chemical inputs have not had ideal trends, and despite the further growth of agricultural production, the quality of the environment has experienced a declining trend."

"The results of the decoupling index related to the use of chemical pesticides and fertilizers in Iran show that during a period of 26–year, only 5 and 4 years of using these inputs have had a sustainable state compared to the production growth; besides, a strong negative decoupling state occurred as the most unsustainable state in relation to chemical fertilizer for 7 years."

"Among the factors affecting the decoupling index, the value-added variable of the agricultural sector has had the most positive effect on this index, and thus, in the long run, it increases the level of pollution in the agricultural sector."

"Adopting effective strategies to improve resource efficiency, planning for the implementation of biotechnological methods, and doing investment for creating green infrastructure in the agricultural sector can be effective in the ideal decoupling of pollution and agricultural economy growth in Iran."

Introduction

"In spite of the high share of the agricultural sector in the spread of pollution and the appropriateness of this method assessing the relationship between the development of the agricultural economy and pollution in this industry, there is a paucity of research in this area (Li and others [110, 111]); besides, in the previous studies, carbon emission have been generally considered as a measure of environmental pollution."

"There is a wide variety of literature on the relationship between economic growth factors, agricultural production, and carbon emissions as an indicator of pollution (Ali and others [112])."

"According to afore mentioned points, the first goal in the current study is to evaluate the relationship between the growth of agricultural production value and pollution caused by the production process in this sector in Iran using agricultural production data and the use of chemical inputs as an indicator of pollution."

Materials and Methods

"In the first phase of the present study, the decoupling index of the relationship between agricultural production growth and pollution caused by the production process in this sector in Iran was first evaluated using the Tapio decoupling model."

"The causal relationship between the decoupling index and the variables affecting that in the agricultural sector was investigated using ARDL model."

"The relationship between the variables of GDP, value added of the agricultural sector, and the area under cultivation of cereal of decoupling index in the agricultural sector of Iran has been investigated."

"It is noteworthy that as the amount of carbon emissions per kilogram of chemical toxins is more than 5.5 times of the amount of chemical fertilizers (West and Marland [113]), to investigate the relationship between chemical toxin and fertilizers' decoupling index in this study, the decoupling index is aggregated by weight average and by giving more weight to the decoupling index, which reflects the consumption status of chemical pesticides and the value of per capita production."

Results

"Except for the coefficient of the variable of value added of the agricultural sector, other estimated coefficients are negative in the short run and indicate a decrease in the decoupling index; therefore, in the short-run GDP, it does not have a significant effect on the decoupling index."

"If the area under cereal cultivation increases, the decoupling index would decrease by the next year; besides, similar to the long-term case, the effect of value added of the agricultural sector on the decoupling index in the short run is positive."

"The estimated ECT coefficient in the model is statistically significant and indicates a relatively low adjustment rate; in fact, according to the coefficient of error correction term, it can be said that in each period, in the agricultural sector, 14% of the variable deviations of the decoupling index disappears, and in case of a shock to the model, short-term equilibrium is approached to long-term equilibrium after more than 7 years."

Discussion

"Examining the long-term model for the estimated model, all variables had a positive and significant effect on the decoupling index; so increasing economic growth, area under cereal cultivation, and value added of the agricultural sector in the long run would lead to an increase in the decoupling index, and thus, the pollution caused by the use of chemical inputs is increased."

"Considering the positive and significant coefficient of GDP variable per capita in the long run, it can be concluded that economic growth in Iran would increase the decoupling index in the agricultural sector and, thus, would reduce the quality of the environment; Therefore, among the variables affecting the changes in the decoupling index in recent years, the second hypothesis of the present study, in which the value added of the agricultural sector will have a greater impact on the rate of this index, is accepted."

Conclusion and Policy Implications

"The decoupling index in relation to chemical inputs has not had an ideal trend; despite the overall growth of Iran's agricultural economy, the quality of the environment has decreased in most years."

"With the growth of value added in the agricultural sector, the decoupling index is strengthened, and as a result, the quality of the environment decreases; therefore, reducing pollution shall be a priority for Iran's agricultural environmental policies."

"Investing in building green infrastructure in the agricultural sector and moving towards the application of low carbon technologies can help coordinate economic growth and environmental quality in Iran."

"Policy-making and planning for the implementation of biotechnological methods with the aim of reducing pollution in the agricultural sector shall be a priority for Iran's environmental policies; besides, the use of a set of incentive and punishment tools making agricultural producers to maintain the quality of the environment can be helpful."

Limitation and Future Research

"Owing to the limitations of the data, only the use of chemical inputs was considered as a polluting variable in relation to the decoupling index in Iran; but considering more indicators such as the use of plastic in agriculture or energy consumption in different stages of crop production can be effective for better depiction of the situation of pollution and production growth."

"Despite the high potentiality of the decoupling index in assessing the balance between production growth and pollution in the agricultural sector, few studies have been conducted in this field; On the other hand, more research is needed to identify and examine the variables affecting changes in the agricultural sector's decoupling index as a comprehensive environmental indicator."

6.2.11 Investigating the Four-Way Linkages Between Energy Consumption, CO_2 Emissions, Exports, and Economic Growth: New Evidence from SAARC Economies [114]

This is a machine-generated summary of:

Ul Habib, Musavir; Tiba, Sofien; Gaies, Brahim; Jahmane, Abderrahman: Investigating the four-way linkages between energy consumption, CO2 Emissions, exports, and economic growth: new evidence from SAARC economies [114].

Published in: Environmental Economics and Policy Studies (2022).

Link to original: https://doi.org/10.1007/s10018-022-00351-z.

If you want to cite the papers, please refer to the original.

For technical reasons we could not place the page where the original quote is coming from.

Abstract-Summary

"This paper examines the relationship between energy consumption and economic growth, and their mutual influences on CO2 emissions and exports in four SAARC (South Asian Association for Regional Cooperation) member countries, Bangladesh, India, Pakistan, and Sri Lanka, from 1971 to 2014."

"We show that energy consumption and exports drive economic growth in the long run while CO2 emissions stifle it."

"The study provides strong evidence of short-run links between GDP per capita and energy, CO2 emissions and energy consumption, and GDP per capita and CO2 emissions."

Introduction

"In this light, assessing the relationship between economic development, energy consumption and pollution is important for developing economies such as those in the SAARC, as they are in the transition phase of their economic development from low-income to middle-income countries."

"It is important to examine the nature of the link between economic growth and CO2 emissions in SAARC countries, taking into account their energy consumption and export specialization."

"This paper contributes to the literature by unifying and extending four broad strands of studies that support different hypotheses regarding the environmental effects of economic growth and its drivers in developing countries, namely the growth hypothesis assuming that energy consumption causes economic growth (e.g., Mohanty and Chaturvedi [115]), the conservation hypothesis assuming that energy conservation does not hinder economic growth (the feedback/bidirectional hypothesis stipulating a two-way linkage between growth and energy consumption and its effects (e.g., Ben Lahouel and others [116] and the neutrality hypothesis that supports that no relationship exists between economic growth and energy consumption and its environmental consequences (Phiri and Bothwell [117])."

"Investigating the short- and long-run interactions between energy consumption, exports, GDP, and CO2 emissions in Bangladesh, India, Pakistan, and Sri Lanka helps to better understand which hypothesis is more plausible in the SAARC countries."

A Brief Literature Review

"Scholars have highlighted that the scarcity of energy resources exerts a restrictive impact on economic growth, especially in industrialized economies (Tiba and others [118]; Ou and others [119]; Bayramoglu and Yildrim [120])."

"This suggests that any disruption in energy supply will disrupt economic growth."

"Over the past three decades, several empirical studies have been conducted to explain the relationship between economic growth and energy use, but they have not yielded conclusive results."

"The extensive literature on energy and economic growth has structured this link into four hypotheses: (i) the growth hypothesis, assuming that energy consumption causes economic growth, as suggested by Lee and Chang [121], Ang [122], Behara [123], and Mohanty and Chaturvedi [115]; (ii) the conservation hypothesis, assuming that energy conservation does not hinder economic growth, as shown by Sari and others [124], Jamil and Ahmad, [125], Hamdi and others [126], and Raza and others (2016); (iii) the feedback/bidirectional hypothesis, assuming that energy consumption promotes economic growth, and economic growth increases energy consumption, as highlighted by Tiba and Frikha [127], Tiba [128], Amar [129], Kurniawan and others [130], Ben Lahouel and others [116]; and finally (iv) the neutrality hypothesis, indicating no significant link between energy consumption and economic growth, as shown by Halicioglu [131], Ozturk and Acaravci [132], Payne and Taylor [133], Bozoklu and Yilanci [134], and Phiri and Bothwell [117]."

"Investigating the four-way links between energy consumption, CO2 emissions, exports, and economic growth using sophisticated econometric techniques, including the Westerlund [135] cointegration test and the CCE approach, is another contribution of this paper compared to the methodologies employed in previous studies."

Model Specification and Data

"Energy consumption is measured in per capita kilograms of oil equivalent."

"Economic growth is indicated by GDP per capita in constant 2015 US$."

"We empirically examine the interactions between energy consumption and economic growth by including exports as a driver of economic performance and the pollution level for a panel of four Asian economies."

"α and β denote the marginal impact of CO2 emissions and total exports on economic growth in the SAARC countries."

"The decomposition of the residual term in the model above yields the following equation, representing the relationship between GDP per capita and energy consumption for SAARC-4 nations."

Econometric Methodology

"The equation of the Augmented Dickey-Fuller (ADF) test adopted by Levin, Lin, and Chu (LLC) can be expressed as follows: The Pedroni [136–138], Johansen Fisher [139], and Kao [140] tests of cointegration are used to check the long-run relationships between variables."

"We employ Westerlund's [135] panel cointegration approach, which addresses the problem of cross-sectional dependency."

"The long-run coefficients of the cointegration equations are calculated using the ordinary least squares method (OLS) and the dynamic ordinary least squares method (DOLS) estimation developed by Pedroni [141]."

"To deal with cross-sectional dependence, heterogeneity, heteroskedasticity, and serial correlation (Phillips and Sul [142]), Pesaran [143] recommended the CCE approach – l which was later improved by Kapetanios and others [144] and Chudik and others [145] (CCE-MG) – to generate long-term estimators that take into account cross-sectional dependence."

"Even in the presence of error correlation, unit-roots in the variables, and cross-sectional dependency, the estimates remain consistent (Chudik and others [145])."

Empirical Analysis

"We reveal a bidirectional causality between exports and GDP, a unidirectional causality from per capita GDP to energy consumption and from CO2 emissions to energy consumption."

"These results imply that GDP growth and CO2 emissions lead to an increase in energy use in the short run, in line with the conservation hypothesis."

"No causality is detected between energy consumption and exports, nor between exports and CO2 emissions in the short run."

"Row 2 shows that per capita GDP significantly affects energy consumption, confirming the results of the cointegration and the causality tests."

"Our results reveal the following empirical facts: (a) CO2 emissions negatively affect per capita GDP; (b) the increase in energy use is enhanced by per capita GDP; (c) CO2 emissions are increased by the heightened demand for energy, and (d) exports increase as long as per capita GDP does."

Conclusion and Policy Implications

"In this backdrop, this study empirically estimates the relationship between energy consumption, economic growth, CO2 emissions, and exports in four selected SAARC countries."

"The results suggest that an increase in energy consumption considerably boosts economic growth, as shown by the long-run DOLS and OLS estimations."

"The policy implications arising from this study confirm the active role of energy consumption as a determinant factor of economic growth in the SAARC-4 as these economies are large consumers of non-renewable energy resources."

"By taking into consideration the important role played by energy in enhancing economic activity, due to the depletion of non-renewable energy resources, policymakers should develop renewable energy infrastructures for their sustainable development."

"Although considering the substantial role of energy in fueling economic growth, these economies should impose energy taxes to curb pollution and introduce mandatory fuel efficiency standards to make their economic development more sustainable."

Bibliography

1. Berg, Andrew G.; Ostry, Jonathan D. Inequality and Unsustainable Growth: Two Sides of the Same Coin?. IMF Economic Review (2017). https://doi.org/10.1057/s41308-017-0030-8
2. Chaudhuri, S. and M. Ravallion, 2006, "Partially Awakened Giants: Uneven Growth in China and India," in Dancing with Giants: China, India, and the Global Economy, ed. by L. A. Winters and S. Yusuf. World Bank, Washington
3. Pritchett, L., 2000, "Understanding Patterns of Economic Growth," World Bank Economic Review, Vol. 14, No. 2, pp. 221–50
4. Khan, Syed Abdul Rehman The nexus between carbon emissions, poverty, economic growth, and logistics operations-empirical evidence from southeast Asian countries. Environmental Science and Pollution Research (2019). https://doi.org/10.1007/s11356-019-04829-4
5. Lau LS, Choong CK, Eng YK (2014) Investigation of the environmental kuznets curve for carbon emissions in Malaysia: do foreign direct investment and trade matter? Energy Policy 68:490–497
6. Aldakhil AM, Nassani AA, Awan U, Abro MMQ, Zaman K (2018) Determinants of Green logistics in BRICS countries: an integrated supply chain model for Green business. J Clean Prod 195:861–868
7. Zaman K, Shamsuddin S (2017) Green logistics and national scale economic indicators: evidence from a panel of selected European countries. J Clean Prod 143:51–63
8. Zhu Q, Sarkis J (2004) Relationships between operational practices and performance among early adopters of green supply chain management practices in Chinese manufacturing enterprises. J Oper Manag 22:265–289
9. Khan SAR, Dong Q (2017) Does national scale economic and environmental indicators spur logistics performance? Evidence from UK. Environ Sci Pollut Res 24(34):26692–26705
10. Zailani S, Govindan K, Iranmanesh M, Shaharudin MR, Chong YS (2015) Green innovation adoption in automotive supply chain: the Malaysian case. J Clean Prod 108:1115–1122
11. Benitez-Amado J, Llorens-Montes FJ, Fernandez-Perez V (2015) IT impact on talent management and operational environmental sustainability. Inf Technol Manag 16(3):207–220
12. Büyüközkan G, çifçi G (2012) Evaluation of the green supply chain management practices: a fuzzy ANP approach. Prod Plan Control 23(6):405–418
13. Zhu Q, Sarkis J, Lai KH (2012) Examining the effects of green supply chain management practices and their mediations on performance improvements. Int J Prod Res 50(5):1377–1394

14. Benitez J, & Walczuch R (2011) IT, proactive environmental strategy and firm performance: a resource-based analysis. Proceedings of the 17th Americas Conference on Information Systems, Michigas, USA: Detroit, pp 1–10
15. Chien MK, Shih LH (2007) An empirical study of the implementation of green supply chain management practices in the electrical and electronic industry and their relation to organizational performances. Int J Environ Sci Technol 4(3):383–394
16. Green KW Jr, McGaughey R, Casey KM (2006) Does supply chain management strategy mediate the association between market orientation and organizational performance? SCMIJ 11(5):407–414
17. Zhang H, Yang F (2016) On the drivers and performance outcomes of green practices adoption: an empirical study in China. Ind Manag Data Syst 116(9):2011–2034
18. Christmann P (2000) Effects of "best practices" of environmental management on cost advantage: the role of complementary assets. Acad Manag J 43(4):663–680
19. Ibrahim MH, Law SH (2014) Social capital and CO2 emission—output relations: a panel analysis. Renew Sust Energ Rev 29:528–534
20. Masron TA, Subramaniam Y (2019) Does poverty cause environmental degradation? Evidence from developing countries. J Poverty 23(1):44–64
21. Khan SAR, Qianli D, SongBo W, Zaman K, Zhang Y (2017) Environmental logistics performance indicators affecting per capita income and sectoral growth: evidence from a panel of selected global ranked logistics countries. Environ Sci Pollut Res 24(2):1518–1531
22. Arellano M, Bond S (1991) Some tests of specification for panel data: Monte Carlo evidence and an application to employment equations. Rev Econ Stud 58(2):277–297
23. Borin N, Mullikin J, Krishanan R (2013) An analysis of consumer reaction to green strategies. J Prod Brand Manag 22(2):118–128
24. Ahmad, Munir; Muslija, Adnan; Satrovic, Elma Does economic prosperity lead to environmental sustainability in developing economies? Environmental Kuznets curve theory. Environmental Science and Pollution Research (2021). https://doi.org/10.1007/s11356-020-12276-9
25. Lieb CM, (2003). The environmental Kuznets curve: a survey of the empirical evidence and of possible causes, Discussion Paper Series, No. 391, University of Heidelberg, Department of Economics, Heidelberg
26. Halliru AM, Loganathan N, Golam Hassan AA, Mardani A, Kamyab H (2020) Re-examining the environmental kuznets curve hypothesis in the economic community of West African states: a panel quantile regression approach. J Clean Prod 276:124247. https://doi.org/10.1016/j.jclepro.2020.124247
27. Dong K, Sun R, Li H, Liao H (2018) Does natural gas consumption mitigate CO2 emissions: Testing the environmental Kuznets curve hypothesis for 14 Asia-Pacific countries. Renew Sust Energ Rev 94:419–429. https://doi.org/10.1016/j.rser.2018.06.026
28. Gormus S, Aydin M (2020) Revisiting the environmental Kuznets curve hypothesis using innovation: new evidence from the top 10 innovative economies. Environ Sci Pollut Res 27:27904–27913. https://doi.org/10.1007/s11356-020-09110-7
29. Dumitrescu E-I, Hurlin C (2012) Testing for granger non-causality in heterogeneous panels. Econ Model 29(4):1450–1460
30. OECD, (2018). OECD ECONOMIC SURVEYS: TURKEY 2018
31. Jalandoni A, (2018). Greenhouse gas emissions in PH rising: report, ABS-CBN News
32. Jabeen G, Ahmad M, Zhang Q (2021) Perceived critical factors affecting consumers' intention to purchase renewable generation technologies : rural-urban heterogeneity. Energy 218:119494. https://doi.org/10.1016/j.energy.2020.119494
33. Capellán-Pérez, Iñigo; Mediavilla, Margarita; de Castro, Carlos; Carpintero, Óscar; Miguel, Luis Javier More growth? An unfeasible option to overcome critical energy constraints and climate change. Sustainability Science (2015). https://doi.org/10.1007/s11625-015-0299-3

34. Capellán-Pérez I, Mediavilla M, de Castro C et al (2014a) Fossil fuel depletion and socio-economic scenarios: an integrated approach. Energy 77:641–666. https://doi.org/10.1016/j.energy.2014.09.063
35. Ayres RU (2006) Turning point: the end of exponential growth? Technol Forecast Soc Change 73:1188–1203. https://doi.org/10.1016/j.techfore.2006.07.002
36. D'Alessandro S, Luzzati T, Morroni M (2010) Energy transition towards economic and environmental sustainability: feasible paths and policy implications. J Clean Prod 18:532–539. https://doi.org/10.1016/j.jclepro.2010.02.022
37. García-Olivares A, Ballabrera-Poy J (2015) Energy and mineral peaks, and a future steady state economy. Technol Forecast Soc Change 90, Part b:587–598. https://doi.org/10.1016/j.techfore.2014.02.013
38. Meadows DH, Meadows DL, Randers J, Behrens WW III (1972) The Limits to Growth. Universe Books, New York
39. Meadows DH, Randers J, Meadows DL (2004) The limits to growth: the 30-year update. Chelsea Green Publishing Company, White River Junction
40. Nel WP, Cooper CJ (2009) Implications of fossil fuel constraints on economic growth and global warming. Energy Policy 37:166–180. https://doi.org/10.1016/j.enpol.2008.08.013
41. De Castro C, Miguel LJ, Mediavilla M (2009) The role of non conventional oil in the attenuation of peak oil. Energy Policy 37:1825–1833. https://doi.org/10.1016/j.enpol.2009.01.022
42. Yücel G, Barlas Y (2010) Dynamics of the North-South welfare gap and global sustainability. Technol Forecast Soc Change 77:594–614. https://doi.org/10.1016/j.techfore.2009.11.004
43. Dale M, Krumdieck S, Bodger P (2012) Global energy modelling—A biophysical approach (GEMBA) Part 2: methodology. Ecol Econ 73:158–167. https://doi.org/10.1016/j.ecolecon.2011.10.028
44. Höök M, Fantazzini D, Angelantoni A, Snowden S (2014) Hydrocarbon liquefaction: viability as a peak oil mitigation strategy. Philos Trans R Soc Math Phys Eng Sci 372:20120319. https://doi.org/10.1098/rsta.2012.0319
45. Ayres RU (2008) Sustainability economics: where do we stand? Ecol Econ 67:281–310. https://doi.org/10.1016/j.ecolecon.2007.12.009
46. Kerschner C (2010) Economic de-growth vs. steady-state economy. J Clean Prod 18:544–551. https://doi.org/10.1016/j.jclepro.2009.10.019
47. Schneider F, Kallis G, Martínez-Alier J (2010) Crisis or opportunity? Economic degrowth for social equity and ecological sustainability. Introduction to this special issue. J Clean Prod 18:511–518. https://doi.org/10.1016/j.jclepro.2010.01.014
48. Kallis G (2011) In defence of degrowth. Ecol Econ 70:873–880. https://doi.org/10.1016/j.ecolecon.2010.12.007
49. Martínez-Alier J, Pascual U, Vivien F-D, Zaccai E (2010) Sustainable de-growth: mapping the context, criticisms and future prospects of an emergent paradigm. Ecol Econ 69:1741–1747. https://doi.org/10.1016/j.ecolecon.2010.04.017
50. Baloch, Amdadullah; Shah, Said Zamin; Noor, Zaleha Mohd; Magsi, Hussian Bakhsh The nexus between income inequality, economic growth and environmental degradation in Pakistan. GeoJournal (2017). https://doi.org/10.1007/s10708-016-9766-3
51. World Bank. (2006). Pakistan-strategic country environmental assessment: Main report. Country environmental analysis (CEA) (Vol. I: Main Re). Washington, DC
52. Baek, J., & Gweisah, G. (2013). Does income inequality harm the environment? Empirical evidence from the United States. Energy Policy, 62(Nov), 1434–1437. https://doi.org/10.1016/j.enpol.2013.07.097
53. Hao, Y., Chen, H., & Zhang, Q. (2016). Will income inequality affect environmental quality? Analysis based on China's provincial panel data. Ecological Indicators, 67(8), 533–542. https://doi.org/10.1016/j.ecolind.2016.03.025

54. Zhang, Y. J., Liu, Z., Zhang, H., & Tan, T. D. (2014). The impact of economic growth, industrial structure and urbanization on carbon emission intensity in China. Natural Hazards, 73(2), 579–595. https://doi.org/10.1007/s11069-014-1091-x
55. Arimoro, Augustine Edobor Beyond Economics! The (Evolving) Role of Law in the Eradication of Extreme Poverty. Journal of Human Rights and Social Work (2023). https://doi.org/10.1007/s41134-023-00247-2
56. Si, S., Ahlstrom, D., Wei, J., & Cullen, J. (2020). Business, entrepreneurship and innovation-toward poverty reduction. Entrepreneurship and Regional Development, 32(1–2), 1–20. https://doi.org/10.4324/9781003176107
57. The World Bank. (2022). Measuring poverty. The World Bank. Retrieved April 17, 2023, from https://www.worldbank.org/en/topic/measuringpoverty
58. Suckling, E., Christensen, Z., & Walton, D. (2021). Poverty trends Global, regional and national. Devenit. https://devinit.org/resources/poverty-trends-global-regional-and-national/#:~:text=In%202021%20an%20estimated%20698,live%20below%20%245.50%20a%20day
59. United Nations. (2012a). Guiding principles on extreme poverty and human rights. OHCHR. https://www.ohchr.org/sites/default/files/Documents/Publications/OHCHR_ExtremePovertyandHumanRights_EN.pdf
60. United Nations. (2012b). Principles and guidelines for a human rights approach to poverty reduction strategies. United Nations. Retrieved April 17, 2023, from https://www.ohchr.org/Documents/Publications/PovertyStrategiesen.pdf
61. Carmona, M. S., & Donald, K. (2015). Beyond legal empowerment: Improving access to justice from the human rights perspective. The International Journal of Human Rights, 19(3), 242–259. https://doi.org/10.1080/13642987.2015.1029340
62. OECD. (2016). Understanding effective access to jusctice. OECD. https://www.oecd.org/gov/Understanding-effective-access-justice-workshop-paper-final.pdf
63. Zhang, Caiqing; Chen, Panyu Industrialization, urbanization, and carbon emission efficiency of Yangtze River Economic Belt—empirical analysis based on stochastic frontier model. Environmental Science and Pollution Research (2021). https://doi.org/10.1007/s11356-021-15309-z
64. Li H, Mu H, Zhang M, Gui S (2012) Analysis of regional difference on impact factors of China's energy - Related CO2 emissions. Energy 39:319–326. https://doi.org/10.1016/j.energy.2012.01.008
65. Wang C, Zhan J, Bai Y, Chu X, Zhang F (2019a) Measuring carbon emission performance of industrial sectors in the Beijing–Tianjin–Hebei region, China: A stochastic frontier approach. Sci Total Environ 685:786–794. https://doi.org/10.1016/j.scitotenv.2019.06.064
66. Lin B, Li Z (2020) Spatial analysis of mainland cities' carbon emissions of and around Guangdong-Hong Kong-Macao Greater Bay area. Sustain Cities Soc 61:102299. https://doi.org/10.1016/j.scs.2020.102299
67. Battese GE, Coelli TJ (1995) A model for technical inefficiency effects in a stochastic frontier production function for panel data. Empir Econ 20:325–332. https://doi.org/10.1007/BF01205442
68. Balli, Esra; Sigeze, Ciler; Ugur, Mehmet Sedat; Çatık, Abdurrahman Nazif The relationship between FDI, CO_2 emissions, and energy consumption in Asia-Pacific economic cooperation countries. Environmental Science and Pollution Research (2021). https://doi org/10.1007/s11356-021-17494-3
69. Westerlund J (2008) Panel cointegration tests of the Fisher effect. J Appl Econ 23:193–233
70. Emirmahmutoglu F, Kose N (2011) Testing for Granger causality in heterogeneous mixed panels. Econ Model 28:870–876
71. Gray KR (2002) Foreign direct investment and environmental impacts-Is the debate over. Rev Eur Comp & Int'l Envtl L 11:306–313
72. Xing Y, Kolstad CD (2002) Do lax environmental regulations attract foreign investment? Environ Resour Econ 21:1–22

73. Mihci H, Cagatay S, Koska O (2005) The impact of environmental stringency on the foreign direct investments of the OECD countries. J Environ Assess Policy Manag 7:679–704
74. He J (2006) Pollution haven hypothesis and environmental impacts of foreign direct investment: the case of industrial emission of sulfur dioxide (SO2) in Chinese provinces. Ecol Econ 60:228–245
75. Wagner UJ, Timmins CD (2009) Agglomeration effects in foreign direct investment and the pollution haven hypothesis. Environ Resour Econ 43:231–256
76. Tang J (2015) Testing the pollution haven effect: does the type of FDI matter? Environ Resour Econ 60:549–578
77. Solarin SA, Al-Mulali U, Musah I, Ozturk I (2017) Investigating the pollution haven hypothesis in Ghana: an empirical investigation. Energy 124:706–719
78. Zhang XP, Cheng XM (2009) Energy consumption, carbon emissions, and economic growth in China. Ecol Econ 68:2706–2712
79. Erdal G, Erdal H, Esengün K (2008) The causality between energy consumption and economic growth in Turkey. Energy Policy 36:3838–3842
80. Apergis N, Payne JE (2009) CO2 emissions, energy usage, and output in Central America. Energy Policy 37:3282–3286
81. Belloumi M (2009) Energy consumption and GDP in Tunisia: cointegration and causality analysis. Energy Policy 37:2745–2753
82. Dagher L, Yacoubian T (2012) The causal relationship between energy consumption and economic growth in Lebanon. Energy Policy 50:795–801
83. Zhixin Z, Xin R (2011) Causal relationships between energy consumption and economic growth. Energy Procedia 5:2065–2071
84. Wang S, Li Q, Fang C, Zhou C (2016) The relationship between economic growth, energy consumption, and CO2 emissions: empirical evidence from China. Sci Total Environ 542:360–371
85. Magazzino C (2016b) The relationship between CO2 emissions, energy consumption and economic growth in Italy. Int J Sustain Energy 35:844–857
86. Antonakakis N, Chatziantoniou I, Filis G (2017) Energy consumption, CO2 emissions, and economic growth: an ethical dilemma. Renew Sust Energ Rev 68:808–824
87. Rahman MM (2021) The dynamic nexus of energy consumption, international trade and economic growth in BRICS and ASEAN countries: a panel causality test. Energy 229:120679
88. Gokmenoglu K, Taspinar N (2016) The relationship between CO2 emissions, energy consumption, economic growth and FDI: the case of Turkey. J Int Trade Econ Dev 25:706–723
89. Refinitiv Eikon Datastream, (2021). Refinitiv. Available. https://www.refinitiv.com/en/products/datastream-macroeconomic-analysis . (Accessed 21 April 2021).
90. Pesaran MH (2004) General diagnostic tests for cross-sectional dependence in panels. Cambridge Working Papers in Economics 0435, Faculty of Economics, University of Cambridge. https://ideas.repec.org/p/cam/camdae/0435.html . Accessed 21 Dec 2020
91. Ho, Sin-Yu; Iyke, Bernard Njindan Finance-growth-poverty nexus: a re-assessment of the trickle-down hypothesis in China. Economic Change and Restructuring (2017). https://doi.org/10.1007/s10644-017-9203-8
92. Ding Z, Tan L, Zhao J (2011) The effect of rural financial development on poverty reduction. Issues Agric Econ 2011(11):1–13
93. Wen T, Ran G, Xiong D (2005) Financial development and the income growth of farmers in China. Econ Res J 2005(09):30–43
94. Nain MZ, Kamaiah B (2014) Financial development and economic growth in India: some evidence from non-linear causality analysis. Econ Change Restruct 47(4):299–319
95. Hao C (2006) Development of financial intermediation and economic growth: the Chinese experience. China Econ Rev 17:347–362
96. Turner G, Tan N, Sadeghian D (2012) The Chinese banking system. Reserve Bank Aust Bull 2012:53–64
97. Ye Q, Xu Z, Fang D (2012) Market structure, performance, and efficiency of the Chinese banking sector. Econ Change Restruct 45(4):337–358

98. Zhang J, Wang L, Wang S (2012) Financial development and economic growth: recent evidence from China. J Comp Econ 40(3):393–412
99. World Development Indicators (2016). http://data.worldbank.org/products/wdi . Accessed 6 July 2016
100. Abosedra S, Shahbaz M, Nawaz K (2016) Modeling causality between financial deepening and poverty deduction in Egypt. Soc Indic Res 126:955–969
101. Seetanah B (2008) Financial development and economic growth: an ARDL approach for the case of the small island state of Mauritius. Appl Econ Lett 15:809–813
102. Uddin GS, Shahbaz M, Arouri M, Teulon F (2014) Financial development and poverty reduction nexus: a cointegration and causality analysis in Bangladesh. Econ Model 36:405–412
103. Husain I (2004) Financial sector reforms and pro-poor growth: case study of Pakistan. Presidential address at the annual general meeting of the institute of bankers Pakistan, Karachi, 21 Feb 2004
104. Sen A (1983) Poor, relatively speaking. Oxf Econ Pap 35:153–169
105. Burgess R, Pande R (2005) Do rural banks matter? Evidence from the Indian social banking experiment. Am Econ Rev 95(3):780–795
106. Pesaran, M. H., Shin, Y., & Smith, R. J. (2001). Bounds testing approaches to the analysis of level relationships. Journal of Applied Econometrics, 16(3), 289–326. https://doi.org/10.1002/jae.616
107. Hsueh SJ, Hu YH, Tu CH (2013) Economic growth and financial development in Asian countries: a bootstrap panel Granger causality analysis. Econ Model 32(294):301
108. Sehrawat M, Giri AK (2016a) Financial development and poverty reduction in India: an empirical investigation. Int J Soc Econ 43(2):106–122
109. Najafabadi, Mostafa Mardani; Mirzaei, Abbas; Laskookalayeh, Somayeh Shirzadi; Azarm, Hassan An investigation of the relationship among economic growth, agricultural expansion and chemical pollution in Iran through decoupling index analysis. Environmental Science and Pollution Research (2022). https://doi.org/10.1007/s11356-022-21004-4
110. Li M, Wang J, Chen Y (2019a) Evaluation and influencing factors of sustainable development capability of agriculture in countries along the Belt and Road route. Sustainability 11(7):2004
111. Li S, Gong Q, Yang S (2019b) Analysis of the agricultural economy and agricultural pollution using the decoupling index in Chengdu, China. Int J Environ Res Public Health 16(21):4233
112. Ali S, Ying L, Shah T, Tariq A, Ali Chandio A, Ali I (2019) Analysis of the Nexus of CO2 emissions, economic growth, land under cereal crops and agriculture value-added in Pakistan using an ARDL approach. Energies 12(23):4590
113. West TO, Marland G (2002) A synthesis of carbon sequestration, carbon emissions, and net carbon flux in agriculture: comparing tillage practices in the United States. Agric Ecosyst Environ 91:217–232
114. Ul Habib, Musavir; Tiba, Sofien; Gaies, Brahim; Jahmane, hv Investigating the four-way linkages between energy consumption, CO2 Emissions, exports, and economic growth: new evidence from SAARC economies. Environmental Economics and Policy Studies (2022). https://doi.org/10.1007/s10018-022-00351-z
115. Mohanty A, Chaturvedi D (2015) Relationship between electricity energy consumption and GDP: evidence from India. Int J Econ Financ 7(2):186–202
116. Ben Lahouel B, Zaied YB, Yang GL, Bruna MG, Song Y (2021) A non-parametric decomposition of the environmental performance-income relationship: evidence from a non-linear model. Annal Oper Res 1–34
117. Phiri A, Bothwell N (2016) Revisiting the electricity-growth nexus in South Africa. Stud Bus Econom 11(1):97–111
118. Tiba S, Omri A, Frikha M (2015) The four-way linkages between renewable energy, environmental quality, trade and economic growth: a comparative analysis between high and middle-income countries. Energy Syst 7:103–144
119. Ou P, Huang R, Yao X (2016) Economic Impacts of Power Shortage. Sustainability 8(7):687

120. Bayramoglu AT, Yildirim E (2017) The relationship between energy consumption and economic growth in the USA: a non-linear ARDL bounds test approach. Energy Power Eng 9(3):170–186
121. Lee CC, Chang CP (2005) Structural breaks, energy consumption, and economic growth revisited: evidence from Taiwan. Energy Econ 27(6):857–872
122. Ang JB (2007) CO2 emissions, energy consumption and output in France. Energy Policy 35(10):4772–4778
123. Behara J (2015) Energy consumption and economic growth in India: A reconciliation of disaggregate analysis. J Energy Technol Policy 5(6):15–27
124. Sari R, Ewing BT, Soytas U (2008) The relationship between disaggregate energy consumption and industrial production in the United States: an ARDL approach. Energy Econom 30(5):2302–2313
125. Jamil F, Ahmad E (2010) The relationship between electricity consumption, electricity prices and GDP in Pakistan. Energy Policy 38(10):6010–6025
126. Hamdi H, Sbia R, Shahbaz M (2014) The nexus between electricity consumption and economic growth in Bahrain. Econ Model 38(1):227–237
127. Tiba S, Frikha M (2018) Income, trade openness and energy interactions: Evidence from simultaneous equation modeling. Energy 147:799–211
128. Tiba S (2019a) A non-linear assessment of the urbanization and climate change nexus: the African context. Environm Sci Pollut Res 26(3):32311–32321
129. Amar AB (2021) Economic growth and environment in the United Kingdom: robust evidence using more than 250 years data. Environml Econ Policy Stud 1–15
130. Kurniawan R, Sugiawan Y, Managi S (2021) Economic growth–environment nexus: An analysis based on natural capital component of inclusive wealth. Ecol Ind 120:106982
131. Halicioglu, F. (2009). An econometric study of CO2 emissions, energy consumption, income and foreign trade in Turkey. Energy Policy, 37(3), 1156–1164. https://doi.org/10.1016/j.enpol.2008.11.012
132. Ozturk I, Acaravci A (2010) CO2 emissions, energy consumption, and economic growth in Turkey. Renew Sustain Energy Rev 14(9):3220–3225
133. Payne JE, Taylor JP (2010) Nuclear Energy consumption and economic growth in the U.S.: an empirical note. Energy Sources Part B Econ Plan Policy 5(3):301–307
134. Bozoklu S, Yilanci V (2013) Energy consumption and economic growth for selected OECD countries: Further evidence from the Granger causality test in the frequency domain. Energy Policy 63:877–881
135. Westerlund J (2007) Testing for error correction in panel data. Oxf Bull Econ Stat 69:709–748
136. Pedroni P (1997) Panel cointegration, asymptotic and finite sample properties of pooled time series tests, with an application to the PPP hypothesis: new results. Economics Working Paper, Indiana University
137. Pedroni P (1999) Critical values for cointegration tests in heterogeneous panels with multiple regressors. Oxford Bull Econ Stat 61:653–678
138. Pedroni P (2004) Panel cointegration: asymptotic and finite sample properties of fooled time series tests with an application to the PPP hypothesis. Economet Theor 20:597–625
139. Johansen S (1988) Statistical analysis of cointegration vectors. J Econ Dyn Control 12(2–3):231–254
140. Kao C, Chiang MH (2001).On the estimation and inference of a cointegrated regression in panel data. In: Baltagi BH, Fomby TB, Hill RC (Eds.), Non Stationary Panels, Panel Cointegration, and Dynamic Panels (Advances in Econometrics, vol. 15. Emerald Group Publishing Limited, pp. 179–222
141. Pedroni P (2001) Fully modified OLS for heterogeneous cointegrated panels. In: Baltagi BH, Fomby TB, Hill RC (Eds.), Non stationary Panels, Panel Cointegration, and Dynamic Panels (Advances in Econometrics, Volume 15). Emerald Group Publishing Limited, pp. 93–130
142. Phillips PCB, Sul D (2007) Bias in dynamic panel estimation with fixed effects, incidental trends and cross section dependence. J Economet 137:162–188

143. Pesaran MH (2007) A simple panel unit root test in the presence of cross-section dependence. J Appl Economet 22:265–312
144. Kapetanios G, Pesaran MH, Yamagata T (2011) Panels with non-stationary multifactor error structures. J Econ 160:326–348
145. Chudik A, Pesaran MH, Tosetti E (2011) Weak and strong cross-section dependence and estimation of large panels. Economet Jl 14:C45–C90

Chapter 7
Decoupling Economic Growth from Natural Resource Consumption

7.1 Introduction by Editor

In the earlier chapter we looked at the unsustainable components of economic growth. The report published by UNEP's International Resource Panel (2011) suggests that by the year 2050, the global economy would have reached a resource consumption of nearly 140 billion tons per year. This includes minerals, fossil fuels and biomass. There is a stark difference in the consumption patterns of developed and developing countries. Citizens of developed nations consume an average of 16 tons per head, of the key resources, while the highest gluttons use up over 40 tons per person. The average Indian consumption is about 4 tons per year. The report warns that resource consumption levels are "far beyond what is likely sustainable". Unless the economic growth rate is "decoupled" from the rate of natural resource consumption, future generations are likely to run out of inputs.

The way out of this predicament, is shown by United Nations. For developing countries, economic growth is inevitable. It contributes to the tax revenues that are important for developmental activities. However, this growth must turn 'greener'. Such economic growth is possible by a carefully planned sustainable use of natural capital. The UN suggests that such growth must be decoupled from two factors:

(i) Economic growth decoupled from natural resource consumption
(ii) Economic growth decoupled from environmental degradation

The premise behind the decoupling argument is that it is possible to grow the economy while consuming lesser resources and lowering the environmental footprint.

Decoupling natural resource use can take up several pathways. Some of these are:

(i) **Raising the value added per use of natural resources**. This is possible in many developing economies like India. There is significant reduction in resource use possible simply by improving efficiency and reducing waste. Furthermore, raising the value from resource use is possible by producing more value added output from the same resource. For example, in rural India, a

G. Nulkar (ed.), *The Economics of Sustainable Development*,
https://doi.org/10.1007/978-981-99-7379-8_7

village producing and selling raw turmeric roots can shift to the production of dry ready-to-use turmeric powder. This can fetch more revenue out of the same resource use. Another potential source of value addition is bioresources. With the huge advances in biotechnology, they offer a huge potential to GDP growth. India's phenomenal biodiversity and deep traditional knowledge provide the perfect setting for a bioeconomy. In the earlier example of turmeric, curcumin extracted from this root can fetch far higher prices than its powder.

(ii) **Increase the number of times value is added per resource us (circular economy)**: This involves using materials more than once, and necessitates designing products for longer life, enabling repair, reuse or recycle. New business models like selling products as a service, and sharing platforms are taking shape in the developed world. Building such a circularity within industry requires the support of policies and tax regimes. The European Union has stepped up their efforts in enabling circularity. This will employ data Geographic Information Systems (GIS), Internet Of Things (IOT) and other technologies for tracking material flows, wastes, making stiffer standard for input materials, operationalizing concepts like 'polluter pays' and 'right to repair'.

(iii) **Increasing the value addition in the renewable resources economy:** Agriculture is undoubtedly the largest renewable resource economy. An agrarian country like India is faced with farmer poverty. Significant changes in the farming sector are needed to improve price realisation for farmers and increase value added produce. Developments in biomass-based energy holds a great promise to increase farmer incomes. Biomass pellets, briquettes, compressed biogas and Hydrogen are growing in demand. India has over 90 million hectares of land that are in degraded condition. On the other hand, the ample sunlight throughout the year, and a monsoon based climate provide compelling arguments to reclaim the degraded lands into the emerging bioeconomy.

All the three paths have a positive effect on the environment. For example, Europe generates 20 percent of the global resource-related value-added but suffers less than 10 percent of the environmental impacts. Compare that to India, where just 4 percent of the resource value-added is generated there. Yet, the nation suffers 24 percent of the global water stress impacts, 8 percent of climate change impacts, 7 percent of pollution impacts, and 7 percent of land use biodiversity impacts.

In this chapter we present various international cases where decoupling has been attempted. The studies present the results of decoupling, on air and water pollution, and economic progress. Research ranges from Pakistan, India, and China to countries in the European Union. The UNEP Report cautions that a decoupling strategy will need significant changes in government policies, enhancements in corporate metrics, and reduction in market consumption. While these changes are not easy, they can be driven by innovations in business models and development of low carbon technologies.

References

UNEP (2011) Decoupling natural resource use and environmental impacts from economic growth, A Report of the Working Group on Decoupling to the International Resource Panel. Fischer-Kowalski, M., Swilling, M., von Weizsäcker, E.U., Ren, Y., Moriguchi, Y., Crane, W., Krausmann, F., Eisenmenger, N., Giljum, S., Hennicke, P., Romero Lankao, P., Siriban Manalang, A., Sewerin, S.

7.2 Machine Generated Summaries

Disclaimer: The summaries in this chapter were generated from Springer Nature publications using extractive AI auto-summarization: An extraction-based summarizer aims to identify the most important sentences of a text using an algorithm and uses those original sentences to create the auto-summary (unlike generative AI). As the constituted sentences are machine selected, they may not fully reflect the body of the work, so we strongly advise that the original content is read and cited. The auto generated summaries were curated by the editor to meet Springer Nature publication standards. To cite this content, please refer to the original papers.

Machine generated keywords: decouple, decoupling, energy, emission, economic growth, growth, carbon emission, intensity, energy use, carbon, energy consumption, indicator, decomposition, wang, pakistan.

7.2.1 Decoupling and Decomposition Analysis of Environmental Impact from Economic Growth: A Comparative Analysis of Pakistan, India, and China [110]

This is a machine-generated summary of:

Ozturk, Ilhan; Majeed, Muhammad Tariq; Khan, Sher: Decoupling and decomposition analysis of environmental impact from economic growth: a comparative analysis of Pakistan, India, and China [110]

Published in: Environmental and Ecological Statistics (2021).

Link to original: https://doi.org/10.1007/s10651-021-00495-3.

If you want to cite the papers, please refer to the original.

For technical reasons we could not place the page where the original quote is coming from.

Abstract-Summary

"This study examines the decoupling the CO_2 emissions from the economic growth through the employment of the Tapio decoupling index and decomposition of CO_2 emissions into its pre-determined factors through the Log Mean Divisia Index (LMDI) decomposition technique for Pakistan, India, and China (PIC) for a time span of 1990–2014."

"The findings of the Tapio elasticity analysis depict that in a few years, environmental impact has been seen to be decoupled from the economic growth in the respective PIC countries."

"The analysis of Tapio decoupling elasticity showed that energy intensity is the key factor supporting the decoupling in PIC countries, while population, affluence (GDP per capita) and energy structure have weakened the progress of decoupling."

"The analysis of the LMDI decomposition suggested that population, energy structure and affluence in PIC countries increase the CO_2 emissions, while energy intensity reduces CO_2 emissions, while mixed effects are reflected by carbon intensity."

Introduction

"Due to primary energy consumption, there is a drastic increase in GHG emissions in which CO_2 emission is on top that leads to a severe threat of climate change in the Asian region."

"The increasing level of global climate change threat has driven our interest to examine the linkage between environmental impact and economic growth and how can we achieve sustainable economic growth/decarbonized economy."

"There are multiple reasons, due to which we specifically analyzed the PIC countries; [1] PIC contributes 40% in the total population of the world (Samir and Lutz [2]). Pakistan, India, and China consumed approximately 29% of world energy (Briefing [3]). The exponential economic growth of China has a significant and effective association with the neighboring and regional nations, such as Pakistan and India, so they will be directly or indirectly affected by the industrial growth of China."

Literature Review

"There is also a vast literature available on the decoupling and decomposition analysis of economic growth from the environmental impact in the case of China at the national level (Wang and others [4]; Zhang and Lahr [5]; Wang and Li [6, 7]; Naminse and Zhuang [8]), provincial level (Jiang and others [9]; Siping and others [10]; Luo and others [11]), regional level (Xu and others [12]; Zhou and others [13]), sectoral level (Zhang and others [14]; Zhao and others [15]; Qi and others [16]; Wang and others

[17]) and industrial level (Zhou and others [13]; Fan and others [18]; Tang and others [19]; Liu and others [20]).”

“The results stated that carbon and energy intensity helped to reduce environmental impact, while the energy structure and economic growth significantly increase CO_2 emissions.”

“Khan and Majeed [21] examined the decoupling of economic growth from CO_2 emissions as well as the decomposition of total environmental impact into pre-determined factors in the case of Pakistan.”

“The results stated that overall, Pakistan experienced expensive negative decoupling; along with it, the LMDI decomposition technique depicts that energy intensity and carbon intensity significantly reduces CO_2 emissions, while population, affluence and energy structure helps to increase CO_2 emissions.”

“Guivarch and Mathy [22] examined the decoupling between economic output and energy sector of India, and the results stated that in the short-run India experienced decoupling between economic output and energy-induced environmental impact.”

Research Methodology and Data Sources

“Tapio developed a decoupling index that provides a more precise and yearly decoupling status between CO_2 emissions and national output, further employing the technique of LMDI decomposition provides a perfect decomposition of CO_2 emissions into pre-determined variables, i.e., relying on the Kaya identity.”

“Based on the goals and requirements of the study, the current study uses the Tapio elasticity assessment and LMDI decomposition approach.”

“The LMDI decomposition technique from the IDA family is used in the current study which is based on the Kaya identity.”

“Kaya identity was firstly presented in IPCC in 1990 by Yoichi Kaya and decomposes the total CO_2 emissions into four factors that are population, affluence, energy and carbon intensity, while we have further used energy structure, which is also a major factor of CO_2 emissions as well as economic growth.”

“The current study uses the annual data of GDP, CO_2 emissions, affluence, population, energy intensity, energy structure and carbon intensity.”

Results and Discussion

“The main reason behind the 1994–1995 strong decoupling status is that the government has taken the bailout package from international financial institutions that lead to the increase of taxation level, resulting in the decrease in economic growth but less than the CO_2 emissions (IMF).”

“The result depicts that energy intensity is the most influential factor which promotes and reinforces the decoupling progress in Pakistan, except in the years, 1992–1994 and 1996–1998.”

"Same as population and affluence effect, energy structure also restricts decoupling progress, but in some years, it helps to decouple economic growth from environmental impact."

"The results stated that population and affluence curbs the decoupling progress in China, while the energy and carbon intensity promotes the decoupling progress in China but, in some years, they are also responsible for restricting the decoupling progress."

Conclusion and Policy Recommendations

"The decomposition of decoupling analysis elaborates the response of Kaya identity factors towards the decoupling of environmental harms and economic growth in PIC."

"According to the empirical results, certain policy recommendations are given in this section, Firstly, the energy structure should be focused on by PIC countries as it's the most influential factor deteriorating the environmental quality of the economy, the shift of the economy's energy structure from emission-intensive towards the less carbon-intensive energy consumption that will promote sustainable economic growth along with the low carbon growth."

"In the context of PIC countries, the current research concentrates on the most debated issue, i.e., the association between the economic growth-environmental quality through the employment of Kaya identity, LMDI decomposition approach, and Tapio decoupling indicator."

7.2.2 Decoupling Economic Growth from Emissions: The Case of Policies Promoting Resource Substitution [111]

This is a machine-generated summary of:

Silva, Susana; Soares, Isabel; Afonso, Oscar: Decoupling economic growth from emissions: the case of policies promoting resource substitution [111]

Published in: Environment, Development and Sustainability (2020).

Link to original: https://doi.org/10.1007/s10668-020-00967-9.

If you want to cite the papers, please refer to the original.

For technical reasons we could not place the page where the original quote is coming from.

Abstract-Summary

"We use a stylized equilibrium model to explore distinct policy instruments under which relative decoupling and absolute decoupling between economic growth and emissions occur."

"The government taxes emissions and subsidizes renewables production."

"Achieving absolute decoupling is harder, but it occurs for a certain interval of the subsidy growth rate."

"The choice of the tax and subsidy growth rates is more important than the choice of the initial levels for these instruments."

"The choice of the subsidy growth rate indicates the prioritization between the economy and the environment."

"A lower subsidy growth rate favors the environment, while a higher one favors the economy."

"The tax growth rate dictates the distance between the output growth rate and the emissions growth rate."

"The higher the tax growth rate, the higher the difference between them."

"The government can decide between attaining higher output growth rates, higher emissions decreasing rates, or an intermediate situation."

Introduction

"A specific branch of the literature focuses on the possibilities of decoupling economic growth and environmental degradation, either in relative or absolute terms."

"An important mechanism to decouple economic growth and emissions, in particular in the energy sector, is resource substitution."

"Other authors specifically referred the need to develop renewable energy sources (RES) to achieve decoupling but did not provide any explicit analysis (e.g., Zhang [23]; OECD [24]; Everett and others [25]; Wang and others [26])."

"Some authors emphasized the importance of environmental policy to achieve decoupling (e.g., Azar and others [27]; Galeotti [28]; van der Voet and others [29]; Næss and Høyer [30]; Everett and others [25]; Ru and others [31]), but did not explore the specific policy instruments required for this goal."

"Despite the relevance encountered on resource substitution (e.g., Bovenberg and Smulders [32]; Tahvonen [33]; Di Vita [34]), these studies do not focus on the issue of decoupling economic growth and emissions, as we do here."

"We contribute to the literature of economic growth-emissions decoupling, by modeling and discussing specific policy instruments and their values to achieve decoupling."

The Model

"Our model is in continuous time and includes three sectors: homogenous final-goods, non-renewable resources or fossil fuels (F), and renewable resources (R)."

"Homogeneous final-goods may be used for consumption, C, and investments in resource extraction/production."

"There are N (n = 1, …, N) final-goods producers who face perfect competition."

"Profits to maximize are given by: In this sector there are Z (z = 1, …, Z) perfectly competitive firms who "extract" resources and sell them to final-goods producers."

"The government's budget is balanced in each period since tax revenues are totally used to finance the subsidy and a lump-sum transfer to consumers when it exists."

"The effect of the subsidy growth rate on the emissions growth rate is not straightforward."

"Absolute decoupling occurs if output grows and emissions decrease in the balanced growth path."

Empirical Application

"The first possibility explored is when the government determines the tax level and then decides the best combination for the tax and subsidy growth rates."

"The second simulation examines results when the government fixes the tax growth rate and then decides the best combination for the tax level and the subsidy growth rate."

"The tax level does not significantly impact output and emissions growth rates."

"The higher the tax growth rate, the stronger the effect on emissions."

"A high subsidy growth rate leads to significant output growth and constant emissions."

"The higher the tax growth rate, the stronger the effect on output."

"Our analysis suggests that, to achieve decoupling, the choice of the tax and subsidy growth rates is more important than the choice of the initial levels for the tax and the subsidy."

Conclusions and Policy Implications

"In terms of decoupling results, if the subsidy growth rate is positive, relative decoupling always occurs, i.e., output grows more than emissions."

"The empirical application to the electricity sector in the USA showed that the government has four policy decisions: tax level, subsidy tax level, tax growth rate, and subsidy growth rate."

"The effect of the subsidy growth on emissions and output is not straightforward since, on one side, it promotes resource substitution, but on the other side, it gives firms extra resources to increase production (or generate more electricity) which is the main driver of emissions."

"The subsidy increases renewables use which decreases emissions directly through resource substitution, but increases them indirectly through higher output."

"Our analysis indicates that, to achieve decoupling, the choice of the tax and subsidy growth rates is more important than the choice of the initial levels for those policy instruments."

7.2.3 The Great Decoupling: Why Minimizing Humanity's Dependence on the Environment May not Be Cause for Celebration [112]

This is a machine-generated summary of:

Shockley, Kenneth: The Great Decoupling: Why Minimizing Humanity's Dependence on the Environment May Not Be Cause for Celebration [112]

Published in: Journal of Agricultural and Environmental Ethics (2018)

Link to original: https://doi.org/10.1007/s10806-018-9736-5.

If you want to cite the papers, please refer to the original.

For technical reasons we could not place the page where the original quote is coming from.

Abstract-Summary

"Characterizations of the Anthropocene often indicate both the challenges that our new epoch poses for human well-being and a sense of loss that comes from a compromised environment."

"The environmental conditions characteristic of the Anthropocene do so, I claim, by compromising flourishing on two fronts."

"The comparatively novel conditions of the Anthropocene risk rupturing our narratives, putting at risk our sense of self and connections to familiar environments."

"This paper argues that to the extent humanity is decoupled from their environments humans are not only less able to access opportunities our understanding of who we are, our identities, and our capacity to make sense of the world around us through those identities is compromised."

Terra Incognita: Our Unfamiliar World

"I will argue that one of the great challenges faced in the Anthropocene is the decoupling of humanity from the ecological systems with which we are familiar, on which we depend, and through which we understand ourselves."

"This paper explores how the decoupling of humans from their surrounding environments compromises our ability to access and actualize our opportunities for flourishing."

"As it is more difficult to access the capabilities necessary for flourishing in an unfamiliar world, decoupling causes separation between the capabilities that have enabled flourishing in the past and the opportunities and available options we face in a world notably divergent from the past."

"Ruptures in narrative, caused by changes to the background conditions through which we partly understand ourselves and one another, make it difficult to access the opportunities and possibilities that, in part, constitute human flourishing."

The Anthropocene

"One vital condition made apparent by the capabilities approach, a condition that will have to be satisfied for any substantive account of flourishing, is that we are able to make connections, both in terms of resources and opportunity, across the transition between our historical relationship to the environment and our uncertain future."

"As an era characterizing a period of comparative ecological instability and representing a substantive break from our historical relationship with the environment, the Anthropocene poses a challenge for human flourishing not merely in the obvious sense that our material and institutional resources may be compromised but in the sense that the decreased familiarity of our environment represents a decreased ability to access the opportunities that constitute flourishing."

"While humans are capable of adapting to substantial changes in their environment, a broad, global sense of nature being unfamiliar is novel."

Narratives at Risk: Understanding Ourselves in an Unfamiliar World

"Our narratives, our modes of self-understanding, blend our reliance on resources, our sense of place, and our familiar environment."

"Our narratives allow us to situate ourselves in a world, a world that makes sense because of a recognizable past (and where we find ourselves situated from, in, and with our environments)."

"Jonathan Lear [35] reminds us of the importance of narrative and background environment to human flourishing through his recounting of the thoughts of Chief Plenty Coups of the Crow Nation, someone who endured catastrophic threats to the stability of his own society and a remarkable, if less global, version of decoupling."

"As the case of Chief Plenty Coups and the Crow makes clear, decoupling humans from their environmental background is highly problematic, anticipated by the way we live from, in, and with the environment."

Flourishing in Unfamiliar Terrain

"The severity of the challenge posed by the Anthropocene can be seen in the interdependence of the two threats to human flourishing described above: identity (narrative self-understanding) and flourishing (manifesting capabilities)."

"We can develop our narratives in ways that make sense of narrative rupture, that allow us to make new ways to flourish in our changing world."

"We saw that the instabilities caused by climate change and other harbingers of the Anthropocene compromise the environmental foundations of our capabilities, and so undermined the options available to us that enable flourishing."

"Our capabilities depend on our ability to make sense of opportunities as live options for human flourishing; live options are typically familiar options."

"What is important for flourishing in the Anthropocene in the face of the Great Decoupling is that we are able to make connections, both in terms of narrative and opportunity, across the transition between our historical relationship to the environment and our uncertain future."

7.2.4 The Causal Link Between Circular Economy and Economic Growth in EU-25 [113]

This is a machine-generated summary of:

Chen, Chun-Chih; Pao, Hsiao-Tien: The causal link between circular economy and economic growth in EU-25 [113]

Published in: Environmental Science and Pollution Research (2022)

Link to original: https://doi.org/10.1007/s11356-022-21010-6.

If you want to cite the papers, please refer to the original.

For technical reasons we could not place the page where the original quote is coming from.

Abstract-Summary

"The selected CE indicators included municipal waste recycling rate, CE-related investment, municipal waste generation per capita, circularity rate, and trade in recyclable raw materials."

"Panel vector error correction model results confirmed that in terms of short-run causality, an increase in material recycling led to a decrease in waste generation, an increase in waste generation led to an increase in CE-related investment, and economic growth led to circular economy growth, but not vice versa."

"This implies that encouraging CE-related innovation investments and promoting material recycling to stimulate the secondary raw material market can help achieve zero waste goals."

"Looking at the long-term causality, the GDP and CE indicators constituted a causal loop, which implies that there is co-evolution between them, although the circular economy is still in its infancy."

Introduction

"Promoting a circular economy requires an innovative business model that closes the loop throughout the life cycle of products, materials, and resources to achieve sustainability and profitability, while being attractive to customers and suppliers."

"To understand the current state of the EU's promotion of a circular economy, we collected six CE-related data for preliminary analysis, including municipal waste recycling rate (RMW), CE-related investment (INV), municipal waste generation per capita (GMWp), circularity rate (CUR), trade in recyclable raw materials (TRM), and real GDP."

"The growth of circular economy indicators and GDP imply that European countries were making efforts to promote efficient resource management and sustainable economic growth."

"Most recent studies indicate that a circular economy is conducive to economic development."

"To fill this gap, this study uses a panel vector error-correcting econometric model (VECM) to examine whether the causal relationship between the circular economy and economic growth is beneficial, inhibitory, neutral, or feedback."

Literature Review

"Each aspect contains some quantitative indicators of circular economy, such as resource productivity (RP), recycling rate of e-products (ReP), municipal solid waste generation (MSWG), and municipal waste generation per capita (GMWp) in the production and consumption aspect; the recycling rate from municipal waste, bio-waste, e-waste, and packaging waste (RMW, RbW, ReW, and RpW) in the waste management aspect; and CE-related private investment (INV) and recycling-related patents (PAT) in the competitiveness and innovation aspect; as well as trade volume of recyclable raw materials (TRM) and circularity rate (CUR) in secondary raw materials aspect."

"Regarding energy resources and economic growth nexus, many historical literatures use panel VECM to analyze the causal relationship between them (i.e., energy–growth causality)."

"Busu [36] found long-run unidirectional causality from renewable energy consumption to growth and bidirectional causality between resource productivity and growth in EU-28."

Model and Methodology

"The variables LGDP, LINV, LGMWp, and LTRM are the natural logarithms of GDP, CE-related investments, generation of municipal waste per capita, and trade volume of recyclable raw materials, respectively."

"The panel unit root test is used to assess for stationary in a time series."

"Time series with unit root is nonstationary and is called integrated of order 1 or I (1)."

"A stationary time series is called integrated of order 0 or I(0)."

"Three panel unit tests, namely Fisher-type ADF (Augmented Dickey–Fuller), PP (Phillips–Perron) (Maddala and Wu [37]; Choi [38]), and LLC (Levin and others

[39]) are used to find the order of integration of LGDP, LINV, LGMWp, LTRM, RMW, and CUR."

"The joint-Wald test for the lag periods of the first-order difference of each explanatory series is to find the short-run causality from the independent variable to the dependent variable."

Results and Discussions

"For the two indicators in the aspect of secondary raw materials, TRM and CUR, the mean of their 8-year average growth rate ((0.284 + 1.264)/2) was less than 1% and the 5-year average growth rate was the lowest, indicating that the EU should actively understand the main barriers that hinder an effective secondary raw materials market, which will make the circular economy more effective."

"In the short-run, (1) the existence of a negative unidirectional causality from TRM to GMWP implied that increased material recycling helped reduce waste generation; (2) the existence of a positive unidirectional causality from GMWp to INV implied that the increase in waste generation stimulated CE-related investments in order to effectively convert waste into gold for more sustainable development; and (3) the existence of a unidirectional causality from GDP to CE indicators without feedback indicated that economic growth promoted circular economy, but not vice versa."

Conclusions, Policy Implications, and Research Limitations

"This research uses panel data from the 25 EU countries from 2010 to 2018 to innovatively explore the causality between circular economy and economic growth (CE–growth causality)."

"Regarding causality, the estimated results of the panel VECM showed that in the short run, the increased in material recycling led to a decrease in waste generation and the increase in waste generation led to an increase in CE-related investment, indicating that EU countries are committed to achieving zero waste environmental benefits through investment in resource efficiency, which should be the effect of the EU's active implementation of its waste policy."

"Based on the Cradle-to-Cradle design concept, the two key elements of the policy should include encouraging CE-related research and innovation investment to stimulate the secondary raw materials market and improve materials recycling efficiency, as well as formulating laws to implement the EU Waste Policy and the European Green Deal."

7.2.5 Economic Growth Versus Climate Balancing: Some Reflections on the Sustainable Management of Forest Resource in India [114]

This is a machine-generated summary of:

Banerjee, Sarmila; Bit, Jayita: Economic growth versus climate balancing: some reflections on the sustainable management of forest resource in India [114]

Published in: DECISION (2015)

Link to original: https://doi.org/10.1007/s40622-015-0090-4.

If you want to cite the papers, please refer to the original.

For technical reasons we could not place the page where the original quote is coming from.

Abstract-Summary

"Designing of a sustainable forest policy is of extreme importance in our present world where the process of rapid economic growth is causing imbalance in the climatic cycles by denying ecosystems sufficient time for adaptation and the socioeconomic systems adequate opportunity for mitigation."

"The forests are considered highly resilient natural resource that plays a major role in reducing the impact of global climate change through carbon sequestration, heat absorption, watershed protection, acid deposition, etc. This paper examines the prospect of sustainable forest management for an emerging economy like India, where forest coverage has gone up over the last three decades in spite of population growth, rapid urbanization, and fast economic growth."

"From the perspective of material balance, the results of structural decomposition analysis reveal increasing dominance of economic growth over other effects indicating necessity of designing intervention to decouple potential future economic growth from forest resources to ensure long run sustainability."

Context

"Our analysis concentrates on the forestry sector as it is a sector of fundamental importance for maintaining the climatic balance and the forest coverage in India has gone up from 63.94 to 68.44 million hectares over the last three decades in spite of her moderate population growth, rapid urbanization, and this fast economic development (http://data.gov.in/sites/default/files/FOREST_COVER_AREA_1.xls)."

"To be economically sustainable, the growth process needs to be decoupled from the forestry sector, and the direct forest intensity of production matrix of the economic system should reveal tendency to-ward substituting forest-dependent products in the production basket by other nonforest-dependent products."

"India's unleashed demand (met through import from outside) would eventually have its impact on reduced stock of natural forest of the other countries and the consequent pressure on the climatic cycle of the planet will not only thwart the process of economic development but mere sustainable existence of the system will be questionable in the longer-run."

Economy–Forest Interactions

"Throughout the history of human civilization forests have been a major source of natural resource products."

"From daily urban life needs to source of livelihood to home for varied habitats, forests' usefulness is almost unaccountable."

"The mix of services that is available to any economy from forests depends, in addition to their biological characteristics, on the nature of economic regime within which they are exploited."

"Others such as nontimber forest products may be extracted under a variety of arrangements, the range varying from open access to common property regimes."

India's Macroeconomic Dependence on Forestry Sector

"For our purpose, the calculation of forward linkage will be more relevant as we are interested in knowing the increase in induced demand for the product of the forestry and logging sector by the rest of the economy following a change in both the production and consumption matrices over the last two decades."

"This exercise will help us to quantify the absorption of the products of forestry and logging sector in the production of other sectors through direct and indirect linkages and will ascertain the technological dependence of the economy on the forest-based products as inputs."

"For the Forestry and Logging sector, the input inverse-based value is greater than the output inverse-based value indicating that the effect of increase in this input on suppliers is greater than that on the users; in contrast, for all other derived industries like wooden furniture and fixture, wood, and wood products; and paper, paper products and newsprint, the impacts on user industries dominate."

Growth of Forest Products in India: Production, Consumption, and Export and Import

"In order to check the domestic usage pattern of forest products, data have been collected on production, consumption, export and import of various forest products from FAO where the major segments are round wood (RW), sawn wood (SW), wood-based panels (WBP), wood pulp (WP), and paper and paperboard (PPB)."

"From the discussion presented in the earlier sections it has been found that on the basis of simple time based forecasting the gap between consumption (demand) and production (supply) is widening for all types of forestry related products."

"The temporal trends in demand–supply gap assessed in terms of raw data are statistically significant for all products and the trend is the maximum for the RW followed by PPB and SW."

"Like China, India is depending on its non-OECD trading partners for the supply of semi-processed wood products, and the important source of manufactured forest products are the OECD countries."

A Structural Decomposition Analysis

"For sustainable management of forest, it would be interesting to isolate the contribution of different drivers like the changes in (i) the technique of production, (ii) the composition of the economic system, and (iii) the scale of the Indian economy in explaining this exponential increase in final consumption of wood-based products and to investigate the ecological pressure imparted on the physical system of the country in terms of the change in the magnitude of direct forest input (DFI) used."

"Structural decomposition analysis (SDA) is generally applied to break down the growth in some variable into the change in its determinants (Dietzenbacher and Los [40]) and has been widely used to assess the environmental stress in terms of material balance for a number of rapidly growing economies."

Policy Challenges

"To conserve forest resource, the growth process needs to be de-coupled from the excessive use of forestry-based products, and therefore, management of this critical resource is not a mere ecological or environmental issue—the social and economic aspects are also of equal importance."

"The National Forest Policy of India, 1988 has focused on meeting national needs, protecting forest at national level with consideration for local livelihood protection by encouraging social forestry."

"The joint forest management (JFM) program considers mostly the economic and social aspects at local or at the most national level."

"Although all these works have gained some partial success in stitching the local-level practices with the national policies, they mostly failed to address the problems of socially equitable, economically efficient, and ecologically balanced management of forest resources."

7.2.6 Using a New Decoupling Indicator (ZM Decoupling Indicator) to Study the Relationship Between the Economic Growth and Energy Consumption in China [115]

This is a machine-generated summary of:

Song, Yan; Zhang, Ming: Using a new decoupling indicator (ZM decoupling indicator) to study the relationship between the economic growth and energy consumption in China [115]

Published in: Natural Hazards (2017)

Link to original: https://doi.org/10.1007/s11069-017-2903-6.

If you want to cite the papers, please refer to the original.

For technical reasons we could not place the page where the original quote is coming from.

Abstract-Summary

"The ZM decoupling indicator is utilized to explore the state of decoupling between economic growth and energy consumption in China."

Introduction

"Among all the existing methods, the decoupling analysis is a useful method to study the relationship between economic growth (GDP) and energy consumption, or environment issue."

"There are two decoupling methods, which were widely used to study the causal relationship between energy use or energy-related environmental degradation and economic growth."

"The Taipo decoupling method has been widely used to study the relationship between economic growth and energy consumption or environment issue (Ren and Hu [41])."

"Climent and Pardo [42] utilized the Taipo decoupling method to investigate the relationship between Spanish economic growth and energy consumption."

"That method was utilized by Freitas and Kaneko [43] to study the occurrence of a decoupling between Brazil's economic growth and energy-related CO_2 emission from 2004 to 2009."

"Based on the LMDI method, Zhang and Wang [44] also defined a decoupling indicator, which was utilized to analyze the decoupling of electricity consumption from economic growth in China."

Methodology and Data

"The IDA decomposition techniques also include two methods: the complete decomposition method and LMDI method."

"By comparing various IDA methods, Ang [45] concluded that the LMDI method was the preferred method."

"Ang and Liu [46] presented eight strategies to handle zero values in LMDI method."

"The LMDI method is used to decompose energy consumption into several influence factors in this paper."

"The detail of the LMDI method can be found in the reference (Ang and Liu [46])."

"All the energy data used in this paper are all collected from various issues of China Energy Statistical Yearbook (CESY [47]–1996, 1997–1999, 2000–2002, 2003, 2004)."

"The whole China economy is divided into three aggregated industries, namely the primary, secondary, and tertiary industry."

Results and Discussion

"Because the secondary industry accounted for more than 75% of total final energy consumption, the energy intensity of the secondary industry played a determined role in the total energy consumption."

"Expansive negative decoupling only appeared in three years: 2003, 2004, and 2008, and the decoupling indicator was 0.62, 0.88, and 0.24, respectively."

"The energy consumption presented weak decoupling with economic growth in the rest 10 years."

"The energy intensity for the secondary industry increased in those three years."

"The slowly decline of the energy intensity was the reason that expansive coupling occurred in those five years."

"Due to the percentage change of GDP is positive, only four decoupling status appeared during the study period: strong decoupling, weak decoupling, expansive coupling, and expansive negative decoupling."

"Weak negative decoupling appeared in the three years under the Tapio decoupling indicator."

Conclusions

"Decoupling theory is a useful tool to study the relation between economic growth (GDP) and energy consumption or environment issue."

"The ZM decoupling indicator was also utilized to explore the state of decoupling between the economic growth and energy consumption in China."

"The main conclusions drawn from the present study are as follows: (1) Along with the rapid economic growth, the final energy consumption in China rose to 2275.02 Mtce in 2012."

7.2.7 Decoupling of Environmental Pressure and Economic Growth: Evidence from High-Income and Nuclear-Dependent Countries [116]

This is a machine-generated summary of:

Pao, Hsiao-Tien; Chen, Chun-Chih: Decoupling of environmental pressure and economic growth: evidence from high-income and nuclear-dependent countries [116]

Published in: Environmental Science and Pollution Research (2019)

Link to original: https://doi.org/10.1007/s11356-019-07122-6.

If you want to cite the papers, please refer to the original.

For technical reasons we could not place the page where the original quote is coming from.

Abstract-Summary

"This paper proposes a three-step process, based on the definition of absolute decoupling (AbsDec), to analyze the role of nuclear energy in the absolute decoupling between environmental pressures (including Energy resource use and CO_2Emissions) and Economic growth (3-E) during the period 1987–2016."

"For the USA, a commensalism of energy-led growth and an amensalism of emissions-limited growth exist."

"The aggregated results indicate that the 4-NDC seem to have achieved an AbsDec between 3-E, the USA appears to achieve an AbsDec between economic growth and emissions while undergoing a relative, and perhaps absolute, decoupling between 3-E. The findings can infer that nuclear power is one of the most important energy sources for achieving absolute decoupling and genuinely sustainable development."

"The policy implication is that measures to reduce energy consumption and control CO_2 emissions may not significantly impair economic growth in countries that rely on nuclear power."

Introduction

"The research objective is to analyze whether a country with a high nuclear power share has reached AbsDec between 3-E. If the AbsDec between 3-E can be achieved in countries with a high nuclear power share, nuclear power will likely become one of the most important energy sources for achieving genuinely sustainable development, and measures to reduce energy consumption and control CO_2 emissions will not significantly damage EG."

"Based on the definition of AbsDec, we propose a three-step process to analyze the role of nuclear energy in the AbsDec between 3-E. First, we select countries with a high nuclear power share as samples to study the impact of nuclear energy on the AbsDec of the country."

"The aggregated answers help us understand the impact of nuclear energy on AbsDec between 3-E. The sample countries in this study include France, Belgium, Sweden, and Switzerland (the 4-NDC), which have the highest share of nuclear power in high-income economies, and the USA, the world's largest nuclear power producer (5-country)."

Literature Review

"To obtain a better understanding of global sustainability issues, numerous studies have explored the EKC/CKC hypothesis, the decoupling relationship of emissions from economic growth, and the dynamic causality between 3-E. Empirical results can be used as a reference for policy development, whose inconsistency may be related to country/market attributes, analysis time period, model structure, or research methods."

"Recent papers on the emissions-economy nexus examined the EKC/CKC hypothesis or the decoupling relationship to verify whether EG can decouple from carbon emissions as countries improve their energy efficiency and technology."

"Since causal analysis helps understand the decoupling relationship, we focused on recent papers of the causal dynamic between 3-E for the 5-country and their related cross-country panels."

"The eighth column shows that the Nuc elasticity of emissions is negative (Pao and Chen [48]; Lapinskienė and others [49]; Lee and others [50]; Marques and others [51]; Onater-Isberk, [52]), implying that nuclear-dependent high-income countries seem to support an inverse relationship between Nuc and carbon emissions."

Data, Model, and Methodology

"To understand the impact of nuclear energy on mitigating climate change and promoting global sustainable development, this study proposed a three-step process to analyze whether countries with a high share of nuclear power will achieve decoupling between 3-E. Out-of-sample predictions were performed for all variables until 2025."

"These are shown below as follows: orwhere i represents the cross section (five nuclear-dependent countries), t denotes the time series, ε_{it} is a serially uncorrelated error term, and time series LCO, LY, LNuc, and LNNuc are the natural logarithms of CO_2 emissions, real GDP, and nuclear and nonnuclear energy consumptions, respectively."

"The model is expressed as three first-order differential equations with nine parameters, which accounts for the logistic population growth, intraspecies interactions, and interspecies interactions, as shown below (Leslie [53]; Liu and Gopalsamy [54]; Kim and others [55]):where X_1, X_2, and X_3 represent populations of emissions, total energy consumption/nuclear energy consumption (TEC/Nuc), and real GDP, respectively."

"There is decoupling between emissions and GDP in countries that are highly dependent on nuclear energy."

Empirical Results

"The (−, 0) of (β_{31}, β_{13}) indicate a commensalism, where real GDP benefits from TEC and TEC are unaffected, implying that decoupling exists between the growth rates of the economy and energy consumption."

"The (−, 0) of (β_{12}, β_{21}) and (β_{13}, β_{31}) indicate commensalisms, where Nuc benefits from both emissions and GDP, and these are unaffected by Nuc, implying that the use of nuclear energy is a key means for dealing with emissions."

"Excessive energy conservation policies may reduce EG; [2] the combination of evidence on emissions limitation growth with negative CAGR in emissions and positive CAGR in GDP indicates an AbsDec of emissions from economic growth; [3] evidence of emissions-driven Nuc implies that nuclear energy use is a key means of dealing with emissions."

"Based on the LV–ECOM for each country, the CAGR of CO_2 emissions is negative, that of energy consumption/nuclear power is negative or lower than 0.5% and that of real GDP is positive."

Conclusion and Policy Implications

"This study selected nuclear-dependent countries as samples to examine the impact of nuclear energy on achieving an AbsDec between EG, EC, and emissions (3-E) by using both a linear CKC model and a nonlinear LV–ECOM over the 1987–2016 period."

"We extended the literature on decoupling analysis between EG and carbon emissions to decoupling analysis between 3-E. The five sample countries are France, Belgium, Sweden, and Switzerland (the 4-NDC), which have the highest share of nuclear power of all high-income economies, and the USA, the world's largest nuclear power producer."

"We conclude that the proposed three-step process can be used to investigate the absolute decoupling between 3-E. Future research should examine the decoupling between 3-E in low-/middle-income countries with a high nuclear power share to further understand the role of nuclear power in achieving absolute decoupling."

7.2.8 Deepening Decoupling for Sustainable Development: Evidence from Threshold Model [117]

This is a machine-generated summary of:

Zeng, LanTing; Ye, Azhong; Lin, Wen: Deepening decoupling for sustainable development: evidence from threshold model [117]

Published in: Energy Efficiency (2022)

Link to original: https://doi.org/10.1007/s12053-022-10031-2.

If you want to cite the papers, please refer to the original.

For technical reasons we could not place the page where the original quote is coming from.

Abstract-Summary

"Based on the theoretical assumption arising from solution to endogenous growth model and using the panel data of 50 countries during the period of 2004–2016, this paper establishes a threshold model to quantify effects of industrial structure, income, urbanization, and other production factors on decoupling of economic growth from energy consumption (decoupling) under the constraint of technological level."

"The impact of industrial structure and income on decoupling has changed in direction and is characterized by inverted U-shaped curve under a single technology threshold."

"The substitution effect from renewable energy rather than production factors is conducive to decoupling, not restricted by the technological level."

"Deepening decoupling should be devoted to exploring the potential of driving factors and refraining from the dragging effects of resisting force."

Introduction

"National wealth especially economic growth depends on material and energy consumption (Grinter [56]), resulting in much more greenhouse gas emission (Warr and others [57]), which is the cause of climate problems."

"OECD officially defined decoupling as a monitoring indicator to ensure sustainable economic growth with high environmental quality, which is commonly adopted in the economic field to measure efficiency of energy use and economic growth, and widely advertised by many countries as great achievements towards the fight against climate change and other environmental impacts (EEA [58])."

"Decoupling economic growth from energy consumption is an important indicator to unlock high-quality economic growth and environmental protection."

"According to the statistics of the International Energy Agency, since 2004, total energy consumption of all the world has increased by 192.72%, while the indicators of sustainable development goals, including decoupling index, have performed well."

"Decoupling usually occurs and fluctuates in many countries, where total energy consumption shows upward trend (Ekins [59])."

Literature Review

"Tapio [60] used elasticity index to measure relationship between energy consumption and growth in transportation sector, and defined eight categories ranging from coupling to decoupling."

"Wu and others [61] employed grey relational analysis (GRA) for the factors on energy consumption change, and found industrial structure and economic growth pattern have greater impact on decoupling than technical progress in developing countries, while it is contrary in developed countries."

"Even various methods successfully measuring decoupling, there are few studies conducting analysis directly on roots of decoupling energy consumption and economic growth, which may be the results of one single index changing or both two."

"In order to obtain more evidence about the changing correlation between energy consumption and economic growth, a direct study on decoupling considering technology threshold should be carried out, especially at the global level."

Theory Hypotheses on Decoupling Driving

"Labor H is divided into two parts, one of which is invested in the final product production department, and the other in the knowledge accumulation department that decides technological progress."

"The larger the human capital is, the higher the production capacity and stock of new knowledge gets."

"When thrust and resistance power coexist, the influence of other input factors on decoupling is not conclusive."

"We can get: Hypothesis 1: Modernity can strengthen decoupling, and the more advanced the economy and social is featured, the stronger the decoupling will become."

"Hypothesis 2: The substitution and complementary effects of other input factors on decoupling is obvious, while inconclusive when put together, and will be coordinated by the input of R & D department."

Data and Methodology

"Gross National Income per capital (gni) is chosen for economic development from the consumers' side, and industry value added of GDP (indus) for industrial structure and urban population of total population for urbanization (urban)."

"Total renewable energy consumption(renew) is employed to represent renewable energy effects on decoupling."

"Data for Y, gni, indus, urban, tech, renew, K, and L are collected from World Bank and International Monetary Fund (IMF)."

"To ensure the common quality of the selected samples, our data covers 50 high-, middle-, and low-income countries during the period of 2004–2016, of which total energy consumption accounted for 88% of the world."

"Technology is set as the threshold with gni as core independent variables representing the different economic development level."

"The significance of the threshold value can be carried out by bootstrapping, and judged by the asymptotic P-value for F with null hypothesis of no threshold, which fits the fixed-effect panel threshold model (Wang [62])."

Empirical Results and Discussion

"In the low-tech regime, 1% increase in gni will lead to 0.000187% decrease in Dr, and decoupling deepened, indicating that the improvement of living standards can alleviate the burden of energy consumption and promote economic growth more."

"In the high-tech regime, a positive gni coefficient shows that 1% increase in income will lead to 0.000523% increase in Dr, indicating that the decoupling driving effect caused by high income has disappeared."

"The significant renew coefficient in the first model shows that 1% increase will cause 0.189% decrease in Dr, indicating that renewable energy can alleviate energy demands and strengthen decoupling without being constrained by technological level."

"As to relationship between capital and decoupling, in low-tech regime, the significant coefficient of K shows that 1% increase in K will lead to Dr up by 1.137%."

Conclusions and Suggestions

"The lack of sustainable and stable decoupling calls on the developing to keep on exploring the potential of deep decoupling for climate governance targets [2]. Technological level shows an obvious breaking point effect in the form of constraining the influences of industrial structure and income on decoupling, featured as inverted U-shaped curve, that is, an active driving power on decoupling strengthened below the breaking point, while a blocking role in decoupling dominates above it."

"Only when social development matches the technological level, economic growth exceeds energy consumption growth, and even energy consumption slows down, leading to decoupling deepened [3]. The effect of modernity on decoupling is composed of the superposition force and pulling force brought by the three indicators."

"Promotion of modernization is not always beneficial to the green development [4]. Production input factors are complementary to energy, showing opposite power on decoupling without being constrained by the technological level."

"Converting energy-intensive capital to technology-intensive, promoting renewable energy, and improving the skilled force are effective strategies for pursuing sustainable decoupling for green development."

7.2.9 On the Examination of the Decoupling Effect of Air Pollutants from Economic Growth: A Convergence Analysis for the US [118]

This is a machine-generated summary of:

Polemis, Michael L.; Fotis, Panagiotis; Tzeremes, Panayiotis G.; Tzeremes, Nickolaos G.: On the examination of the decoupling effect of air pollutants from economic growth: a convergence analysis for the US [118]

Published in: Letters in Spatial and Resource Sciences (2022)

Link to original: https://doi.org/10.1007/s12076-022-00320-w.

If you want to cite the papers, please refer to the original.

For technical reasons we could not place the page where the original quote is coming from.

Abstract-Summary

"Three alternative decoupling indicators associated with the related global (carbon dioxide emissions) and local air pollutants (e.g., sulfur dioxide and nitrogen oxide emissions) for the US regions are estimated over the period 1990–2017 based on eight related decoupling criteria."

"By employing the Phillips and Sul methodology, we examine for the first time in the literature whether the three decoupling indicators converge among the US regions."

"The findings indicate that for the carbon dioxide decoupling indicator three (merged) US regions converge to a steady state, while for sulfur dioxide and nitrogen oxide, two and seven convergence (merged) clubs are present."

Introduction

"The term decoupling was first adapted to literature by Zhang [23] and presented as an indicator by the OECD [63] report, which distinguishes between two types of decoupling effect: the absolute (when the environmental variable moves in the opposite direction from the economic growth at a stable or decreasing trend) and the relative (when the environmental variable is positive but at a slower rate than the growth of economic activity) decoupling effect."

"According to UNEP [64], the impact of decoupling is referring to a growth path under which a region can increase its economic growth by having specific policies under which the environmental pressures will deteriorate."

"The growth of the variables under scrutiny (environmental & economic activity) can be positive or negative, expressed as expansive coupling (growing link) and recessive coupling (recession link), while decoupling may be divided into three categories such as weak, strong, and recessive decoupling."

Data Description and Methodology

"This section describes the methodology we use to examine the convergence analysis developed by Phillips and Sul [65, 66] alongside the sample and variable description."

"In our analysis, we use emission and GDP data for the U.S. regions over the period 1990–2017."

"If among two time periods we obtain $\Delta CO_2 > 0$, $\Delta GDP > 0$, and the DI has a value greater than 1.2, then the state of the region is facing an "Expansionary Negative Decoupling" of CO_2 emissions."

"If a region among two time periods has $\Delta SO_2 < 0$, $\Delta GDP > 0$, and the DI value less than 0, then the state of the region is facing a "Strong Decoupling" of SO_2 emissions."

"We utilize Phillips and Sul [65, 66] methodological framework, to explore the existence of convergence clubs among U.S. regions' estimated decoupling indices over the examined period."

Empirical Findings

"Club 2 has 38 members (Alaska, Alabama, Arizona, California, Colorado, Connecticut, Georgia, Hawaii, Iowa, Illinois, Kansas, Kentucky, Louisiana, Massachusetts, Maryland, Maine, Michigan, Missouri, Montana, North Carolina, North Dakota, Nebraska, New Hampshire, Ohio, Oklahoma, Oregon, Pennsylvania, Rhode Island, South Carolina, South Dakota, Tennessee, Texas, Utah, Virginia, Washington, Wisconsin, West Virginia, and Wisconsin), while Club 3 consists only of two regions, namely Denver and Nevada."

"Only the initial convergence Club 1 (Idaho, Vermont, Louisiana, Alaska, Florida, New Jersey, Montana, New Mexico, Oregon, Missouri, Nebraska, Tennessee, West Virginia, Oklahoma, Maryland, Georgia, Utah, Arizona, South Carolina, Hawaii, Indiana, Alabama, Texas, Kansas, Pennsylvania, Michigan, Denver, Massachusetts, Iowa, Maine, Illinois, Wyoming, Washington, North Dakota, California, North Carolina, Ohio, Kentucky, Virginia, Colorado, Nevada, Minnesota, and New Hampshire) and Club 2 (Arkansas, Connecticut, South Dakota, and Wisconsin) can be merged into one and the relevant t-statistic (0.989) is larger than the critical value of -1.65 failing to reject the null hypothesis."

Conclusions

"The objective of this study is to examine the convergence patterns of decoupling factors of three environmental hazards (CO_2, SO_2, and NO_X) from economic growth across the U.S. regions over the period 1990–2017."

"On the opposite, the other local environmental hazard (NO_X emissions) seems to deviate from the "concentrated" spatial pattern, since eleven primary convergence clubs are detected across the US territory."

"This heterogeneity sheds some light on the future direction of the environmental policy that must be pursued by government officials and regulators to combat climate change and successfully decouple the NO_X emissions from the level of regional economic growth."

"The empirical analysis in this paper could be useful for government officials and policymakers in their efforts to combat environmental degradation alongside economic growth."

"Policymakers and government officials should seriously address these issues since the role of decoupling and the (regional/federal) environmental policies to achieve this, is one of the most challenging issues."

7.2.10 *Decomposing the Decoupling Relationship Between Energy-Related CO_2 Emissions and Economic Growth in China [119]*

This is a machine-generated summary of:

Li, Wei; Sun, Shuang; Li, Hao: Decomposing the decoupling relationship between energy-related CO_2 emissions and economic growth in China [119]

Published in: Natural Hazards (2015)

Link to original: https://doi.org/10.1007/s11069-015-1887-3.

If you want to cite the papers, please refer to the original.

For technical reasons we could not place the page where the original quote is coming from.

Abstract-Summary

"In order to explore the decoupling relationship and its influence factors between economic growth and carbon emissions in China, the decoupling elasticity decomposition quantitative model of carbon emissions based on extended Log-Mean Divisia Index and Tapio decoupling models is established in this paper."

"The carbon emissions induced by household (HOU) sector ranked fifth among the nine sectors; hence, we analyzed the HOU sector with the production (PRO) sector together."

"The economic output effect is the main contributor, and the energy intensity effect is the major inhibitor factor to the carbon emissions in both the PRO sector and HOU sector, respectively."

"The impact factors of the carbon emissions and the decoupling elasticity values are in complete agreement."

Introduction

"No one takes the life department into account when researching the decoupling relationship between energy-related carbon emissions and economic growth."

"To address these gaps, we established a Tapio model based on extended Kaya identity to study the decoupling relationship between energy-related carbon emissions and economic growth in both production sector and household sector."

"There are three primary objectives of this study: (1) apply Log-Mean Divisia Index (LMDI) model decomposing energy-related carbon emissions into energy mix, energy intensity, industry structure, economic output and population-scale effects and further analyze the reasons for each effect in depth; (2) employ Tapio decoupling model based on extended LMDI method calculating the decoupling elasticity values and ascertain the reasons of decoupling state in PRO sectors and HOU sector; (3) provide some useful recommendations for policy makers."

Methodology

"Tapio (2005) decoupling model, which is developed on the basis of the OECD decoupling model, has been widely utilized to research the relationship between economic growth and carbon emissions (Wang and Yang [67]; Gray and others [68])."

"The Tapio model is as follows:where $D_{c,g}$ is the decoupling factor; C is the carbon emissions for current year; ΔC is the variation of carbon emissions on current time compared with the base period; G is for gross domestic product in current year; and ΔG is the variation of gross domestic product on current time compared with the base period."

"The original Kaya identity is as follows.where C denotes carbon emission; E is total energy consumption; GDP is gross domestic product (GDP); and P is population scale."

"The Kaya identity is established to research the relationship between population, economic policy and the CO_2 emission caused by human activities and is widely used in the field of energy economy and environmental economy (Wang and others [69]; Liu and others [70]; Xu and others [71])."

Data Sources and Processing

"The data of energy consumption are collected from China Energy Statistical Yearbook (CESY)."

"Added value of industrial sectors, GDP and population data were collected from China Statistical Yearbook (CSY)."

"We used the data of GDP and industrial value added at 1996 price."

Results and Discussion

"The development of economy in turn contributes to increase in energy consumption and carbon emissions."

"In consistent with the production sector, the main inhibitory factor of energy-related carbon emissions of HOU is also energy intensity effect."

"If other energy consumption stays unchanged, a decrease in the proportional consumption of coal or an increase in the proportion of natural gas, carbon emissions would reduce (Xu and others [71])."

"Despite the fact that the energy intensity decoupling values have been negative in HOU sector, it showed an increasing tendency."

"Because decoupling value is an elasticity value, the annual growth rate of expenditure of HOU sector increases faster than that of the energy intensity effect, which is 8.27 and 4.90%, respectively."

"The key point to achieve decoupling of carbon emissions and economic growth in HOU sector is to decrease the energy intensity."

Conclusions and Policy Recommendations

"Based on LMDI method and Tapio model, this paper constructed an extended decomposition model for decoupling elasticity index between energy-related carbon emissions and economic growth in both production sector and household sector."

"The direct energy-related carbon emissions from HOU ranked fifth of the nine sectors."

"Energy intensity is the dominated factor in reducing carbon emissions, which results in carbon emission decline by 3.52 billion tons in PRO sector and 0.30 billion tons in HOU sector."

"Economic output and energy intensity are the first inhibiting and first promoting factor to decoupling the relation between carbon emissions and economic growth in both PRO sector and HOU sector."

"The factors to increase (reduce) carbon emissions and the decoupling elasticity values are in complete agreement, in both PRO sector and HOU sector."

7.2.11 Analysing Energy Intensity Trends and Decoupling of Growth from Energy Use in Indian Manufacturing Industries During 1973–1974 to 2011–2012 [120]

This is a machine-generated summary of:

Dasgupta, Shyamasree; Roy, Joyashree: Analysing energy intensity trends and decoupling of growth from energy use in Indian manufacturing industries during 1973–1974 to 2011–2012 [120]

Published in: Energy Efficiency (2016)

Link to original: https://doi.org/10.1007/s12053-016-9497-9.

If you want to cite the papers, please refer to the original.

For technical reasons we could not place the page where the original quote is coming from.

Abstract-Summary

"This paper has two research objectives: first, it derives and analyses energy intensity trends for seven energy intensive manufacturing industries and the aggregate manufacturing sector in India for the period 1973–1974 to 2011–2012 and compares the same with best practice benchmarks."

"The study finds faster decline in energy intensity in all the seven industries during the recent years (1998–1999 to 2011–2012)."

"Aluminium, cement and fertilizer industries are found to operate close to the global best-practice energy intensities with transformational changes in process technology."

"Declining energy intensity has been able to neutralize a major portion of the growing energy demand resulting in decoupling trends, especially in recent years, with more energy efficiency-related voluntary and mandatory policies in place)."

Introduction

"There is a range of studies dealing with energy efficiency trends of industries in the context of OECD and other developed countries (Howarth and others [72]; Farla and others [73]; Worrell and others [74]; Schipper and others [75, 76]; Sun [77]; Liaskas and others [78]; Miketa [79]; Phylipsen and others [80]; Allcott and Greenstone [81]; IEA [82]; Voigt and others [83])."

"In India, while there are some studies dealing with the energy intensity of the overall manufacturing industries (Dasgupta and Roy [84–86]; Bhaduri and Chaturvedi [87]; Kumar [88]; Mukherjee [89]; Goldar [90]) and their disaggregate counterparts including aluminium, iron and steel, cement, paper and textile industries (Sahu and Narayanan [91]; Reddy and Ray [92, 93]), there is dearth of comprehensive literature analysing long-term trends in energy intensity comparable with the best available technology in the world for a comprehensive set of industries at a disaggregate level."

"Against this backdrop, the present study provides a comprehensive analysis of energy intensity (EI) trends of these manufacturing industries prior to implementation of PAT and draws a comparison with respective best practice EIs (BPEI) acting as an international benchmark."

"It contributes to the energy efficiency and decoupling literature by providing evidence based on EI trends for a set of significant energy intensive manufacturing industries in India for a period of almost four decades (1973–1974 to 2011–2012)."

Analytical Framework

"EI in industrial sector (i.e. energy use per unit of production) has been perceived in many ways: measuring both energy inputs and activity in monetary units (EI_1); measuring both energy inputs and activity in physical units (e.g. GJ/ton) (EI_2); measuring activity in monetary terms while measuring energy use in physical units (e.g. MJ per thousand money value of sales or production) (EI_3)."

"For decomposition analysis, the coverage of industries, timeframe considered and output and energy variables are similar to that of EI_1."

"For the first group, the structural change represents change in the share of PAT and other energy intensive industries in the manufacturing sector."

"For the second group, the structural change refers to changing share of relatively higher energy intensive PAT industries such as cement, iron and steel as compared to the others."

Results and Discussion

"This trend is important given the fact that the industrial policies adopted since early 1990s encouraged production growth in these energy intensive manufacturing industries but there was no thrust on any mandatory energy efficiency policy per se."

"The declining trend of EI during the second sub-period clearly suggests that the massive growth in production in these industries during the era of economic liberalization was accompanied by gains in energy efficiency."

"There was a penetration of energy efficient natural gas-based technology in the industry with increasing efficiency of the plants."

"While EI trends of Indian manufacturing industries (both EI_1 and EI_2) exhibit steady decline with potential for further improvements, the question still remains to what extent such trend has contributed to decoupling of energy demand from industrial activity growth."

"This period witnessed negligible decline in EI resulting in coupled growth of activity and energy use in aggregate manufacturing sectors as well as in PAT industries."

Conclusion

"For the industries operating close to the external benchmark, the question that immediately follows is how they can manage their energy use more efficiently, even they met the world benchmark."

"Mandatory energy efficiency policies such as PAT encouraging the industrial units to achieve the 'internal benchmark' can play a crucial role here."

"The transition pathway of cement industry clearly indicates that policy has an important role to play when energy efficiency potential lies beyond process upgradation."

"There is a clear need for comprehensive policy guidelines on how this industry can realize and achieve further energy efficiency through recovery, use and management of waste which can contribute to the goal of reduction of energy demand."

"Given this background, it will be interesting to observe how the mandatory energy efficiency policy PAT can contribute in enhanced energy efficiency gains in these industries to reinforce the decoupling trend."

7.2.12 Assessing Pakistan's Energy Use, Environmental Degradation, and Economic Progress Based on Tapio Decoupling Model [121]

This is a machine-generated summary of:

Yasmeen, Humaira; Tan, Qingmei: Assessing Pakistan's energy use, environmental degradation, and economic progress based on Tapio decoupling model [121]

Published in: Environmental Science and Pollution Research (2021)

Link to original: https://doi.org/10.1007/s11356-021-15413-x.

If you want to cite the papers, please refer to the original.

For technical reasons we could not place the page where the original quote is coming from.

Abstract-Summary

"A comprehensive assessment of energy use, environmental degradation, and economic progress can play a significant role in transition towards low-carbon economy, and it can serve as a reference for the green economic development for the rest of the developing world."

"Following this, we have analyzed the decoupling relation among environmental degradation, energy use, and economic progress in Pakistan."

"The study adopted the comprehensive data from year 1972–2017 and applied Tapio decoupling method to explore the decoupling status of environmental degradation, energy use, and economic progress in Pakistan."

"From the decoupling point of view, we found the decoupling linkage between energy use and carbon emissions that is growth negative decoupling, whereas a weak decoupling relation has been observed among carbon emissions and economic progress which means that in most of the year's county has achieved more economic growth compared with the carbon emissions."

"The similar weak decoupling relationship was found among energy use and economic progress."

Introduction

"To develop low-carbon economy and advocate energy conservation and emission reduction is one of the key matters of investigation in the light of economic progress and environmental degradation."

"Rehman and others [94] employed NARDL approach and studied the role of renewable energy toward carbon emissions; their finding also suggest that renewable energy is a way forward to improve environmental quality."

"The comprehensive assessment of year over year to explore the linkage of energy use, economic progress, and environmental degradation is extremely important."

"In view of the said research questions, this study aims to conceptualize the role of various kinds of energy consumption toward carbon emissions in Pakistan over the years."

"Following this, we explored the role of various kinds of energy use toward carbon emissions in Pakistan over the years to provide in-depth information about policy formation."

Literature Review

"Over the past several decades, various governments have introduced distinguished policies for the energy sector which significantly influenced the environment, in this regard, a large number of studies have explored the linkage of energy consumption and other economic factors with carbon emissions."

"Shaheen and others [95] employed ARDL model to study the role of energy use and GDP toward carbon emissions in Pakistan; they found the GDP growth and energy use intensify carbon emissions."

"None of the study in the past employed Tapio decoupling model in the context of Pakistan to understand the linkage of energy and environment over the years."

"Mahmood and others [96] measured the role of energy and industrial production toward environment using ARDL method."

"Khan and others [97, 98] employed ARDL model on the data of Pakistan and found that financial development stimulates carbon emissions in the country."

Research Methods

"The appraisal of the recent environmental pressure and economic development status lacks reference value, whereas Tapio decoupling analysis selects the previous year as the base period for the calculation, and it can also evaluate the decoupling of emission reduction under the decoupling elasticity index of energy consumption, carbon emissions, and economic growth."

"This article uses the Tapio decoupling elastic analysis method to analyze the decoupling relationship between energy consumption, carbon emissions, and economic progress in Pakistan."

"This paper studies the decoupling relationship between energy consumption, carbon emissions, and economic progress in Pakistan and chooses to analyze it by constructing a Tapio decoupling model."

"Convert the decoupling elasticity between the country's energy consumption, carbon emissions, and economic progress."

Results and Findings

"In years 1976, 2004, 2011, and 2013, it is seen a negative growth rate in term of carbon emissions which is similar with the growth trend of energy consumption."

"In few years, such as 1974, 1991, 1997, and 2002, there was a decline in decoupling and a connection between the energy consumption and carbon emissions in Pakistan."

"We found the decoupling linkage between energy consumption and carbon emissions is growth negative decoupling, whereas, in very few years, a growth link and also recession decoupling has been seen which is almost similar to the actual situation of the county."

"In few years, in which we found the growth link between energy consumptions and carbon emissions, it indicates that the growth rate of energy consumption value of Pakistan is similar to that of carbon emissions."

Discussions

"Many studies in context of Pakistan have been done which emphasize on overall energy consumptions and carbon emissions such as the studies of M. Hussain and others [99], Mirza and Kanwal [100], and Lin and Raza [101]."

"None of the study explored how oil consumption contributes toward carbon emissions in the country."

"For the first time, we have checked in context of Pakistan that coal consumption along with other kinds of energy consumptions such as oil, natural gas, and coal contributes toward carbon emissions."

"We have used Tapio decoupling model to perform decoupling analysis between energy consumption, carbon emissions, and economic progress in Pakistan."

"It was inferred that there is a positive association between energy consumption and carbon emissions."

"Mirza and Kanwal [100] used dynamic causality analysis and confirmed the positive relation of energy and carbon emissions in Pakistan."

Conclusion and Implications

"We have tried to first investigate the relationship between various kinds of energy consumptions and carbon emissions in Pakistan."

"Following this, we checked the decoupling relation of energy consumption, economic progress, and carbon emissions."

"Three kinds of energy consumptions (oil, natural gas, and coal) were analyzed to find any relation with carbon emissions, and we excluded renewable energy while calculating carbon emissions from energy use."

"The results indicate that coal consumption in the country positively stimulates carbon emissions."

"The study assesses and conceptualizes various kinds of energy consumptions, economic growth, and carbon emissions in Pakistan."

"Of all, I indicate that, due to data availability and other issues, only the energy consumption of coal, oil, and natural gas was used for calculation of carbon emissions."

7.2.13 Spatiotemporal Dynamics and Decoupling Mechanism of Economic Growth and Carbon Emissions in an Urban Agglomeration of China [122]

This is a machine-generated summary of:

Hu, Han; Lv, Tiangui; Zhang, Xinmin; Fu, Shufei; Geng, Can; Li, Zeying: Spatiotemporal dynamics and decoupling mechanism of economic growth and carbon emissions in an urban agglomeration of China [122]

Published in: Environmental Monitoring and Assessment (2022)

Link to original: https://doi.org/10.1007/s10661-022-10195-52.

If you want to cite the papers, please refer to the original.

For technical reasons we could not place the page where the original quote is coming from.

Abstract-Summary

"Carbon emissions and economic growth are two contradictions in urban development, and their decoupling is related to the sustainable development of cities."

"Low level and medium level areas continue to decrease, and relatively high level area gradually become dominant [2]. Spatially, carbon emissions are in a pattern of middle-hot and east-cold."

"Jiangxi is in the sub-cold and coldspot area, while the hotspot area is driven by the transformation from Wuhan's single-core to Wuhan and Changsha's dual-core [3]. Since 2010, most cities have been in a good decoupling state, and weak decoupling cities have risen from 35.5% in the initial period to 87.1% in 2010–2011, but the decoupling situation of industrial cities with more high-energy-consuming industries still rebounded slightly [4]. The economic level and energy intensity effect had the most significant impact on the economic decoupling of carbon emissions, whose absolute contribution rates were greater than 35%."

"Urbanization and economic level both play a positive role in promoting carbon emissions, and the energy intensity plays a negative role in retarding carbon emissions."

"We suggest that decoupling carbon emissions from economic growth requires developing green urbanization and a decarbonized economy, optimizing the structure of energy consumption and guiding rational population flow."

Introduction

"Many scholars have focused their attention on the modeling of the drivers of carbon emission decoupling, trying to reveal the carbon emission growth code to help countries formulate reasonable economic and energy policies (Liu & Xiao [102])."

"Wang and others [103] used the decoupling model and the LMDI model to study the decoupling performance of the industrial economic output and carbon emissions and explored its decoupling trends and effects."

"Other scholars have also paid attention to the decoupling of carbon emissions and economic growth in the agricultural sector."

"Looking at the existing research, the current academic research on carbon emissions and economic growth has become mature, but the research on the factors affecting the decoupling of carbon emissions and the economy is more focused on the perspective of economy, energy, and population."

Conceptual Framework

"Economic growth mainly affects carbon emissions through three paths, which are the energy transformation process, economic development process, and population migration process."

"These processes include positive factors that promote the decoupling of carbon emissions caused by energy reform and negative factors that hinder the decoupling of carbon emissions brought about by economic development."

"On the bright side, behind the energy transformation process are technological progress and fuel substitution, which is the primary driver of carbon emissions decoupling."

"The economic development process is the most essential activity that hinders carbon emission reduction."

"Economic growth has promoted an increase in carbon emissions, which is mainly due to more energy consumption and greenhouse gas emissions brought about by urbanization and the improvement of the economic level (Ortega-Ruiz and others [104]; Song and others [105])."

Study Area and Methodology

"Taking the UAMRYR as the study area to carry out research is of great necessity and practical significance for achieving carbon emission reduction targets in central China under the background of high-quality development."

"Spatial data analysis combines data visualization and statistical methods to analyze spatial data, which can more intuitively reflect the spatial heterogeneity and spatial connection pattern of carbon emissions."

"The Kaya model is suitable for studying the relationship between economy, technology, population, and carbon emissions."

"To analyze the impact of urbanization and economic level on carbon emissions, this paper expands the Kaya model and incorporates urbanization, per capita industrial output value, energy intensity, carbon emissions density, and population into the influencing factor analysis system."

Results

"The carbon emission hotspot area is basically located in the circle [2]. The carbon emission center has always been in Xianning, but it has also experienced a continuous migration to the southeast, with a total offset distance of 26.64 km, of which the largest offset occurred between 2010 and 2013, with a moving distance of 14.05 km, which is mainly due to Hubei's enterprises continue to transform and upgrade, and Jiangxi has undertaken a large number of high-input industries in coastal areas in recent years [3]. From the perspective of the difference in direction angle, taking 2006–2010 as the benchmark, the center of the standard deviation ellipse rotates approximately 4.26° to the southeast, which shows that the carbon emissions in the southeast of the middle triangle increase faster than those in the northwest [4]. The expansion of flattening reflects a more obvious carbon emission trend."

Discussion and Policy Implications

"Since 2011, the UAMRYR has issued a series of policies to vigorously promote energy conservation, emissions reduction, and consumption reduction, carry out low-carbon development pilot projects, and promote the development of the carbon trading market and industrial transformation and upgrading and green energy development in resource-based cities such as Xinyu."

"In terms of influencing factors, many scholars have analyzed the mechanism of carbon emission decoupling by factors such as industrial structure, imports and exports, population, and energy (Piłatowska & Włodarczyk [106]; Tenaw & Hawitibo [107]), but relatively few studies have introduced urbanization into the analysis."

"Judging from the research results, the identified key factors and their mechanisms of decoupling carbon emissions are consistent with the existing research, all of which warn us to pay attention to the effects of economic level and energy intensity (Liu and others [108]; Wang and others [109])."

Conclusions and Policy Implications

"The development of green urbanization with low energy consumption, low emissions, and low pollution will contribute to carbon emission reduction."

"Promoting the upgrading of the industrial structure and developing a decarbonization economy are powerful measures to effectively solve the decoupling problem between carbon emissions and the economy."

"The government also needs to speed up the research on low-carbon key technologies such as industrial energy conservation and carbon reduction technology, zero carbon emissions, and negative carbon emissions in old industrial cities such as Wuhan, and pay attention to the upgrading and transformation of environmental protection equipment."

"With respect to decoupling carbon emissions from economic growth, we suggest that the government can develop green urbanization and a decarbonized economy, and make changes from the perspectives of optimizing the structure of energy and guiding the rational flow of the population."

Bibliography

1. Chang CC (2010) A multivariate causality test of carbon dioxide emissions, energy consumption and economic growth in China. Appl Energy 87(11):3533–3537
2. Samir KC, Lutz W (2017) The human core of the shared socioeconomic pathways: population scenarios by age, sex and level of education for all countries to 2100. Glob Environ Chang 42:181–192

3. Briefing US (2013) International energy outlook 2013. US Energy Information Administration
4. Wang C, Chen J, Zou J (2005) Decomposition of energy-related CO2 emission in China: 1957–2000. Energy 30(1):73–83
5. Zhang H, Lahr ML (2014) China's energy consumption change from 1987 to 2007: a multi-regional structural decomposition analysis. Energy Policy 67:682–693
6. Wang Q, Li R (2016a) Journey to burning half of global coal: trajectory and drivers of China's coal use. Renew Sustain Energy Rev 58:341–346
7. Wang Q, Li R (2016b) Drivers for energy consumption: a comparative analysis of China and India. Renew Sustain Energy Rev 62:954–962
8. Naminse EY, Zhuang J (2018) Economic growth, energy intensity, and carbon dioxide emissions in China. Pol J Environ Stud 27(5):2193–2201
9. Jiang JJ, Ye B, Zhou N, Zhang XL (2019) Decoupling analysis and environmental Kuznets curve modelling of provincial-level CO2 emissions and economic growth in China: a case study. J Clean Prod 212:1242–1255
10. Siping J, Wendai L, Liu M, Xiangjun Y, Hongjuan Y, Yongming C, Haiyun C, Hayat T, Alsaedi A, Ahmad B (2019) Decoupling environmental pressures from economic growth based on emissions monetization: Case in Yunnan, China. J Clean Prod 208:1563–1576
11. Luo Y, Long X, Wu C, Zhang J (2017) Decoupling CO2 emissions from economic growth in agricultural sector across 30 Chinese provinces from 1997 to 2014. J Clean Prod 159:220–228
12. Xu SC, He ZX, Long RY, Chen H, Han HM, Zhang WW (2016) Comparative analysis of the regional contributions to carbon emissions in China. J Clean Prod 127:406–417
13. Zhou X, Zhang M, Zhou M, Zhou M (2017) A comparative study on decoupling relationship and influence factors between China's regional economic development and industrial energy-related carbon emissions. J Clean Prod 142:783–800
14. Zhang X, Zhao Y, Wang C, Wang F, Qiu F (2019) Decoupling effect and sectoral attribution analysis of industrial energy-related carbon emissions in Xinjiang, China. Ecol Ind 97:1–9
15. Zhao X, Zhang X, Li N, Shao S, Geng Y (2017) Decoupling economic growth from carbon dioxide emissions in China: a sectoral factor decomposition analysis. J Clean Prod 142:3500–3516
16. Qi T, Weng Y, Zhang X, He J (2016) An analysis of the driving factors of energy-related CO2 emission reduction in China from 2005 to 2013. Energy Econ 60:15–22
17. Wang WW, Zhang M, Zhou M (2011) Using LMDI method to analyze transport sector CO2 emissions in China. Energy 36(10):5909–5915
18. Fan T, Luo R, Xia H, Li X (2015) Using LMDI method to analyze the influencing factors of carbon emissions in China's petrochemical industries. Nat Hazards 75(2):319–332
19. Tang Z, Shang J, Shi C, Liu Z, Bi K (2014) Decoupling indicators of CO2 emissions from the tourism industry in China: 1990–2012. Ecol Ind 46:390–397
20. Liu LC, Fan Y, Wu G, Wei YM (2007) Using LMDI method to analyze the change of China's industrial CO2 emissions from final fuel use: an empirical analysis. Energy Policy 35(11):5892–5900
21. Khan S, Majeed MT (2019) Decomposition and decoupling analysis of carbon emissions from economic growth: a case study of Pakistan. Pak J Commer Soc Sci 13(4):868–891
22. Guivarch C, Mathy S (2012) Energy-GDP decoupling in a second best world—a case study on India. Clim Change 113(2):339–356
23. Zhang, Z. (2000). Decoupling China's carbon emissions increase from economic growth: An economic analysis and policy implications. World Development, 28(4), 739–752
24. OECD (2002). Indicators to measure decoupling of environmental pressure from economic growth. SG/SD(2002)1/FINAL. Retrieved from http://www.oecd.org/officialdocuments/publicdisplaydocumentpdf/?cote=sg/sd(2002)1/final&doclanguage=en
25. Everett, T., Ishwaran, M., Ansaloni, G., & Rubin, A. (2010). Economic growth and the environment. Defra Evidence and Analysis Series Paper 2
26. Wang, W., Liu, R., Zhang, M., & Li, H. (2013). Decomposing decoupling of energy-related CO2 emissions and economic growth in Jiangsu Province. Energy for Sustainable Development, 17, 62–71

27. Azar, C., Holmberg, J., & Karlsson, S. (2002). Decoupling—past trends and prospects for the future. Stockholm: Environmental Advisory Council
28. Galeotti, M. (2003). Environment and Economic Growth: Is Technical Change the Key to Decoupling? FEEM Working Paper No. 90.2003
29. van der Voet, E., van Oers, L., Moll, S., et al. (2005). Policy review on decoupling: development of indicators to assess decoupling of economic development and environmental pressure in the EU-25 and AC-3 countries. Centre of Environmental Sciences (CML), Leiden
30. Næss, P., & Høyer, K. (2009). The emperor's green clothes: Growth, decoupling, and capitalism. Capitalism Nature Socialism, 20(3), 74–95. https://doi.org/10.1080/10455750903215753
31. Ru, X., Chen, S., & Dong, H. (2012). Na empirical study on relationship between economic growth and carbon emissions based on decoupling theory. Journal of Sustainable Development, 5(8), 43–51
32. Bovenberg, A. L., & Smulders, S. (1995). Environmental quality and pollution-augmenting technological change in a two-sector endogenous growth model. Journal of Public Economics, 57, 369–391
33. Tahvonen, O. (1997). Fossil fuels, stock externalities, and backstop technology. The Canadian Journal of Economics, 30(4a), 855–874
34. Di Vita, G. (2006). Natural resources dynamics: Exhaustible and renewable resources, and the rate of technical substitution. Resources Policy, 31, 172–182
35. Lear, J. (2008). Radical hope: Ethics in the face of cultural devastation. Cambridge: Harvard University Press
36. Busu M (2020) Analyzing the impact of the renewable energy sources on economic growth at the EU level using an ARDL model. Mathematics 8:1367. https://doi.org/10.3390/math8081367
37. Maddala GS, Wu S (1999) A comparative study of unit root tests with panel data and a new simple test. Oxford Bull Econ Stat 61:631–652
38. Choi M (2001) Threat effect of foreign direct investment on labor union wage premium. https://doi.org/10.2139/ssrn.335480 Accessed 15 May 2021.
39. Levin A, Lin CF, Chu CSJ (2002) Unit root tests in panel data: asymptotic and finite-sample properties. J Econ 108(1):1–24
40. Dietzenbacher E, Los B (1998) Structural decomposition techniques: sense and sensitivity. Econ Syst Res 10(4):307–324
41. Ren SG, Hu Z (2012) Effects of decoupling of carbon dioxide emission by Chinese nonferrous metals industry. Energy Policy 43:407–414
42. Climent F, Pardo A (2007) Decoupling factors on the energy-output linkage: the Spanish case. Energy Policy 35:522–528
43. Freitas, L., & Kaneko, S. (2011). Decomposing the decoupling CO2 emissions and economic growth in Brazil. Ecological Economics, 70, 1459–1469
44. Zhang M, Wang WW (2013b) Decoupling analysis of electricity consumption from economic growth in China. J Energy South Afr 24:57–66
45. Ang, B. (2004). Decomposing analysis for policy making in energy: Which is the preferred method? Energy Policy, 32, 1131–1139
46. Ang BW, Liu N (2007) Handling zero values in the logarithmic mean Divisia index decomposition approach. Energy Policy 35:238–246
47. CESY (China Energy Statistical Yearbook) (1991–1996, 1997–1999, 2000–2002, 2003, 2004. 2005, 2006, 2007, 2008, 2009, 2010, 2011, 2012) National Bureau of Statistics of China, National Development and Reform Commission
48. Pao HT, Chen CC (2019) Decoupling strategies: CO2 emissions, energy resources, and economic growth in the group of twenty. J Clean Prod 206:907–919 https://doi.org/10.1016/j.jclepro.2018.09.190
49. Lapinskienė G, Peleckis K, Radavičius M (2015) Economic development and greenhouse gas emissions in the European Union countries. J Bus Econ Manag 16(6):1109–1123. https://doi.org/10.3846/16111699.2015.1112830

50. Lee S, Kim M, Lee J (2017) Analyzing the impact of nuclear power on CO2 emissions. Sustainability 9(8):1428. https://doi.org/10.3390/su9081428
51. Marques AC, Fuinhas JA, Nunes AR (2016) Electricity generation mix and economic growth: what role is being played by nuclear sources and carbon dioxide emissions in France? Energ Policy 92:7–19. https://doi.org/10.1016/j.enpol.2016.01.027
52. Onater-Isberk E (2016) Environmental Kuznets curve under noncarbohydrate energy. Renew Sust Energ Rev 64:338–347. https://doi.org/10.1016/j.rser.2016.06.022
53. Leslie P (1985) A stochastic model for studying the properties of certain biological systems by numerical methods. Biometrika 45:16–31. https://doi.org/10.2307/2333042
54. Liu P, Gopalsamy K (1997) On a model of competition in periodic environments. Appl Math Comput 82(2–3):207–238. https://doi.org/10.1016/S0096-3003(96)00044-6
55. Kim J, Lee DJ, Ahn J (2006) A dynamic competition analysis on the Korean mobile phone market using competitive diffusion model. Comput Ind Eng 51(1):174–182. https://doi.org/10.1016/j.cie.2006.07.009
56. Grinter, A. (2019). Energy and the wealth of nations: an introduction to biophysical economics. Economic Record, 95 (308), 167–168
57. Warr, B., Ayres, R., & Eisenmenger, N. (2010). Energy use and economic development: A comparative analysis of useful work supply in Austria, Japan, the United Kingdom and the US during 100 years of economic growth. Ecological Economic, 69(10), 1904–1917
58. EEA. (2016). Trends in energy intensity, gross domestic product and gross inland energy consumption. European Environment Agency.
59. Ekins, P. (2009). The economic growth engine: how energy and work drive material prosperity. International Journal of Sustainable Engineering, 3(3), 228–228
60. Tapio, P. (2005). Towards a theory of decoupling: degrees of decoupling in the EU and the case of road traffic in Finland between 1970 and 2001. Transport Policy, 12(2), 137–151
61. Wu, Y., Zhu, Q., & Zhu, B. (2018). Comparisons of decoupling trends of global economic growth and energy consumption between developed and developing countries. Energy Policy, 116, 30–38
62. Wang, X. (2018). Technology introduction, absorptive capacity and China's economic growth: Theoretical logic research and threshold regression analysis of Provincial Panel Data. International Business, 2, 45–57
63. OECD: Indicators to measure decoupling of environmental pressure from economic growth. OECD report. OECD, Paris (2002)
64. United Nations Environment Programme (UNEP): Decoupling Natural Resource Use and Environmental Impacts from Economic Growth (2011). http://wedocs.unep.org/handle/20.500.11822/9816
65. Phillips, P.C., Sul, D.: Transition modeling and econometric convergence tests. Econometrica 75(6), 1771–1855 (2007)
66. Phillips, P.C., Sul, D.: Economic transition and growth. J. Appl. Econ. 24(7), 1153–1185 (2009)
67. Wang ZH, Yang L (2015) Delinking indicators on regional industry development and carbon emissions: Beijing-Tianjin-Hebei economic band case. Ecol Indic 48:41–48
68. Gray D, Anable J, Illingworth L (2006) Decoupling the link between economic growth, transport growth and carbon emissions in Scotland. Technical Report, Robert Gordon University
69. Wang ZH, Liu W, Yin JH (2014) Driving forces of indirect carbon emissions from household consumption in China: an input-output decomposition analysis. Nat Hazards. https://doi.org/10.1007/s11069-014-1114-7
70. Liu LC, Wei YM, Fan Y, Wu G (2007) The impact of lifestyle on energy use and CO2 emission: an empirical analysis of China's residents. Energy Policy 35:247–257
71. Xu SC, He ZX, Long RY (2014) Factors that influence carbon emissions due to energy consumption in China: decomposition analysis using LMDI. Appl Energy 127:182–193
72. Howarth, R. B., Schipper, L., & Anderson, B. (1993). The Structure & Intensity of energy use: trends in five OECD nations. The Energy Journal., 14(2), 27–45

73. Farla, J., Blok, K., & Schipper, L. (1997). Energy efficiency developments in the paper and pulp industry. Energy Policy, 25(7–9), 745–758
74. Worrell, E., Price, L., Martin, N., Farla, J., & Schaeffer, R. (1997). Energy intensity in the iron and steel industry: a comparison of physical and economic indicators. Energy Policy, 25(7–9), 727–744
75. Schipper, L., Unander, F., Marie-Lilliu, C. (1998). Energy Use in Canada in an International Perspective. Comparison of trends through the early 1990s. International Energy Agency
76. Schipper, L., Unander, F., Marie-Lilliu, C., Walker, I., Gorham, R. (2000). Energy use in Australia in an International Perspective: Comparison of trends through the mid 1990s. International Energy Agency, Paris and Canberra- Department of Industry, Science and Resources
77. Sun, J. W. (1999). Decomposition of aggregate CO2 emissions in the OECD. 1960–1995. The Energy Journal, 20(3), 147–155
78. Liaskas, K., Mavrotas, G., Mandaraka, M., & Diakoulaki, D. (2000). Decomposition of industrial CO2 emissions: the case of European Union. Energy Economics, 22(4), 383–394
79. Miketa, A. (2001). Analysis of energy intensity developments in manufacturing sectors in industrialized and developing countries. Energy Policy, 29(2001), 769–775
80. Phylipsen, D., Blok, K., Worrell, E., & de Beer, J. (2002). Benchmarking the energy efficiency of Dutch industry: an assessment of the expected effect on energy consumption and CO2 emissions. Energy Policy, 30(2002), 663–679
81. Allcott, H., & Greenstone, M. (2012). Is there an energy efficiency gap? Journal of Economic Perspectives, 26(1), 3–28
82. IEA (2012). World energy outlook 2012. Paris: International Energy Agency
83. Voigt, S., De Cian, E., Schymura, M., & Verdolini, E. (2014). Energy intensity developments in 40 major economies: structural change or technology improvement? Energy Economics, 41, 47–62
84. Dasgupta, M., & Roy, J. (2000). Manufacturing energy use in India: a decomposition analysis. Asian Journal of Energy and Environment, 1(3), 223–247
85. Dasgupta, M., Roy, J. (2002a). Energy Consumption in India: an Indicator Analysis. Development Alternatives, October, pp 12–13. http://www.devalt.org/newsletter/oct02/of_6.htm Accessed 22 July 2016
86. Dasgupta, M., Roy, J. (2002b). Energy consumption in India: An indicator analysis. Development Alternatives, (October), 12–13. http://www.devalt.org/newsletter/oct02/of_6.htm
87. Bhaduri, S. N., & Chaturvedi, R. K. (2002). Decomposition of India's industrial energy use, a case study using energy intensity approach. International Journal of Global Energy Issues, 17(1), 92–105
88. Kumar, A. (2003). Energy Intensity: A quantitative exploration for Indian manufacturing. Working Paper, Indira Gandhi Institute of Development Research, Mumbai. http://papers.ssrn.com/sol3/papers.cfm?abstract_id=468440 . Accessed 22 July 2016
89. Mukherjee, K. (2008). Energy use efficiency in the Indian manufacturing sector: an interstate analysis. Energy Policy, 36(2008), 662–672
90. Goldar, B. (2010). Energy Intensity of Indian Manufacturing Firms: Effect of Energy Prices, Technology and Firm Characteristics. Institute of Economic Growth, Delhi, India. http://www.mse.ac.in/Frontier/m13%20Goldar%20A.pdf Accessed 22 July 2016
91. Sahu, S., Narayanan, K., (2009). Determinants of Energy Intensity: A Preliminary Investigation of Indian Manufacturing. Indian Institute of Technology, Mumbai, MPRA Paper no. 16606
92. Reddy, S., & Ray, B. K. (2010). Decomposition of energy consumption and energy intensity in Indian manufacturing industries. Energy for a Sustainable World, 14(8), 35–47
93. Reddy, S. B., & Ray, B. K. (2011). Understanding industrial energy use: physical energy intensity changes in Indian manufacturing sector. Energy Policy, 39(11), 7234–7243
94. Rehman A, Ma H, Ahmad M, Ozturk I, Işık C (2021a) An asymmetrical analysis to explore the dynamic impacts of CO2 emission to renewable energy, expenditures, foreign direct investment, and trade in Pakistan. Environmental Science and Pollution Research:1–13

95. Shaheen A, Sheng J, Arshad S (2020a) The dynamic linkage between income, energy consumption, urbanization and carbon emissions in Pakistan. Polish Journal of Environmental Studies 29(1)
96. Mahmood MT, Shahab S, Hafeez M (2020) Energy capacity, industrial production, and the environment: an empirical analysis from Pakistan. Environmental Science and Pollution Research 27(5):4830–4839
97. Khan MK, Teng J-Z, Khan MI (2019a) Effect of energy consumption and economic growth on carbon dioxide emissions in Pakistan with dynamic ARDL simulations approach. Environmental Science and Pollution Research 26(23):23480–23490
98. Khan MK, Teng J-Z, Khan MI, Khan MO (2019b) Impact of globalization, economic factors and energy consumption on CO2 emissions in Pakistan. Science of The Total Environment 688:424–436
99. Hussain M, Irfan Javaid M, Drake PR (2012) An econometric study of carbon dioxide (CO2) emissions, energy consumption, and economic growth of Pakistan. International Journal of Energy Sector Management 6(4):518–533
100. Mirza FM, Kanwal A (2017) Energy consumption, carbon emissions and economic growth in Pakistan: Dynamic causality analysis. Renewable and Sustainable Energy Reviews 72:1233–1240
101. Lin B, Raza MY (2019) Analysis of energy related CO2 emissions in Pakistan. Journal of cleaner production 219:981–993
102. Liu, D., & Xiao, B. (2018). Can China achieve its carbon emission peaking? A scenario analysis based on STIRPAT and system dynamics model. Ecological Indicators, 93, 647–657
103. Wang, Q., Zhao, M., & Li, R. (2019). Decoupling sectoral economic output from carbon emissions on city level: A comparative study of Beijing and Shanghai, China. Journal of Cleaner Production, 209, 126–133
104. Ortega-Ruiz, G., Mena-Nieto, A., & García-Ramos, J. (2020). Is India on the right pathway to reduce CO2 emissions? Decomposing an enlarged Kaya identity using the LMDI method for the period 1990–2016. Science of the Total Environment, 737, 139638
105. Song, X., Jia, J. S., Chen, C. D., & Chen, J. H. (2020). Spatio-temporal characteristics, decoupling relation and its driving factors of the carbon emission from energy consumption in underdeveloped Jiangxi Province. Acta Ecologica Sinica, 40, 7451–7463. (in Chinese)
106. Piłatowska, M., & Włodarczyk, A. (2018). Decoupling economic growth from carbon dioxide emissions in the EU countries. Montenegrin Journal of Economics, 14, 7–26
107. Tenaw, D., & Hawitibo, A. (2021). Carbon decoupling and economic growth in Africa: Evidence from production and consumption-based carbon emissions. Environment, Development and Sustainability, 6, 100040
108. Liu, X., Gao, C., Song, Y., Zhang, Y., Su, Q., & Tian, Y. (2017). Temporal and spatial pattern and driving factors of carbon emissions from fossil energy consumption in Hunan. Acta Ecologica Sinica, 37, 2476–2487. (in Chinese)
109. Wang, W., Kuang, Y., Huang, N., & Zhao, D. (2014). Empirical research on decoupling relationship between energy-related carbon emission and economic growth in Guangdong province based on extended Kaya identity. The Scientific World Journal, 11, 782750
110. Ozturk, Ilhan; Majeed, Muhammad Tariq; Khan, Sher Decoupling and decomposition analysis of environmental impact from economic growth: a comparative analysis of Pakistan, India, and China. Environmental and Ecological Statistics (2021). https://doi.org/10.1007/s10651-021-00495-3
111. Silva, Susana; Soares, Isabel; Afonso, Oscar Decoupling economic growth from emissions: the case of policies promoting resource substitution. Environment, Development and Sustainability (2020). https://doi.org/10.1007/s10668-020-00967-9
112. Shockley, Kenneth The Great Decoupling: Why Minimizing Humanity's Dependence on the Environment May Not Be Cause for Celebration. Journal of Agricultural and Environmental Ethics (2018). https://doi.org/10.1007/s10806-018-9736-5
113. Chen, Chun-Chih; Pao, Hsiao-Tien The causal link between circular economy and economic growth in EU-25. Environmental Science and Pollution Research (2022). https://doi.org/10.1007/s11356-022-21010-6

114. Banerjee, Sarmila; Bit, Jayita Economic growth versus climate balancing: some reflections on the sustainable management of forest resource in India. DECISION (2015). https://doi.org/10.1007/s40622-015-0090-4
115. Song, Yan; Zhang, Ming Using a new decoupling indicator (ZM decoupling indicator) to study the relationship between the economic growth and energy consumption in China. Natural Hazards (2017). https://doi.org/10.1007/s11069-017-2903-6
116. Pao, Hsiao-Tien; Chen, Chun-Chih Decoupling of environmental pressure and economic growth: evidence from high-income and nuclear-dependent countries. Environmental Science and Pollution Research (2019). https://doi.org/10.1007/s11356-019-07122-6
117. Zeng, LanTing; Ye, Azhong; Lin, Wen Deepening decoupling for sustainable development: evidence from threshold model. Energy Efficiency (2022). https://doi.org/10.1007/s12053-022-10031-2
118. Polemis, Michael L.; Fotis, Panagiotis; Tzeremes, Panayiotis G.; Tzeremes, Nickolaos G. On the examination of the decoupling effect of air pollutants from economic growth: a convergence analysis for the US. Letters in Spatial and Resource Sciences (2022). https://doi.org/10.1007/s12076-022-00320-w
119. Li, Wei; Sun, Shuang; Li, Hao Decomposing the decoupling relationship between energy-related CO_2 emissions and economic growth in China. Natural Hazards (2015). https://doi.org/10.1007/s11069-015-1887-3
120. Dasgupta, Shyamasree; Roy, Joyashree Analysing energy intensity trends and decoupling of growth from energy use in Indian manufacturing industries during 1973–1974 to 2011–2012. Energy Efficiency (2016). https://doi.org/10.1007/s12053-016-9497-9
121. Yasmeen, Humaira; Tan, Qingmei Assessing Pakistan's energy use, environmental degradation, and economic progress based on Tapio decoupling model. Environmental Science and Pollution Research (2021). https://doi.org/10.1007/s11356-021-15416-x
122. Hu, Han; Lv, Tiangui; Zhang, Xinmin; Fu, Shufei; Geng, Can; Li, Zeying Spatiotemporal dynamics and decoupling mechanism of economic growth and carbon emissions in an urban agglomeration of China. Environmental Monitoring and Assessment (2022). https://doi.org/10.1007/s10661-022-10195-5

Chapter 8
Accounting Stocks and Flows of Natural Capital

8.1 Introduction by Editor

Natural capital is the foundation of the human economy. Everything that is made, served or sold in markets consumes natural resources. Fossil fuel, land, water, gases, minerals, each of these are supplied by nature, and form the 'Natural Capital'. The human economy is the system of production, distribution, and consumption of goods that are made from this natural capital. The marketplace converts the manufactured goods and services into money.

Natural ecosystems provide the foundation needed for human economy to function. The physical and biological diversity of the planet generates ecosystems services that are vital for survival and the functioning of our economic engine. The hydrological cycle, food production, ground water recharge, carbon cycle, atmospheric gas balance, these and many other ecosystem services are critical for the economy.

However, much of this natural capital is not accounted for in national accounts. Economic production and social consumption lead to pollution, and drives changes in land use. This leads to deforestation, degradation of soil fertility, and reduction in biological diversity. The impacts have economic consequences, such as reduced crop yields, increased healthcare costs, price fluctuations in raw materials, and loss in tourism revenues. The pressures of economic growth have led to a sharp decline in stocks of non-renewable resources, and pollution of renewable resources. Is it possible to account for the depleted stock of non-renewable resources, degradation of ecosystem services, and pollution of renewable resources?

Robert Costanza et al. (1997) estimated the economic value of 17 ecosystem services for 16 biomes, based on a synthesis of published studies and some original calculations. They estimated the value, for the entire biosphere, to be in the range of \$16 - 54 trillion/yr., with an average of \$33 trillion/yr (Costanza & al, 1997). Most of this value, they estimated, was outside of markets, meaning it was not accounted for. Since 1997 there has been significant development in this topic.

G. Nulkar (ed.), *The Economics of Sustainable Development*,
https://doi.org/10.1007/978-981-99-7379-8_8

Recognizing the importance of natural capital, several countries and organizations are working out ways to embed natural resources accounting into national accounting frameworks. The European Union, for example, is developing experimental natural capital accounts, which aim to measure the quantity and value of natural assets and the ecosystem services they provide. In India there have been efforts to develop a "green GDP" measure that accounts for the depletion and degradation of natural resources. The United States Bureau of Economic Analysis has explored methods for incorporating natural capital into the national accounts, though progress has been slower compared to the EU and India (Partyn et al., 2020; Koshy et al., 2019; Costanza et al., 2009; Costanza et al., 2013). Nonetheless, the growing recognition of the need to move beyond traditional GDP as a measure of economic progress has spurred further research and innovation in this area. For individual companies, accounting for natural capital can provide valuable insights into their dependencies and impacts on the environment, and help them prepare for future risks and opportunities. One Swedish company in the paper and pulp business has already begun to develop "natural capital statements" that report on their use of natural resources and the associated costs and benefits. This approach can help companies better understand their relationship with the natural environment and make more informed decisions about their operations and investments.

Accounting for natural capital is necessary to maintain sustained economic growth and development of a nation. The process of building up and prudently managing a portfolio of assets, including manufactured capital, natural capital, and human and social capital, is at the foundation of sustainable development. Nobel laureate Joseph Stiglitz had famously remarked that corporate performance is evaluated on its income and balance sheet, however, nations only generate an income statement (GDP) while leaving out accounting of the natural capital.

The Economics of Ecosystems and Biodiversity (TEEB) and the United Nations Natural Capital Accounting and Valuation of Ecosystem Services (NCAVES) are two initiatives that are aimed at capturing the value of ecosystem services and incorporating them into national accounting decisions (Bouma & Beukering, 2015). The ecosystem services approach aims to establish the value that society derives from the natural world, such that the true cost of proposed development actions becomes apparent to decision-makers. The Economics of Ecosystems and Biodiversity (TEEB) initiative was launched in 2007 with the intention of informing the global economic benefits of biodiversity and to highlight the growing costs of biodiversity loss and ecosystem degradation (Bouma & Beukering, 2015). It presents an approach to help decision-makers recognize, demonstrate, and capture the values of ecosystem services and biodiversity. This initiative has been influential in shifting the discourse around conservation from a purely ecological perspective to one that also considers the economic implications of environmental degradation.

Similarly, the United Nations Natural Capital Accounting and Valuation of Ecosystem Services (NCAVES) project, launched in 2017, aims to advance the theory and practice of ecosystem accounting. This project provides a framework for integrating the value of ecosystem services into national accounting systems, allowing policymakers to better understand the trade-offs between economic development and environmental protection (Lafortezza & Chen, 2016; Bouma & Beukering, 2015).

By quantifying the economic value of ecosystem services, both the TEEB and the NCAVES have the potential to disclose the social costs or benefits that would otherwise remain hidden. With these frameworks, it would be easier to capture information that otherwise be excluded from economic decision-making processes.

In this chapter, we present studies on putting biodiversity into national accounts, soil as a natural capital, quantifying waste and pollution, and how green recovery can be achieved with natural accounting. The studies range from India to European Union, and present readers with compelling arguments to measure and account for natural capital.

References

Bouma, J., & Beukering, P V. (2015, January 29). Ecosystem services: from concept to practice. Cambridge University Press, 3–22. https://doi.org/10.1017/cbo9781107477612.002

Brandt, N., Schreyer, P., & Zipperer, V. (2016, August 25). Productivity Measurement with Natural Capital. Wiley-Blackwell, 63(s1). https://doi.org/10.1111/roiw.12247

Costanza, R., Hart, M., Posner, S., & Talberth, J. (2009, January 1). Beyond GDP: The Need for New Measures of Progress. https://open.bu.edu/bitstream/2144/22665/1/PP-004-GDP.pdf

Koshy, A., Raynaud, J., Özdemiroğlu, E., & Provins, A. (2019, July 3). Natural Capital Statements: a case study on SCA, a Swedish paper and pulp company. Taylor & Francis, 8(4), 394–412. https://doi.org/10.1080/21606544.2019.1635917

Lafortezza, R., & Chen, J. (2016, May 1). The provision of ecosystem services in response to global change: Evidences and applications. Elsevier BV, 147, 576–579. https://doi.org/10.1016/j.envres.2016.02.018

Partyn, H., Zahorodniy, A., Pylypenko, L., & O.V., O. (2020, July 1). Natural Capital: Essence, Types and Evaluation Procedure., VIII(Issue 3), 202–214. https://doi.org/10.35808/ijeba/496

Troy, A., & Wilson, M A. (2006, December 1). Mapping ecosystem services: Practical challenges and opportunities in linking GIS and value transfer. Elsevier BV, 60(2), 435–449. https://doi.org/10.1016/j.ecolecon.2006.04.007

8.2 Machine Generated Summaries

Disclaimer: The summaries in this chapter were generated from Springer Nature publications using extractive AI auto-summarization: An extraction-based summarizer aims to identify the most important sentences of a text using an algorithm and uses those original sentences to create the auto-summary (unlike generative AI). As the constituted sentences are machine selected, they may not fully reflect the body of the work, so we strongly advise that the original content is read and cited. The auto generated summaries were curated by the editor to meet Springer Nature publication standards. To cite this content, please refer to the original papers.

Machine generated keywords: natural capital, capital, natural, accounting, ecosystem, ecosystem service, wealth, asset, account, service, information, type, inclusive, soil, flow.

8.2.1 Putting Biodiversity into the National Accounts: Creating a New Paradigm for Economic Decisions [1]

This is a machine-generated summary of:

Vardon, Michael; Keith, Heather; Obst, Carl; Lindenmayer, David: Putting biodiversity into the national accounts: Creating a new paradigm for economic decisions [1]

Published in: Ambio (2018)

Link to original: https://doi.org/10.1007/s13280-018-1114-z.

If you want to cite the papers, please refer to the original.

For technical reasons we could not place the page where the original quote is coming from.

Abstract-Summary

"Integrating the environment into national accounts has been suggested as a way to improve information but so far impact on decision making is limited."

"Using examples from Australia and Botswana, we show how integrating information on biodiversity, resource use and the economy via accounting can help create a new decision-making paradigm and enable a new policy framing with spending on biodiversity conservation and sustainability seen as an investment, not a cost."

Introduction

"With human use of the environment reaching dangerous levels (Rockström and others [2]), better ways need to be found to achieve environmental, economic and social objectives."

"For economic policy, the standard framing involves the assessment of future benefits obtained from investments now—understanding the different returns to investment choices."

"Seeing decisions as investment choices is second nature for economists, financiers and accountants, but is largely missing from policies on the environment where the validity of outcomes, such as conservation or sustainability, is taken as given."

"Such a framing also allows recognition of the importance of ecosystem services which make explicit the dependencies of human economic activity, and human well-being more broadly, on ecosystems and biodiversity and the wide variety of priced and unpriced services they provide (e.g. Boyd and Banzhoff [3]; Milner-Gulland and others [4])."

"The importance of ecosystem services and integrating them into economic decisions has been reviewed and acknowledged at other levels (e.g. Bateman and others [5])."

The Role of Accounting in the Paradigm Shift

"Standard economic and financial accounts encompass information on returns (incomes, expenditure, profit and loss) and underlying assets (manufactured and financial)."

"Since 2012, the System of Environmental-Economic Accounting (SEEA; United Nations and others [6]) has provided international statistical standards for governments, while from 2016, the Natural Capital Protocol (NCP; Natural Capital Coalition [7]) has provided an approach for business."

"These platforms for institutionalizing the use of environmental information in economic and environmental decision making are available now."

"The use of extended accounting for decision making remains limited and accounting has struggled to make the transition from theory to practice and make real impact on government or business strategic planning or management processes (Vardon and others [8])."

"We use biodiversity accounting and cases from Australia and Botswana to show how accounting can help make an urgently needed paradigm shift in the planning processes and management of natural resources by government and business."

Using Accounting for Biodiversity Conservation

"The Aichi Target 2 states: "By 2020, at the latest, biodiversity values have been integrated into national and local development and poverty reduction strategies and planning processes and are being incorporated into national accounting, as appropriate, and reporting systems". (Emphasis added) Aichi Target 2 provides a clear entry point for biodiversity and accounting experts to work together."

"The valuation work within the Intergovernmental Science-Policy Platform on Biodiversity and Ecosystem Services (IPBES) has focused on the different conceptions of value with respect to the environment (an important discussion) but is only now progressing to consider the value of biodiversity specifically or the integration of environmental and economic values into national accounting, which requires monetary valuation."

"This requires not just reporting on biodiversity loss but also on connecting to data on the extent and condition of terrestrial ecosystems (e.g. forests) and enabling assessment of levels of use of environmental stocks relative to the regenerative capacity of those stocks and the economic revenue."

Example of Ecosystem Accounts for Forest Management in Australia

"It was demonstrated that if logging native forests ceased, then USD$10.9 million of industry value added (IVA; the contribution to GDP) would be lost directly but that this would be offset by expected gains to the water supply industry (USD$ 7.4 million, IVA) through higher water yield, plus the ability to enter the carbon market (USD$ 11.5 million IVA), plus possible increases in tourism and plantation timber production, plus the non-monetary benefit of reducing the extinction risk of species (all values for 2013–2014, average AUD$ to USD$ for 2013–2014 was 0.91)."

"The monetary information compiled from an accounting perspective was more readily amendable to making policy tradeoffs, than the previously estimated, standalone value of between USD$53–110 million for Leadbeaters' Possum using contingent valuations (Jakobsson and Dragun [9]), reflecting social welfare values that include consumer surplus."

Example of Ecosystem Accounts for Wildlife Management in Botswana

"The allocation can be viewed as both an investment choice to support tourism-based industries rather than alternative land and water uses (and especially agriculture) as well as a broader statement of the value of wildlife not captured via standard economic approaches."

"The presentation of information on multiple conceptions of value allow governments, businesses and individuals to express their preferences through economic means (e.g. spending on ecosystem management, accommodation, tours and travel to and from national parks, etc. as well as purchases of land) as well as other means (e.g. time spent appreciating wildlife, voting for candidates that support policies that protect wildlife)."

"In these conceptions of value, biodiversity assets (i.e. elephants, other species and their habitats) are recognized as delivering services that support economic activity, such as tourism, which can be compared to alternative uses of water and land that generate income (e.g. agriculture), as well as services that do not need to be monetized to be included in management and investment decisions (e.g. cultural and recreational services)."

Some Lessons from Australia and Botswana

"These examples from Australia and Botswana show that accounting for biodiversity can provide new information to support decision making by government and business."

"In the water accounts developed until then (Department of Water Affairs [10]), wildlife use of water had not been considered."

"With the new information on both water use by wildlife and the economic contribution of tourism to the Botswana economy, the Department of National Parks and Wildlife is able to argue that more water should be made available to wildlife."

"While these case studies are positive, for biodiversity accounting to become a mainstreamed part of decision making, accounts must be expanded to cover larger areas (e.g. river basins, countries), include more comprehensive metrics describing biodiversity, and be updated regularly, ideally annually."

"Consistent, long-term, nationwide monitoring of biodiversity at the species or ecosystem level seldom occurs."

Making the Connections: Business, Government and Academia

"International organizations and multi-stakeholder coalitions have established processes to develop environmental-economic accounting platforms for governments via the SEEA (United Nations and others [6]), and for businesses via the NCP (Natural Capital Coalition [7])."

"Business and government linkages will be essential if a new paradigm of decision making, recognizing our dependence on ecosystems and biodiversity, not just the human impacts, is to be established to meet the real and present economic, social and environmental challenges of the modern world."

"With greater understanding of the tradeoffs between the environment and the economy revealed by environmental-economic accounts, different parts of society are able to regularly review and modify their activities."

"It also enables different parts of society to hold each other accountable and for business and government to make institutional changes to enable information to be analysed more effectively and translated into policy and management responses."

8.2.2 Managing Soil Natural Capital: A Prudent Strategy for Adapting to Future Risks [11]

This is a machine-generated summary of:

Cong, Rong-Gang; Termansen, Mette; Brady, Mark V.: Managing soil natural capital: a prudent strategy for adapting to future risks [11]

Published in: Annals of Operations Research (2015)
Link to original: https://doi.org/10.1007/s10479-015-2066-3.

If you want to cite the papers, please refer to the original.

For technical reasons we could not place the page where the original quote is coming from.

Abstract-Summary

"Managing soil natural capital—the capacity of the soil to generate ecosystem services of benefit to farmers—has been proven to generate the double dividend: increasing farm profit and reducing associated risk."

"We explore whether managing soil natural capital has a third dividend: reducing the downside risk (increasing the positive skewness of profit)."

"This we refer to as the prudence effect which can be viewed as an adaptation strategy for dealing with future uncertainties through more prudent management of soil natural capital."

"We do this by developing a dynamic stochastic portfolio model to optimize the stock of soil natural capital—as indicated by soil organic carbon (SOC) content—that considers the mean, variance and skewness of profits from arable farming."

"We find that managing soil natural capital can, not only improve farm profit while reducing the risk, but also reduce the downside risk."

"Prudent adaptation to future risks should therefore consider the impact of current agricultural management practices on the stock of soil natural capital."

Introduction

"Managing soil natural capital might therefore be an additional option for adapting to production risk because higher abundances and diversity of soil organisms can increase both the generation and reliability of soil ecosystem services (Koellner and Schmitz [12])."

"We aim to investigate whether managing soil natural capital (as represented by SOC concentration) can even provide farmers with prudent adaption to multiple agricultural risks (such as those presented by accelerating climate change, and increasing volatility of input and output prices) as represented by the skewness of farm profits."

"About soil natural capital, Brady et al. [13] examined the effect of soil capital (using SOC as an indicator) on crop yield, input use and farmers' profits based on a production-function approach."

"Cong et al. [14] further develop a dynamic portfolio model combined with crop production functions to examine the effect of managing SOC on profits and risk."

The MVS (Mean–Variance–Skewness) Portfolio Model

"The farmer has three objectives: increase the expectation and skewness of profit, and reduce its variance."

"The land that is not used for growing cash crops, i.e. if the land constraint is slack, is assumed to be put under a SOC replenishing land use such as a multi-year grass ley, with zero profit and variance."

"In the MV portfolio model, the combinations of means and variances of profits below the efficient frontier are considered inefficient."

"Different from the static MVS portfolio model, the land covered with the replenishing use is assumed to rejuvenate SOC which will increase land productivity in the following periods (Nandwa [15]; Ajayi et al. [16])."

"The risk constraint in the dynamic model restricts the variance of the expected net present value (NPV) of accumulated profits over T periods to a certain level."

"The dynamic MVS portfolio model optimizes farmers' land use over a finite rather than infinite time horizon."

Data Sources and Sub-models

"This section briefly describes the case-study region, the sources of data and sub-models (Copula models and crop production functions) and the site-specific settings, which will be jointly used to generate inputs for the portfolio models."

"It has a temperate European climate (the mildest climate in Sweden) where precipitation and temperature are the two most important climatic factors affecting crop yields (De Wit et al. [17]; Schlenker and Roberts [18])."

"The main barrier for modelling the climatic risks (i.e. the uncertainties in precipitation and temperature) and price risks is the potential interdependence among risk factors (Richardson [19]; Dai et al. [20])."

"The interdependence among price series of the four cash crops and inputs (i.e. energy and fertilizer) is mainly because of their internal economic linkages."

"The four cash crops (or inputs) studied here are imperfect substitute goods (Gill et al. [21]; Li et al. [22]) and hence their prices are interrelated as measured through cross-price elasticities (Gardner [23]; Zhu et al. [24])."

"For the price series of the cash crops, fertilizer and energy inputs, we used data from 1966 to 2010 (SCB [25])."

Empirical Results: Optimal Crop Diversification and Soil Capital Management Within the MVS Portfolio Framework

"We find that among the four cash crops, winter wheat is most important for increasing the skewness of profit and hence farmers' prudence (e.g. the area of wheat increases when the profitability constraint is relaxed (i.e. from 100 to 70% of max profit) while the area of other crops usually decreases)."

"Compared with cash crops, the rejuvenating land use (grass ley) is always effective in increasing farmers' prudence (i.e. from 100 to 10% max profit)."

"To evaluate the optimal SOC management strategies for increasing the prudence, the state of SOC is compared for different combinations of the profitability and risk constraints."

"We examine whether farmers can both increase profits and prudence when reducing risk through the dynamic management of SOC."

"Showing that through active management of SOC, the triple dividend can be achieved: increasing profits and prudence while reducing risks."

Discussion

"The main innovation of this research is that we demonstrate that management of SOC (our indicator of soil natural capital) together with crop diversification can increase the farmers' profit and prudence and decrease their risk, the so-called triple dividend, thanks to increasing flows of supporting soil ecosystem services."

"There could be two reasons for this phenomenon: (1) Farmers are not well informed of the capacity of soil natural capital to increase profits and prudence, and reduce risk (according to our knowledge, this study is the first one to explore the potential of soil capital to adapt to agricultural risks); (2) Farmers could be myopic—they value current profit more than the long-term risk-adjusted profit while ignoring the future risks (i.e. they are too optimistic)."

"We lack policy instruments to give farmers incentives to manage soil capital for adapting to future risks."

Conclusion

"We show that these three objectives can be improved simultaneously through dynamic management of soil natural capital and associated supporting ecosystem services."

"Beyond the economic benefits and risk reduction, we show that managing of soil natural capital can also reduce the downside risk (losses) through the prudence effect, which can be viewed as a natural self-insurance for agricultural risks."

"We suggest that prudent adaptation to future risks should therefore consider the impact of current agricultural management practices on the stock of soil natural capital."

8.2.3 Path Toward Net-Zero Buildings: A Natural Capital Assessment Framework [26]

This is a machine-generated summary of:

Hossaini, Navid; Hewage, Kasun; Sadiq, Rehan: Path toward net-zero buildings: a natural capital assessment framework [26]

Published in: Clean Technologies and Environmental Policy (2017)
Link to original: https://doi.org/10.1007/s10098-017-1468-z.

If you want to cite the papers, please refer to the original.

For technical reasons we could not place the page where the original quote is coming from.

Abstract-Summary

"There is a gap in the literature for linking natural capital assessment and net-zero buildings."

"This paper presents a methodology for assessment of natural capital to achieve net-zero targets in buildings."

"The contribution of water, energy and building materials is discussed, and various scenarios are evaluated on how the building can become net-zero over its life cycle by using local natural resources."

Introduction

"A net-zero building is a building system that uses sustainable materials and produces its needed energy and water throughout its life cycle (Hossaini and others [27])."

"Among all of the sustainability assessment tools and rating systems, Living Building Challenge® (LBC) is the only rating system requiring net-zero targets (Bendewald and Zhai [28])."

"LBC does not provide a clear methodology on how a building can achieve net-zero."

"There is a gap in the literature for linking natural capital assessment and net-zero buildings."

"This paper presents a methodology for assessment of natural capital to achieve net-zero targets in buildings."

"Energy, water and materials have been selected as three main elements of net-zero buildings."

"The methodology evaluates how local natural resources can be used by the building to become self-sufficient with net-zero impact."

"The methodology includes an off-site carbon offset program for buildings that cannot reach net-zero from the on-site natural resources."

Background

"These studies often develop natural capital based models to study sustainability of global economy."

"The study also focuses on how to reduce the natural capital appropriation of building construction by using renewable energy resources and 'environmentally inexpensive' materials."

"Other studies that focus on net-zero buildings or near-zero buildings mainly focus on energy as the main component of natural capital."

"There is a clear gap in the literature for use of natural capital in assessing buildings potential to move toward net-zero targets."

"The idea presented in this paper focuses on assessing natural capital to achieve net-zero targets in buildings."

"Natural capital can be strategically used in buildings to reduce their environmental footprint and help them be more sustainable."

"If carefully planned and managed, 'natural capital' is enough for buildings to become self-sufficient and have a net-zero impact on the environment."

Natural Capital Methodology for Net-Zero Buildings

"A system boundary for the building is defined and demand assessment is performed to evaluate buildings energy and water needs."

"Natural capital assessment is then conducted for three modules: energy, water and building materials."

"All buildings need to consume energy to function properly, so to achieve net-zero objective buildings should be designed to consume less energy (via efficient design and systems) and offset that energy use by generating renewable energy."

"It is important to consider energy use of the building over its full life cycle including, construction, operation and demolition phases."

"These parameters include building location, geometry, construction materials, electrical system, mechanical system and renewable energy generation systems."

"To achieve net-zero, a building should generate renewable energy to offset the energy it uses over its lifetime."

"To reach net-zero energy, the building's potential to generate renewable energy from all applicable forms should be evaluated."

Results and Discussions

"The contribution of water, energy and materials is discussed, and various scenarios are evaluated on how the building can become net-zero."

"The building energy and water demand assessment was completed by professional consultant engineers during the design phase of the building and was evaluated during the operational phase."

"Feasibility of wind energy mainly depends on the wind speed at the building's location and area availability for the installation of the wind turbine."

"The building can capture 21,900 m^3 (15,525 + 6375 m^3) of water on-site from precipitation and groundwater sources."

"The LBC building is not net-zero impact over its life cycle after integrating the three main systems (water, energy and building materials)."

"The result emphasizes the importance of an integrated pre-design sustainability assessment that evaluates building materials, energy systems and water systems."

"Solar energy accounts for 51% of total cost savings and has the highest potential for the net-zero building (Scenario 3)."

Conclusions

"Current net-zero building rating systems do not consider full life cycle impact of buildings in their assessment."

"This paper fills the gap by proposing a methodology for assessment of natural capital to achieve net-zero targets in buildings."

"The presented methodology evaluates the location of the building for available natural resources and evaluates how buildings can achieve net-zero targets."

"The result of the paper indicates that the LBC design is not net-zero CO_2e over the life of the building."

"The methodology described in this paper considers the entire life cycle of buildings and evaluates if the building is net-zero in this timeframe."

"Various scenarios are considered to bring the building to net-zero CO_2e emission over its life cycle."

"The result shows that the building can become net-zero CO_2e emission over its full life cycle by investing more on renewable energy and with the help of materials recycling and off-site carbon offset program."

8.2.4 Assessing the Capacity of Ecosystems to Supply Ecosystem Services Using Remote Sensing and an Ecosystem Accounting Approach [29]

This is a machine-generated summary of:

Vargas, Leonardo; Willemen, Louise; Hein, Lars: Assessing the Capacity of Ecosystems to Supply Ecosystem Services Using Remote Sensing and An Ecosystem Accounting Approach [29]

Published in: Environmental Management (2018)

Link to original: https://doi.org/10.1007/s00267-018-1110-x.

If you want to cite the papers, please refer to the original.

For technical reasons we could not place the page where the original quote is coming from.

Abstract-Summary

"Ecosystems contribute to economic development through the supply of ecosystem services such as food and fresh water."

"Ecosystem accounting includes the compilation of an ecosystem services supply and use account, which reflects actual flows of ecosystem services, and the ecosystem capacity account, which reflects the capacity of ecosystems to sustainably supply ecosystem services."

"A capacity assessment requires detailed data on ecosystem processes, which are often not available over large scales."

"We examined how net primary productivity derived from remote sensing can be used as an indicator to assess changes in the capacity of ecosystems to supply services."

"We compared ecosystem capacities with the level of ecosystem service supply to assess a sustainable use of ecosystems."

"Our study provides insights on how the capacity of ecosystems can be quantified using remote sensing data in the context of ecosystem accounting."

Introduction

"A key innovation in ecosystem accounting is the inclusion and guidance on the spatially explicitly measurement of stocks by assessing changes ecosystems in terms of extent, condition and capacity to supply ecosystem services, and the measurement of flows of ecosystem services (United Nations and others [30])."

"Different spatially explicit methods can be used to assess the capacity of an ecosystem to supply ecosystem services, including biophysical models, static land cover-based look-up tables, remote sensing and direct measurements (Bagstad and others [31]; Schröter and others [32]; Willemen and others [33])."

"Because ecosystem accounting requires environmental data, spatially explicit, repeatable and accessible, suitable to assess large areas in various accounting periods, this study explored the use of remote sensed data to assess the capacity of an area to sustainable deliver ecosystem services."

"The objective of this study is to explore if and how remote sensing spectral information can be used to assess the capacity of ecosystems to supply ecosystem services following the ecosystem accounting guidelines."

Methods

"Of the 11 grouped land cover types, only six land cover types (forest, grassland, savannah, woody savannah, natural mixed and oil palm plantations) were relevant for the selected ecosystem services: oil palm FFB, grazing pastures for cattle, timber and carbon sequestration."

"We link the selected ecosystem services with the six EA: oil palm FFB are supplied by oil palm plantations, grazing pastures are supplied by four ecosystems: grassland, savannah, woody savannah and natural mixed ecosystem (which are pastures and trees), and timber is supplied by forest ecosystem (see Supplemental Materials for more details on ecosystem services and EA)."

"Ecosystem functioning indicators such as Net Primary Productivity (NPP) have been used in earlier assessments of ecosystem change and ecosystem services supply (Costanza and others [34]; van Oudenhoven and others [35])."

"We selected NPP as an indicator of the capacity of ecosystems to supply ecosystem services because of two aspects."

Results

"With an annual mean of 11 ton of carbon per hectare oil palm plantations had the highest capacity (compared to other ecosystems) to sequester carbon for the period 2001 to 2014."

"Forests, grassland and savannah ecosystem have a similar capacity to sequester carbon (in ton per hectare per year), however forests and woody savannah will provide more benefits as more carbon will be kept in the ecosystem for longer periods of time."

"A substantial portion of the capacity to supply pastures is removed by grazing cattle in grassland (40%) and natural mixed (41%) ecosystems, because conditions (e.g. water availability, soil type, palatability) and management (e.g. infrastructure) in these two ecosystems are more favourable to graze cattle."

"Although adult oil palms have a high capacity to supply FFB biomass (16 ton of carbon per hectare per year), the harvesting of FFB removes 56% of this capacity (9 ton of carbon per hectare per year)."

Discussion

"Ecosystem accounting is spatially explicit approach and includes an assessment of the capacity of ecosystems to supply ecosystem services (United Nations and others [30])."

"Our study explored the feasibility of compiling remote sensed spatially explicit information following ecosystem accounting guidelines to assess the capacity of ecosystems to supply ecosystem services at the level of a river basin."

"Our results also show that remote sensed information can potentially be used to assess the capacity of ecosystems to supply ecosystem services following the ecosystem accounting guidelines."

"Our study included MODIS NPP to assess capacity by assessing the amount of aboveground biomass that is used to supply an ecosystem service."

"MODIS NPP is a powerful tool to assess the spatial variation of the capacity of ecosystems to supply ecosystem services in large areas, however, can be limited in contexts where a high level of detail is required such as municipality and local level, e.g. in Remme and others [36] and Villamagna and others [37]."

Conclusion

"Ecosystem accounting includes the assessment of ecosystems' capacity to supply multiple ecosystem services (United Nations and others [30])."

"Our study provided insights on (i) how the capacity of ecosystems to supply ecosystem services can be assessed and (ii) how remote sensing can be used to support this assessment in large areas."

"We proposed NPP as a suitable indicator to assess the capacity of ecosystems to supply timber, pastures for grazing cattle, oil palm FFB and to sequester carbon because NPP is sensitive to changes in ecosystem condition and changes in primary productivity that affect the supply of ecosystem services."

"The presented approach used for the assessment of the capacity of ecosystems to supply ecosystem services is a basis for further refinements that will allow developing capacity–supply models for ecosystem accounting and other applications."

8.2.5 Stakeholder Perspectives on the Value Proposition of Enterprise-Level Natural Capital Accounting for Three Primary Industries [38]

This is a machine-generated summary of:

van Putten, E. Ingrid; Pinkard, Elizabeth; O'Grady, Anthony; Schmidt, Rebecca K.; Cresswell, Ian; Raoult, Vincent; Taylor, Matt D.: Stakeholder perspectives on the value proposition of enterprise-level natural capital accounting for three primary industries [38]

Published in: Environment Systems and Decisions (2021)

Link to original: https://doi.org/10.1007/s10669-021-09817-z.

If you want to cite the papers, please refer to the original.

For technical reasons we could not place the page where the original quote is coming from.

Abstract-Summary

"Many primary industries and enterprises are highly dependent on renewable and non-renewable natural capital, but there has been little focus on measuring and monitoring the flows benefits from natural capital."

"To assist future implementation of natural capital accounting we assess the value proposition perceived by primary industry stakeholders."

"Interviewees generally suggested natural capital accounting is of public benefit by improving environmental outcomes."

"There were some concerns across the industries about the potential costs natural capital accounting may add to normal business operations and the lack of consistency in practices."

"Mainstreaming natural capital accounting across primary industries will mean that the value propositions and the implementation barriers need to be addressed."

"The value proposition of implementing natural capital accounts at the enterprise level must stack up favourably against its costs."

Introduction

"The value proposition of implementing natural capital accounts at the enterprise level must stack up favourably against its costs."

"User perceptions about the value proposition for natural capital accounting will likely vary, not only because some users further up the supply chain will have different needs."

"The value proposition for global brands seeking to use natural capital accounting often revolves around transparency across their supply chains, enabling more informed decisions about the environmental sustainability of their businesses (Modak and others [39])."

"We undertook stakeholder analysis to provide insights into the value proposition and usefulness of enterprise-level natural capital accounts for three primary industries."

"The aim was to understand the exact nature and role of the economic/financial and non-economic factors in the overall value proposition of implementing natural capital accounting."

Case Study Description

"The industry is an important regional employer in inland regions of the states of the New South Wales (NSW) and Queensland, and to a lesser extent in Victoria and Western Australia."

"The cotton industry has received much attention globally for its potential to negatively impact natural capital."

"Key issues in Australia for consumers and the community are loss of biodiversity associated with land clearing for cotton production, over-extraction of water from rivers and aquifers, and issues related to the use of agrichemicals (Wall [40])."

"Wild-harvest fisheries are an extractive primary industry that targets free-ranging aquatic organisms (Hart and Reynolds [41])."

"Recreational fisheries that occur in the same ecosystems harvest over 8500 tonnes per year in NSW (Henry and Lyle 38), with state-wide expenditure by anglers estimated to be approximately $1.626 b per year (McIlgorm and Pepperell [42])."

"In Australia, plantation forestry plays an important role in rural and regional economies, contributing to employment, skills development and broader economic activity (Schirmer and others [43]; Brand [44])."

Methods

"The survey was implemented as semi-structured interviews, to gather quantitative and qualitative data that was analysed to understand the value propositions as perceived by different stakeholders (e.g. growers, processors, managers, scientists) in three key primary industries (plantation forestry; cotton; and fisheries)."

"To recruit an adequate number of key-informant respondents for the interviews, the primary industries and their stakeholder groups were first mapped out."

"The mapping of stakeholders was carried out in consultation with relevant industry organisations and it involved firstly determining all groups that may have an interest and secondly, listing the names of potential individuals that could be approached for an interview."

"Potential participants for stakeholder interviews were determined from project partner networks, registrants at the Australian Cotton Conference 2018, and recommendations from interviewees (snowball sampling)."

"Potential participants for stakeholder interviews were identified through discussions with fishery and natural resource managers, project partner networks and recommendations from interviewees (snowball sampling)."

Results

"For the cotton industry the direct value of natural capital accounting was perceived to be market access."

"To plantation forestry and fisheries, respondents representing the cotton industry did specifically mention the value of natural capital accounting on their operations and business model."

"The respondent sample contained fewer females [36] than males [45], but there was no distinguishable difference in their choice of value proposition of natural capital accounting (i.e. environmental, market access and operations and business model were ranked most frequently as value propositions)."

"Even though these respondents indicated that they had low knowledge, their contribution to the total number of perceived value propositions of natural capital accounting was proportional to the size of this group (i.e. 26% of respondents had low knowledge and they contributed 25% of value propositions) and there is no significant difference (Kruskal–Wallis chi-squared = 2.8837, p = 0.4099) between knowledge level."

Discussion

"Our approach provided insights into the different value propositions and usefulness of enterprise-level natural capital accounts for three primary industries."

"Although there was seeming agreement over the most obvious value propositions – environmental benefits, market access, and operations and business models – there were some differences in perceptions of benefits between different stakeholder groups, notably the perceived role of natural capital accounting in accessing markets."

"The variable set of value propositions for natural capital accounting identified by the different industries also highlights the 'early days' of implementing natural capital accounts (O'Grady and others [46]), and their current usefulness at the local level."

"Even though the level of knowledge of natural capital accounting is not a key factor that affects the perceived value propositions (as we found in this study), greater involvement of stakeholder groups, supported by further education and outreach, could help clarify needs of different industries and potential lack of, or incorrect, information, which in turn would improve the usefulness of accounts."

Conclusion

"Natural capital accounting can better account for management of natural capital in business practices."

"Our study shows that primary industries (cotton, fisheries, and plantation forestry) have different value propositions for implementing natural capital accounting but that stakeholder groups largely seem to agree over the top value propositions."

"Important value propositions for implementing natural capital accounts are economic and financial (i.e. market access) but environmental benefits are higher on the list for some industries."

"Co-production of industry-specific information and communication around natural capital accounting is needed to ensure these differences are recognised."

8.2.6 Quantifying the Effect of Waste on Soil Health in European Union: What Are the Roles of Technology, Natural Capital, and Institutional Quality? [47]

This is a machine-generated summary of:

Shah, Muhammad Ibrahim; Zakari, Abdulrasheed; Kumar, Siddharth; Abbas, Shujaat; Sheraz, Muhammad: Quantifying the effect of waste on soil health in European Union: what are the roles of technology, natural capital, and institutional quality? [47]

Published in: Environmental Science and Pollution Research (2022)

Link to original: https://doi.org/10.1007/s11356-022-20909-4.

If you want to cite the papers, please refer to the original.

For technical reasons we could not place the page where the original quote is coming from.

Abstract-Summary

"Taking this into account, this study presents the effect of hazardous and non-hazardous waste, plastic waste, and electronic waste on soil health for 24 European Union (EU) countries during 2004–2018 period."

"The impacts of several other variables such as technological innovation, ICT, natural capital, fossil fuel energy consumption, and institutional quality on soil health are also examined."

"Technological innovation, ICT, and institutional quality, as well as fossil fuel energy consumption, have positive impacts on soil health."

"Natural capital moderates the effect of plastic and electronic waste on soil health."

Introduction

"To be more specific, Germany and France were both the top countries in generating plastic and electronic waste, and Italy occupied the 4th position in the EU in total waste generation during 2018."

"Another reason is the rise of hi-tech productions, causing electronic waste to increase in the EU (Almendro-Candel and others [48])."

"Given the increase in the per capita waste production in the EU countries (Tiseo [49]), we argue that EU countries are more appropriate to study because of their vulnerability to waste generation."

"We examine the effects of the waste product on soil fertility or soil health for 24 EU countries from 2004 to 2018 using panel data."

"To capture these effects, the study incorporates the interaction of technology and ICT with wastes and analyzes whether they can moderate the effect of waste on soil quality."

Literature Review

"Both soil fertility and agricultural productivity can be used interchangeably throughout the paper, and as a result, in this section, we discuss the relationship of waste, human capital, natural capital, quality of institutions, and technology with agricultural productivity or the agricultural sector in general."

"They found that industrial waste from enzyme production is associated with increasing soil quality, particularly in the grassland with low fertility, thus improving agricultural proceeds."

"Carmo and others [50] found that the impact of waste product may differ according to the soil's nature."

"Merga and others [51] argues that brewery waste or by-products of the brewing industry may enhance agricultural productivity by improving soil fertility, which can be reflected in an upward improvement in crop production."

"The authors found that waste product has a negative relationship with soil quality."

Data and Methodology

"This study uses the following base equation: where NB represents the nutrients balance, a proxy for soil health, which is the dependent variable of our study, whereas HNHW, PW, EW, TECH, ICT, NC, INSTQ, and FEC represent hazardous and non-hazardous waste, plastic waste, electronic waste, technological innovation, information and communication technology, natural capital, institutional quality, and fossil fuel energy consumption, respectively."

"To see the moderating impacts of technological innovation, ICT, natural capital, and quality of institution with hazardous and non-hazardous waste on nutrient balance, the model 2 is formed as follows: In model 3, the moderating impacts of technological innovation, ICT, natural capital, and quality of institutions with plastic waste on nutrient balance are formed as follows: Furthermore, to see whether technological innovation, ICT, natural capital, and quality of institutions can moderate the impacts of electronic waste on nutrient balance, a fourth model is formed: Following Al-Busaidi and Ahmed [52], the relationship between nutrient balance and different types of waste is expected to be positive."

Driscoll and Kraay's Robust Test (D-K)

"The D-K technique resolves the issues of cross-sectional dependence for the robust estimator."

"Following Hafeez and others [53] and Sheraz and others [54], the current study employs the DK technique with standard errors for pooled OLS (ordinary least square) for panel data while having a linear functional form as in the following equations: Here, NB indicates the dependent variable of the study, Z represents control variables of the study, t shows the sample time period ($t = 10$), whereas i shows the cross-sections of sample = 1, 2, 3, … 24 (24 European Union countries)."

"For robustness analysis, another static model known as panel-corrected standard error (PCSE) technique is employed which an alternative estimation technique of feasible generalized least square method and can fit cross-sectional linear time series model on the basis of the assumption that errors are contemporary correlated across the panels and heteroscedastic by default (Nathaniel and Adeleye [55])."

Empirical Results

"After describing the variables and checking for multicollinearity, we move on to the analysis where using the Driscoll-Kraay standard error model, the impact of waste (hazardous and non-hazardous, plastic, electronic), ICT, technology, fossil fuel energy consumption, and institutional quality on nutrient balance is studied."

"Model 1 shows the direct impact of hazardous and non-hazardous waste, plastic waste, electronic waste, technological innovation, ICT, natural capital, Institutional quality, and fossil fuel energy consumption on the nutrient balance of the soil in the 24 EU countries."

"In model 2, the interaction of hazardous and non-hazardous waste is interacted with ICT, natural capital, institutional quality, and technological innovation."

"Model 4 demonstrates how the moderating effects of ICT, natural capital institutional quality, and technological innovation work with electronic waste."

"Although technology and natural capital can offset the effect of plastic waste, institutional quality cannot do that to improve soil health."

Conclusion and Policy Implications

"The nature of the waste can impact not only the agricultural productivity but also soil health."

"The present study sets out to unravel the role of waste on soil health in 24 European Union countries during 2004–2018."

"The findings of the study reveal that soil health gets deteriorated by electronic waste, while technology, institutional quality, and natural resources have been successful in the prevention of deteriorating soil health."

"The findings would help the policymakers to focus on containing the adverse impacts of hazardous and non-hazardous waste on the nutrient balance of the soil and as a result, on the agricultural productivity."

"Production leading to low soil nutrient balance will jeopardize the food security in the countries selected for the study."

8.2.7 Natural Capital for Nature's Contributions to People: The Case of Japan [56]

This is a machine-generated summary of:

Kumagai, Junya; Wakamatsu, Mihoko; Hashimoto, Shizuka; Saito, Osamu; Yoshida, Takehito; Yamakita, Takehisa; Hori, Keiko; Matsui, Takanori; Oguro, Michio; Aiba, Masahiro; Shibata, Rei; Nakashizuka, Tohru; Managi, Shunsuke: Natural capital for nature's contributions to people: the case of Japan [56]

Published in: Sustainability Science (2021)

Link to original: https://doi.org/10.1007/s11625-020-00891-x.

Copyright of the summarized publication:

If you want to cite the papers, please refer to the original.

For technical reasons we could not place the page where the original quote is coming from.

Abstract-Summary

"Other studies have investigated the value of various types of ecosystem services and natural capital by integrating a number of findings using meta-analyses at the global level."

"Although these studies have provided information on either the global value of natural capital or the local value of specific subjects, there is little evidence on the country-specific values of natural capital in Japan, which will provide useful information for national environmental policies."

"We investigated the perceived values of terrestrial and marine natural capital in Japan using internet surveys and payment card methods."

"We conducted future predictions of terrestrial and marine natural capital using a scenario developed in a previous study."

"Our results indicate that Japan should follow a population-dispersed scenario for the sustainable management of natural capital up to 2050."

Introduction

"A number of studies have investigated the values of natural capital by focusing on ecosystem services."

"We focus on the values and determinants of terrestrial and marine natural capital in Japan."

"We predict the future of terrestrial natural capital in Japan at the municipality level to measure future sustainability up to 2050 based on a scenario suitable to natural capital projection (Saito and others [57])."

"Some studies have focused on the value and future of natural capital in Japan at the prefectural level (Fujii and others [58]; Ikeda and Managi [59])."

"A limited number of studies predict the future of natural capital at the municipality level."

"This subsection clarifies how much future natural capital will be affected by the type of population and land use scenarios Japan will follow, considering the differences between urban and rural areas."

WTP Estimation for Terrestrial and Marine Natural Capital in Japan

"We additionally asked the respondents about their WTP to maintain overall natural environment in Japan to estimate the maximum amount of money that an average Japanese household can pay for the entire environmental conservation project in Japan."

"The coefficients of the eight region dummies are mostly negative, where some are statistically significant, which indicates that the respondents living in the baseline area, i.e., Hokkaido, are more likely to be WTP to maintain natural capital than the residents in other areas."

"The coefficient estimates in the outcome equation reveal that the variables of gender, graduate, household size, household income, visit frequency, perceived importance, and area of residence can be the determinants for the WTP amounts to maintain much of the natural capital."

"The WTP for coral reefs, seaweed beds, and mangrove forests were higher if the respondents had children while the other natural capital categories were not related to having children."

Natural Capital Prediction and its Importance to Sustainability

"We calculated the increase in the monetary value of natural capital stock generated by the conversion of abandoned agricultural land into forests, considering five scenarios developed by PANCES."

"We calculated the increase in monetary value of terrestrial natural capital by converting abandoned agricultural lands into forests for each of the five PANCES scenarios and the three conversion scenarios developed above."

"We estimated the increases in the monetary value of natural capital for each scenario by converting abandoned land into forest."

"In terms of the differences between the PANCES scenarios, the values of the increases in terrestrial natural capital are higher in the produced capital-based scenarios compared with the natural capital-based scenarios because a larger area of abandoned agricultural lands remains in the produced capital-based society in Japan."

Discussion

"Residents in urban areas are less likely to pay for some natural capital than those in rural areas."

"It implies that a policy aimed at population dispersion to rural areas can lead to better results in terms of coastal natural capital maintenance in Japan."

"More terrestrial and marine natural capital can be maintained by implementing policies aimed at population dispersion and attracting people to rural areas."

"Based on these aspects, promoting population dispersion to maintain more natural capital in Japan should be effective."

"This implies that the spread of autonomous vehicles has significant potential to promote optimal population distributions for the sustainable management of natural capital in rural areas."

"Population dispersion is one of the key measures to maintain natural capital, especially in rural areas."

"To sustainably manage natural capital in Japan, the policies mentioned above should be implemented to maintain the optimal population of municipalities in rural areas."

8.2.8 From COVID-19 to Green Recovery with Natural Capital Accounting [60]

This is a machine-generated summary of:

Vardon, Michael; Lucas, Paul; Bass, Steve; Agarwala, Matthew; Bassi, Andrea M.; Coyle, Diane; Dvarskas, Anthony; Farrell, Catherine A.; Greenfield, Oliver; King, Steven; Lok, Martin; Obst, Carl; O'Callaghan, Brian; Portela, Rosimeiry; Siikamäki, Juha: From COVID-19 to Green Recovery with natural capital accounting [60]

Published in: Ambio (2022)

Link to original: https://doi.org/10.1007/s13280-022-01757-5.

If you want to cite the papers, please refer to the original.

For technical reasons we could not place the page where the original quote is coming from.

Abstract-Summary

"Economic recovery is now an opportunity to rebuild natural capital alongside financial, physical, social and human capital, for long-term societal benefit."

"Current decision-making is dominated by economic imperatives and information systems that do not consider society's dependence on natural capital and the ecosystem services it provides."

"New international standards for natural capital accounting (NCA) are now available to integrate environmental information into government decision-making."

Introduction

"Dasgupta [61] highlighted that investment in natural capital is an investment in the economy and society, but governments have not yet translated this understanding into economic recovery spending."

"Around 24% of announced recovery spending is 'green' (contributing to environmental objectives) and most is targeted at climate change mitigation, with 3% positive for natural capital, and up to 17% negative (O'Callaghan and Murdock [62])."

"NCA serves as a bridge between economists and environmental scientists, enabling better decision-making for sustainable development, by including information on the impacts and dependencies of society on natural capital (Ruijs and others [63])."

"While business has a role in Green Recovery, and has adopted various forms of sustainability reporting (e.g. International Integrated Reporting Framework and Natural Capital Protocol), the focus of this paper is on public policy and how to improve current national information systems that do not properly account for natural capital, hindering decision-making for sustainable development."

Green Recovery and Natural Capital Approaches

"By bringing human and environmental challenges, such as climate change and biodiversity loss, to the forefront, Green Recovery can combine short-term socio-economic recovery with medium- to long-term transitions to address persistent human and environmental challenges."

"Despite some differences, each Green Recovery strategy is concerned with maintaining or enhancing natural capital for the benefit of current and future generations."

"They recognise that protecting, sustainably managing and restoring natural capital provides short-term employment and boosts economic growth, while at the same time delivering social benefits (such as improving health and food security), improving ecosystem services (e.g. flood control and carbon sequestration), reducing physical risks (e.g. flooding and storm-related damage) and helping to prevent future pandemics (Cohen-Shacham and others [64]; IPBES [65]; Seddon and others [66]; WWF and ILO [67]; Dobson and others [68]; Palomo and others [69])."

"Both components provide short-term economic benefits, while at the same time enhancing natural capital in the long term."

Natural Capital Accounting

"These frameworks integrate environmental data with the economic data from the System of National Accounts (SNA) that has played such a key role in decision-making."

"Having integrated and harmonised environmental and economic data in regularly updated accounts enables decision-makers to move beyond traditional siloed measures of economic success, notably GDP, which is based on obsolete economic theory and a mid-twentieth Century world-view that barely considered the environment (Stiglitz and others [70]; Coyle [71]; Hamilton and Hepburn [72]; Dasgupta [61])."

"With growing exper-tise, experience, access to online data and modelling platforms (e.g. ARIES for SEEA, EO4EA and InVEST), it is increasingly possible to rapidly produce basic accounts (Lucas and Vardon [73]; World Bank [74])."

"While the use of models for examining environmental or economic issues is not new, the availability of integrated environmental and economic data from NCA makes it easier to feed models and analyse the interrelationships between the economy and the environment (Banerjee and others [75]; Lucas and Vardon [73])."

A Green Recovery Through Natural Capital Accounting

"By providing integrated economic and environmental data NCA can improve decision-making for a Green Recovery."

"It brings together often disparate actors, supporting the alignment of multiple Green Recovery perspectives, shaping recovery packages so they build or improve natural capital and creating incentives for actors to cooperate (World Bank [74]) with a compelling example from Ireland (Farrell and others [76])."

"NCA has been used by several countries for analysing issues aligned with Green Recovery, including biodiversity conservation and restoration (Coates and others [77]; King and others [78]; Farrell and others [76]), tackling climate change (Pizarro [79]), integrated land management (Meijer and others [80]) and SDG monitoring (Ruijs and others [81]), with many examples from across the world (Vardon and others [82, 83]; Vardon and Bass [84])."

NCA Informing Biodiversity Conservation

"NCA has been used to integrate economic development with biodiversity conservation in Rwanda, through the means of nature-based tourism."

"NCA played a prominent role in developing Rwanda's recovery plan, providing the evidence needed to ensure the protection of ecosystems while demonstrating their role in economic development (Benitez and others [85])."

"With the evidence from NCA, the Rwandan Government estimated an investment of US$3.9 billion was required to maintain the environment to ensure that nature-based tourism can return to pre-COVID-19 levels and continue to grow, while also providing ecosystem services like carbon sequestration and soil retention (Benitez and others [85])."

"At a global level, the role of NCA is prominent in the first draft of the CBD Post-2020 Global Biodiversity Framework, which recognises the importance of embedding the value of biodiversity in decision-making and promotes transparency with the implementation of a natural capital accounting framework."

NCA Informing Climate Action

"Natural capital investments provide benefits for climate change mitigation and adaptation (Klenert and others [86])."

"Many ecosystems sequester and store carbon while some ecosystem types provide resilience against climate change (e.g. mangroves provide coastal protection from storm surges)."

"Investments in green space also provide benefits to climate change."

"Heris and others [87] used NCA to assess the economic benefits of urban trees in US cities for two ecosystem services: (1) cooling and mitigating climate change, and thereby reducing the need for air conditioning, and (2) rainfall interception providing improved water quality and flood mitigation."

"In up to 11% of these cities, investing in natural capital for climate mitigation and adaptation (green infrastructure) was determined to reduce costs (51% less), and to generate additional benefits (28% more) when compared to equivalent "grey" infrastructure (produced capital; Bassi and others [88])."

NCA Informing Achievement of the Sustainable Development Goals (SDGs)

"The SDGs are "integrated and indivisible and balance the three dimensions of sustainable development: the economic, social and environmental"."

"NCA also provides the information necessary to target and monitor the structural reforms needed for achieving the SDGs."

"Implementation of NCA is itself an indicator for SDG target 17.9 on capacity building and for SDG target 17.19 on supporting statistical capacity building in developing countries."

"Other countries are using NCA to help achieve particular SDGs."

NCA Informing Finance and Macroeconomic Policy

"On the expenditure side, accounts record spending on environmental protection and restoration, resource management and subsidies (or foregone income) which may harm or benefit nature."

"Such expenditure, by both the public and private sector, may be considered investments in natural capital."

"The cost of restoration (offsetting the depreciation in the condition of natural capital or reducing the debt) may be calculated from the expenditures and subsidies recorded."

"The UK's natural capital accounts show that COVID-19 travel restrictions reduced fuel duty revenues by 19% from 2019 to 2020 (ONS [89])."

"Natural capital investments that improve health (e.g. by improving air quality) can spur additional tax revenues due to increased labour productivity, while simultaneously reducing expenditure on treating respiratory illnesses."

"Natural capital accounts provide evidence for assessing if the investments financed by green bonds lead to the expected environmental benefits, helping reduce the risk of "greenwashing"."

Conclusions and Next Steps

"As the examples presented in this paper show, NCA can assist the development and implementation of policies, programs and financing needed for Green Recovery."

"Authorities in charge of recovery stimulus and reform instruments can use NCA across the policy cycle (from policy design, planning and financing to implementation and monitoring) to 'reset' the economy to deliver Green Recovery, simultaneously reinvigorating economic growth, arresting environmental decline and achieving a more equitable society."

"All parts of government have a stake in Green Recovery and NCA can provide them with consistent and regularly updated information for planning, implementation and monitoring."

"To increase awareness, the NCA and Green Recovery communities should continue collaborating, engaging with decision-makers to help them better understand, use and trust NCA so they can be applied to Green Recovery."

"With NCA government agencies responsible for assessing, funding and implementing recovery programs will have the information needed to plan and implement Green Recovery."

8.2.9 Contributions of Financial, Social and Natural Capital to Food Security Around Kanha National Park in Central India [90]

This is a machine-generated summary of:

Neelakantan, Amrita; DeFries, Ruth; Sterling, Eleanor; Naeem, Shahid: Contributions of financial, social and natural capital to food security around Kanha National Park in central India [90]

Published in: Regional Environmental Change (2020)

Link to original: https://doi.org/10.1007/s10113-020-01589-7.

If you want to cite the papers, please refer to the original.

For technical reasons we could not place the page where the original quote is coming from.

Abstract-Summary

"We examine the contributions of financial, social, and natural capital to household food access of 883 households around Kanha National Park (KNP) in central India over three seasons."

"We use regression trees and mixed effects models to identify associations between natural, social, and financial capital indicators and household food access (an indicator of well-being)."

"Financial capital (e.g., salaried jobs and proximity to towns for all seasons) is most prominently associated with higher food consumption scores."

"Households supplement incomes by converting social capital (e.g., 28% of surveyed households access "food in lieu of work or credit" in monsoon) and natural capital (e.g., 14% of surveyed households sold forest products in summer) to financial capital seasonally."

"Financial capital dwarfs contributions of social and natural capital around KNP, in contrast to other studies, which suggest that gains from natural capital are essential for well-being of people around protected areas."

"Food insecurity in the KNP landscape primarily relates to the lack of financial capital."

Introduction

"The expansion of protected areas has had quantifiable and large detrimental impacts on vulnerable human communities by restricting access to natural capital (ecosystem goods and services; Brooks and others [91]; West and Brockington [92]; Torri [93])."

"To displacement, costs of restricted access to natural capital by protected areas to local people include restricted access to wild foods, bush-meat, and non-timber forest products (Coad and others [45])."

"We show that using a globally recognized framework allows for increased applicability of local case-specific information in conservation landscapes where natural and social capital have been documented to play important roles to people's livelihoods and well-being."

"Natural capital provides benefits to people living around protected areas through goods (e.g., wild foods, bush-meat) and ecosystem services for agricultural production and human health (directly or indirectly by conversion to other types of capital; Zhang and others [94]; Golden and others [95]; Johnson and others [96])."

Methods

"We conducted semi-structured interviews, at each survey household, to collect data on socio-economic status, food access measurements, and interactions with forests."

"Our primary survey instrument to measure the contributions of natural, social, and financial capital to food security was a semi-structured interview at each household repeated in each survey season (summer, monsoon, and winter)."

"We use validated standardized multi-dimensional metrics measured at the household level with a 7-day recall time-period to assess food security in the KNP landscape (Jones and others [97])."

"We use household Food Consumption Scores (FCSs) and Coping Strategy Indices (CSIs) in every survey season (summer, monsoon, and winter)."

"We also estimated tree cover within various use radii (1 km, 2 km, 4 km, 6 km, and 8 km) as a proxy for unmeasured natural capital gains from forests that contribute to household food security."

Results

"Labor incomes are positively associated with FCSs in summer but negatively associated with FCSs in monsoon and winter."

"Food in lieu of work or credit" is negatively associated with FCSs in summer and household asset indices in summer as well as in monsoon."

"Sending household members to eat elsewhere is not significantly associated with FCSs or asset indices in our study."

“Households with lower social capital and labor incomes form a distinct subset (n = 320) and the households with highest and second highest FCSs all have incomes that were either agricultural or from salaries.”

“Households with agricultural and salaried incomes with higher asset indices partition into a distinct group (n = 50).”

“Households with agricultural and salaried incomes with lower asset indices partition according to social capital—those with higher social capital (n = 30) and those with lower social capital (n = 228).”

Discussion

“Considering that most households report agricultural and labor incomes, we expected poor households (low assets and no salaried incomes) to use opportunities for additional incomes from access to natural capital and social capital (coping strategies) to compensate in times of household stress and food scarcity.”

“In the KNP landscape, the link between social capital and labor is most intense in monsoon—where many households are food scarce and provide labor on other’s farms in return for food.”

“Proximity to built-up areas is associated with higher asset indices (increased financial capital) as urban centers in the KNP landscape might provide more income avenues year-round.”

“We were not able to explore mechanistic gains from forest available to households and their food security but significant positive associations between per capita forest availability (unmeasured natural capital) and FCSs in monsoon suggest households might somehow be reliant on forests in times of food scarcity.”

Conclusions

“Our results suggest that increased financial capital through salaried incomes in the KNP landscape could allow households to have higher FCS, especially by creating additional employment during monsoon.”

“The dominating influence in our measured indicators of financial capital relative to natural and social capital to household food security around KNP is in contrast to many studies around the world that report direct benefits from natural capital for well-being of people around protected areas (Sunderlin and others [98]; Bharucha and Pretty [99]; Golden and others [95]).”

“We suggest that more studies recognize that increasingly in landscapes where populations are less directly reliant on protected areas (for goods and services), people are likely to depend on financial capital to meet most of their well-being needs (Thulstrup [100]; Asih and Klasen [101]).”

“While this study concluded that financial capital is highly relevant for food security in the KNP landscape, other landscapes around tiger reserves might indicate the importance of natural, social or human capital.”

8.2.10 *The Trade-Off Between Natural Capital and Human Capital in Pakistan [102]*

This is a machine-generated summary of:

Islam, Moinul; Zhang, Bingqi; Managi, Shunsuke: The trade-off between natural capital and human capital in Pakistan [102]

Published in: Sustainability Science (2022)

Link to original: https://doi.org/10.1007/s11625-022-01143-w.

If you want to cite the papers, please refer to the original.

For technical reasons we could not place the page where the original quote is coming from.

Abstract-Summary

"Pakistan is the fifth largest populous nation in the world, and it is endowed with significant wealth in natural resources."

"This article calculates the natural capital (NC), human capital (HC), produced capital (PC) and inclusive wealth (IW) of Pakistan from 1992 to 2019 at the national level, geographic grid level, and provincial level."

"According to the disaggregated grid cell level data analysis, HC and PC of Pakistan grow positively due to the different measures of government."

"NC in Pakistan decreases in the areas where the reserve of fossil fuels decreases, and natural land cover is converted to other uses in the process of urbanization."

"Whether Pakistan's economy can continue to grow sustainably largely depends upon its NC base."

"The IW estimates for Pakistan offer policymakers a framework to look at three main issues for sustainable development: first, understanding trade-offs across different asset bases and time when making sustainability decisions is critically important; second, understanding social prices and market prices and their deviations; third, the asset side of the national balance sheet is now expanded to include HC and NC."

Introduction

"The UNEP's Inclusive Wealth Index (IWI) and the World Bank's Changing Wealth of Nations are two such examples that have demonstrated that it is possible to assess changes in natural capital (NC), human capital (HC), and produced capital (PC) across all countries, regardless of income level."

"Results indicated that it is possible to trace the changes of the components of wealth by country and link these to economic growth by highlighting that the impact of declines or increases in NC can be significant to the economic productive base (UNU-IHDP and UNEP [103])."

"The IWI looks at a country's well-being through a sustainability lens and measures its wealth in terms of social, environmental, and economic assets, the three key pillars of sustainability."

"During the period 1990–2014, Pakistan suffered an average loss in NC per capita of 5%, the steepest decline of the 26 Asian countries included in the IWI calculations (Managi and Kumar [104])."

Methodology

"IW includes three different categories of capital assets: natural capital (NC), human capital (HC) and produced capital (PC)."

"Assets in all three categories of capital provide the inputs for production, and thus are referred to as the productive base of the economy."

"Produced capital is perhaps the easiest to conceptualize, as it encompasses assets that have been historically referred to as capital by economists."

"Some wealth estimates include heath metrics into human capital accounts, but the scope of this paper allows for education-related human capital only."

"For estimating natural capital, the current accounting addresses sub-soil non-renewable resources, forests, and agricultural land."

"They are formally defined bygiven a forecast of how produced, human, and natural capitals and other flow variables evolve in the future in the economy in question."

Results

"The growth of the IW of Pakistan is consistently positive, and the growth rates of HC and PC are the main driving forces of this positive growth of wealth overall."

"Punjab and Sindh regions of Pakistan observed higher values of IW over the time period of analysis, and compared to the year 2000, the IW of these regions increased in years 2010 and 2019."

"From the grid level data analysis, it is clearly evident that Punjab, Sindh and Khyber Pakhtunkhwa contribute to the increase in HC of Pakistan."

"Similar to the HC distribution pattern, PC is concentrated in the Punjab, Sindh and Khyber Pakhtunkhwa areas of Pakistan."

"Sindh and Khyber Pakhtunkhwa provinces follow Punjab as the second and third wealthiest in HC stocks, with Balochistan showing the least amount of development in this area."

Conclusion

"IW represents the combined value of HC, PC and NC."

"In Pakistan, the relative share of HC, PC and NC has changed over time."

"This indicates that there have been clear trade-offs between the PC and NC of Pakistan, where the investment of PC has been prioritized over investment in protecting and improving NC."

"The trade-offs between the PC creation and NC degradation are often overlooked."

"In terms of development potential, there are clear trade-offs among the provinces of Pakistan."

"Understanding trade-offs across different asset bases and across time when making decisions on sustainability is critically important."

"Considering the importance of analyzing the interrelationships between HC, NC and PC, future studies should apply statistical methods (see Kumar and others [105]; Fujii and Managi [106]) to identify the precise relationship of these resources in different administrative and geographic levels."

8.2.11 Distributional Issues in Natural Capital Accounting: An Application to Land Ownership and Ecosystem Services in Scotland [107]

This is a machine-generated summary of:

Atkinson, Giles; Ovando, Paola: Distributional Issues in Natural Capital Accounting: An Application to Land Ownership and Ecosystem Services in Scotland [107]

Published in: Environmental and Resource Economics (2021)

Link to original: https://doi.org/10.1007/s10640-021-00613-6.

If you want to cite the papers, please refer to the original.

For technical reasons we could not place the page where the original quote is coming from.

Abstract-Summary

"Accounting for ecosystems is increasingly central to natural capital accounting."

"We examine the significance of the ownership of land on which ecosystem assets (or ecosystem types) is located in the context of natural capital accounting."

"We illustrate this in an empirical application to two ecosystem services and a range of ecosystem types and land ownership in Scotland, a context in which land reform debates are longstanding."

"Our results indicate the relative importance of private land in ecosystem service supply, rather than land held by the public sector."

"The contributions in this paper, we argue, represent important first steps in anticipating distributional impacts of natural capital (and related) policy in natural capital accounts as well as connecting these frameworks to broader concerns about wealth disparities across and within countries."

Introduction

"Natural capital accounting (NCA) describes a body of statistical work that seeks to construct better metrics of nature for policy."

"While there is no hard and fast definition of what NCA is, at its core is an emphasis on measuring flows—typically called 'ecosystem services'—as well as the underlying 'natural capital' stocks giving rise to these outputs."

"In the current paper we focus on a different spatial dimension that has so far been neglected in NCA: how natural capital is distributed amongst economic actors, i.e. people and organizations."

"We account for the use and supply distribution of two ecosystem services (climate change regulation through carbon sequestration and air pollution removal) across private and publicly owned land on which natural capital is located."

"How natural capital value is distributed by relative size of privately owned landholding depends on the particular ecosystem service examined."

Distributional Issues and Natural Capital Accounting: Concepts and Framework

"These ecosystem assets will be located on land in some way, and natural production and ecosystem service supply will be militated by how that land is managed by a range of different economic actors."

"Our empirical focus is on the physical and monetary flow of ecosystem services, as well as ecosystem extent, disaggregated within NCA to look at distributional issues."

"UN [108] does briefly anticipate such issues in suggesting, for example, that: "... data from cadastres showing the sector of ownership or the nature of the tenure can be linked to data on ecosystem assets, and hence provide a basis for monitoring the effects of land management policies within a given region, e.g., a water catchment" (para."

"What is changed is the lower portion of the panel, which now describes how ecosystem service supply can be attributed to natural capital on land which has an identifiable landholder such as a legal owner."

Spatial Distribution of Landownership and Ecosystem Services in Scotland

"We set out our data sources, data analysis and analytical criteria used for our NCA distributional analysis with its focus on carbon sequestration and air pollution removal services as well as extent of ecosystem asset or types."

"We also observe that where there are data about landownership in North-East and South-East Scotland properties tend to be small to medium size in areas dominated by arable land used for cereals, other crops, horticulture, or crop-based farming while in areas with limited agricultural value (grassland, rough grazing) and woodland, properties are of a medium to very large size class."

"These are also public ecosystem services: i.e. flows which are (largely) consumed by economic units other than those using or owning land (e.g. such as households in the case of air pollution removal services)."

"We now summarize the data we use for our assessment of the spatial distribution of air pollution removal and carbon sequestration services."

Results

"Aggregated natural capital accounts for carbon sequestration in the UK, indicate that about 71% of the physical and monetary value associated with this ecosystem service flow was mainly provided by woodland in 2016."

"Woodland is also associated with 28% of air pollution removal service, while enclosed farm, mountainous and moorland areas contribute to 23% each and semi-natural grassland to 17% of this service (ONS [109])."

"The monetary values (for flow and asset) associated with air pollution removal would be lower when air quality is higher (and vice versa), other things being equal."

"We find, however, a relatively weak positive relationship between gross income and total air pollution removed, which indicates a higher air quality control services in areas of higher incomes, which in Scotland correspond to more densely populated areas."

"In aggregated terms, air pollution removal had a monetary value of £39 million for the areas covered in this study (2016 prices)."

Discussion and Conclusions

"Natural capital accounting (NCA) provides a powerful framework for organizing otherwise voluminous information on ecosystem assets—and natural capital more generally—as well as the flow of services that these assets produce."

"We have couched this in terms of identifying distributional issues in the supply and use of ecosystem services and location of ecosystem assets (defined as ecosystem types)."

"Who owns, and manages, natural capital might be a secondary concern so long as public policy regimes succeed in provides sufficient flows of public ecosystem services to users, perhaps further ensuring these flows are reasonably equitably distributed in use."

"Extending distributional issues in supply to ecosystem condition accounts might, in addition, start to connect NCA with questions addressed elsewhere about the connection between land inequality and ecosystem loss or degradation (e.g. McDonald and others [110]; Albuquerque Sant'Anna [111]; Ceddia [112])."

8.2.12 The Impact of Renewable Versus Non-renewable Natural Capital on Economic Growth [113]

This is a machine-generated summary of:

Gasmi, Farid; Recuero Virto, Laura; Couvet, Denis: The Impact of Renewable Versus Non-renewable Natural Capital on Economic Growth [113]

Published in: Environmental and Resource Economics (2020)
Link to original: https://doi.org/10.1007/s10640-020-00495-0

If you want to cite the papers, please refer to the original.

For technical reasons we could not place the page where the original quote is coming from.

Abstract-Summary

"In a dataset on 83 countries covering the years 1960–2009, we find a negative indirect effect of the share of renewable natural capital in wealth on economic growth transmitted through demographic factors, more specifically, population fertility."

"In countries with lower income inequality and higher institutional quality, the share of non-renewable natural capital in wealth has a direct positive impact on growth."

"We also find that countries with higher income per capita, human development, and institutional quality have a higher share of renewable natural capital per capita, but a lower share of renewable natural capital in wealth."

Introduction

"We use data on some renewable and non-renewable natural capital variables to develop a comparative analysis, based on cross-country estimations (World Bank [114]), of the impact of these two types of natural capital on economic growth."

"In order to understand how natural capital contributes to economic growth, it is necessary to make a distinction between renewable and non-renewable natural capital, as well as between direct and indirect effects on economic growth including the effects that work through international trade."

"We take the view that renewable and non-renewable natural capital impact economic growth in different ways, in particular, through variables that relate to various dimensions of human well-being as well as institutions, among other factors."

"This article is an original contribution to the understanding of the role played by natural capital in economic growth while distinguishing the direct and indirect effects of its renewable and non-renewable components."

Related Literature

"Several articles have suggested that the negative relationship between natural resource dependence and economic growth can be reversed if institutional quality is sufficiently high (Boschini and others [115])."

"Although natural resource dependence, measured by the share of exports of primary products in GDP or other ratio indicators, may slow economic growth, natural resource abundance, often measured by per capita value of production and reserves or stocks, may have no significant impact on, or affect positively, economic growth."

"As to the issue of accounting for the presence of international heterogeneity when analyzing the impact of natural resources on economic growth, the academic literature based on fixed-effects estimation provides evidence that dependence variables based on natural resource exports are not significant determinants of economic growth (Mansano and Ribogon [116]; Lederman and Maloney [117])."

Econometric Methodology

"Our economic growth regressions include proxies for both proximate and fundamental theories of growth."

"Proximate theory proxy variables can only have direct impacts on economic growth whereas fundamental theory proxy variables can have direct and/or indirect impacts."

"If proximate variables or fundamental variables were significant, this would imply that the corresponding theories have a direct explanatory power of economic growth."

"To uncover indirect impacts of fundamental theories' proxy variables on economic growth, growth regressions should only include fundamental theories' proxy variables."

"In the introduction the BMA estimation approach that we use allows us to estimate regressions aimed at testing alternative theories of economic growth while explicitly accounting for the associated models' uncertainty and the uncertainty related to the specification of proxy variables for each theory to be used as regressors (Brock and Durlauf [118]; Brock and others [119])."

Data and Preliminary Analysis

"For the sub-sample above the median cut-off point in investment in physical capital, the share of renewable natural capital is correlated with the demography proximate theory (fertility and life expectancy variables)."

"For the sub-sample below the median cut-off point in investment in physical capital, the share of renewable natural capital is correlated with neoclassical (population growth rates), demography (fertility), regional heterogeneity (Sub-Saharan Africa) proximate theories Building on these results, we analyze whether or not the fundamental theories just mentioned, religion, renewable natural capital, fractionalization, and institutions have some explanatory power in the growth regressions, beyond the influence exerted through proximate theories' variables."

"According to this result, the economic growth regressions are developed for the whole sample and for the two sub-samples defined by the cut-off point in the median level of investment in physical capital, which determines the two economic growth regimes."

BMA Estimation Results

"To improve the interpretation of our results, we end this section with an exploration of how natural capital is related to the standard variables affecting economic growth, namely, income per capita, income inequality, human well-being, especially human development, and institutional quality."

"We focus on income per capita, income inequality, human development, and institutional quality to distinguish different growth regimes while analyzing the role of natural capital in economic growth."

"These results hold for two different sub-samples, namely, those characterized by lower income inequality and by higher institutional quality, given that natural capital is a robust determinant of growth when both proximate and fundamental theories are analyzed together."

"To further interpret our results, we investigate the issue of how natural capital is related to the different variables affecting economic growth, namely, income per capita, income inequality, human well-being, especially human development, and institutional quality."

Conclusion

"When using the whole sample in the BMA estimation and assuming a unique growth regime, we find that the natural capital theory, as reflected in natural capital dependence, proxied by share of natural capital in wealth, and natural capital abundance proxied by natural capital per capita, is not a robust direct explanatory theory of growth, i.e., such proxy variables do not have a direct significant impact on economic growth proxied by GDP per capita rate of growth."

"Enough though, we find that the share of renewable natural capital in wealth affects economic growth negatively and indirectly through its influence on proximate theories proxy variables, more specifically, through its effect on fertility."

"We find a direct positive impact of the share of non-renewable natural capital in wealth on economic growth when exploring proximate and fundamental theories together in the sub-samples corresponding to lower income inequality and higher institutional quality."

Bibliography

1. Vardon, Michael; Keith, Heather; Obst, Carl; Lindenmayer, David Putting biodiversity into the national accounts: Creating a new paradigm for economic decisions. Ambio (2018). https://doi.org/10.1007/s13280-018-1114-z
2. Rockström, J., W. Steffen, K. Noone, Å. Persson, F. S. Chapin, III, E. Lambin, T. M. Lenton, M. Scheffer, et al. 2009. Planetary Boundaries: Exploring the Safe Operating Space for Humanity. Ecology and Society 14: 32. http://www.ecologyandsociety.org/vol14/iss2/art32/
3. Boyd, J., and S. Banzhoff. 2007. What are Ecosystem Services? The Need for Standardized Environmental Accounting Units. Ecological Economics 63: 616–626
4. Milner-Gulland, E.J., J.A. Mcgregor, M. Agarwala, G. Atkinson, P. Bevan, T.J. Clements, T. Daw, K. Homewood, et al. 2014. Accounting for the Impact of Conservation on Human Well-Being. Conservation Biology 28: 1160–1166
5. Bateman, I.J., G.M. Mace, C. Fezzi, G. Atkinson, and K. Turner. 2010. Economic Analysis for Ecosystem Service Assessments. Environmental & Resource Economics 48: 177–218
6. United Nations, European Commission, Food and Agriculture Organization, International Monetary Fund, Organisation for Economic Co-operation and Development, World Bank. 2014. System of Environmental-Economic Accounting 2012—Central Framework. New York: United Nations
7. Natural Capital Coalition. 2016. Natural Capital Protocol Principles and Framework. London: Natural Capital Coalition. http://naturalcapitalcoalition.org/protocol/
8. Vardon, M., P. Burnett, and S. Dovers. 2016. The Accounting Push and the Policy Pull: Balancing Environment and Economic Decisions. Ecological Economics 124: 145–152
9. Jakobsson, K.M., and A.K. Dragun. 2001. The Worth of a Possum: Valuing Species with the Contingent Valuation Method. Environmental Research and Economics 19: 211–227
10. Department of Water Affairs. 2017. Botswana Water Accounting Report 2015/16. Gaborone, Department of Water Affairs, Botswana. https://www.wavespartnership.org/en/knowledge-center/botswana-water-accounting-report-201516

11. Cong, Rong-Gang; Termansen, Mette; Brady, Mark V. Managing soil natural capital: a prudent strategy for adapting to future risks. Annals of Operations Research (2015). https://doi.org/10.1007/s10479-015-2066-3
12. Koellner, T., & Schmitz, O. J. (2006). Biodiversity, ecosystem function, and investment risk. BioScience, 56(12), 977–985
13. Brady, M., Hedlund, K., Cong, R.-G., Hemerik, L., Hotes, S., Machado, S., et al. (2015). Valuing supporting soil ecosystem services in agriculture: A natural capital approach. Agronomy Journal, 107(5), 1809–1821. https://doi.org/10.2134/agronj14.0597
14. Cong, R.-G., Hedlund, K., Andersson, H., & Brady, M. (2014a). Managing soil natural capital: an effective strategy for mitigating future agricultural risks? Agricultural Systems, 129, 30–39
15. Nandwa, S. (2001b). Soil organic carbon (SOC) management for sustainable productivity of cropping and agro-forestry systems in Eastern and Southern Africa. In C. Martius, H. Tiessen, & P. L. G. Vlek (Eds.), Managing organic matter in tropical soils: Scope and limitations. Developments in Plant and Soil Sciences (Vol. 93, pp. 143–158). Netherlands: Springer
16. Ajayi, O., Franzel, S., Kuntashula, E., & Kwesiga, F. (2003). Adoption of improved fallow technology for soil fertility management in Zambia: Empirical studies and emerging issues. Agroforestry Systems, 59(3), 317–326
17. De Wit, Ad, Boogaard, H., & Van Diepen, C. (2005). Spatial resolution of precipitation and radiation: The effect on regional crop yield forecasts. Agricultural and Forest Meteorology, 135(1), 156–168
18. Schlenker, W., & Roberts, M. J. (2009). Nonlinear temperature effects indicate severe damages to US crop yields under climate change. Proceedings of the National Academy of Sciences, 106(37), 15594–15598
19. Richardson, C. W. (1981). Stochastic simulation of daily precipitation, temperature, and solar radiation. Water Resources Research, 17(1), 182–190
20. Dai, A., Trenberth, K. E., & Karl, T. R. (1999). Effects of clouds, soil moisture, precipitation, and water vapor on diurnal temperature range. Journal of Climate, 12(8), 2451–2473
21. Gill, S., Vasanthan, T., Ooraikul, B., & Rossnagel, B. (2002). Wheat bread quality as influenced by the substitution of waxy and regular barley flours in their native and extruded forms. Journal of Cereal Science, 36(2), 219–237. https://doi.org/10.1006/jcrs.2001.0458
22. Li, Y. L., McAllister, T. A., Beauchemin, K. A., He, M. L., McKinnon, J. J., & Yang, W. Z. (2011). Substitution of wheat dried distillers grains with solubles for barley grain or barley silage in feedlot cattle diets: Intake, digestibility, and ruminal fermentation. Journal of Animal Science, 89(8), 2491–2501. https://doi.org/10.2527/jas.2010-3418
23. Gardner, B. L. (1976). Futures prices in supply analysis. American Journal of Agricultural Economics, 58(1), 81–84
24. Zhu, Y., Ghosh, S. K., & Goodwin, B. K. (2008). Modeling dependence in the design of whole-farm insurance contract: A copula-based model approach. In Annual meetings of the American Agricultural Economics Association, Orlando, FL, 2008 (pp. 27–29)
25. SCB. (2012). Jordbrukstatistik årsbok 2012 (Yearbook of agricultural statistics 2012). Örebro: Statistics Sweden
26. Hossaini, Navid; Hewage, Kasun; Sadiq, Rehan Path toward net-zero buildings: a natural capital assessment framework. Clean Technologies and Environmental Policy (2017). https://doi.org/10.1007/s10098-017-1469-z
27. Hossaini N, Hewage K, Sadiq R (2015) Spatial life cycle sustainability assessment: a conceptual framework for net-zero buildings. Clean Technol Environ Policy. https://doi.org/10.1007/s10098-015-0959-0
28. Bendewald M, Zhai ZJ (2013) Using carrying capacity as a baseline for building sustainability assessment. Habitat Int 37:22–32
29. Vargas, Leonardo; Willemen, Louise; Hein, Lars Assessing the Capacity of Ecosystems to Supply Ecosystem Services Using Remote Sensing and An Ecosystem Accounting Approach. Environmental Management (2018). https://doi.org/10.1007/s00267-018-1110-x

30. United Nations, European Commission, Food and Agriculture Organization of the United Nations, International Monetary Fund, Organisation for Economic Co-operation and Development, World Bank (2014b) System of environmental economic accounting 2012: experimental ecosystem accounting. United Nations, New York, NY
31. Bagstad KJ, Semmens DJ, Waage S, Winthrop R (2013) A comparative assessment of decision-support tools for ecosystem services quantification and valuation. Ecosyst Serv 5:27–39. https://doi.org/10.1016/j.ecoser.2013.07.004
32. Schröter M, Remme RP, Sumarga E, Barton DN, Hein L (2015) Lessons learned for spatial modelling of ecosystem services in support of ecosystem accounting. Ecosyst Serv 13:64–69. https://doi.org/10.1016/j.ecoser.2014.07.003
33. Willemen L, Burkhard B, Crossman N, Drakou EG, Palomo I (2015) Editorial: best practices for mapping ecosystem services. Ecosyst Serv 13:1–5. https://doi.org/10.1016/j.ecoser.2015.05.008
34. Costanza R, Fisher B, Mulder K, Liu S, Christopher T (2007) Biodiversity and ecosystem services: a multi-scale empirical study of the relationship between species richness and net primary production. Ecol Econ 61:478–491
35. van Oudenhoven AP, Petz K, Alkemade R, Hein L, de Groot RS (2012) Framework for systematic indicator selection to assess effects of land management on ecosystem services. Ecol Indic 21:110–122
36. Remme RP, Edens B, Schröter M, Hein L (2015) Monetary accounting of ecosystem services: A test case for Limburg province, the Netherlands. Ecol Econ 112:116–128. https://doi.org/10.1016/j.ecolecon.2015.02.015
37. Villamagna AM, Angermeier PL, Bennett EM (2013) Capacity, pressure, demand, and flow: a conceptual framework for analyzing ecosystem service provision and delivery. Ecol Complex 15:114–121. https://doi.org/10.1016/j.ecocom.2013.07.004
38. van Putten, E. Ingrid; Pinkard, Elizabeth; O'Grady, Anthony; Schmidt, Rebecca K.; Cresswell, Ian; Raoult, Vincent; Taylor, Matt D. Stakeholder perspectives on the value proposition of enterprise-level natural capital accounting for three primary industries. Environment Systems and Decisions (2021). https://doi.org/10.1007/s10669-021-09817-z
39. Modak P, Mathur A, Vaidyanathan K, et al (2019) Mainstreaming natural capital valuation. Paper presented at the T20 Summit 2019: International finsancial architecture for Stability and Development / Crypto-assets and Fintech, Tokyo, Japan. https://www.gatewayhouse.in/wp-content/uploads/2019/04/t20-japan-tf2-14-mainstreaming-natural-capital-valuation.pdf
40. Wall J (2019) Biodiversity assets of NSW and Qld cotton growing areas. Draft report to the CRDC, Cotton Research and Development Corporation, Cotton Research and Development Corporation
41. Hart PJ, Reynolds JD (2008) Handbook of fish biology and fisheries: fisheries. Wiley, New York
42. Mcilgorm A, Pepperell J (2013) Developing a cost effective state wide expenditure survey method to measure the economic contribution of the recreational fishing sector in NSW in 2012, Australian National Centre for Ocean Resources and Security (ANCORS), University of Wollongong, Australian National Centre for Ocean Resources and Security (ANCORS), University of Wollongong, Report to the NSW Recreational Fishing Trust, NSW Department of Primary Industries
43. Schirmer, J., Mylek, M., Magnusson, A., et al. (2017), Socio-economic impacts of the forest industry: Green Triangle, Forest and Wood Products Australia, Forest and Wood Products Australia
44. Brand D (2019) Transforming the forestry asset class, New Forests, New Forests
45. Coad L, Campbell A, Miles L, Humphries K (2008) The costs and benefits of protected areas for local livelihoods: a review of the current literature. UNEP World Conservation Monitoring Centre. UK, Cambridge
46. O'Grady AP, Smith GS, Ascui F et al (2020) The rise and rise of natural capital: what role for forestry? Aust for 83:103–106

47. Shah, Muhammad Ibrahim; Zakari, Abdulrasheed; Kumar, Siddharth; Abbas, Shujaat; Sheraz, Muhammad Quantifying the effect of waste on soil health in European Union: what are the roles of technology, natural capital, and institutional quality?. Environmental Science and Pollution Research (2022). https://doi.org/10.1007/s11356-022-20909-4
48. Almendro-Candel MB, Navarro-Pedreño J, Lucas IG, Zorpas AA, Voukkali I, Loizia P (2019) The use of composted municipal solid waste under the concept of circular economy and as a source of plant nutrients and pollutants. In: (ed) Municipal solid waste management. IntechOpen. https://doi.org/10.5772/intechopen.83386 . Accessed 22 Mar 2022
49. Tiseo I (2019) Global waste generation per person outlook by region 2050. https://www.statista.com/statistics/233624/forecast-of-per-capita-waste-generation-worldwide-by-region/ . Accessed 23 Mar 2022
50. Carmo DLD, Lima LBD, Silva CA (2016) Soil fertility and electrical conductivity affected by organic waste rates and nutrient inputs. Revista Brasileira de Ciência do Solo, 40
51. Merga B, Mohammed M, Ahmed A (2020) Socio-economic impacts of possible brewery waste recycling in agriculture. Cogent Environ Sci 6(1):1732112
52. Al-Busaidi A, Ahmed M (2016) Treated municipal wastes: are they contaminating or enriching the soil?. In: Larramendy ML, Soloneski S (eds) Soil contamination - current consequences and further solutions. IntechOpen. https://doi.org/10.5772/64962 . Accessed 22 Mar 2022
53. Hafeez M, Yuan C, Shahzad K, Aziz B, Iqbal K, Raza S (2019) An empirical evaluation of financial development-carbon footprint nexus in One Belt and Road region. Environ Sci Pollut Res 26(24):25026–25036. https://doi.org/10.1007/s11356-019-05757-z
54. Sheraz M, Deyi X, Ahmed J, Ullah S, Ullah A (2021) Moderating the effect of globalization on financial development, energy consumption, human capital, and carbon emissions: evidence from G20 countries. Environ Sci Pollut Res 28(26):35126–35144
55. Nathaniel SP, Adeleye N (2021) Environmental preservation amidst carbon emissions, energy consumption, and urbanization in selected African countries: implication for sustainability. J Clean Prod 285:125409
56. Kumagai, Junya; Wakamatsu, Mihoko; Hashimoto, Shizuka; Saito, Osamu; Yoshida, Takehito; Yamakita, Takehisa; Hori, Keiko; Matsui, Takanori; Oguro, Michio; Aiba, Masahiro; Shibata, Rei; Nakashizuka, Tohru; Managi, Shunsuke Natural capital for nature's contributions to people: the case of Japan. Sustainability Science (2021). https://doi.org/10.1007/s11625-020-00891-x
57. Saito O, Kamiyama C, Hashimoto S, Matsui T, Shoyama K, Kabaya K, Uetake T, Taki H, Ishikawa Y, Matsushita K, Yamane F, Hori J, Ariga T, Takeuchi K (2019) Co-design of national-scale future scenarios in Japan to predict and assess natural capital and ecosystem services. Sustain Sci 14(1):5–21. https://doi.org/10.1007/s11625-018-0587-9
58. Fujii H, Sato M, Managi S (2017) Decomposition analysis of forest ecosystem services values. Sustainability (Switzerland). https://doi.org/10.3390/su9050687
59. Ikeda S, Managi S (2019) Future inclusive wealth and human well-being in regional Japan: projections of sustainability indices based on shared socioeconomic pathways. Sustain Sci 14(1):147–158. https://doi.org/10.1007/s11625-018-0589-7
60. Vardon, Michael; Lucas, Paul; Bass, Steve; Agarwala, Matthew; Bassi, Andrea M.; Coyle, Diane; Dvarskas, Anthony; Farrell, Catherine A.; Greenfield, Oliver; King, Steven; Lok, Martin; Obst, Carl; O'Callaghan, Brian; Portela, Rosimeiry; Siikamäki, Juha From COVID-19 to Green Recovery with natural capital accounting. Ambio (2022). https://doi.org/10.1007/s13280-022-01757-5
61. Dasgupta, P. 2021. The Economics of Biodiversity: The Dasgupta Review. London: HM Treasury
62. O'Callaghan, B. J. and Murdock, E. 2021. Are We Building Back Better? Evidence from 2020 and Athways to Inclusive Green Recovery Spending. United Nations Environment Programme (UNEP)
63. Ruijs A, Vardon M, Bass S et al (2019) Natural capital accounting for better policy. Ambio 48:714–725

64. Cohen-Shacham, E., G. Walters, C. Janzen, and S. Maginnis. 2016. Nature-Based Solutions to Address Global Societal Challenges. Gland: IUCN
65. IPBES. 2019. Global assessment report on biodiversity and ecosystem services of the Intergovernmental Science-Policy Platform on Biodiversity and Ecosystem Services, Bonn, Germany: IPBES Secretariat
66. Seddon, N., S. Sengupta, M. García-Espinosa, I. Hauler, D. Herr, and A.R. Rizvi. 2019. Nature-Based Solutions in Nationally Determined Contributions: Synthesis and Recommendations for Enhancing Climate Ambition and Action by 2020. Oxford: IUCN and University of Oxford
67. WWF and ILO. 2020. Nature hires: How Nature-based Solutions can power a green jobs recovery. World Wide Fund For Nature (WWF) and International Labour Organization (ILO)
68. Dobson, A.P., S.L. Pimm, L. Hannah, L. Kaufman, J.A. Ahumada, A.W. Ando, A. Bernstein, J. Busch, et al. 2020. Ecology and economics for pandemic prevention. Science 369: 379–381
69. Palomo, I., B. Locatelli, I. Otero, M. Colloff, E. Crouzat, A. Cuni-Sanchez, E. Gómez-Baggethun, et al. 2021. Assessing Nature-Based Solutions for Transformative Change. One Earth 4: 730–741
70. Stiglitz, J., A. Sen, and J.-P. Fitoussi. 2010. Report by the Commission on the Measurement of Economic Performance and Social Progress. Paris: The Commission on the Measurement of Economic Performance and Social Progress
71. Coyle, D. 2014. GDP: A Brief But Affectionate History. Princeton: Princeton University Press
72. Hamilton, K., and C. Hepburn. 2017. National Wealth: What is Missing, Why It Matters. Oxford University Press
73. Lucas, P. L. and Vardon, M. 2021. Greening the recovery to make it last: the role of Natural Capital Accounting. The Hague: PBL Netherlands Environmental Assessment Agency
74. World Bank. 2021b. From Accounts to Policy: WAVES Closeout Report Wealth Accounting and Valuation of Ecosystem Services Global Partnership (2012–2019). Washington DC: World Bank
75. Banerjee, O., N. Crossman, R. Vargas, L. Brander, P. Verburg, M. Cicowiez, J. Hauck, and E. Mckenzie. 2020. Global Socio-economic Impacts of Changes in Natural Capital and Ecosystem Services: State of Play and New Modeling Approaches. Ecosystem Services 46: 101202
76. Farrell, C.A., J. Aronson, G.C. Daily, L. Hein, C. Obst, P. Woodworth, and J.C. Stout. 2022. Natural Capital Approaches: Shifting the UN Decade on Ecosystem Restoration from aspiration to reality. Restoration Ecology. https://doi.org/10.1111/rec.13632
77. Coates, D., J.Y. Chan, and C. Hill. 2020. Natural Capital Accounting for Integrated Biodiversity Policies. New York: United Nations
78. King, S., M. Vardon, H.S. Grantham, M. Eigenraam, S. Ferrier, D. Juhn, T. Larsen, C. Brown, and K. Turner. 2021. Linking Biodiversity into National Economic Accounting. Environmental Science & Policy 116: 20–29
79. Pizarro, R. 2020. Natural Capital Accounting for Integrated Climate Change Policies. New York: United Nations
80. Meijer, J., E. Berkhout, C.J. Hill, and M. Vardon. 2020. Integrated Landscape Management and Natural Capital Accounting: Working Together for Sustainable Development. The Hague: PBL Netherlands Environmental Assessment Agency
81. Ruijs, A., M. Van Der Heide, and J. And Van Den Berg. 2018. Natural Capital Accounting for the Sustainable Development Goals. The Hague: PBL Netherlands Environmental Assessment Agency
82. Vardon, M., Bass, S., Ahlroth, S. and Ruijs, A. 2017. Forum on Natural Capital Accounting for Better Policy Decisions: Taking Stock and Moving Forward. Washington DC: the World Bank
83. Vardon, M., Bass, S. and Ahlroth, S. 2019. Natural Capital Accounting for Better Policy Decisions: Climate Change and Biodiversity. Proceedings and Highlights of the 3rd Forum on Natural Capital Accounting for Better Policy Decisions. Washington DC: World Bank WAVES

84. Vardon, M. and Bass, S. 2020. Natural Capital Accounting for Better Policy Decisions: Measuring and Valuing Natural Capital to Improve Landscape Management and Governance. In Proceedings and Highlights of the 4th Forum on Natural Capital Accounting for Better Policy Decisions. Washington DC: World Bank WAVES
85. Benitez, P., Blignaut, J., Kalisa, J., Katanisa, P. and Rutebuka, E. 2021. Nature-smart post-pandemic economic recovery in Rwanda: A Natural Capital Accounting approach with an emphasis on the natural resources sectors. Word Bank
86. Klenert, D., F. Funke, L. Mattauch, and B. O'callaghan. 2020. Five Lessons from COVID-19 for Advancing Climate Change Mitigation. Environmental and Resource Economics 76: 751–778
87. Heris, M., K.J. Bagstad, C. Rhodes, A. Troy, A. Middel, K.G. Hopkins, and J. Matuszak. 2021. Piloting Urban Ecosystem Accounting for the United States. Ecosystem Services 48: 101226
88. Bassi, A.M. 2021. Policy Scenario Analysis using SEEA Ecosystem Accounting. New York: United Nations
89. ONS. 2021. UK Environmental Accounts: 2021. London: UK Office for National Statistics
90. Neelakantan, Amrita; DeFries, Ruth; Sterling, Eleanor; Naeem, Shahid Contributions of financial, social and natural capital to food security around Kanha National Park in central India. Regional Environmental Change (2020). https://doi.org/10.1007/s10113-020-01589-7
91. Brooks TM, Bakarr MI, Boucher T, Da Fonseca GA, Hilton-Taylor C, Hoekstra JM, Moritz T, Olivieri S, Parrish J, Pressey RL, Rodrigues AS (2004) Coverage provided by the global protected-area system: is it enough? BioScience 54(12):1081–1091. https://doi.org/10.1641/0006-3568(2004)054[1081:CPBTGP]2.0.CO;2
92. West P, Brockington D (2006) An anthropological perspective on some unexpected consequences of protected areas. Conserv Biol 20:609–616. https://doi.org/10.1111/j.1523-1739.2006.00432.x
93. Torri M (2011) Conservation, relocation and the social consequences of conservation policies in protected areas: case study of the Sariska Tiger Reserve, India. Conserv Soc 9:54. https://doi.org/10.4103/0972-4923.79190
94. Zhang W, Ricketts TH, Kremen C, Carney K, Swinton SM (2007) Ecosystem services and dis-services to agriculture. Ecol Econ 64:253–260. https://doi.org/10.1016/j.ecolecon.2007.02.024
95. Golden CD, Fernald LCH, Brashares JS, Rasolofoniaina BJ, Kremen C (2011) Benefits of wildlife consumption to child nutrition in a biodiversity hotspot. Proc Natl Acad Sci U S A 108:19653–19656. https://doi.org/10.1073/pnas.1112586108
96. Johnson KB, Jacob A, Brown ME (2013) Forest cover associated with improved child health and nutrition: evidence from the Malawi demographic and health survey and satellite data. Glob Health Sci Pract 1:237–248. https://doi.org/10.9745/GHSP-D-13-00055
97. Jones AD, Ngure FM, Pelto G, Young SL (2013) What are we assessing when we measure food security? A compendium and review of current metrics. Adv Nutr:481–505. https://doi.org/10.3945/an.113.004119
98. Sunderlin WD, Angelsen A, Belcher B, Burgers P, Nasi R, Santoso L, Wunder S (2005) Livelihoods, forests, and conservation in developing countries: an overview. World Dev 33:1383–1402. https://doi.org/10.1016/j.worlddev.2004.10.004
99. Bharucha Z, Pretty J (2010) The roles and values of wild foods in agricultural systems. Philos Trans R Soc Lond Ser B Biol Sci 365:2913–2926. https://doi.org/10.1098/rstb.2010.0123
100. Thulstrup AW (2015) Livelihood resilience and adaptive capacity: tracing changes in household access to capital in Central Vietnam. World Dev 74:352–362. https://doi.org/10.1016/j.worlddev.2015.05.019
101. Asih DN, Klasen S (2017) Improving food security? Setting indicators and observing change of rural household in Central Sulawesi. Forest and Society 1:154–161 https://doi.org/10.24259/fs.v1i2.2099
102. Islam, Moinul; Zhang, Bingqi; Managi, Shunsuke The trade-off between natural capital and human capital in Pakistan. Sustainability Science (2022). https://doi.org/10.1007/s11625-022-01143-w

103. UNU-IHDP and UNEP (2012) Inclusive Wealth Report 2012: Measuring progress toward sustainability. Cambridge University Press, Cambridge
104. Managi S, Kumar P (eds; 2018) Inclusive Wealth Report 2018: measuring progress towards sustainability. Routledge Taylor and Francis Group, Oxford
105. Kumar S et al (2015) Substitute or complement? Assessing renewable and non-renewable energy in OECD Countries. Appl Econ 47(14):1438–1459
106. Fujii H, Managi S (2016) An evaluation of inclusive capital stock for urban planning. Ecosyst Health Sustain 2(10):1–11. https://doi.org/10.1002/ehs2.1243
107. Atkinson, Giles; Ovando, Paola Distributional Issues in Natural Capital Accounting: An Application to Land Ownership and Ecosystem Services in Scotland. Environmental and Resource Economics (2021). https://doi.org/10.1007/s10640-021-00613-6
108. UN (2021) System of environmental-economic accounting – ecosystem, accounting. final draft. United Nations, SEEA. Available online: (https://seea.un.org/ecosystem-accounting). last accessed 13 Apr 2020
109. ONS (2020a) Scottish Natural Capital Accounts: 2020, UK Office for National Statistics
110. McDonald RI, Forman RTT, Kareiva P (2010) Open space loss and land inequality in United States' Cities, 1990–2000. PLoS ONE 5(3):e9509
111. Albuquerque Sant'Anna, A. (2016) Land inequality and deforestation in the Brazilian amazon. Environ Dev Econ 22:1–25
112. Ceddia MG (2019) The Impact of income, land, and wealth inequality on agricultural expansion in Latin America. Proc Natl Acad Sci 116(7):2527–2532
113. Gasmi, Farid; Recuero Virto, Laura; Couvet, Denis The Impact of Renewable Versus Non-renewable Natural Capital on Economic Growth. Environmental and Resource Economics (2020). https://doi.org/10.1007/s10640-020-00495-0
114. World Bank (2006) A guide to valuing natural resources wealth. Environment Department, World Bank, USA
115. Boschini A, Pettersson J, Roine J (2013) The resource curse and its potential reversal. World Dev 43:19–41
116. Mansano O, Ribogon R (2001) Resource curse or debt overhang? National Bureau of Economic Research Working Paper No. 8390
117. Lederman D, Maloney W (2002) Open questions about the link between natural resources and economic growth: Sachs and Warner revisited. Central Bank of Chile Working Paper 141
118. Brock W, Durlauf SN (2001) Growth empirics and reality. World Bank Econ Rev 15(2):229–272
119. Brock W, Durlauf SN, West K (2003) Policy analysis in uncertain economic environments (with discussion). Brook Pap Econ Act 1:235–322

Chapter 9
Green Jobs

9.1 Introduction by Editor

According to the International Labour Organisation (ILO), Green Jobs are defined as employment opportunities that meet the criteria of being both decent and conducive to the preservation or repair of the natural environment. Skills can be cultivated within conventional industries like manufacturing and construction, as well as within developing sectors focused on sustainability, such as renewable energy and energy efficiency. Green jobs have the potential to facilitate the widespread adoption of renewable energy sources, enhance material utilisation and efficiency, mitigate pollution, reduce greenhouse gas emissions, safeguard and rehabilitate natural ecosystems, and provide assistance in adapting to and mitigating the impacts of climate change.

The topic of transitioning to a more environmentally sustainable economy has garnered significant attention in contemporary discourse. The objective of this notion is to supplant the prevailing "business as usual" paradigm with an alternative framework that holds promise for the preservation and restoration of natural ecosystems. The approach encompasses several strategies such as providing incentives and implementing legislative frameworks to encourage environmentally friendly investments, fostering innovation, promoting sustainable consumption practices, and facilitating information exchange among firms. The concept of a greening of the economy has garnered interest from both governmental and corporate entities due to its potential to simultaneously tackle economic and environmental challenges. The incorporation of strategies pertaining to the green economy is an intrinsic component of the trajectory towards achieving sustainable development. The United Kingdom's 'Build Back Greener' scheme aims to create 2,000 new construction jobs to manufacture offshore wind turbines and upgrade ports. The program outlines strategies and recommendations for decarbonizing all UK economic sectors in order to achieve our net zero aim by 2050. According to the Institute for Public Policy Research, 70,000 employment opportunities could be produced in offshore wind energy, and over 200,000 jobs might be created in energy efficiency.

G. Nulkar (ed.), *The Economics of Sustainable Development*,
https://doi.org/10.1007/978-981-99-7379-8_9

The European Union (EU) has adopted the Circular Economy Action Plan in 2015.[1] This plan sets ambitious targets for increasing resource productivity and recycling rates, both of which have shown to have a positive impact on economic growth and employment generation. One bibliometric analysis highlights the progress made by the EU in waste management, resource efficiency, and the integration of circular economy principles into various sectors (Pascale et al., 2021). Another study found that the implementation of circular economy practices in the EU could create up to 700,000 new jobs by 2030 (Mazur-Wierzbicka, 2021).

The United States, however, has not kept pace with the EU. They are yet to adopt a circular economy approach, but are focusing primarily on the green jobs in renewable energy, energy efficiency, and sustainable transportation. Government initiatives such as the Green New Deal have the potential to accelerate the transition towards a more sustainable economy, creating new economic opportunities and addressing environmental concerns.

India, on the other hand, has been taking some steps in promoting resource efficiency and circular economy principles. The country's National Action Plan on Climate Change and the establishment of eco-industrial parks have demonstrated a commitment to sustainable development. As highlighted by Ali et al., the integration of green growth strategies in China's sports industry has proven effective in cutting pollution and conserving materials, suggesting that similar approaches could be successful in other emerging economies like India.

Even as the green businesses provide a significant opportunity for low-carbon economic growth, a key aspect of the shift is the need to address the skills gap and facilitate the transition of workers from traditional, carbon-intensive industries to the green sectors (Zhou et al., 2023). There is a need in existing occupations to retrain, repurpose and upgrade skills (Jeon, 2012). To ensure a just transition, it is crucial to focus on segments of the population that may be disproportionately affected, such as those working in carbon-intensive industries, and to provide them with the necessary skills and support to adapt to the changing labor market.

As the reserves of fossil fuels diminish, various industries will face mounting pressure to transition towards production processes that are less reliant on them. This necessitates the advancement of energy-efficient technologies and the adoption of sustainable practices. The assertion holds particular significance in the context of emerging economies such as India, which mainly depend on imported fossil fuels in for their energy demands. Based on the findings of the CEEW research (Tyagi, 2021), it is indicated that in order to achieve its target of attaining 500 GW of non-fossil energy generation capacity by 2030, India must augment its solar capacity by 238 GW and its wind capacity by 101 GW. The prospective outcome of this initiative has the capacity to generate approximately 3.4 million employment opportunities in both the immediate and extended periods. Approximately one million workers have the potential to occupy these environmentally friendly positions, considering that a single worker may be capable of fulfilling many roles. The International Labour

[1] First circular economy action plan—European Commission (europa.eu) Last accessed 21 June 2024.

Organisation (ILO) has identified other industries in India that are conducive to the creation of green jobs. These sectors include forestry, watershed development, mass transport, and ecological restoration.

This chapter provides an overview of many studies that demonstrate the potential of circular economies in generating employment opportunities, securing venture investment for environmentally sustainable firms, and the current status of green recruitment practices and human resource management. The chapter provides a discussion of the reasons supporting the creation of an environment that fosters the growth of green employment and green jobs.

References

Mazur-Wierzbicka, E. Towards Circular Economy—A Comparative Analysis of the Countries of the European Union. Resources 2021, 10, 49. https://doi.org/10.3390/resources10050049.

De Pascale, A., et al. A systematic review for measuring circular economy: The 61 indicators November 2020. Journal of Cleaner Production 281(3):124,942. https://doi.org/10.1016/j.jclepro.2020.124942.

Tyagi, A. C. (2021). India's Expanding Clean Energy Workforce. New Delhi: Council on Energy, Environment and Water, Natural Resources Defense Council, and Skill Council for Green Jobs. Retrieved from https://www.ceew.in/sites/default/files/Green-Jobs-Report-Jan27.pdf

Zhou, L., Zhao, K., & Waqas, M. (2023, July 1). Beyond the Arena: How sports economics is advancing China's sustainable development goals. Elsevier BV, 9(7), e18074–e18074. https://doi.org/10.1016/j.heliyon.2023.e18074.

9.2 Machine Generated Summaries

Disclaimer: The summaries in this chapter were generated from Springer Nature publications using extractive AI auto-summarization: An extraction-based summarizer aims to identify the most important sentences of a text using an algorithm and uses those original sentences to create the auto-summary (unlike generative AI). As the constituted sentences are machine selected, they may not fully reflect the body of the work, so we strongly advise that the original content is read and cited. The auto generated summaries were curated by the editor to meet Springer Nature publication standards. To cite this content, please refer to the original papers.

Machine generated keywords: green, job, labor, innovation, company, external, regulation, enterprise, business, economy, creation, environmental, environmentally, capital, human capital.

9.2.1 *The Circular Economy and the Green Jobs Creation [60]*

This is a machine-generated summary of:

Sulich, Adam; Sołoducho-Pelc, Letycja: The circular economy and the Green Jobs creation [60]

Published in: Environmental Science and Pollution Research (2021).

Link to original: https://doi.org/10.1007/s11356-021-16562-y.

If you want to cite the papers, please refer to the original.

For technical reasons we could not place the page where the original quote is coming from.

Abstract-Summary

"This pro-environmental economic model induces visible changes in the labor market which are Green Jobs (GJs)."

"This paper is focused on the creation of Green Jobs in the CE."

"Some Sustainable Development Goals (SDGs) and their measure were used as main indicators reflecting changes in the labor market."

"Results are presented as a model indicating which of the SDGs can support CE and enhance a number of the Green Jobs."

"Presented results contribute to the science because combine factors influencing GJs creation in EGSS, in a CE perspective."

"This study underlines a lack of uniform methods for measuring and forecasting the effects of Green Jobs creation and indicates future research directions."

Introduction

"Striving for sustainable development (SD) is at the basis of creating a contemporary socio-economic policy of the EU and many countries around the world (Ledoux and others 1; Tortorella and others 2)."

"The EU's documents concern not only the transition to the CE but also Green Jobs creation (Friant and others 3)."

"The research gap identified in this paper is the lack of the Green Jobs creation econometric model connections with the CE."

"Such a model should describe between theoretical assumptions of CE and empirical implications in the emerging GJs using indicators constructed upon the Sustainable Development Goals (SDGs)."

"This research intends to fulfill indicated gap by investigating the EGSS in EU countries from 2009 to 2019."

"After the research gap and aim of this paper are presented in the introduction, the themes crucial for this research are discussed in the literature review in the second point."

Literature Review

"The EGSS is the sector where Green Jobs are created next to the EP activities (Ernst and others 4), which are categorized according to the Classification of Environmental Protection Activities (CEPA)."

"The EGSS is related to the CE because closing linear chains into loops creates direct and indirect (induced) Green Jobs based on technology development."

"This initiative has a set of goals: "to promote awareness and dialog; to identify and respond to knowledge gaps; to facilitate a "just transition" that reflects the environmental, economic and social pillars of sustainable development; to promote policies and measures to achieve Green Jobs; to catalyze employment and poverty alleviation within climate mitigation and adaptation programs; and to strengthen collaboration between UNEP/ILO/ITUC, within the UN system and with the international business community to establish a common "Green Jobs" definition" (UNEP 5)."

Research Design

"Data obtained from the Eurostat database for all 28 EU countries in years, 2009–2019 were used, because of their reliability (Eurostat 6)."

"The EGSS data are made available every year, and these data are also part of the sustainable development indicators (more than 100 overlapping variables in a very wide context), gathered in the Eurostat database grouped in 17 sustainable development goals (SDGs) (Schroeder and others 7; Rincón-Moreno and others 8)."

"In the research method, they are considered as the describing independent variables (with symbols from CE_1 to CE_15) to the variable with symbol GJs—Green Jobs (described variable)."

"The presented division of the variables is in order with the method adopted by the Eurostat (Eurostat 6)."

"The performed multiple regression aim was to create the econometric model to cover described in the introduction section research gap and to propose a reliable model based on the variables proposed by Eurostat."

Data Analysis and Discussion

"The main research assumption is the correlation examination of variables possesses a sense only a cause-and-effect relation is present."

"The simplified equation was obtained, representing the relations between GJs and its describing variables CE_14, CE_15, and C_9."

"This equation presents regression results, reducing the number of variables that were used in the final model."

"Regression allowed the description of statistically significance relations between GJs (Employment in the Environmental Goods and Services Sector) and CE_14; CE_15; CE_9."

"Obtained results prove the unequivocal character of the three variables with described variable GJs."

"CE_14 (Private investments, jobs, and gross value added related to circular economy sectors) represents crucial expenditures related to the development of EGSS."

"Significant for the model is variable CE_15 (Patents related to recycling and secondary raw materials), which also indicates that technological progress combined with investments can influence CE_9 (Recycling of biowaste)."

Discussion

"The GJs creation process in the literature is described mostly qualitatively what includes sectors of the economy (Bruyère and Filiberto 9; Conlon and others 10): 1."

"This list of the areas or factors influencing the GJs creation process then is not used in a mathematical method to predict or calculate the number of the GJs but to prove transformation toward the CE (McMahon and others 11; Rojas Arboleda and others 12)."

"Contrary to Luca and others (13), this paper covers 2009 and 2019 for all EU member states, and based on secondary data from Eurostat, the linear model of the Green Jobs creation process has been proposed."

"This research proposed the Green Jobs creation econometric model connected to the CE."

"This research contributes to science by identifying the green indicators of CE required in the GJs creation."

"This model summarizes our research as it identifies three of the most important variables for Green Jobs creation."

Conclusions and Implications

"The main findings revealed in the proposed explanatory model of GJs creation are reflected in striving to implement the idea of CE."

"The model is indicating which strategic management fields can support a CE and enhance employment in EGSS which in this paper is described as GJs."

"Presented calculations are rare in the field of the CE scientific discussion about factors of GJs creation."

"The limitation of this study lies in the initial number of variables describing the CE proposed by Eurostat and adopted to this study."

"When pointing to the possibilities of implementing the CE model in practice, it should be considered that the GJs are next to the pillar of this concept, and they are necessary to achieve SDGs."

"The GJs importance is not only based on protecting and developing the natural environment."

9.2.2 Is Green the New Gold? Venture Capital and Green Entrepreneurship [61]

This is a machine-generated summary of:

Mrkajic, Boris; Murtinu, Samuele; Scalera, Vittoria G.: Is green the new gold? Venture capital and green entrepreneurship [61]

Published in: Small Business Economics (2017).

Link to original: https://doi.org/10.1007/s11187-017-9943-x.

If you want to cite the papers, please refer to the original.

For technical reasons we could not place the page where the original quote is coming from.

Abstract-Summary

"We test whether born-to-be-green represents a signal toward potential venture capital (VC) investors on a sample of Italian, independent, unlisted, high-tech entrepreneurial firms."

"Our results show that neither running a business based on green technologies nor positioning a business in a green sector per se are strongly correlated with the likelihood to get VC."

"We find that born-to-be-green can be a reliable signal for investors only when entrepreneurs perform activities based on green technologies/products and position their business in a green sector, at the same time."

"Further, we present three contingencies that moderate the association between green business propositions and the likelihood to get VC, namely the technical/scientific education of the founder(s), the origin of the firm as academic spin-out, and the presence of corporate shareholders into the venture's equity."

Introduction

"Green ventures represent a way through which entrepreneurship can fulfill the call for a greener and more environment-friendly business orientation, providing practical and innovative solutions to environmental problems (Criscuolo and Menon 16; Esty and Winston 17)."

"Even though policy makers have implemented an array of public policies to subsidize and foster the development of environment-friendly businesses (Bürer and Wüstenhagen 18; Criscuolo and Menon 16), born-green new ventures still face severe capital constraints, at least comparable to the ones of all the other new high-tech innovative ventures."

"Our work contributes to the literature at the nexus between entrepreneurship and finance (e.g., Bürer and Wüstenhagen 18; Criscuolo and Menon 16; del-Palacio and others 19), aiming at understanding whether born-to-be-green represents a signal toward potential VCs."

"Leveraging theoretical arguments from entrepreneurial finance and (green) entrepreneurship literature, we argue that, for high-tech ventures, born-to-be-green may theoretically entail either a positive or a negative signal toward VCs."

Green Ventures and VC Financing

"We refer to green ventures as high-tech ventures using green technologies in their businesses or offering green products/services."

"This financing issue appears to be extremely relevant because the born-green entrepreneurial ventures are capital constrained, in spite of favoring regulations and active governmental support directed to environment-friendly businesses (Bürer and Wüstenhagen 18; Criscuolo and Menon 16)."

"The notorious information asymmetries between high-tech ventures and capital markets (Hall and Lerner 20), mainly the debt providers, are particularly hefty in the case of born-green ventures."

"VC is thus considered to be the most suitable source of capital for high-tech ventures, including the born-green ones."

"In the case of green high-tech ventures, ambiguity is coupled with considerable uncertainty derived from the extremely high (technological and managerial) complexity, the longer investment duration, and the strong reliance on strictly regulated markets characterized by regulatory volatility and unstable prices (Amore and Bennedsen 21; Demirel and Parris 22; Petkova and others 23, 24)."

Hypotheses Development

"The above arguments lead us to posit that for high-tech ventures, the born-to-be-green signal may be positive, enabling them to an easier access to the VC market compared to the other high-tech ventures."

"Apart from the typical issues new highly-innovative ventures face to obtain external funding caused by the capital market imperfections (Carpenter and Petersen 25)—, information asymmetry, highly variable returns, and intangible knowledge-based assets—there are peculiarities of green ventures, which might additionally hinder their odds of obtaining external funding, such as VC."

"Fifth, the success of new green ventures strongly depends on the governmental support, which introduces regulatory uncertainty in the investment equation for VC firms (Criscuolo and Menon 16; Wüstenhagen and Teppo 26)."

"Relying on these arguments, we claim that born-to-be-green may also represent a negative signal toward VC investments, considering the higher risk, uncertainty, and complexity entailing the green businesses, compared to other high-tech ventures."

Data and Methodology

"We use the following probit model specification: Pr(VC) is a dummy variable that equals one if the venture has been VC-backed after venture foundation."

"The vector Controls includes industry dummies; further, it accounts for time effects, which is measured by the foundation year, being every year potentially affected by different shocks influencing both the formation of green ventures and the likelihood to get VC."

"As a first step to control for such endogeneity, we exploit venture-level information provided by the RITA dataset and directly related to the "active search for VC financing" and include such information to preliminary isolate the possible endogeneity effect."

"As exclusion restriction, we use a variable sourced from the World Values Survey (WVS) and related to the awareness about environmental issues of the local community where the focal venture is located."

Results

"We find a positive and statistically significant (at 1% confidence level) impact of TechEducation on the likelihood to get VC: the marginal effect ranges between + 3.52 (significant at 5%) and + 4.76 (significant at 1%)."

"FuctionalHeterogeneity leads to a positive and statistically significant (at 5%) contribution to the likelihood to access VC: the marginal effect at means is estimated to range between + 9.47 and + 18.83%."

"Quite surprisingly, we find that the presence in venture teams of previously self-employed founders or founders with a prior management experience shrinks the likelihood to get VC: marginal effects range from − 2.16% in column II (significant at 1%) to − 3.14% in column I (significant at 5%)."

"Regarding our main independent variable, our estimates show that running a green business per se is not correlated with the likelihood to receive VC, and its marginal effect is not statistically significant."

Discussion and Conclusions

"This twofold view, both horizontal and vertical, enables us to better distinguish the signaling effect that may stem, on the one hand, from the adoption of a green technology (or the offer of a green product/service) and, on the other hand, from the entrepreneurial ability to position the venture within a green sector."

"Re-adapting the idea of Heil and Robertson (27) that "a new product signal will create more perceived sender commitment if the sender has potentially available production capacity" (p. 408), we argue that the alignment between the core green business and the ability of the entrepreneur to position her/his venture within such business makes the "sender of the green signal" more committed to the eyes of potential VC investors."

"In order to be effective, the born-to-be-green signal needs to be properly communicated by the entrepreneurial ability to position the business and also needs to be aligned with the actual venture's underlying technology/product portfolio."

"The question regarding the effectiveness of the born-to-be-green signal to access the VC market should be studied in a cross-country setting."

9.2.3 *How Does Green Entrepreneurship Affect Environmental Improvement? Empirical Findings from 293 Enterprises [62]*

This is a machine-generated summary of:

Zeng, Juying; Ren, Jun: How does green entrepreneurship affect environmental improvement? Empirical findings from 293 enterprises [62]

Published in: International Entrepreneurship and Management Journal (2022).

Link to original: https://doi.org/10.1007/s11365-021-00780-4.

If you want to cite the papers, please refer to the original.

For technical reasons we could not place the page where the original quote is coming from.

Abstract-Summary

"Enterprises tend to change their production and emission behaviours to foster green entrepreneurship under endogenous innovation development pressure and exogenous environmental regulation."

"This research innovatively establishes a theoretical framework on the dual dynamic linkage of environmental regulation and green entrepreneurship."

"The dynamic regression identification then validates the stimulatory effect of environmental regulation on green entrepreneurship cultivation through internal competitiveness enhancement."

"With the application of the triple difference method (DDD), this research verifies that green entrepreneurship is positive in cultivating technological innovation and significantly facilitates environmental regulation to achieve better regulation effects through overall enterprise pollution emissions, total sulfur dioxide emissions, and total nitrogen oxide emissions."

Introduction

"Such regulations facilitate green and sustainable development and provide a new ideal that is compatible with environmental protection incentives for industries since all industries tend to set up entry barriers through high waste emission standards (Hilmersson & Hilmersson, 28)."

"Concerning environmental regulations, the Environmental Bureau of Zhejiang has successively issued the "Action Plan for Pollution Prevention and Control of Large Enterprises (2013–2017)" and "Action Plan for Comprehensive Transformation and Upgrading of Local Coal-Fired Cogeneration Industry in Zhejiang Province" since 2013 to guarantee the emission reduction of heavy-polluting industries, such as the thermoelectric industry."

"Several cities in Zhejiang Province (Hangzhou, Huzhou, Lishui, etc.) are taken as pilot cities for green and environmentally friendly development, and their environmental emission regulations are stricter than those of other cities."

"Environmental regulation can promote the green development of enterprises by formulating emission standards, increasing punishment for enterprises with excessive emissions, and forcing enterprises to replace backward processes with green innovative technologies."

Literature Review and Hypothesis Development

"Existing studies have mostly separately focused on waste emission decreases or technological innovation due to environmental regulation and various definitions of entrepreneurship; however, solid evidence on how environmental regulation cultivates green entrepreneurship through detailed channels for micro-level enterprises and regions is lacking."

"To overcome the listed defects, this research innovatively constructs a comprehensive indicator system of green entrepreneurship with both internal and external competitiveness, summarizes the possible relevant impacts of environmental regulation on green entrepreneurship through entrepreneurs' innovation and waste emission reduction, and finally investigates how green entrepreneurship facilitates environmental regulation to achieve better environmental improvement with the application of several data-driven econometric models."

"The research proposes H1 concerning the development of green entrepreneurship as follows: Second, the literature review concerning enterprise reactions to environmental regulation shows that environmental regulation has either positive or negative impacts on the innovation behaviours of enterprises."

Data Illustration and Model Specification

"With the reference of existing studies, the research takes emissions of sulfur dioxide and nitrogen oxides by heavily polluting industries as explained variables of environmental quality."

"This research takes policy regulation and green entrepreneurship as the two main indicators for analysis."

"According to the discreteness of sample index data, this research uses the entropy method to evaluate the green entrepreneurship of 11 cities in Zhejiang Province."

"This research mainly studies the impact of environmental regulation on green entrepreneurship through both internal and external competitiveness."

"Zhejiang Province is the pilot city in environmental regulation and green development."

"The research assumes that enterprises can cultivate green entrepreneurship and bring spillover effects on environmental improvement, such as a reduction in waste emissions."

"Regional green entrepreneurship could make local heavy pollution enterprises more active and effective under environmental regulation constraints."

Green Entrepreneurship in Zhejiang Province

"The research employs the data-driven entropy weight method to measure green entrepreneurship without environmental regulation in Zhejiang Province with the sample data during the period from 2016 to 2019."

"The estimation results validate Hypothesis 2 (H2), i.e., environmental regulation stimulates green entrepreneurship cultivation in Zhejiang Province."

"The regression coefficients of environmental regulation intensity (lnpolicy) and quadratic regulation intensity ($(\ln\text{policy})^2$) on the internal competitiveness of green entrepreneurship are 0.265 and 0.217 at the significance level of 1%, respectively."

"Environmental regulation intensity does not exert a significant role on the external competitiveness of green entrepreneurship, and all the regression coefficients were insignificant."

"The conducive role of environmental regulation on the internal competitiveness of green entrepreneurship witnessed an increasing rate from 2016 to 2019."

Empirical Evidence of Green Entrepreneurship on Environmental Improvement

"Model 6 further controls the economic variables based on Model 5, and the joint effect of the interaction terms environmental regulation, green entrepreneurship and time on the overall waste emissions is -14.881."

"Model 6 further controls the economic variables based on Model 5, and the joint effect of the interaction term of environmental regulation, green entrepreneurship and time on total sulfur dioxide emissions is -18.616."

"Driven by environmental regulation, green entrepreneurship promotes the green innovation level of enterprises, significantly controls the sulfur dioxide emissions of enterprises, and further achieves the effect of sulfur dioxide emission control."

"Model 6 further controls the economic variables based on Model 5, and the joint effect of the interaction term environmental regulation, green entrepreneurship and time on total sulfur dioxide emissions is -19.189."

Conclusion and Remarks

"This research investigates how environmental regulation stimulates green entrepreneurship and how green entrepreneurship conversely facilitates environmental improvement based on panel data from 293 thermoelectric industry enterprises in Zhejiang, China, during 2017–2020."

"The research first constructs a comprehensive indicator system of green entrepreneurship and applies the entropy method to measure regional green entrepreneurship for 11 prefecture-level cities in Zhejiang; then, it measures environmental regulation intensity and its role in stimulating green entrepreneurship with a dynamic regression equation model; and finally, it applies DDD to measure how cultivated green entrepreneurship facilitates environmental regulation to achieve environmental improvement through the channels of various pollutants."

"Enterprises should increase investments in the research and development of energy conservation and emission reduction technologies and make use of clean energy and technologies for production, which will not only improve the environmental quality but also contribute to enhancing the market competitiveness of enterprises. (3) Regional green entrepreneurship has different impacts on the emission of different pollutants, so it is necessary to implement differentiated governance approaches."

9.2.4 The Complexity of Green Job Creation: An Analysis of Green Job Development in Brazil [63]

This is a machine-generated summary of:

Dordmond, Gertjan; de Oliveira, Heder Carlos; Silva, Ivair Ramos; Swart, Julia: The complexity of green job creation: An analysis of green job development in Brazil [63]

Published in: Environment, Development and Sustainability (2020).

Link to original: https://doi.org/10.1007/s10668-020-00605-4.

If you want to cite the papers, please refer to the original.

For technical reasons we could not place the page where the original quote is coming from.

Abstract-Summary

"Transforming current economic structures into a green economy is a complex process."

"The creation of green jobs is one of the components of the "going green" process."

"Using occupational data for 27 Brazilian states between 2003 and 2013, this paper examines whether the level of economic complexity of a state explains why one state has a greener occupation space than another."

"We construct a "green jobs index" which quantitatively measures the greenness of the occupation space of a state."

"We then empirically show that economic complexity has a positive explanatory power when it comes to differences in this green jobs index."

"States which had a low green jobs index became relatively greener than states that had initially a high green jobs index."

Introduction

"This paper adopts the definitions of a green economy and job proposed by the Occupational Information Network Resource Center (O*NET, a program of the US Department of Labor/Employment and Training Administration)."

"They use the following definition of green jobs: "The "greening" of occupations refers to the extent to which green economy activities and technologies increase the demand for existing occupations, shaping the work and worker requirements needed for occupational performance, or generate unique work and worker requirements."

"The creation of green jobs is closely related to the occupational structure of a country and is a logical consequence of a transition toward a green economy."

"The characteristics of this occupational structure change in a significant way when new green jobs enter the economy."

"This paper sets out to examine whether the level of economic complexity of an economy plays an important explanatory role in the transition toward a green economy through the channel of green job creation."

Theoretical Framework

"Several countries have explicitly incorporated green growth in their policy objectives (OECD 29)."

"Green growth theories correct this market failure by including natural capital as a factor of production and lead to an optimal outcome."

"Comparative advantage in such sectors can lead to growth as the world transitions into green economies (Jacobs 30)."

"He shows that China's model of the strategic pursuit of green growth is delivering economic benefits, creating jobs in great numbers, and reducing carbon emissions at the same time."

"Many developing countries already struggle with internal inequality, and a green growth process that does not consider regional differences could lead to undesirable outcomes."

"This creates the possibility of tailor-made policies, rather than a generic green growth policy for the entire country, which is what most of the existing literature suggests."

"They have developed a view of economic growth and development that gives a central role to the complexity of a country's economy."

Methodology

"The second stage consists of an empirical analysis of the impact of the economic complexity of the states and various control variables on the capability of an economy creates green jobs."

"In "Theoretical framework," based on the literature we expect the level of economic complexity of an economy to have a positive effect on the occupational greenness of a state."

"The green jobs index is the main dependent variable, and the following model is estimated: We test the null hypothesis that β_1 and β_2 are zero."

"The newly constructed green jobs index variable allows us to empirically investigate the relationship between greening the occupational space and economic complexity."

"We expect this variable to have a positive effect on the green jobs index because economies that are well developed and have a lot of green jobs are likely to have a lot of green jobs in the next period as well."

Results and Discussion

"The green jobs index shows how the greenness of the occupational spaces of the 27 Brazilian states has developed between 2003 and 2013."

"As for the main dependent variable, our hypothesis is that we expect the level of economic complexity of a state to have a positive effect on the greenness of the occupational space of a location."

"This variable has a positive impact on the green job index, and it is significant at the 1% level."

"All dummies have the expected sign, being in the North or the Northeast has a negative impact on the green jobs index of a state, while being in the South or the Southeast has a positive impact, relative to being in the West-Center region."

"These results allow us to conclude that we accept the hypothesis that economic complexity has a positive effect on green job development in the Brazilian states."

Conclusion

"It was constructed the green jobs index for the Brazilian states and the results show that São Paulo is not only the economic capital of Brazil but also its green capital, as no other state even comes close to the green jobs index of São Paulo."

"The second thing that is clear is that the green jobs index of states changes very slowly on a regional level, not a single state experienced considerable change between 2003 and 2013."

"Brazil seems to become greener; however, the green jobs index growth seems to respond to the state of the economy."

"Knowing in which of these jobs, that are close to being green, a state has a relative comparative advantage is useful for policymakers who want to make their economy greener, as it shows which sectors have potential for that state."

9.2.5 Born to Be Green: New Insights into the Economics and Management of Green Entrepreneurship [64]

This is a machine-generated summary of:

Demirel, Pelin; Li, Qian Cher; Rentocchini, Francesco; Tamvada, J Pawan: Born to be green: new insights into the economics and management of green entrepreneurship [64]

Published in: Small Business Economics (2017).

Link to original: https://doi.org/10.1007/s11187-017-9933-z.

If you want to cite the papers, please refer to the original.

For technical reasons we could not place the page where the original quote is coming from.

Abstract-Summary

"The papers in this special issue shed light on these issues by underscoring the role of several factors, such as industry life cycles, knowledge spillovers, institutions, and availability of external finance, in shaping decision-making and firm behaviour in green start-ups."

Green Entrepreneurship: An Emerging Field of Research

"Notwithstanding the conceptual appeal of the "green wave" of creative destruction and the predictions that green markets are on track to reach a "critical mass", the scholarly literature exploring entrepreneurial firms that are "born green" has emerged only more recently at the intersection between innovation and entrepreneurship literatures (see, Hall and others 31, for a review)."

"Born green firms, used interchangeably with "green start-ups" here, follow an entrepreneurial path that fulfils the need for a greener and more environmentally friendly approach to business, providing practical and innovative solutions for social and environmental concerns (Criscuolo and Menon 16; Esty and Winston 32)."

"This special issue attempts to integrate the economics and entrepreneurship literatures on new green ventures to advance the scholarship on the role of born green firms."

"The special issue is organised around four main themes where the published papers place their contribution: (i) industry and corporate life cycle, (ii) role of institutional structure and government, (iii) access to finance, and (iv) geographical clusters and regional drivers of green start-ups."

Key Insights from the Literature and Areas for Development

"Little empirical evidence exists to unravel the role of age and industry life cycle for the evolution of green start-ups' innovation and financial performance across different green sectors."

"Young born green companies entering markets with their innovative products and services are likely to face significantly different innovation challenges compared to the more established firms that adopt environmentally friendly innovations as a reaction to the changing industry landscape."

"Despite a much improved understanding of issues around access to external finance by green innovators (Criscuolo and Menon 16; Meyskens and Carsrud 33), there is still a dearth of knowledge around the overlap and timing of distinct financing modes along the business life cycle, where distinct logics apply at different phases, and little is known about the existence and efficacy of the "green" signals from these new ventures for attracting external funding."

Contribution of the Articles in this Special Issue

"The first group of papers in the special issue investigate the relevance of firm and technology life cycle for green start-ups."

"Colombelli and Quatraro (34, this issue) combine the knowledge spillovers theory of entrepreneurship with the literature on sustainable entrepreneurship and analyse how the properties of the regional knowledge stock relate to the creation of green start-ups at the regional level."

"Corradini (35, this issue) examines the location determinants of new green tech firms across European regions."

"The fourth contribution on the role of region by Sunny and Shu (36, this issue) examines the impact of clusters on entry by green tech firms across different regions of the USA."

"Put together, these four papers present some of the first insights into the role of knowledge spillovers, regional entrepreneurial capabilities, awareness, and social norms as well as policies that have an impact on green entrepreneurship and significantly extend the current research linking green entrepreneurship and regional location."

Concluding Remarks and Potential Areas for Future Research

"Relatedly, the second area for future research concerns the relevant methodological issues that can provide a grounded understanding of green entrepreneurship."

"A third future challenge refers to the integration of green entrepreneurship into related research fields such as entrepreneurship, innovation studies, and finance in order to explore unanswered research questions at the intersections of these fields."

"The creation of more comprehensive databases that capture a broad range of green entrepreneurs across different countries (from both developed and developing background) is essential for researchers to conduct rigorous empirical studies on green entrepreneurship."

"A more coordinated effort and collection of larger-scale data will benefit the cause of advancing green entrepreneurship research."

"The growing use of machine learning methods are likely to complement these efforts and open up new avenues of research (Varian 37; Székely and Vom Brocke 38), provided green entrepreneurship and green activities in general can be conceptualised and defined consistently."

9.2.6 *Think Green: The Eco-innovative Approach of a Sustainable Small Enterprise [65]*

This is a machine-generated summary of:

Campo, Raffaele; Trio, Oronzo: Think Green: the Eco-innovative Approach of a Sustainable Small Enterprise [65]

Published in: Journal of the Knowledge Economy (2021).

Link to original: https://doi.org/10.1007/s13132-021-00833-7.

If you want to cite the papers, please refer to the original.

For technical reasons we could not place the page where the original quote is coming from.

Abstract-Summary

"The objective of this research is to analyze the implementation of eco-innovation as a tool to improve the quality of products, in order to comprehend the benefits deriving from its application."

"This research is aimed at exploring the way eco-innovation can be implemented in small- and medium-sized enterprises (SMEs)."

"The analysis highlights that the case of Ecobabydesign embodies the total application of ecological innovation."

"It is an example of innovation which is generated firstly in its mission and then in its high-quality products, whose characteristics derive from its green choice."

Studies on Eco-innovation

"The Eco-innovation Observatory Report (39) defines it as "the introduction of any new or significantly improved product (good or service), process, organizational change or marketing solution that reduces the use of natural resources (including materials, energy, water, and land) and decreases the release of harmful substances across the whole life-cycle." (p. 8); Klemmer and others (40) highlight that it is linked to ideas, behaviors, products, and processes which contribute to reducing environmental burdens or ecological targets, while Fussler and James (41), Wolff and others (42) and Oltra and Saint Jean (43) note that it is the development of new products/processes/services which provide value to customers and business, with a lower environmental impact."

"In terms of determinants of eco-innovations Horbach and others (44) and Doran and Ryan (45) underline that regulations (for reducing air and water emissions and for increasing recyclability), cost savings (for reducing energy and material use), and customer benefits (with regard to material efficiency, reduction of energy consumption, waste and the use of dangerous substances) affect companies and their environmental choices."

"With reference to SMEs (which is an emerging research topic according to de Jesus Pacheco and others, 46), Triguero and others (47) find that those companies collaborating with universities, research institutes, and agencies in trying to increase demand for green products are more active in terms of eco-innovations; moreover, supply-side factors are important drivers for environmental processes and organizational innovations."

Eco-innovation in Practice: A Case Study

"Since 2010, Ecobabydesign has developed a project of technological innovation by basing its exclusive range of eco-design models on technologies, processes, treatment methodologies, and productive practices falling within the fields of ecology and sustainability."

"One of the key concepts that inspired Ecobabydesign and its choice of sustainability is the principle in which the material should permit doing more with less; from this perspective, a first benefit deriving from the use of PMMA is the quantity of energy required for its production, which is lower than that used to produce glass."

"Ecobabydesign wants to be perceived in the minds of consumers/potential customers as the brand par excellence that embodies luxury design, eco-sustainability, made in Italy quality, whose products are specifically designed and tailored to the needs of children, in other words, a brand to which potential customers immediately refer when they have to satisfy their needs."

Conclusions

"Conceiving eco-innovation as a moral duty and as a source of economic benefits represents a strength for Ecobabydesign; this enterprise has a clear vision of the opportunities provided by the ecological choice, inasmuch as they declare that sustainability creates a virtuous circle among investments, benefits for the environment and for the company balance sheet, in terms of: - improvement of the ecological footprint of products and services; - savings in terms of energy consumption and resources; - differentiation of products and services, a strength in comparison to competitors; - acquisition of new customers, sensitive and attentive to the environmental impact of products and services; - possibility of having an economic contribution made available by the public administration in order to incentivize a sustainable change."

9.2.7 *Greening Work: Labor Market Policies for the Environment [66]*

This is a machine-generated summary of:

Bohnenberger, Katharina: Greening work: labor market policies for the environment [66]

Published in: Empirica (2022).

Link to original: https://doi.org/10.1007/s10663-021-09530-9.

If you want to cite the papers, please refer to the original.

For technical reasons we could not place the page where the original quote is coming from.

Abstract-Summary

"This paper assesses how labor market policies can green employment."

"The paper analyses the potential effects of eight different policy strategies on four dimensions of the Taxonomy of Sustainable Employment: conversion of plants and businesses, environmental labor law, climate decommodification, socio-ecological job guarantee, vocational guidance and retraining, distribution of employment time, alternative income sources, and equalization of income."

"In the light of environmental crises, the results suggest widening the toolbox of labor market policies for a green and just transition."

Introduction

"Given the importance of jobs and high levels of employment for household income, state budgets, stable social security systems, and social participation in communities, it is not surprising that the threat of unemployment has been the most powerful argument against adequate environmental policies, particularly climate policies that entail a profound change of production processes (Vona 48)."

"Using a Taxonomy of Sustainable Employment, this paper analyses how eight different strategies in employment and social policies contribute towards environmental sustainability."

"With organizations like the International Labor Organization, the social dimension of employment has already been elevated to an established position, while the environmental dimension of work is largely disregarded in assessments of employment."

"Existing labor policies focus on employment and largely disregard the domain of unpaid work."

"As this paper seeks to analyze the institutionalizing and potentially transformative power of labor (market) policies, it focuses on the domain of work mostly structured by labor (market) policies: paid, formal employment."

Theoretical Approaches to the Greening of Work: The Sustainable Employment Taxonomy

"The indirect and micro-effects assess the consequences of employment relationships for environmentally friendly lifestyles beyond the workplace (Work-lifestyles approach)."

"Statistics using the sectoral approach approximate green jobs through the environmental goods and services sector, like wastewater treatment, renewable energies, or end-of-pipe technologies (Janser 49, p. 80)."

"The output type approach of the taxonomy of sustainable employment, therefore, defines jobs as sustainable which provide goods or services that serve human needs (and not only their desires) in an environmentally generalizable way."

"The occupation approach of the taxonomy of sustainable employment evaluates jobs by the greenness of the activities people do at workplaces, which covers the tasks of the job description and other activities that are taking place at the workplace or in preparation for it."

"The third dimension of the Taxonomy of Sustainable Employment analyses the effects of jobs and employment relationships on lifestyles beyond the workplace (work-lifestyles approach)."

Data and Method

"The previous section presented the Taxonomy of Sustainable Employment with its four distinct dimensions that matter for the evaluation of employment in respect to its environmental sustainability."

"This paper only seeks to provide an overview of the different policy strategies that exist, discuss how they can promote sustainable employment, and elaborate on what needs to be considered to maximize the environmental benefits of the policies."

"Three steps were undertaken to identify and select the eight major policy strategies: First, relevant literature in the domain of labor markets and social policy was scanned for policy proposals that claim to promote the environment."

"Eight different strategies in the field of labor market policies could be identified."

"The next section describes the strategies, provides examples, and evaluates the contribution of the strategies on the four dimensions of the Taxonomy of Sustainable Employment."

Results and Discussion: Eight Labor Market Strategies and Their Contribution to Sustainable Employment

"Environmental labor law mainly addresses the tasks and activities of workers and to a smaller extent the outcome efficiency dimension of green jobs and the work-lifestyles dimension."

"The lifestyle dimension is least affected by climate decommodification—although altered income and time wealth during participation in the program could have a positive impact for sustainable lifestyles and the option of leaving an unsustainable job could make workers in unsustainable sectors more open to reconsidering their lifestyles."

"The smallest, yet still positive impact will be in the dimensions outcome efficiency and work-lifestyles: During educational leave, people do not contribute to the production process and when employed in brown jobs their absence might lower the production of an unsustainable sector during the time of their leave."

"Another variant of work-sharing could be an intra-family shift of work time: Employees in exnovation sectors could reduce employment hours and increase care work, enabling their partners in more sustainable jobs to reduce care work and increase employment hours."

Conclusion

"This paper analyzed how eight different labor market strategies contribute to the greening of employment."

"The eight identified policy strategies cover a wide range of instruments: conversion of plants and businesses, environmental labor law, climate decommodification, socio-ecological job guarantee, vocational guidance and retraining, distribution of employment time, alternative income sources, and equalization of income."

"All eights strategies have the potential of greening employment but vary in their intensities."

"How these strategies can be implemented in the labour market policies of countries depend on the legal and institutional specifics of the welfare state, labour market institutions and industrial relations."

"Many strategies, e.g. work time distribution, require changes in the social security institutions, such as unemployment insurances or tax and labour law Future research is required to explicate country- and time-specific conditions and design options for translating the eight different classes of strategies into specific design."

9.2.8 Green New Hiring [67]

This is a machine-generated summary of:

Darendeli, Alper; Law, Kelvin K. F.; Shen, Michael: Green new hiring [67]

Published in: Review of Accounting Studies (2022).

Link to original: https://doi.org/10.1007/s11142-022-09696-y.

If you want to cite the papers, please refer to the original.

For technical reasons we could not place the page where the original quote is coming from.

Abstract-Summary

"We propose a new measure, Green Score, to capture firms' investment in green human capital based on the concentration of green skills required in firms' job postings."

"We find that firms that increase their Green Score have higher future profitability."

"Traditional ratings widely used to evaluate firms' environmental efforts do not consider firms' Green Score."

Introduction

"We exploit firms' demand for green skills in their job postings as a proxy for their investment in green human capital – a type of environmental effort."

"We propose a new measure, Green Score, based on the concentration of green skills required in firms' job postings."

"A firm that increases its Green Score invests more in green human capital and puts in more environmental efforts."

"Our measure of Green Score is based on the skills mentioned in firms' job postings."

"We find that firms' investment in green human capital, proxied by the concentration of green skills required in firms' job postings, is associated with future profitability after we control for KLD ratings."

"By using firms' job postings data to measure firms' investment in green human capital, our paper complements their findings."

Testable Hypothesis, Data, and Research Design

"We conjecture that the concentration of green skills required in firms' job postings is a proxy for firms' investment in green human capital."

"We conjecture that the concentration of green skills, not green jobs, is a meaningful proxy for firms' investment in green human capital."

"Firms that post jobs with a higher concentration of green skills could also be developing new services or products built around environmental considerations."

"The concentration of green skills required in job postings could reflect firms' commitment to environmental issues beyond just compliance."

"We state our main testable hypothesis as follows: H1: Firms with job postings that require a higher concentration of green skills have higher future profitability."

"Burning Glass is an employment data analytics firm that provides real-time data on online job postings and skills in demand."

"We identify all of a firm's jobs that require green skills in a given year."

Main Results

"In column 1, we find that the estimate of Green Score is positively and statistically significant when we control for Firm Fixed Effects and Year Fixed Effects."

"In column 2, we continue to find that the estimate of Green Score is positively and statistically significant, controlling for a comprehensive set of control variables and Firm Fixed Effects and Year Fixed Effects."

"A one standard deviation increase in Green Score is associated with a 3.6% (= 0.345 × 0.104) increase in future Net Profit Margin."

"The analyses above suggest that an increase in sales and net profit margin are the primary drivers of the improved profitability for firms that increase Green Score."

"The results in column 1 show that firms that increase their Green Score generate more patents."

"Firms that increase their Green Score have more innovation activities and subsequently have better financial performance (i.e., profitability, sales, and net profit margin)."

[Section 4]

"To test this conjecture, we estimate the following regression specification: The dependent variable is the one-year-ahead Green Score, which captures the extent of green skills in firm j's jobs in year t + 1."

"In column 3, Green Score (Fiscal) is the average Green Score per Job across all of a firm's jobs that require green skills in a fiscal year period."

"In column 4, Green Score (First Instance: Year) is the average Green Score per Job across a firm's jobs that require green skills in the first three months of a given year."

"In column 7, Green Score (All Jobs) is the average Green Score per Job across all of a firm's jobs (not just green jobs that require green skills) in a given year, and our main results remain robust."

[Section 5]

"We offer a simple application of our Green Score to detect greenwashing firms."

"Such greenwashing improves a firm's environmental rating without requiring a corresponding increase in the firm's investment in green human capital."

"To help identify greenwashing firms, we construct Greenwashing, an indicator variable equal to one if the increase in Green Score is less than the increase in KLD strengths."

"Greenwashing captures the difference between the concurrent increases in firms' environmental ratings and investment in green human capital."

"A firm that improves its environmental rating without a corresponding increase in its green human capital is likely to engage in greenwashing and have a higher value of Greenwashing."

"We find that the main effect is only observed when firms greenwash their environmental strengths (column 3) but improve their other strengths (column 4)."

"The results in this section suggest that Green Score can detect firms' greenwashing behaviors."

Conclusion

"We exploit firms' demand for green skills in their job postings as a proxy for their investment in green human capital – a type of environmental effort."

"We propose a new measure, Green Score, based on the concentration of green skills required in firms' job postings."

"We find that firms that increase their Green Score have higher future profitability."

"We show that the KLD ratings that are widely used to evaluate firms' environmental efforts do not consider firms' Green Score."

"We find that firms' investment in green human capital proxied by the concentration of green skills required in firms' job postings is associated with future profitability after we control for KLD ratings."

"Our results will inform firm executives – especially those with the worst environmental track records – by suggesting that firms could step up their investment in green human capital more proactively and thereby increase their future profitability."

9.2.9 Green Human Resource Management Research in Emergence: A Review and Future Directions [68]

This is a machine-generated summary of:

Ren, Shuang; Tang, Guiyao; E. Jackson, Susan: Green human resource management research in emergence: A review and future directions [68]

Published in: Asia Pacific Journal of Management (2017).

Link to original: https://doi.org/10.1007/s10490-017-9532-1.

If you want to cite the papers, please refer to the original.

For technical reasons we could not place the page where the original quote is coming from.

Abstract-Summary

"GHRM research raises new, increasingly salient questions not yet studied in the broader human resource management (HRM) literature."

"This paper, seeking to better understand research opportunities and advance theoretical and empirical development, evaluates the emergent academic field of GHRM with a narrative review."

"Going beyond a function-based perspective that focuses on specific HRM practices and building on advances in the strategic HRM literature, we discuss possible multi-level applications, the importance of employee perceptions and experiences related to GHRM, contextual and cultural implications, and alternative theoretical approaches."

"The detailed and focused review provides a roadmap to stimulate the development of the GHRM field for scholars and practicing managers."

Defining and Measuring Green Human Resource Management

"The conceptualization of GHRM thus involves not only traditional HRM functions that contribute to environmental goals, but also factors of employment and labor force that emerge or change as organizations adopt EM practices (Gholami and others, 50)."

"This definition adopts the meaning of an HRM system provided by Jackson and others (51: 3–4), and thus casts GHRM as an organization's aspiration to design and implement an HRM system that supports a proactive and positive approach to addressing environmental concerns by (1) formulating an overarching HRM philosophy that reflects green values; (2) promulgating formal HRM policies that state the organization's intent and serve to direct and partially constrain the green behavior of employees; (3) actively ensuring actual green HRM practice (by ensuring the daily enactment of green HRM philosophies and policies); and (4) using green technological processes for designing, implementing, evaluating, and modifying GHRM philosophies, policies, and practices as they evolve."

Theoretical Foundations

"While the strategic HRM literature has witnessed an array of theoretical perspectives (e.g., resource-based view, human capital theory, and the behavioral perspective), the GHRM research to date has mostly adopted the behavioral perspective (e.g., Jackson & Seo, 52; Dubois & Dubois, 53)."

"The behavioral perspective of strategic HRM was systematically adopted in Renwick and others's (54) seminal review that argued that specific GHRM practices can shape the competencies, motivations, and opportunities available in the organization's workforce."

"The abilities-motivation-opportunities model has since become the most widely used theoretical perspective for GHRM scholarship (e.g., Guerci and others, 55, 56; Pinzone and others, 57)."

"Seldom specified, the resource-based view lays a foundation for theorizing about the integration of GHRM (e.g., Antonioli and others, 58) with other management functions (such as green supply chain management and product marketing) and is supported by an emerging research stream that takes a multidisciplinary approach to understanding green organizations (e.g., Jabbour & Jabbour, 59)."

"Alternative theoretical foundations have also begun to take shape in the GHRM literature."

A Framework for Understanding Antecedents, Consequences, and Contingencies of GHRM

"While the empirical work linking GHRM to team-level outcomes is quite limited, it is reasonable to assume that HRM practices that influence the types of employees placed in leadership positions will also influence the green behaviors of other team members (see Kim and others, 2017)."

"While the majority of speculation about GHRM outcomes concerns desirable effects such as improved environmental performance, an emerging stream of research links EM with a broader array of employee attitudes and behaviors beyond those specifically considered "green" and directly relevant to the environment (e.g., Bode, Singh, & Rogan, 60)."

"Whereas the hard targets of GHRM include direct control of employee behaviors that are expected to affect environmental performance outcomes, the soft consequences of GHRM include improved employee attitudes toward their employing organization more generally."

Research Agenda

"Building upon the concept of HRM strength, future research could also investigate the extent to which employees perceive GHRM practices as fostering a strong and persuasive situation leading to pro-environmental attitudes and behavior—the strength of GHRM systems potentially having both direct and moderating effects on employees' attitudes and behavior."

"Building on the concept of HRM attribution, future research might investigate how employees understand and make sense of organizations' motivations, as such attributions are likely to influence how employees respond to GHRM policies and practices."

"One opportunity for future research is to investigate the processes through which alignment of GHRM and other management functions can best be achieved; it is likely that the effectiveness of GHRM partly depends on how well the system used all of an organization's functional areas to target environmental concerns."

Implications for Practice

"To foster GHRM, our framework suggests that organizations need to pay attention to both organization-initiated changes related to EM and employee perceptions and interpretations of GHRM."

"Organizations could invest in employee engagement surveys to understand how their employees perceive the alignment between GHRM practices and "best HRM practices."

"Organizations could also encourage information flow to understand how employees perceive the environmentally focused messages that GHRM conveys."

"Given that every organization faces resource constraints, it is advised that GHRM systems include the objective of encouraging employee-driven initiatives."

Conclusion

"Owing to the relatively young age of GHRM as a field, there is not yet sufficient variety in the theoretical perspectives utilized to assess which are likely to prove most useful for future development."

"Strategic HRM seems to be the dominant meta-theory for the foundation of GHRM, yet the field of strategic HRM itself has been criticized as lacking theoretical depth and sophistication (e.g., Guest, 55, 56)."

"This weakness creates various opportunities for innovative and important research in the GHRM field."

"Our proposed framework highlights and organizes several likely antecedents, consequences, and contingencies of GHRM and positions GHRM systems within an organization's wider management context."

"With the continued awareness of environmental sustainability, GHRM is now clearly a legitimate field of academic pursuit."

[Section 7]

"Research on GHRM introduces ideas and issues that are only beginning to be studied by human resource management (HRM) scholars as they realize the strategic importance of EM in building sustainable organizations."

"The originality of this current review lies in: (1) an integrated approach to understanding the antecedents, contingencies, and outcomes of GHRM from the strategic HRM perspective, which in turn enables assessing the state of GHRM research vis-à-vis recent strategic HRM research advances (e.g., an employee-centric perspective and multi-level modeling); (2) a critical analysis of GHRM concerning its conceptualization, measurement, and theoretical basis; and (3) articulation of contextual factors that should be addressed in future GHRM research, including cultural influences that have been largely overlooked."

Bibliography

1. Ledoux L, Mertens R, Wolff P (2005) EU sustainable development indicators: an overview. Nat Resour Forum 29:392–403. https://doi.org/10.1111/j.1477-8947.2005.00149.x
2. Tortorella MM, Di Leo S, Cosmi C, et al (2020) A methodological integrated approach to analyse climate change effects in agri-food sector: the TIMES Water-Energy-Food Module. Int J Environ Res Public Health 17. https://doi.org/10.3390/ijerph17217703
3. Friant MC, Vermeulen WJV, Salomone R (2020) A typology of circular economy discourses: navigating the diverse visions of a contested paradigm. Resour Conserv Recycl 161:104917. https://doi.org/10.1016/j.resconrec.2020.104917
4. Ernst C, Brizuela ASR, Epifanio D (2019) Green jobs in Argentina: opportunities to move forward with the environmental and social agenda. CEPAL Rev 2019:51–71. https://doi.org/10.18356/280A25AA-EN
5. UNEP (2008) Green jobs: towards decent work in a sustainable, low-carbon world
6. Eurostat (2020c) Main tables -circular economy - Eurostat. In: Circ. Econ. https://ec.europa.eu/eurostat/web/circular-economy/indicators/main-tables . Accessed 22 Dec 2020
7. Schroeder P, Anggraeni K, Weber U (2019) The relevance of circular economy practices to the sustainable development goals. J Ind Ecol 23:77–95. https://doi.org/10.1111/jiec.12732
8. Rincón-Moreno J, Ormazabal M, Álvarez MJ, Jaca C (2020) Shortcomings of transforming a local circular economy system through industrial symbiosis: a case study in Spanish SMEs. Sustainability 12. https://doi.org/10.3390/su12208423
9. Bruyère SM, Filiberto DM (2013) The green economy and job creation : inclusion of people with disabilities. Int J Green Econ 7:257–275. https://doi.org/10.1504/IJGE.2013.058151.This
10. Conlon K, Jayasinghe R, Dasanayake R (2019) Circular economy: waste-to-wealth, jobs creation, and innovation in the global south. World Rev Sci Technol Sustain Dev 15:145–159. https://doi.org/10.1504/WRSTSD.2019.099377
11. McMahon K, Ryan-Fogarty Y, Fitzpatrick C (2021) Estimating job creation potential of compliant WEEE pre-treatment in Ireland. Resour Conserv Recycl 166. https://doi.org/10.1016/j.resconrec.2020.105230, 105230
12. Rojas Arboleda M, Pfeiffer A, Bezama A, Thrän D (2021) Anticipatory study for identifying the key influential factors of the biogas system in Germany contributing to the energy system of 2050. Futures 128:. https://doi.org/10.1016/j.futures.2021.102704, 102704
13. Luca FA, Epuran G, Ciobanu CI, Horodnic AV (2019) Green jobs creation - main element in the implementation of bioeconomic mechanisms | Crearea de locuri de muncă ecologice - componentă de bază pentru implementarea mecanismelor bioeconomice. Amfiteatru Econ 21 (50): 60–74. https://doi.org/10.24818/EA/2019/50/60
14. Ayyagari, M., Demirgüç-Kunt, A., & Maksimovic, V. (2011). Firm innovation in emerging markets: The role of finance, governance, and competition. Journal of Financial and Quantitative Analysis, 46(6), 1545–1580.
15. Chen, F., Hope, O.-K., Li, Q., & Wang, X. (2011). Financial reporting quality and investment efficiency of private firms in emerging markets. Accounting Review, 86(4), 1255–1288.
16. Criscuolo, C., & Menon, C. (2015). Environmental policies and risk finance in the green sector: Cross-country evidence. Energy Policy, 83, 38–56.
17. Esty, D., & Winston, A. (2009). Green to gold: How smart companies use environmental strategy to innovate, create value, and build competitive advantage. Hoboken: John Wiley & Sons.
18. Bürer, M. J., & Wüstenhagen, R. (2009). Which renewable energy policy is a venture capitalist's best friend? Empirical evidence from a survey of international cleantech investors. Energy Policy, 37(12), 4997–5006.
19. del-Palacio, I., Zhang, X. T., & Sole, F. (2012). The capital gap for small technology companies: Public venture capital to the rescue? Small Business Economics, 38(3), 283–301.
20. Hall, B. H., & Lerner, J. (2010). Handbook of the economics of innovation. Handbook of the Economics of Innovation, 1(31), 609–639.

21. Amore, M. D., & Bennedsen, M. (2016). Corporate governance and green innovation. Journal of Environmental Economics and Management, 75, 54–72.
22. Demirel, P., & Parris, S. (2015). Access to finance for innovators in the UK's environmental sector. Technology Analysis & Strategic Management, 27(7), 782–808.
23. Petkova, A. P., Wadhwa, A., Yao, X., & Jain, S. (2011). Being first or making a difference? A study of U.S venture capital investments into the "clean" energy sector. Frontiers of Entrepreneurship Research, 31(2), 49–61.
24. Petkova, A. P., Wadhwa, A., Yao, X., & Jain, S. (2014). Reputation and decision making under ambiguity: A study of U.S. venture capital firms' Investments in the Emerging Clean Energy Sector. Academy of Management Journal, 57(2), 422–448.
25. Carpenter, R. E., & Petersen, B. C. (2002). Capital market imperfections, high-tech investment, and new equity financing. The Economic Journal, 112(477), F54–F72.
26. Wüstenhagen, R., & Teppo, T. (2006). Do venture capitalists really invest in good industries? Risk-return perceptions and path dependence in the emerging European energy VC market. International Journal of Technology Management, 34(1/2), 63–87.
27. Heil, O., & Robertson, T. S. (1991). Toward a theory of competitive market signaling: A research agenda. Strategic Management Journal, 12, 403–418.
28. Hilmersson, F. P., & Hilmersson, M. (2021). Networking to accelerate the pace of SME innovations. Journal of Innovation & Knowledge, 6(1), 43–49. https://doi.org/10.1016/j.jik.2020.10.001
29. OECD. (2012). OECD work on green growth. Retrieved April, 2018, from http://www.oecd.org/greengrowth/oecdworkongreengrowth.htm .
30. Jacobs, M. (2012). Green growth: Economic theory and political discourse. Centre for Climate Change Economics and Policy. Working paper number 92.
31. Hall, J. K., Daneke, G. A., & Lenox, M. J. (2010). Sustainable development and entrepreneurship: Past contributions and future directions. Journal of Business Venturing, 25(5), 439–448.
32. Meyskens, M., & Carsrud, A. L. (2013). Nascent green-technology ventures: A study assessing the role of partnership diversity in firm success. Small Business Economics, 40(3), 739–759.
33. Colombelli, A. & Quatraro, F. (2017). Green startups and local knowledge spillovers from clean and dirty technologies. Small Business Economics, forthcoming in this issue.
34. Corradini, C. (2017). Location determinants of green technological entry: evidence from European regions. Small Business Economics, forthcoming in this issue.
35. Sunny, S. A. & Shu, C. (2017). Investments, incentives and innovation: geographical clustering dynamics as drivers of sustainable entrepreneurship. Small Business Economics, forthcoming in this issue.
36. Varian, H. R. (2014). Big data: new tricks for econometrics. Journal of Economic Perspectives, 28(2), 3–28.
37. Székely, N., & Vom Brocke, J. (2017). What can we learn from corporate sustainability reporting? Deriving propositions for research and practice from over 9,500 corporate sustainability reports published between 1999 and 2015 using topic modelling technique. PloS One, 12(4), e0174807. https://doi.org/10.1371/journal.pone.0174807 .
38. Eco-innovation Observatory Report. (2012). Methodological Report, available at http://www.ecoinnovation.eu/index.php/reports/methodological-report (last access 01/14/2019).
39. Klemmer, P., Lehr, U., Lobbe, K. (1999). Environmental innovation: Incentives and barriers; Analytica: Berlin.
40. Fussler, C., & James, P. (1996). Driving eco-innovation: A breakthrough discipline for innovation and sustainability. Pitman Publishing.
41. Wolff, S., Brönner, M., Held, M., & Lienkamp, M. (2020). Transforming automotive companies into sustainability leaders: A concept for managing current challenges. Journal of Cleaner Production, in press.
42. Oltra, V., & Saint Jean, M. (2009). Sectoral systems of environmental innovation: An application to the French automotive industry. Technological Forecasting & Social Change, 76, 567–583.

43. Horbach, J., Rammer, C., & Rennings, K. (2012). Determinants of eco-innovations by type of environmental impact — The role of regulatory push/pull, technology push and market pull. Ecological Economics, 78, 112–122.
44. Doran, J., & Ryan, G. (2012). Regulation and firm perception, eco-innovation and firm performance. European Journal of Innovation Management, 15, 421–441.
45. De Jesus Pacheco, D. A., Caten, C. S., Jung, C. F., Ribeiro, J. L. D., Navas, H. V., Cruz-Machado, V. A. (2017). Eco-innovation determinants in manufacturing SMEs: Systematic review and research directions. Journal of Cleaner Production, 142(part 4), 2277–2287.
46. Triguero, A., Moreno-Mondéjar, L., & Davia, M. A. (2013). Drivers of different types of eco-innovation in European SMEs. Ecological Economics, 92, 25–33.
47. Vona F (2019) Job losses and political acceptability of climate policies: why the 'job-killing' argument is so persistent and how to overturn it. Clim Policy 19(4):524–532. https://doi.org/10.1080/14693062.2018.1532871
48. Janser M (2018) The greening of jobs: empirical studies on the relationship between environmental sustainability and the labor market. University of Bamberg
49. Gholami, H., Rezaei, G., Saman, M. Z. M., Sharif, S., & Zakuan, N. 2016. State-of-the-art green HRM system: Sustainability in the sports center in Malaysia using a multi-methods approach and opportunities for future research. Journal of Cleaner Production, 124: 142–163.
50. Jackson, S. E., Schuler, R. S., & Jiang, K. 2014. An aspirational framework for strategic human resource management. Academy of Management Annals, 8(1): 1–56.
51. Kim, A., Kim, Y., Han, K., Jackson, S. E., & Ployhart, R. E. 2017. Multilevel influences on voluntary workplace green behavior: Individual differences, leader behavior, and coworker advocacy. Journal of Management, 43(5): 1335–1358.
52. Dubois, C. L. Z., & Dubois, D. A. 2012. Strategic HRM as social design for environmental sustainability in organization. Human Resource Management, 51(3): 799–826.
53. Renwick, D. W. S., Jabbour, C. J. C., Müller-Camen, M., Redman, T., & Wilkinson, A. 2016. Contemporary developments in green (environmental) HRM scholarship. International Journal of Human Resource Management, 27(2): 114–128.
54. Guest, D. E. 1997. Human resource management and performance: A review and research agenda. International Journal of Human Resource Management, 8(3): 263–276.
55. Guest, D. E. 2011. Human resource management and performance: Still searching for some answers. Human Resource Management Journal, 21(1): 3–13.
56. Pinzone, M., Guerci, M., Lettieri, E., & Redman, T. 2016. Progressing in the change journey towards sustainability in healthcare: The role of green HRM. Journal of Cleaner Production, 112: 201–211.
57. Antonioli, D., Mancinelli, S., & Mazzanti, M. 2013. Is environmental innovation embedded within high-performance organisational changes? The role of human resource management and complementarity in green business strategies. Research Policy, 42: 975–988.
58. Jabbour, C. J. C., & Jabbour, A. B. L. D. S. 2016. Green human resource management and green supply chain management: Linking two emerging agendas. Journal of Cleaner Production, 112: 1824–1833.
59. Bode, C., Singh, J., & Rogan, M. 2015. Corporate social initiatives and employee retention. Organization Science, 26(6): 1702–1720.
60. Sulich, Adam; Sołoducho-Pelc, Letycja The circular economy and the Green Jobs creation. Environmental Science and Pollution Research (2021). https://doi.org/10.1007/s11356-021-16562-y
61. Mrkajic, Boris; Murtinu, Samuele; Scalera, Vittoria G. Is green the new gold? Venture capital and green entrepreneurship. Small Business Economics (2017). https://doi.org/10.1007/s11187-017-9943-x
62. Zeng, Juying; Ren, Jun How does green entrepreneurship affect environmental improvement? Empirical findings from 293 enterprises. International Entrepreneurship and Management Journal (2022). https://doi.org/10.1007/s11365-021-00780-4
63. Dordmond, Gertjan; de Oliveira, Heder Carlos; Silva, Ivair Ramos; Swart, Julia The complexity of green job creation: An analysis of green job development in Brazil. Environment, Development and Sustainability (2020). https://doi.org/10.1007/s10668-020-00605-4

64. Demirel, Pelin; Li, Qian Cher; Rentocchini, Francesco; Tamvada, J. Pawan Born to be green: new insights into the economics and management of green entrepreneurship. Small Business Economics (2017). https://doi.org/10.1007/s11187-017-9933-z
65. Campo, Raffaele; Trio, Oronzo Think Green: the Eco-innovative Approach of a Sustainable Small Enterprise. Journal of the Knowledge Economy (2021). https://doi.org/10.1007/s13132-021-00833-7
66. Bohnenberger, Katharina Greening work: labor market policies for the environment. Empirica (2022). https://doi.org/10.1007/s10663-021-09530-9
67. Darendeli, Alper; Law, Kelvin K. F.; Shen, Michael Green new hiring. Review of Accounting Studies (2022). https://doi.org/10.1007/s11142-022-09696-y
68. Ren, Shuang; Tang, Guiyao; E. Jackson, Susan Green human resource management research in emergence: A review and future directions. Asia Pacific Journal of Management (2017). https://doi.org/10.1007/s10490-017-9532-1

Chapter 10
Bioeconomy

10.1 Introduction by Editor

The understanding that "business as usual" will exacerbate climate change is now widely acknowledged. The escalating levels of greenhouse gas emissions, dwindling reserves of fossil fuels, and the deterioration of land have become impossible to disregard within the realm of conventional economics. Extensive research is currently being conducted on various alternatives, with some promising options emerging that have the ability to replace or significantly modify the existing high emission intensity economic model. One of the prospective opportunities is that of the field of Bioeconomy. The theoretical framework of a circular bioeconomy encompasses the use of renewable natural resources in a manner that is both efficient and sustainable, hence facilitating transformative changes in various areas including land use, food production, health, and industry. The primary goal is to achieve long-term economic well-being while simultaneously fostering a mutually beneficial coexistence with the natural environment.

Many countries are now assessing the potential of bioeconomy as a means to address the challenges of climate change, resource depletion, and environmental degradation. Bioeconomy includes the production and utilization of renewable biological resources, such as biomass, to create a wide range of products, including biofuels, biochemicals, and biomaterials. This shift towards a bioeconomy has the potential to transform the way we produce and consume goods, moving away from a linear, fossil-fuel-based economy towards a more sustainable, circular model. The transition to a bioeconomy is driven by several key factors like advancements in biotechnology, the need to reduce greenhouse gas emissions, and the desire to develop more sustainable and resilient economic systems. Biotechnology has enabled the development of a range of innovative processes and products, from the production of advanced biofuels to the creation of renewable, biodegradable materials (Gajula & Reddy, 2021) (Scarlat et al., 2015). At the same time, the urgent need to address climate change and reduce our reliance on fossil fuels has spurred the development

G. Nulkar (ed.), *The Economics of Sustainable Development*,
https://doi.org/10.1007/978-981-99-7379-8_10

of biomass-based energy sources, which can offer a more sustainable alternative to traditional energy production.

The potential of bioeconomy to generate economic and environmental benefits is huge. Estimates suggest that the current bioeconomy market is worth approximately €2. 4 billion, employing 22 million people globally (Scarlat et al., 2015). The transition to a "Billion Ton bioeconomy" in the United States could expand direct bioeconomy revenue to nearly $259 billion and create 1. 1 million jobs by 2030 (Rogers et al., 2016). Beyond the economic benefits, the bioeconomy also offers the opportunity to reduce greenhouse gas emissions, improve waste management, and promote the sustainable use of natural resources.

According to Martinez-Hernandez (2018), in order to achieve the transition to a bioeconomy, it is necessary to utilise natural resources and produce goods by effectively reclaiming waste materials. This endeavour necessitates a thorough comprehension of process engineering. As an illustration, India generates a substantial quantity of agricultural waste. The residual biomass has the potential to be effectively exploited through the conversion processes that transform it into biofuels such as Compressed Biogas or biomass briquettes. Nevertheless, the successful implementation of this requires the advancement of economically feasible technologies and the enhancement of people resources with the necessary skills to effectively run these systems. It is worth noting that natural sunlight serves as the most abundant energy source for India. This presents a significant opportunity for the cultivation of biomass on soils that have been degraded. India has over 90 million hectares of land that has been subjected to degradation. Reclaiming these lands into the production of biomass for biofuels is a viable approach. This offers a dual advantage of facilitating carbon sequestration on previously unutilized areas, and substituting fossil fuels, that can improve India's foreign account shortfall. In order to effectively gain from this opportunity, it is imperative for India to install measures that promote the growth of biomass production. This can be achieved through the provision of policy support and market incentives, while ensuring that such efforts do not result in a conflict between the utilisation of land for food and nonfood biomass production.

The concept of a bioeconomy extends beyond the mere substitution of renewable energy sources for fossil fuels. It also encompasses the replacement of materials that are reliant on fossil fuels. Therefore, the substitution of plastics, concrete, and synthetic fabrics with other materials is necessary. There is a pressing need for substantial investments in the study and experimentation of renewable biobased materials, as well as the development of technologies capable of effectively harnessing these low-carbon intensity materials. Achieving this outcome is feasible by means of collaborative endeavours among several stakeholders, including industry, research institutes, and universities.

Muscat et al. (2021) suggest five ecological principles that can offer a framework for directing the utilisation of biomass towards the establishment of a circular bioeconomy. These principles encompass the following: the preservation and restoration of the well-being of our (agro)ecosystems; the avoidance of non-essential products and the wastage of essential ones; the prioritisation of biomass streams for fundamental human necessities; the utilisation and recycling of by-products derived from (agro)ecosystems; and the adoption of renewable energy sources while simultaneously minimising overall energy consumption. The implementation of these principles necessitates a comprehensive overhaul of our existing economic system, encompassing substantial modifications to legislation, technologies, organisations, social behaviour, and markets.

This chapter provides an analysis of the idea of biocircularity, as well as an examination of the associated issues. Research conducted by the European Union has demonstrated a growing trend among governments to adopt a circular framework and promote the use of biomaterials within the industrial sector.

References

Elias Martinez-Hernandez, J. S. (2018). Chapter 24 - Process Design and Integration Philosophy for Competitive Waste Biorefineries: Example of Levulinic Acid Production From Representative Lignocellulosic Biomasses. In A. P.-J. Thallada Bhaskar, Waste Biorefinery (pp. 695–725). Elsevier.

Gajula, S., & Reddy, C R K. (2021, May 4). More sustainable biomass production and biorefining to boost the bioeconomy. Wiley, 15(5), 1221–1232. https://doi.org/10.1002/bbb.2227.

Muscat, A. d.-B. (2021). Principles, drivers and opportunities of a circular bioeconomy. Nature Food, 2, 561–566. doi: https://doi.org/10.1038/s43016-021-00.340-7.

Rogers, J N., Stokes, B J., Dunn, J B., Cai, H., Wu, M., Haq, Z U., & Baumes, H S. (2016, November 21). An assessment of the potential products and economic and environmental impacts resulting from a billion ton bioeconomy. Wiley, 11(1), 110–128. https://doi.org/10.1002/bbb.1728.

Scarlat, N., Dallemand, J., Monforti-Ferrario, F., & Viorel, N. (2015, July 1). The role of biomass and bioenergy in a future bioeconomy: Policies and facts. Elsevier BV, 15, 3–34. https://doi.org/10.1016/j.envdev.2015.03.006.

10.2 Machine Generated Summaries

Disclaimer: The summaries in this chapter were generated from Springer Nature publications using extractive AI auto-summarization: An extraction-based summarizer aims to identify the most important sentences of a text using an algorithm and uses those original sentences to create the auto-summary (unlike generative AI). As the constituted sentences are machine selected, they may not fully reflect the body of the work, so we strongly advise that the original content is read and cited. The auto generated summaries were curated by the editor to meet Springer Nature publication standards. To cite this content, please refer to the original papers.

Machine generated keywords: bioeconomy, circular, european, bioeconomy strategy, biobase, biomass, strategy, biological, concept, european commission, transformation, justice, commission, circular economy, German.

10.2.1 Why is Communicating the Circular Bioeconomy so Challenging? [129]

This is a machine-generated summary of:

Giurca, Alexandru: Why is Communicating the Circular Bioeconomy so Challenging? [129]

Published in: Circular Economy and Sustainability (2022).

Link to original: https://doi.org/10.1007/s43615-022-00.229-y.

If you want to cite the papers, please refer to the original.

For technical reasons we could not place the page where the original quote is coming from.

Abstract-Summary

"This commentary synthesizes what is known, presumed, and still unknown about how to effectively communicate about the circular bioeconomy."

"It postulates that the circular bioeconomy communication challenge appears due to three main reasons: (i) one rooted in differing conceptualizations and competing discourses; (ii) one rooted in normativity; and (iii) due to several knowledge gaps in the communication process."

"It provides a series of lessons learned and some suggestions for future research on circular bioeconomy communication."

"The insights provided here are useful for environmental communications, and the society and technology scientific communities, as well as for policymakers, practitioners, and communication experts interested in effectively communicating the circular bioeconomy."

The Communication Challenge

"Despite growing political and academic backing, one problem persists: the CBE largely remains an obscure concept for the broader public."

"A growing number of perception studies point to a gap in our understanding regarding CBE awareness."

"Farmers in Austria tend to be more critical of the bioeconomy concept than other stakeholder groups [1]."

"Citizen perceptions seem to depend on specific CBE technologies and products and on how these are communicated [2]."

"The literature is far from conclusive on matters of CBE perceptions, acceptance of bio-based products, and lifestyle changes."

"Some studies focus on specific bio-based products, others on lifestyle changes, and the broader bioeconomy concept."

"Despite a lack of a comparable, comprehensive meta-analysis focusing on public perceptions of CBE, the existing literature tells us that perceptions vary depending on how this complex concept is constructed and communicated."

Differing Conceptualizations and Discourses

"The dominant way of framing the CBE is through the lens of existing policies, particularly the OECD's bioeconomy policy agenda [3] and the European Union's bioeconomy strategy [45]."

"The bioeconomy was not always the political concept concocted by the OECD and the European Commission."

"Only in the last decades did the bioeconomy concept gain traction in the European Union, particularly through biotechnology-related programs [6]."

"This brief historical detour of the origins of the CBE concept is important as it shows how the growing rift between different conceptualizations, expectations, and imaginaries came to be."

"Bioeconomy proponents emphasize economic growth enabled by the use of biotechnology using and processing large amounts of biomass (forests, crops, agricultural residues, etc.) [e.g., 6–8]."

"Bioeconomics proponents consider that the concept has been "hijacked" from Georgescu-Roegen's "bioeconomics program" [9] and argue for de-growth, new social organization, and low-tech innovations [e.g., 9, 10], critiquing the supposed circularity as only being related to targets of "reusing and recycling" which are ultimately aimed at economic growth [11]."

Normativity

"Participation in the CBE has been mainly framed from the consumer perspective, expecting citizens to buy (new) bio-based products and adopt significant lifestyle changes to reduce climate and environmental footprints [12]."

"This is why the specialty literature often points to regulatory and normative policies necessary for eliciting changes in everyday behaviors and new societal values [13, 14]."

"Envisioned CBE "lifestyle changes" are presented as fundamental pattern changes in consumption behavior which may include new dietary habits (e.g., vegetarian or vegan diets), food waste reduction, consumption, circularity, leisure activities, or mobility [15, 16]."

"Whereas bioeconomy and biotechnology proponents are normative in the sense of encouraging and nudging citizens to buy into the new bio-based market and opt into a lifestyle enabled by neoliberalist "green growth", ecological bioeconomics is intrinsically normative by advocating for radical lifestyle changes related to consumption and production patterns, sufficiency, entropy, and ecological sustainability within planetary boundaries [17]."

The Communication Process

"Basic questions related to "who" communicates (i.e., who the messenger is) and "to whom" it is communicated (i.e., what the target group is) need to be answered first."

"We have seen that different groups have different levels of knowledge regarding bioeconomy, and interpretations of the concept differ widely depending on who communicates and what is communicated."

"As to the channels of communication (the "how" question), messengers have employed different approaches with so far little known effects."

"Despite CBE communication campaigns being initiated in different European countries, so far, there is no research on the circulation and framing of the concept on social media."

"There is a research gap in terms of the different strategies employed, the role and interests of the different messengers, their strategies, and the effectiveness of these communication approaches in terms of public and/or consumer awareness."

Conclusion and the Way Forward

"Building communication bridges between science, policymaking, and society is paramount for a successful CBE transformation."

"For future communication and research endeavors, the following key aspects are important: There is no single, generally accepted CBE conceptualization."

"CBE communication has much to learn from climate change communication [e.g., 18, 19–20]."

"Whereas there are certainly many important lessons to be learned from decades of climate change communication, one concept embodies a scientific reality, the other a (political) solution."

"Just as the scientific expertise generated by CBE researchers elicits different imaginaries and policies, so too has the scientific expertise around climate change eventually translated into various normative mitigation policies [21]."

"Just as Maslow's "law of the instrument" warns against the overreliance on a preferred tool, so too must policymakers and communicators avoid presenting the CBE as the silver bullet to all our sustainability problems."

10.2.2 Biocircularity: A Framework to Define Sustainable, Circular Bioeconomy [130]

This is a machine-generated summary of:

Holden, Nicholas M.; Neill, Andrew M.; Stout, Jane C.; O'Brien, Derek; Morris, Michael A.: Biocircularity: a Framework to Define Sustainable, Circular Bioeconomy [130]

Published in: Circular Economy and Sustainability (2022)

Link to original: https://doi.org/10.1007/s43615-022-00.180-y.

If you want to cite the papers, please refer to the original.

For technical reasons we could not place the page where the original quote is coming from.

Abstract-Summary

"Bioeconomy will have to embrace circularity to enable production of renewable feedstocks in terms of both biomass yield and maintaining essential natural capital."

"The concept of biocircularity is proposed as an integrated systems approach to the sustainable production of renewable biological materials focusing on extended use, maximum reuse, recycling, and design for degradation from polymers to monomers, while avoiding the "failure" of end of life and minimizing energy demand and waste."

"Challenges are discussed including sustainable production and consumption; quantifying externalities; decoupling economic growth from depletion; valuing natural ecosystems; design across scales; renewable energy provision; barriers to adoption; and integration with food systems."

"Biocircularity offers a theoretical basis and measures of success, for implementing sustainable circular bioeconomy."

Introduction

"Circular bioeconomy "…focus[es] on the sustainable, resource-efficient valorization of biomass in integrated, multi-output production chains (e.g., biorefineries) while also making use of residues and wastes and optimizing the value of biomass over time via cascading…" [22]."

"Circular bioeconomy is an approach that can provide society with a sustainable future by allowing economic growth within the planetary boundaries and decoupling demand from resource extraction, but fears have been expressed that it will increase pressure on natural ecosystems [23]."

"The only way to achieve a truly sustainable, circular bioeconomy is to embrace all elements of circularity including eco-design of products, use of processes, and services that drive integrated systemic thinking, targeting sustainable production and consumption of renewable biological materials and prioritizing maintenance and enhancement of natural capital."

The Concept of Biocircularity

"The processing system that creates objects must be designed to do more than merely cascade the biomass [22]; rather, it must minimize energy demand, losses, emissions, material degradation, mass loss during use cycles, and minimum loss/energy demand during the transition from one material/product to the next."

"This means that a feedstock and the primary materials must be developed using a holistic design that considers the primary use of the object and the secondary use of the materials from which it is made."

"The difference between market demand and sustainable primary feedstock supply provides an indication of how much mass would need to remain in closed-loop circulation within the economy, which would allow calculation of the time materials should stay in circulation, or the number of cycles required, prior to energy extraction and end of life."

The Challenges

"A route to a solution for this challenge includes (i) finding and eating diets that minimize the land area needed to feed the global population a nutritious, healthy diet with minimum impact on natural capital (to meet the demands of SDG targets 2.1, 2.2, 2.4, 12.8, 12.C); (ii) estimating the prioritization of non-food feedstocks to provision society with its non-food needs (to meet the demands of SDG target 12.2); (iii) estimating the maximum provisioning from the food system within the limits of sustainable agriculture and minimization of wasted food (to meet SDG target 12.3); (iv) for the remaining land area and considering co-product streams from the food land area, estimating the maximum optimum provisioning that can be achieved within the limits of sustainable agriculture; (v) estimating global demand for non-food materials that can be provided by the bioeconomy; and (vi) calculating the amount of fossil-derived material currently in the technosphere that will have to pump-prime circularity in order for human society to be provisioned within the limits of end-of-life losses (related to SDG target 12.5) and sustainable production."

Concluding Remarks

"For sustainable, circular bioeconomy to work a balance is needed between the production capacity provided by the natural environment (the biogenic carbon feeding the circular bioeconomy [24]) and demand for materials, energy, and products."

"Society lacks the data and means of analysis to manage transition to sustainable circular bioeconomy."

"The concept of biocircularity brings the challenges faced to achieve a sustainable, circular bioeconomy into sharp focus."

"Governance, policy, resources, research, and human effort should be devoted to achieving successful sustainable, circular bioeconomy by using biocircularity as the framework for measuring success."

"To find a route to a sustainable future, society needs to adopt a rigorous and realistic understanding of sustainable, circular bioeconomy, and the need to use quantitative tools to plan for and design reliable approaches to harnessing natural resources."

10.2.3 'The Good Economy': a Conceptual and Empirical Move for Investigating How Economies and Versions of the Good Are Entangled [131]

This is a machine-generated summary of:

Asdal, Kristin; Cointe, Béatrice; Hobæk, Bård; Reinertsen, Hilde; Huse, Tone; Morsman, Silje R.; Måløy, Tommas: 'The good economy': a conceptual and empirical move for investigating how economies and versions of the good are entangled [131]

Published in: BioSocieties (2021)

Link to original: https://doi.org/10.1057/s41292-021-00.245-5.

If you want to cite the papers, please refer to the original.

For technical reasons we could not place the page where the original quote is coming from.

Abstract-Summary

"This paper suggests the concept of 'the good economy' as an analytical tool for investigating how economic practice is entangled in versions of the good."

"Empirically, the paper teases out how what is made to be good in the bioeconomy is radically different than in economies of the recent past."

"While 'the good economy' of the early oil and aquaculture economy concerned how to insert this economy into society in a good manner, society is surprisingly absent in the contemporary bioeconomy."

"The bioeconomy is enacted as an expert issue, pursued by the tools of economic valuation, and based in the unquestioned idea that 'the bio' makes any economy good."

Introduction

"As this paper will show, the various policy strategies enact the new bioeconomy in different ways, yet they all seem to share a strong normative basis: a vision of a shift to a new economy that is innovative, sustainable, responsible and environmentally friendly—in short, novel, ethical and good."

"The analysis implies 'troubling' the good and to pursue a critical reading of the relations between 'the bio' and 'the economy' of the bioeconomy."

"And related to this, rather than taking at face value that the new bioeconomy is a normatively better (or worse) one than the economies it seeks to replace, we argue that questions of 'the good' has in fact always been entangled with the economy—but in highly different ways."

"And most importantly, 'the good economy' concept enables us to grasp that there are more relations at stake in the bioeconomy than bioeconomy relations alone."

Theoretical resources and approaches to 'the good economy'

"Expanded understanding of economy–normativity relations, we build the concept of 'the good economy' as a heuristic and analytical tool for investigating the ongoing discussions of the bioeconomy."

"This leads us to the article's empirical section, in which we employ and further develop our analytical framework within four key versions of economies: first, the obvious choice of the ongoing bioeconomy policy initiatives of the EU, the OECD and national governments in Europe; second, the related—but diverging—version of 'the blue bioeconomy', which is based upon ocean industries and marine resources; third, the establishment of aquaculture as an industry in Norway in the 1970s and fourth, the establishment of a petroleum industry in Norway, also in the 1970s."

"After this empirical exposé, we draw together the different versions of the economy and conclude by discussing how the concept of 'the good economy' may better help us make sense of co-existing forms of bioeconomies and their diverging modes of enacting the good—and ultimately, how this also carries implications not only for bioeconomy relations but also for biosociety relations."

From a protestant ethic, a moral economy and market truths to 'the good economy'

"Even more so, perhaps, this is the case for E. P. Thompson who, in his seminal work on the English working class, suggested the notion of 'moral economy' to capture the tensions and conflicts emerging from the transition to a capitalist economy (Thompson 25)."

"Thompson's point was not to argue for a particular moral in—or integral to—the economy, nor, as in the case of Weber, an ethics spearheading the drive for surplus."

"Thompson's 'moral economy' was rather the moral of the English working class and the normativities embedded in their ways of living and doing their economy, a moral economy which was set aside and suppressed by the transformation to a capitalist economy."

"In this turn to a market society there is no space for moral sentiments or economies beyond the market price, which does not make a difference between the correct price and that which is good."

Processes of economization versus 'versions of the economy'

"Our notion of 'the good economy' still takes inspiration from her term, while also addressing normativities beyond ethics."

"These have not addressed the economy in relation to the explicit issue of 'the good', but rather pursued the valuation approach as more of a value-neutral term, so to speak."

"The concept of 'the good economy' draws from these contributions, but extends them towards analysing the choreographies and compositions of the 'versions of economies' in question (cf. Asdal 26) and the entanglements between the economy and 'the good'."

"Rather than a label of what the economy is, the 'good economy' concept is intended as an investigative tool to question and trouble the composition of the so-called bioeconomy and its entanglements with 'the good', including 'the good' that it performs and invests in."

The good green bioeconomy: troubling the inherent good of the 'bio'

"The European Commission, in the above-mentioned document Innovating for Sustainable Growth: A Bioeconomy for Europe (EU 27), delineate the bioeconomy as follows: "The bioeconomy encompasses the production of renewable biological resources and their conversion into food, feed, bio-based products and bioenergy" (EU 27, p. 16)."

"While all the bioeconomy policy documents emphasize the need to expand the 'bio' component of the economy—whether biomass or biotech—they diverge in how they position themselves vis-à-vis fossil fuels and the petroleum economy."

"The Norwegian bioeconomy strategy (NFD 28) similarly adopts a clear transition narrative, but rather than highlighting the need to move away from fossil fuels, what is emphasized is rather what the bioeconomy moves toward."

"In the documents themselves, these are not addressed or acknowledged—neither in the founding documents from the supranational institutions, such as the OECD and the EU, nor in the national documents in which these versions are bound together, such as the Norwegian bioeconomy strategy."

The good blue economy: enabling a fully economized nature by bringing in economics and its 'tools of valuation'

"While fostering growth has indeed been a key preoccupation of all the 'blue documents' from the EU's Blue Growth strategy onwards (EU 29), the OECD's Ocean Economy project went further than conventional estimates of potential industrial growth."

"Being based in the OECD's Futures Project and employing policy foresight techniques, it sought to advance the methods for assessing potential industrial growth by integrating the value of the ocean's ecosystems directly into the calculation."

"Another blue economy report, this time from the Norwegian context, offers a particularly strong example of how the promise of economizing future bio-values may be used, not to preserve ecosystems, but to leverage political will for rapid, large-scale industrial growth."

"A report titled Value Created from Productive Oceans in 2050 calculated the Norwegian marine industries to have the potential of at least a fivefold growth in value creation by the year 2050 (SINTEF 30)."

The good aquaculture economy: justifying expansion by staying small and local

"The first decisive action taken by (a unanimous) Parliament in 1973 was to limit the possibilities for an industrial path of growth in fish farming, by making any further expansion contingent upon production licences, with a maximum volume per fish farm (PNP 31)."

"This growing, new sector of the economy is justified by reference to the overall aims it should help fulfil, and the ways it might fit together with and complement already existing activities: increased food production, rural development and new opportunities specifically directed to vulnerable, coastal districts, and lastly, employment and wages comparable to fisheries and agriculture."

"For aquaculture to help achieve such aims, it needed to be developed in a certain way: industrial growth—understood as large, mechanized units and 'external' capital—was curtailed in favour of smaller, geographically dispersed labour-intensive fish farms with local ownership and locally available labour and natural resources."

The good oil economy: fostering 'a qualitatively better society'

"At the time of discovering oil, Norwegian politicians were cautious that the potential new wealth would not automatically be positive for Norwegian society and its economy."

"The Norwegian government issued a white paper which explicitly warned that uncontrolled use of the oil revenues would have severe consequences for Norwegian society and its economy."

"The guidelines that are drawn up for the petroleum activities and the use of revenues must therefore be part of a planned transformation of Norwegian society. [...] Democratic institutions must gain actual government [herredømme] of the development in increasingly more areas."

"Quite the contrary, the white paper repeatedly asserts that the infusion of oil into the economy might produce multiple "problems of transition" (Ministry of Finance 32, p. 6)."

"The white paper's key concern was to outline how a smooth and beneficial transition into an oil economy might be done."

"If the coming oil economy is governed in an active, deliberate manner, it might become 'a good economy'."

Drawing versions of 'the good economy' together

"Whereas the early oil and aquaculture economies were problematized in terms of how these new economic activities were being inserted into society, society does not exist in the same way in the policy documents on the bioeconomy."

"Rather than an economy inserting itself into society, the bioeconomy is oriented towards inserting 'the bio' into the economy."

"And importantly, this insertion of 'the bio' into the economy is not envisioned to happen by societal means, but by way of the tools of valuation of economics, either neoclassical economics, such as ecosystem services, or business school models and strategies."

"While comparing and contrasting with other versions of the economy, what becomes apparent is how society is curiously lacking in the bioeconomy policy programmes."

"This version of the good economy is one where 'the bio'—the nature upon which the bioeconomy rests—is fully economized, thus, transformed into an economic object."

Conclusion

"The overriding objective of this paper is to suggest a recasting of the bioeconomy, most notably by putting forward the notion of 'the good economy' as an analytical working tool made for opening up and investigating this economy further."

"And as a more overriding point, the newness seems related to the normativities that are involved: the bioeconomy steps forward as a new form of good economy—as the 'good economy'."

"The concept of the good economy is not put forward in order to analyse the bioeconomy exclusively."

"This conceptual move is done in order to work as a heuristic tool for analysing good economy relations more broadly, and also in other versions of the economy."

"We also need to include the 'tools of valuation' that are involved in doing such good economy relations."

10.2.4 Conceptual Evolution of the Bioeconomy: A Bibliometric Analysis [132]

This is a machine-generated summary of:

Mougenot, Benoit; Doussoulin, Jean-Pierre: Conceptual evolution of the bioeconomy: a bibliometric analysis [132]

Published in: Environment, Development and Sustainability (2021)

Link to original: https://doi.org/10.1007/s10668-021-01.481-2.

If you want to cite the papers, please refer to the original.

For technical reasons we could not place the page where the original quote is coming from.

Abstract-Summary

"The bioeconomy suggests a model based on the production of renewable biological resources and the conversion of these resources into value-added products."

"The main aim of this article is to assess the impact of the bioeconomy on the scholar."

"This is one of the first research which analyzes the literature pathways of the bioeconomy issue using a computational analysis."

"Our article concludes that the principles of the bioeconomy have a strong potential to address these related challenges to manage and maintain the environment."

Introduction

"Brazil, Russia, India, China, and South Africa (BRICS), as well as other African countries, have incorporated this new economic development based on bioeconomy-based strategies in energy production, agriculture, and biomass (Bambo & Pouris, 33; Louvel & de Gromard, 34; Virgin & Morris, 35)."

'Bugge and others (36) performed a bibliometric literature indexed analysis of 453 articles from 2005 to 2014 concluding that the importance of bioeconomy had increased."

"The aim of the bioeconomy is to build an economy based on the consumption and production of goods and services from the direct use of biological resources and its sustainable transformation (European Commission, 37)."

"The importance of this study concerns the first bibliometric research that analyzes the literature pathways of the bioeconomy issue using Biblioshiny a web interface for bibliometrix analysis based in R package (Gagolewski, 38; Moral-Muñoz and others, 39; Saikia and others, 40)."

"The increased demand for bioeconomy issues justifies the need for more effective economic policies with a focus on sustainability."

Materials and methods

"The complete records for each publication retrieved during the search were converted as a Scopus BibTex file and imported into Bibliometrix and Biblioshiny."

"Bibliometrix allows completing the full process of scientific literature analysis and data process."

"Biblioshiny captures the core Bibliometrix code and creates an online data analysis framework (Aria & Cuccurullo, 41)."

"Using the Bibliometrix and the Biblioshiny packages, the research allows showing bibliometric indicators on bioeconomy such as publication volume in number of articles, citation count, and keywords."

Results and discussion

"The country published the National Bioeconomy Blueprint, a plan document to reinforce its activities around the bioeconomy and bio-based products (McCormick & Kautto, 42)."

"This significant increase in the term bioeconomy is driven by policies proposed in countries such as Germany, Spain, and Finland."

"The analysis showed in which research topics of the bioeconomy concept the authors of bioeconomy publications had explored and which sources they had most often published."

"The analysis of keywords, top authors, and sources indicated that four authors (i.e., Sanders JPM, Patel MK, and Birch K) and five sources (i.e., Journal of Cleaner Production, Biofuels, Bioproducts and Biorefining; Industrial; Industrial Biotechnology, Journal of Chemical Technology and Biotechnology and Sustainability) were closely linked to the main research topics of bioeconomy ("bioeconomy," "biorefinery," "bio-based economy" and "biomass")."

Conclusions

"While advances in environmental bibliometric research are often seen as performing a critical function by focusing on important and contemporary issues, for instance, environmental conservation, environmental management accounting, biomass energy and the environment, the article has indicated that the bioeconomy establishes a useful examination field; despite the fact that everything considered, the examination analyzed in this survey has probably been engaged in related fields previously, or in a comparative exploration under various headings, for example, the sustainable production and consumption."

"The first record on bioeconomy appeared in 1979 (Ikeda, 43), but it was only after 2000 that the concept attracted researchers and since then the number of publications in this field has increased to 1732 documents published between 1999 and 2018."

"The country that has done much of the bioeconomy research is the USA (568 publications, 7851 citations, and 13.82 average article citations)."

10.2.5 Conceptualizing Controversies in the EU Circular Bioeconomy Transition [133]

This is a machine-generated summary of:

Starke, Jan R.; Metze, Tamara A. P.; Candel, Jeroen J. L.; Termeer, Catrien J. A. M.: Conceptualizing controversies in the EU circular bioeconomy transition [133]

Published in: Ambio (2022)

Link to original: https://doi.org/10.1007/s13280-022-01.730-2.

If you want to cite the papers, please refer to the original.

For technical reasons we could not place the page where the original quote is coming from.

Abstract-Summary

"However, it remains underexplored in the CBE literature how controversies develop throughout a transition process."

"We suggest that controversies can be situated on and between different system layers within a transition."

"These controversies can have both productive and unproductive outcomes while they unfold throughout a transition."

"We illustrate this understanding with the example of biorefineries as CBE key technology and discuss a research agenda on controversies in sustainability transitions."

Introduction

"Biorefineries are a key technology in the CBE transition because they convert bio-based resources into materials such as chemicals, plastics or feed products (Cherubini 44)."

"Examples of controversies in the CBE transition with relevance for biorefineries include food-feed-fuel (Muscat and others 45), green growth versus degrowth (D'Alessandro and others 46), globalization versus regionalization (Priefer and others 47, pp."

"Controversies are particularly relevant in the EU CBE transition, as both of the concepts forming the transition's goal, circularity and bioeconomy, are contested (Bugge and others 48; Kirchherr and others 49; Bauer 50) and thus prone to conflicting interpretations."

"We therefore aim to advance the understanding of how controversies develop throughout the transition towards a CBE in the EU."

"We argue that current perspectives on controversies in the CBE transition could be advanced in three regards."

"We situate controversies within the transition process."

"We highlight the outcomes of controversy changes on the transition process."

Current perspectives on conflict in the CBE literature

"Despite acknowledging the relevance of conflict, the ways in which controversies develop as well as their outcomes on an unfolding transition process remain largely underexplored in the state-of-the-art literature on the CBE."

"We synthesized the literature into two overarching perspectives on the role of conflicts and controversies in the CBE transition: a perspective on conflict as a design fault of novel technologies and supply chains and a perspective focusing on framing conflicts to (de)legitimize transition pathways and visions."

"We contribute to both perspectives by contextualizing controversies as an intractable type of conflict in the transition process and conceptualize how such controversies change throughout a transition."

"Although we acknowledge the relative stability of actor groups, we add to this perspective by explicitly conceptualizing how and why storylines and associated actor groups change over time, particularly in long-term controversies, and what this means for the overall transition process."

Situating controversies in the transition process

"Controversies are contextualized in these loci, which are specific locations in a transition process with particular involved (groups of) actors, frames and communication rules."

"Micro-level controversies concern alternative, regime-challenging ways of thinking, doing and organizing (Van Der Minne and others 51)."

"Whereas engineers might develop a new biorefinery design on the micro level, policymakers craft strategic decisions on the future of the CBE in the EU on the macro level, and the installation ultimately has to be located in a municipality, concerning the meso level."

"A controversy could develop between an innovative micro-level expert team proposing a biorefinery using genetically engineered algae as feedstock and facing meso-level regime regulations impeding the use of this feedstock."

"The algae controversy could be salient between micro- and macro-level actors in the design phase of a biorefinery and pop up later in the form of meso- and macro-level citizen concerns about locating the new facility."

Changes of controversies due to dynamic storylines

"We conceptualize these actor groups as dynamic discourse coalitions (Metze and Dodge 52) around common storylines, which involved actors reproduce and shape."

"Actors in different discourse coalitions frame knowledge and experiences divergently because they make sense of new information and select, name, and categorize aspects strategically to build their storylines (Van Hulst and Yanow 53)."

"Storylines and surrounding discourse coalitions develop throughout the transition process."

"Discourse coalitions change over time through processes of confirmation (strengthening of a storyline), integration (connection of storylines), disintegration (contestation of a storyline within the discourse coalition itself) and polarization (reconfirming the differences in competing discourse coalitions' storylines) (Metze and Dodge 52, p. 4)."

"Because of changes in storylines, the surrounding discourse coalitions are also in constant flux: new actors join the coalition along the transition process, others leave it."

"A discourse coalition can confirm its underlying storylines."

Unproductive and productive outcomes of controversies

"Controversies evolve during the transition process in both unproductive and productive ways."

"These developments affect the inter-personal relationships of actors shaping both transition vision and pathways in dynamic discourse coalitions, while the transition process develops."

"We propose criteria on how to differentiate unproductive and productive evolutions of controversies while they unfold in a transition process."

"Another sign of a controversy becoming unproductive is when key actors manage to actively exclude actors with a different perspective from decision-making processes."

"Actors talking past each other without regarding the arguments of their adversaries is an unproductive form of controversy (cf. Van Eeten 54)."

"Controversies can evolve in productive ways, meaning that they improve the CBE transition process."

"Controversies can thus bring together previously unconnected actors in new groups with new resources and new power relations."

Discussion and research agenda

"Controversies in the CBE transition cannot be 'out-designed'."

"We have provided conceptual steps towards understanding how these controversies change during a transition and how controversies shape a transition process."

"This temporal contextualization provides analysts with a lens to assess how controversies have developed, what controversial aspects of the CBE transition have surfaced before and could reappear in the future."

"Future empirical research could apply our conceptualization to identify different controversies and their dynamics in other sustainability transitions."

"Understanding changes of controversies in the context of a transition helps to illuminate the discursive dimension of responsible research and innovation, as is recently called for (Jakobsen and others 55)."

"Conceptualizing the changes of controversies in the CBE transition is a first step towards designing governance interventions to stimulate productive forms of controversy."

10.2.6 Environmental and Socioeconomic Footprints of the German Bioeconomy [134]

This is a machine-generated summary of:

Bringezu, Stefan; Distelkamp, Martin; Lutz, Christian; Wimmer, Florian; Schaldach, Rüdiger; Hennenberg, Klaus Josef; Böttcher, Hannes; Egenolf, Vincent: Environmental and socioeconomic footprints of the German bioeconomy [134]

Published in: Nature Sustainability (2021)

Link to original: https://doi.org/10.1038/s41893-021-00.725-3.

If you want to cite the papers, please refer to the original.

For technical reasons we could not place the page where the original quote is coming from.

Abstract-Summary

"The agricultural biomass FP is dominated by animal-based food consumption, which is slightly decreasing."

"The forestry biomass FP of consumption could potentially shift from net import to total supply from domestic territory."

"Agricultural land use for consumption is triple that of domestic agricultural land (which covers half of Germany) and induced substantial land use change in other regions from 2000–2015."

"The climate FP of BE contributes 18–20% to the total climate FP of domestic consumption, while employment makes up 10% and value added only 8% of the total German economy."

"These findings imply that sufficient monitoring of the BE needs to consider both production and consumption perspectives, as well as global FPs of national economies."

Main

"The intention was mainly to replace fossil- and mineral-based resources with biomass, and there have been high hopes with regards to additional income for farmers and foresters through bioenergy and biomaterials, sustaining rural development, and economic growth through innovations in biotechnology [56]."

"While policy strategies have been implemented and billion-dollar heavy industries (such as the biofuel industry) have been created, comprehensive monitoring of the actual performance of the BE regarding the achievement of sustainability goals, including the effects of past and ongoing policies, has been largely lacking [57, 58]."

"In this Article, we provide a comprehensive analysis of past and projected resource and climate footprints (FPs) of the German BE and compare them with established indicators of socioeconomic performance, such as employment and value added."

"Building on conceptual [59, 60] and empirical [61, 62] work, we concentrate on materials (forest and agricultural biomass), agricultural land, irrigation water withdrawals and climate FPs as essential parts of a future monitoring system."

Results

"The environmental FPs of the German BE have been determined on the basis of global, life cycle-wide requirements of fully or partly biomass-based products annually produced and consumed in Germany, for agricultural raw materials (agriculture biomass FP), primary timber (forestry biomass FP), agricultural land (land FP), irrigation water withdrawals (water FP) and GHG emissions (climate FP)."

"The past level of the consumption FP of 95.1 Mm^3 roundwood equivalents on average could have been completely supplied from Germany's domestic territory as it lay within the potential sustainable national roundwood production range of 86–114 Mm^3 roundwood equivalents in 2015 and 79–118 Mm^3 roundwood equivalents in 2030 [63]."

"While ongoing trends indicate that, as a consequence of climate change, global irrigation water withdrawals might grow by up to 21% from 2015–2030, the German water FP is expected to decline by 28% (from 15.8–11.5 km^3), mainly due to changing import structure."

Discussion

"Our study has shown that the German BE has contributed substantially to land transformation in other regions, that it might continue to add to the scarcity of water in arid areas, that it uses timber from elsewhere that could be produced at home and that the GHG emissions of its consumption are much higher than is reported on the basis of the Kyoto protocol."

"In terms of sustainability, a first orientation on whether a country exerts disproportionate pressure by its domestic consumption of biomass-based products is possible by comparing FP levels with global or continental averages."

"On a per-person basis, the agricultural biomass consumption of Germany exceeds the global average; consumption of primary timber is at a level that can be supplied sustainably at a European scale; agricultural land use has exceeded the global average in the past; and cropland use may still be beyond average and sustainable levels in 2030 without additional measures."

Methods

"BE-related FPs of production and consumption (including capital formation), as well as those for exports, were determined for GHG emissions and resource use (biotic materials, agricultural land and irrigation water withdrawals)."

"Compared with other approaches, such as by the International Resource Panel [64] including the German details, our analysis had two distinct differences: (1) our study focused on the BE part of the whole economy, which required appropriate differentiation of bio-based parts; and (2) the International Resource Panel specified some impacts on a cradle-to-material basis, meaning that use of the materials in the form of products was not included (as a result, for example, GHG emissions from the burning of fossil fuels and biomass were not considered); in contrast, our analysis included basically the full life cycle."

"For the years 2020, 2025 and 2030, land use maps were generated in scenario mode by allocating as much crop cultivation and grazing area as was needed to reach the prescribed quantities of production specific to all crops and livestock in individual countries."

10.2.7 The Bioeconomy in Spain as a New Economic Paradigm: The Role of Key Sectors with Different Approaches [135]

This is a machine-generated summary of:

Ferreira, Valeria; Pié, Laia; Mainar-Causapé, Alfredo; Terceño, Antonio: The bioeconomy in Spain as a new economic paradigm: the role of key sectors with different approaches [135]

Published in: Environment, Development and Sustainability (2023)

Link to original: https://doi.org/10.1007/s10668-022-02.830-5.

If you want to cite the papers, please refer to the original.

For technical reasons we could not place the page where the original quote is coming from.

Abstract-Summary

"Its importance has been demonstrated by its inclusion as a priority in specific and related policies such as the bioeconomy strategy, the Sustainable Development Goals, the European Green Deal, and the Next Generation recovery plan."

"Spain has not lagged behind and considers the bioeconomy as a priority in its policy strategies to achieve a more sustainable economy."

"To carry out policy-relevant impact assessment in support of bioeconomy development, specific databases describing bio-based products are required."

"Analysis results reveal promising products as key wealth generators and growth promoters and allow to identify the most suitable to be stimulated with policies to promote the development of the Spanish bioeconomy."

Introduction

"With the aim of analysing the potential of bioeconomy products in Spain and promote its development, this article presents on the one hand, a multisectoral structural analysis based on two different and complementary methods: the classical methodology and the Hypothetical Extraction Method (HEM); and, on the other, some policy implications derived from this analysis."

"The objective of the article is to analyse the bioeconomy structure in Spain using different methodologies and to identify those promising sectors as key wealth generators and growth promoters that will be suitable to stimulate with policies to promote the development of the Spanish bioeconomy."

"The BioSAM provides a comprehensive economy-wide frameworks that explicitly represent the linkages between the bioeconomy and the broader macroeconomy, and the combination of methodologies applied enables a more in-depth analysis of these linkages and therefore, a better interpretation of those most suitable sectors for promoting the bioeconomy (Iráizoz, 70)."

Data and methodologies

"The idea behind structural analysis is to understand the structure of an economy by analysing the linkages between its sectors to identify those that are highly interconnected and to determine the importance of each sector."

"The HEM determines the importance of a sector, considering the differences in output upon hypothetically eliminating it from the economy (Dietzenbacher and others, 71)."

"Analysis of the BL serves to indicate the impact of extracting a sector that no longer consumes inputs for its production on the other sectors in terms of reduction in output."

"Sectors are extracted one by one, obtaining new production values for each case and detailing the results of each output in a new matrix."

"The difference between the case with the extracted sector and the original situation reflects the economic losses of the rest of the sectors of the economy without the offer of the extracted sector."

Results

"Within the bioeconomy, the products included in the agriculture, food industry and biomass groups stand out."

"The analysis of the results of the FL considering Ghosh's inverse matrix indicates that there are 10 key products, of which 9 belong to the bioeconomy, highlighting those included in agriculture and the food industry."

"Considering this relative value, non-bioeconomy products do not stand out."

"This means that the highest values are related to the products with the highest output, and the results considering the average of all sectors indicate that only non-bioeconomy products stand out."

"If the results are analysed considering the average of the bioeconomy products, those within agriculture stand out, including cereals, other crops, and intensive livestock; within the food industry, animal feed, beverages and tobacco, white meat, dairy, other food, and wine; and in the bioindustry, biochemicals, textiles, and wood products."

Discussions

"In the bioeconomy in Spain, the products related to the food industry stand out for their quantitative importance and their links with the other accounts."

"Considering the average of those belonging to the bioeconomy, the results with the HEM show that various products mainly included in the food industry and bioindustry sectors, and one product in the agricultural sector, are important as much for their representative values in the bioeconomy as for their links with the other accounts."

“Exclusively analysing their links with the other accounts reveals that only wood products are considered as a product with important relations driving the growth of the economy to an above average extent in terms of sectors, with the influence of work and capital standing out.”

Policy implications

“The analysis of the linkages shows the importance of the bioeconomy products’ relationships with the other accounts because most of them can be considered as driver, base or key products within the economy.”

“This article concludes that to promote the Spanish bioeconomy, the products with the greatest influence identified in this study as “key sectors" by some methodology must be stimulated.”

“The promotion of these products will mean that they will represent a larger quantity in the bioeconomy and that their backward and forward linkages promote the development of other activities within it, promoting economic growth.”

“It will be important to invest not only in the traditional sectors previously mentioned that stand out as important in the economy, but also in those that are important at strategic levels because they enable the expansion of the products belonging to the bioeconomy, thus aiding its development in Spain.”

Conclusions

“Considering the bioeconomy sectors, this article focuses on a new perspective and compliment the structural analysis applying different methodologies with the purpose of examining in-depth the relations within the sectors.”

“Different methodologies have been applied to the BioSAM to analyse the structure of the bioeconomy in Spain and to understand the potential of each bio commodity to generate influence in the rest of the economy, providing a complete analysis of Spanish Bioeconomy.”

“The descriptive analysis applied allowed us to identify which products stand out most for their economic importance and/or their link with the other accounts, promoting the development of the rest, and also the products whose economic maturity is not yet relevant.”

“The results indicate that the more traditional bioeconomy products (such as primary sector) have the potential to promote economic growth in the rest of the economy sectors since they stand out for their importance and/or their relations with the other accounts.”

10.2.8 A Comparative Analysis of Bioeconomy Development in European Union Countries [136]

This is a machine-generated summary of:

Dolge, Kristiāna; Balode, Lauma; Laktuka, Krista; Kirsanovs, Vladimirs; Barisa, Aiga; Kubule, Anna: A Comparative Analysis of Bioeconomy Development in European Union Countries [136]

Published in: Environmental Management (2022)

Link to original: https://doi.org/10.1007/s00267-022-01.751-3.

If you want to cite the papers, please refer to the original.

For technical reasons we could not place the page where the original quote is coming from.

Abstract-Summary

"This paper presents the results of a cross-country comparison of 22 EU countries applying the Complex Bioeconomy Sustainability Index for 2012–2018."

"Results were compared with results of the TOPSIS multi-criteria decision-making method on national bioeconomy strategies in Finland, France, Spain, Italy, Ireland, and Latvia."

"The research aims to combine a quantitative assessment of bioeconomy output indicators with a qualitative evaluation of bioeconomy strategies of EU Member States."

Introduction

"It is expected that the monitoring system will serve as a reasonable way to manage the progress of the bioeconomy sectors based on measurable, comparable indicators and help solve the current uncertainties concerning bioeconomy progress evaluation that many authors point to in the academic environment."

"The authors highlight the need for more research on methods and indicators to measure the bioeconomy, considering the diversity of bioeconomy policies and strategies worldwide (Ferreira and others 72)."

"This is one of the few studies that offer an indicator-based evaluation and comparison of bioeconomy progress between almost all EU countries."

"As the literature review shows, previous studies have focused on assessing individual indicators or their groups (i.e., based on the sustainability dimension), bioeconomy policy, or case studies of specific countries."

"Zihare and others proposed an indicator-based approach to measure the sustainable development of the bioeconomy considering the aspects of policy, research and innovations, and technology."

Methodology

"We studied the bioeconomy strategies of selected countries using the systematic literature review and the TOPSIS multi-criteria decision-making method to collect in-depth insight into synergies with the EU 2018 Bioeconomy Strategy and Action Plan (European Commission, 73, 74)."

"This study constructed a Composite Bioeconomy Sustainability Index to monitor performance and evaluate and compare the bioeconomy development trends among EU member states."

"The proposed Composite Bioeconomy Sustainability Index combines 13 indicators grouped into four main dimensions, namely, (i) socio-economic impact, (ii) resource sustainability, (iii) innovation and technology, and (iv) financing and government support, as similarly seen in other studies (Martinez-Hernandez and others 75)."

"The interviewed experts had to evaluate which of the five goals and three areas of action of the EU Bioeconomy Strategy (European Commission 73) are the most important to achieve rapid development of the bioeconomy."

Results

"For Latvia, a substantial decrease in the Bioeconomy Sustainability Index value is explained by a significant decline in innovation and technology dimension indicators (development of environment-related technologies (% all technologies), relative advantage in environment-related technology, development of environment-related technologies (inventions per capita), biotechnology patent applications to the EPO per billion EUR total government budget allocations for R&D) as well in financing and governmental support dimension indicators (environmentally related government R&D budget (% total government R&D))."

"The Netherlands had the 2nd highest score in the composite index among the nine countries and ranked 3rd among the national bioeconomy strategy result assessment."

"While there is no direct correlation between the results of the two methods, the three countries that scored highest on the national strategy scores also scored highest on the Bioeconomy Sustainability Index among the nine countries."

"Despite Spain and Italy's lowest national bioeconomy strategy impact scores, both countries achieved remarkably high results in the Bioeconomy Sustainability Index."

Discussion

"Evaluation of national bioeconomy strategies with a mixed methods approach, combining the systematic literature review with TOPSIS, showed that the method can be used to assess whether and to what extent the priorities set in higher-level policy planning documents are implemented in hierarchically descended documents."

"While this study attempts to identify both the five objectives and the three action lines from the EU Bioeconomy Strategy in the national bioeconomy strategies, for a more in-depth exploration and assessment of possible gaps in policy coherence, it could be suggested narrowing the scope of the study and looking specifically at how the sub-activities of each action line are implemented in practice."

"Marcone and others compared the bioeconomy strategies of nine EU countries to evaluate synergies of national and urban circular bioeconomy transitions looking for indicator sets and found that only Italy and Germany provide bioeconomy assessment framework with a high overall quality rating (Marcone and others 76)."

Conclusions and Policy Implications

"A broadly described strategic direction and objectives alone do not ensure growth in the bioeconomy, as the comparison of the bioeconomy sustainability index with the results of national bioeconomy strategies shows: a well-defined national strategy does not guarantee high performance in this area."

"It is demonstrated by Denmark, which ranks highest in the Bioeconomy Sustainability Index but does not have a national bioeconomy strategy, and Austria, which has developed a nuanced strategy with different priorities, is only sixth in the Bioeconomy Sustainability Index."

"A comparison of the two methods shows that there is no direct link between the Bioeconomy Sustainability Index and the results obtained from the analysis of national bioeconomy strategies, most likely because the composite index analyses data from 2016–2018 and most of the national bioeconomy strategies analyzed were adopted in this period or later."

10.2.9 Energizing a Transformation to a Circular Bioeconomy: Mechanisms to Spread, Deepen and Broaden Initiatives [137]

This is a machine-generated summary of:

Schagen, O. M.; Metze, T. A. P.; de Olde, E. M.; Termeer, C. J. A. M.: Energizing a transformation to a circular bioeconomy: mechanisms to spread, deepen and broaden initiatives [137]

Published in: Sustainability Science (2022).

Link to original: https://doi.org/10.1007/s11625-022-01.249-1.

If you want to cite the papers, please refer to the original.

For technical reasons we could not place the page where the original quote is coming from.

Abstract-Summary

"These emerging circular initiatives, so-called small wins, continuously spread, deepen and broaden and as such contribute to transformative change."

"A small win spreads by becoming larger and more numerous, deepens by becoming more radical and circular, and broadens by connecting with other themes and domains."

"This paper explores how mechanisms drive these processes and, more in particular, how circular initiatives stay or become more transformative while developing."

"By building on existing literature about mechanisms that contribute to the development of initiatives—we extend the small wins framework by focusing on the interplay between mechanisms and spreading, deepening and broadening."

"We empirically explored how the interplay of mechanisms constitutes continuous transformative change."

"Both case studies indicate that the energizing and logic of attraction mechanisms are key in continuous transformative change."

Introduction

"These seeds of transformative change are also referred to as the small wins, initiatives showing concrete in-depth change (Termeer and Metze 77)."

"We approach transformative change as a continuous process through which emerging circular initiatives, so-called small wins, develop (Bours and others 78; Connor and Keil 79; Termeer and Metze 77; Termeer and Dewulf 80; Termeer and others 81)."

"In this paper, small wins are also characterized by in-depth change that alters current practices; positive contribution to a transformative ambition (in our case a circular bioeconomy); having overcome barriers and resistance; and connecting technical and societal change (Termeer and Metze 77; Termeer and Dewulf 80)."

"The concept of mechanisms offers an entry point for studying how initiatives continuously develop in a transformative way."

"We selected Kipster and Herenboeren for an in-depth study, as these initiatives both have a strong circular ambition, comply to the indicators of a small win and are rapidly developing in their own way."

Conceptual framework

"The literature on how mechanisms trigger the development of initiatives into systemwide transformative change is growing and has resulted in various lists of identified mechanisms (von Wirth and others 82; Gorissen and others 83; Lam and others 84; Loorbach and others 85; Termeer and Metze 77) (see Appendix 1)."

"The spreading, deepening, and broadening of small wins are not the final stages and are never completed, as learning involves continuous development by reinforcing mechanisms through positive feedback loops."

"This mechanism concerns the process of financial and human resources flowing toward winners, in this case circular initiatives considered successful (Termeer and Dewulf 80; Weick 86)."

"It also remains unclear how mechanisms constitute the spreading, deepening, and broadening of initiatives."

"The interplay of mechanisms leads to an initiative spreading, deepening, and broadening, potentially generating positive feedback through which mechanisms are strengthened."

"Mechanisms do not necessarily lead to transformative change of initiatives."

Transformative change in circular agriculture: the Kipster and Herenboeren cases

"We present the cases of the circular laying hen farm (Kipster) and community farming (Herenboeren) as small wins and we illustrate how the dynamics of deepening, broadening and spreading manifest in practice."

"The Kipster concept is an integrated farm with attention to key environmental, economic, and social challenges in the laying hen sector, such as circularity, feed-food competition, farm income and animal welfare."

"The Herenboeren approach is showing signs of embedding; the concept is presented as a successful initiative of circular agriculture, including on the government website that gathers examples of circular farming: platformkringlooplandbouw.nl."

"Aligning the wishes of the members, the possibilities of the local farm and the farmer (learning by doing) and sharing this knowledge on the Herenboeren Nederland level (professionalization), a new educational program was introduced."

Discussion

"The aim of this paper was to apply and further develop the small wins framework to understand how circular initiatives in society remain and become transformative through spreading, deepening and broadening."

"The mechanism of stabilizing provides the foundation for an initiative's further development in deepening, broadening or spreading."

"These cycles are sustained by the interactions of mechanisms that create deepening, broadening and spreading and the positive feedback of energizing and logic of attraction that derives from this."

"Our empirical findings indicate that energizing and logic of attraction function as positive feedback on other mechanisms through which the development of initiatives is sustained."

"We need to better understand the complexities at work in processes of transformative change and develop more adaptive 'buttons' for initiatives and all involved actors to foster deepening, broadening and spreading of initiatives."

Conclusion

"Alongside a variety of these small wins transformation is sustained through their further deepening, broadening and spreading."

"We applied and further developed the small wins framework in two cases, circular laying hen farm Kipster and community farming initiative Herenboeren, to illustrate how the interaction of mechanisms constitutes the spreading, deepening and broadening of these small wins."

"Our study shows that focusing on the dynamic of mechanisms visible within the development of small wins provides a better understanding on how these can maintain continuous transformative change."

"The development of small wins in a transformative way implies the continuous (re)activation of energizing and logic of attraction by spreading, deepening and broadening, and preserving these developments."

Appendix 2

"Origin of definitions of the eight mechanisms."

"The definitions for each mechanism were either literary taken from a reference (column 3), or combined and adjusted from multiple references (column 4)."

"A definition can relate to similar identified mechanism by other references (column 5)"

10.2.10 Bioeconomy and Circular Economy: Implications for Economic Evaluation in the Post-COVID Era [138]

This is a machine-generated summary of:

Viaggi, Davide; Zavalloni, Matteo: Bioeconomy and Circular Economy: Implications for Economic Evaluation in the Post-COVID Era [138]

Published in: Circular Economy and Sustainability (2021)

Link to original: https://doi.org/10.1007/s43615-021-00.113-1.

If you want to cite the papers, please refer to the original.

For technical reasons we could not place the page where the original quote is coming from.

Abstract-Summary

"The objective of this paper is to review selected insights about the current economic research on the Bioeconomy and circular economy, with a particular focus for the role of primary sector, and to derive implications for organisation, evaluation and valuation practice in the context of the post-COVID era."

"The paper argues that these issues will become more important in the post-COVID era, due to the plea for increasing Bioeconomy resilience."

"Evaluation, on the other hand, will need to be embedded even more in the decision-making processes, in spite of the increasing uncertainty and difficulties in evaluation."

Introduction and Objectives

"The objective of this paper is to review selected insights about the current economic research on the Bioeconomy and CE, with a particular focus on the role of the primary sector, and to derive implications for organisation, evaluation and valuation practice in the context of the post-COVID era through the proposal of a conceptual economic framework of the CBE."

"Based on the integration of these two sections, in the "Towards a Framework for Circular Bioeconomy Organisation and Evaluation in the Post-COVID Era" section, we provide a framework for the analysis of circular Bioeconomy, focusing in particular on market and organisation, evaluation and valuation practice in the context of the post-COVID era."

"The outcome is presented in the "Towards a Framework for Circular Bioeconomy Organisation and Evaluation in the Post-COVID Era" section, and it is the main result of this work."

Main Conceptual, Technological and Organisational Issues in the Bioeconomy

"While there are many aspects of the Bioeconomy that can be relevant for the objective of this paper, in this section, we focus on six interconnected issues, roughly organised in the terms of increasing scale: The relevance of innovation processes The increased technological separability of processes The importance of flexibility and cascading Consumer's acceptability and market segmentation The (partly consequent) increase in the complexity of the value chains The growing relevance of system-level organisational concepts The relevance of innovation for the Bioeconomy is paramount [87, 88]."

"The increased separability of processes emphasises a direction of change of technology that is already under way, allowing higher specialisation of production and also higher complexity of the value chains."

"The fact that many products of the Bioeconomy may bring process innovations with uncertain acceptability reflects in high instability of preferences and market, and an important role of information."

Bioeconomy and Circular Economy

"The Bioeconomy idea puts a stronger emphasis on the use of renewable resources as both input and energy source in the industrial production process."

"In this prospect, an integration with the Bioeconomy perspective would be useful in adding two dimensions to the problem of circularity: (1) the nature of the raw material (whether is renewable or not) and (2) the costly research and development activities needed to develop the technologies required."

"Moving from the theoretical analysis to the actions taken by actual firms, the need for the Bioeconomy to embed the notion of circularity at the core of the CE concept is highlighted by, e.g. Lokesh and others [89]."

"Business models have been reviewed by Salvador and others [90], who highlight several issues and hint at how to develop them for firms working in the circular-bioeconomy."

Towards a Framework for Circular Bioeconomy Organisation and Evaluation in the Post-COVID Era

"We point here to three main relevant issues: The need to consider circular processes in the appropriate way, including related economic aspects The need to cast circularity in the increased complexity and interconnection of downstream and upstream markets The need for new forms of interaction between demand and supply These topics have taken increasing importance over time for bio-based systems."

"The need to consider circular processes through a proper economic analysis relates to the issue of addressing appropriately the economic aspects of circularity in the valuation."

"That higher circularity has a higher marginal cost (higher marginal costs of recycling technology and lower marginal environmental cost due to decrease pressure on the environment), it is likely that there will be a non-trivial issue in understanding what level of circularity is optimal and how it evolves over time."

Discussion and Final Remarks

"Moving to CBE also requires attention to appropriate economic organisation concepts and evaluation procedures, allowing to address novel concepts such as flexibility, resilience and complex systems rather than mere advocacy and description of value chains."

"In this direction, it is also apparent how evaluation will become more important, but also more complex and difficult, due to the need to account for dynamic processes and a larger variety of interconnected markets."

"These issues are particularly relevant for the primary sector, because the increased flexibility and variety of processing approaches also reflect in higher substitutability among biomass sources, hence high competition and volatility of markets, as well as the need of providing clear anchoring between territory and chains based on socio-economic structures rather than physical constraints."

"This process is probably further promoted by the effects of the pandemics, calling for an increase in flexibility, resilience and preparedness to future (unknown) shocks."

10.2.11 The Bioeconomy and Its Untenable Growth Promises: Reality Checks from Research [139]

This is a machine-generated summary of:

Eversberg, Dennis; Holz, Jana; Pungas, Lilian: The bioeconomy and its untenable growth promises: reality checks from research [139]

Published in: Sustainability Science (2022)

Link to original: https://doi.org/10.1007/s11625-022-01.237-5.

If you want to cite the papers, please refer to the original.

For technical reasons we could not place the page where the original quote is coming from.

Abstract-Summary

"We argue that this stepwise process of moderation is partly due to a series of 'reality checks' to which various strands of research on the bioeconomy have (willingly or unwillingly) subjected these promises, forcing governments to move away from visions exposed as unrealistic and to adopt more humble ones."

"We argue that bioeconomy research should adopt a broader perspective that considers transitions toward bio-based processes and resources as but one element in a comprehensive social–ecological transformation of current modes of production and living, and that understanding the dynamics of societal conflict around that transformation is crucial for assessing the social possibility of bioeconomy visions."

Introduction

"Many national governments and the EU soon followed suit and adopted strategies promising growth from bioeconomy (European Commission 91; BMBF 92; see Fund and others 93)."

"Apart from socio-political contestation by NGOs and other critical actors, which has been relatively weak in a policy field that was long hardly known to the broader public (Lühmann 94; Riemann and others 12), what led to the newer strategy papers' more moderate rhetoric was a whole series of research-based 'reality checks' to the growth promises."

"Resulting from work in a broad variety of disciplines, focusing on widely different aspects of the bioeconomy, based on all kinds of conceptual and methodical approaches, and sometimes even contrary to researchers' own intentions, these reality checks have exposed the promised growth as either impossible or undesirable, or both."

Growth and bioeconomy policies in Europe: moderating expectations

"Ongoing criticism, particularly from the newly founded European Technology Platform 'TP Organics', a stakeholder forum of the organic sector officially recognized as an advisory body to EU research policy, significantly influenced the EU's 2012 bioeconomy strategy 'Innovating for Sustainable Growth: A Bioeconomy for Europe' (European Commission 91; Levidow and others 95)."

"The EU strategy's 2018 update 'A Sustainable Bioeconomy for Europe: strengthening the connection between economy, society and the environment' (European Commission 96) marks a further step in at least rhetorically moderating the promises.'"

"Instead of the growth promise having been abandoned altogether, what has occurred may be more accurately described as an inversion of the terms of the growth-sustainability connection: whereas earlier, growth had appeared as the central promise, made all the more attractive by the expected sustainability gains, the core promise now is that the bioeconomy can help make society sustainable, and bioeconomic growth is framed as a necessary element and condition for achieving that aim."

Reality checks: what they are and how we identified them

"Since their inception, the promissory visions of bioeconomy concepts and strategies have been challenged not only by political critics, but also by research findings from a range of fields of inquiry, including research funded by the same governments that pinned high hopes on the bioeconomy's growth potentials."

"To discuss the reality checks on the bioeconomy's growth promises in a systematic way, we have ordered existing research on the relation between bioeconomy and growth into several categories, each of which we found to have contributed its own type of reality check."

"As a selection criterion for relevant literature, we sought to identify research that in some way addresses the relation between bioeconomy and economic growth, and that explicitly addresses 'bioeconomy' as a comprehensive economic or even societal 'model', rather than merely discussing specific processes or solutions without regard for their possible function within a broader context."

Challenging the growth promises: four core reality checks from bioeconomy research

"Analyses of bioeconomy discourses or policy frameworks make this point by highlighting differences and similarities in rhetoric or priorities of political bioeconomy strategies, with regard to the definitions of bioeconomy deployed, the implied relations between sustainability and growth, etc. To Backhouse and others (97), most bioeconomy strategies from the Global North and South rely on a shared belief in technology and 'innovation' as drivers of growth and means to overcome both social and environmental crises."

"The recurring story in accounts of bioeconomy discourses and knowledges is one of a hierarchical tension between dominant conceptions promising to reconcile economic growth and (weak) ecological sustainability through technological innovation, and dominated, marginalized socio-ecological ideas seeking (strong) sustainability based on principles of agro-ecology and sufficiency."

Conclusion: from empty promise to societal conflict

"We have shown here how research on the bioeconomy's social and ecological preconditions and implications has challenged the claims to the possibility and desirability of such growth by confronting them with irritating or irreconcilable realities on several levels."

"It has highlighted the contested nature of promissory visions in bioeconomy discourses and knowledge production, illuminated the interests, strategies and power relations at play in advancing the promissory policy narrative, exposed expectations of jobs growth and broad-based wealth creation through bioeconomy as hardly realistic in the face of its economic reality, and shown that the provision of biological resources cannot expand at a scale even remotely sufficient to allow for overall economic growth."

"This wider context of societal conflicts and struggles deserves greater attention within bioeconomy research, as it will eventually be crucial for what kinds of bio-based economic practices find democratic support and prove socially (or in Critical Realist terms, 'empirically') possible."

10.2.12 Situating a Sustainable Bioeconomy Strategy on a Map of Justice: A Solution and Its Problems [140]

This is a machine-generated summary of:

Häyry, Matti; Laihonen, Maarit: Situating a sustainable bioeconomy strategy on a map of justice: a solution and its problems [140]

Published in: Environment, Development and Sustainability (2022).

Link to original: https://doi.org/10.1007/s10668-022-02.720-w.

If you want to cite the papers, please refer to the original.

For technical reasons we could not place the page where the original quote is coming from.

Abstract-Summary

"The European Union's 2018 updated bioeconomy strategy A Sustainable Bioeconomy for Europe: Strengthening the connection between economy, society and the environment aims to fulfill the requirements of sustainability and justice while transitioning economy from fossil-based to bio-based."

"A map of justice that shows the tensions and alliances between key interpretations of equality, identify the definition of sustainability informing the strategy, and show how this definition steers the strategy toward a limited view of justice that is geared to ignore social and ecological concerns."

"The governance framework that the strategy adheres to is the 2015 United Nations Sustainable Development Goals approach, which is an outgrowth of the 1987 Brundtland Commission's view on sustainability."

Introduction: justice and a bioeconomy strategy based on sustainable development goals

"Many policies are open to the accusation that they ignore, in practice, the requirements of social or environmental justice."

"Although fairness, inclusiveness and responsiveness have been identified as crucial factors in successful environmental governance, significant dimensions of justice have been practically ignored in many central policy frameworks and in analyses of them."

"Our research question is: To what degree, how, and why do social and environmental dimensions of justice remain underrepresented despite the proclaimed realization that justice is paramount in international agreements and policies?"

"We then scrutinized the European Union's (EU) 2018 bioeconomy strategy (European Commission, 98) as a representative example of a sustainability-oriented international policy; noted its proclaimed aims (environmental, social, and economic); unearthed its actual agenda (technology and business innovation); and identified potential discrepancies between these sets of goals as well as their connections to the prevailing sustainability ideas."

The map of justice

"Positionalists favor spontaneity in the formation of social practices, immaterial (spiritual, esthetic) values, collectivity, tradition, special relationships assigning people their dues, a holistic take on understanding, weight on emotional responses, intersectionality in the acknowledgement of moral agents and patients, and identity as a basis for recognition."

"Universalists champion calculation in the design of laws and social policies, material (tangible, measurable) values, individualism, reform superseding tradition, impartiality in the assignment of duties and entitlements, an analytic approach to understanding, weight on intellectual considerations, cosmopolitanism in the endowment of moral status, and individuality instead of group memberships as the basis of recognition (Stevenson and others, 99; Hänninen & Aaltola, 100)."

"The other prefers impartially assessed good, scorns tradition for its own sake, and advocates rational reforms aimed at the wellbeing of as many sentient beings as possible regardless of their nationality, ethnicity, species, or other differentiating but morally irrelevant features (e.g., Takala & Häyry, 101)."

"Care and relations ethics and the capability approach both champion equal opportunities, but assign them on dissimilar grounds."

The European bioeconomy strategy

"Our focus is on the Strategy's updated version A Sustainable Bioeconomy for Europe: Strengthening the Connection between Economy, Society and the Environment (European Commission, 98), and we confine our scrutiny to an explicit interpretation of the document."

"Global challenges like climate change, land and ecosystem degradation, coupled with a growing population force us to seek new ways of producing and consuming that respect the ecological boundaries of our planet."

"We start in the first column with limited resources, climate change, land and ecosystem degradation, growing population, the unsustainability of current production and consumption, and concern about the prosperity of EU citizens."

"The overall solutions offered by the strategy emphasize changes and growth in production and consumption, the modernization of industries, boosting EU businesses in the global market, and innovating ecosystem failure away by consuming and producing more but better."

The meaning of sustainability in the European bioeconomy strategy

"The Brundtland Commission's task was to show how, by using a combination of economic freedom, prompts for structural changes, and international regulation, the wealthy in the First World can keep living like they do while people in the Third World can improve their situation without causing permanent environmental damage."

"The Brundtland commission ignored the conceptual precaution of their predecessors and went, in their own Report, for a combined account, in which economic and social development are amalgamated with environmental targets."

"While the Brundtland Report emphasized the development and equity dimensions of sustainability, keeping an eye on environmental concerns, the Rio Summit shifted the focus more decidedly on the environment (Drexhage & Murphy, 102)."

"Another point of contention has been the business-as-usual approach that had, according to some observers, by 2015 become the UN hallmark in matters concerning climate change and environmental and social matters more generally (Norren, 103)."

Justice in the European bioeconomy strategy based on SDGs

"The 17 SDGs are: (1) end poverty, (2) end hunger, (3) ensure health and wellbeing for all, (4) ensure quality education for all, (5) achieve gender equality and empower all women and girls, (6) ensure water and sanitation for all, (7) ensure affordable and clean energy for all, (8) promote economic growth and decent work for all, (9) build infrastructure, promote industrialization, and foster innovation, (10) reduce inequality, (11) make human settlements inclusive and safe, (12) ensure sustainable production and consumption, (13) take action to combat climate change, (14) conserve oceans and seas, (15) protect terrestrial ecosystems and protect biodiversity, (16) promote peace and justice, and (17) strike global partnerships (United Nations, 104, p. 14)."

"This distinction is not immediately visible, if we concentrate on the desired outcomes, because both utilitarian and care-and-relations approaches can, in their own ways, give weight to nature, nonhuman animals, and the environment."

Conclusions and a way back

"Sustainability and justice are important principles in global environmental governance and its assessment."

"We situated the professedly sustainable European Union bioeconomy strategy on a map of justice as interpretations of equality."

"We concluded that the Strategy observes the sustainable development ideology that was formulated in the Brundtland Report in 1987 and given its current form in the SDGs agreed upon in Paris and New York in 2015."

"This is also stressed by the compromise views on justice by Rawls and Habermas, flagged up in our account of interpretations of equality."

"Justice does not properly prevail unless their call for ecologico-social equality is heeded."

"The European bioeconomy strategy is an example of environment-related international agreements that do not heed the call."

10.2.13 Envisioning just Transformations in and Beyond the EU Bioeconomy: Inspirations from Decolonial Environmental Justice and Degrowth [141]

This is a machine-generated summary of:

Ramcilovic-Suominen, Sabaheta: Envisioning just transformations in and beyond the EU bioeconomy: inspirations from decolonial environmental justice and degrowth [141]

Published in: Sustainability Science (2022)

Link to original: https://doi.org/10.1007/s11625-022-01.091-5.

If you want to cite the papers, please refer to the original.

For technical reasons we could not place the page where the original quote is coming from.

Abstract-Summary

"This paper applies theoretical ideas from decolonial environmental justice and degrowth, placed in the wider context of transformations, to analyse the EU bioeconomy policy within the global context, and to draw lessons and recommendations for just transformations in the EU bioeconomy policy."

"I identify five dominant logics and approaches in the EU bioeconomy that act as barriers for just transformations and propose alternative ones that can support such transformations."

"Barriers and alternatives include (1) framing 'nature' as a resource and service provider for humans, who are seen as separate from nature, and the need to abandon human–nature duality; (2) dominance of economic green growth and technoscientific policy solutions, and the need to place planetary justice at the centre of tackling socioecological crises; (3) a limited approach to justice, and the need to act upon climate and epistemic justice, including self-determination and self-governing authority; (4) the EU's ambition for global leadership and competitiveness in global bioeconomic markets and governance, and the need to redefine global governance towards partnerships based on the principles of solidarity, mutual respect, reconciliation and redistribution of power and wealth; (5) hegemonic politico-economic structures and actor coalitions in charge of the EU bioeconomy, and the need for decentralized bottom-up leadership coalitions that promote direct democracy, local autonomy and sovereignty beyond state."

Socioecological emergencies, global inequalities, the pandemic, decarbonization, bioeconomy, degrowth, environmental justice: what common threads?

"These socioecological and existential crises, including climate change, are increasingly recognized as a global injustice problem due to (neo)colonial and imperial relations of extraction of labour and resources in the GDP-poor countries (many of which are in the 'Global South') to foster economic growth in the GDP-rich countries ('Global North') (Hickel 105, 106; Hickel and others 107; Martinez-Alier 108)."

"To build more ecologically sound and socially just bio-based societies and economies, the EU needs to look beyond affirmative reformist approaches that rely on the existing political and economic institutions, rules, roles, knowledge and relations, which have contributed to socioecological degradation and inequalities (Büscher and others 109; Gills and Morgan 110; Martin and others 111; Nightingale and others 112; Temper and others 113), and seek to embrace a decolonial and pluralist approach to justice, environment and development more broadly."

"To emphasize the importance of global socioecological justice implications in the context of the EU bioeconomy policy, I review some of the more recent empirical findings that emphasize the socioecological costs and creation of green sacrifice zones associated with transition to green and renewable resources and energy systems."

From a policy of green growth to a practice of green sacrifice zones

"The dominant EU bioeconomy visions and discourses are closely associated with industry, private sector, intensified biomass harvest, technological innovations, and production of novel bio-based materials."

"The bioeconomy policy as a whole has been framed as technology-driven and biomass-based green growth bioeconomy (Bugge and others 48; Levidow and others 114), elite-driven neoliberalization of nature, which remains silent on the politics of bioeconomy (Birch and others 115; Levidow and others 114)."

"While the changes in the revised EU bioeconomy strategy are a step in the right direction, the revised strategy remains problematic in two major ways: first, the framing of challenges vis-à-vis solutions; second, the ignorance of justice and inequalities."

"The proposed actions concerning all aspects related to growth and the EU's competitiveness (e.g. bio-based industry, innovation, green jobs) are elaborated more concretely and include non-voluntary policy packages."

"Concerning justice and equality, the revised strategy is devoid of concerns and issues related to climate justice, carbon inequality, and different carbon footprints and resource use between different sectors of society in the EU, let alone EU ecological footprint and global extraction."

Decoloniality, decolonial environmental justice and degrowth as tenants of transformations

"Unlike decoloniality of mind perspectives, indigenous perspectives frame decoloniality in more material and concrete terms and aims, which include ending structural systemic oppression and violence of colonialism, as well as repatriation or restitution of indigenous land (Tuck and Young 116)."

"An important scholarly debate on decoloniality concerns the questions of what is to be decolonized (i.e. the object of decolonialisation) and whether decoloniality should be used as a means to address a range of socio-political issues (i.e. decoloniality as a means), or an end in itself (i.e. decoloniality as an end) have emerged among decolonial, postcolonial and degrowth scholars (see for example D'Alisa and others 117; Deschner and Hust 118; Feola 119; Garcia-Arias and Schöneberg 120; Tuck and Yang 116)."

"Some scholars argue for a use of decoloniality strictly in relation to undoing colonial histories and injustices (decoloniality as an end in itself), while others approach it in the context of, and for addressing a range of socio-political issues, including justice and education for example (decoloniality as a means)."

Drawing on DEJ and degrowth to identify barriers to and alternatives for socioecological justice and transformations

"To avoid reproduction of green sacrifice zones (Zografos and Robbins 121), EU policies with relations to and implications for the Global South need to change course radically, and here are just a few suggestions: (1) reinvent EU relations with and role in the world, with the humility of a former coloniser, where responsibilities for historical/colonial and modern/neo-colonial debt, burdens and injustices are acknowledged and acted upon; (2) reform EU policy and institutional framework with an aim to fostering reconciliation, reciprocity, respect, care and solidarity with the Global South and marginalized groups across the peripheries, as opposed to exploitation and violence; (3) abandon the 'imperial mode of living' by tackling non-proportional and wasteful consumption and extraction and thus alleviate pressure on lands, peoples, and 'natures' at home and elsewhere; (4) implement reforms for redistribution of wealth and power in a way that addresses current inequalities and repairs historic, as well as contemporary injustices and violence associated with the EU corporate extraction and dispossession; (5) implement policy reforms that allow the EU to operate in full recognition of domestic and, where relevant, local indigenous and customary territorial rules and rights, including cultural and political self-determination and self-governing authority."

Transformations, co-optation, and politics of change: EU bioeconomy and beyond

"The EU bioeconomy policy primarily aims to advance economic sustainability, green growth, and EU global competitiveness, by substituting non-renewables for renewables, modernizing production and consumption, boosting technology, research and innovation, and promoting circular and cascade use of materials and products."

"Rather than proposing far-fetched policy recommendations for more transformational and more just politics in the EU bioeconomy policy domain, in what follows I reflect upon the process of change and risks of co-optation, with a hope to inspire decolonial and just socioecological transformations in and beyond bioeconomy."

"The concepts and processes of decoloniality, degrowth and transformations are crucial for the radical changes, as argued above."

"While co-optation from the outside is the most dominant across all three concepts and political agendas, it currently appears to be the most extreme in the case of transformations, where the term is used to describe and therefore legitimize sociotechnical and eco-modernist solutions as 'transformative' (Kruus and Hakala 122; Tome Valencia Hamilton and Ramcilovic-Suominen forthcoming; UNECE 123)."

10.2.14 How to Implement the Urban Bioeconomy: Insights from Leipzig's Current Policy Mix for Transitioning Towards a Bio-principled City [142]

This is a machine-generated summary of:

Boldt, Christin; Thrän, Daniela: How to implement the urban bioeconomy: insights from Leipzig's current policy mix for transitioning towards a bio-principled city [142]

Published in: Urban Transformations (2023)

Link to original: https://doi.org/10.1186/s42854-023-00.052-0.

If you want to cite the papers, please refer to the original.

For technical reasons we could not place the page where the original quote is coming from.

Abstract-Summary

"Integrating biological principles, such as metabolism, ecosystems and cycles, into urban planning and city life, and promoting a so-called "bio-principled city" provides significant potential for shaping more sustainable urban development."

"This study therefore draws attention to the existing policy mix of the city of Leipzig and evaluates in how far it already supports the city's transformation into a bio-principled city."

"The analysis shows that the current policy framework for a bio-principled city in Leipzig is complex, uncoordinated and lacks in definitive actions."

"By highlighting its strengths and corresponding limitations, we provide a starting point for deliberate interventions to further promote the vision of a bio-principled city politically and thus realize its implementation."

Policy and practice recommendations

"The federal state and local policy levels need to be more explicitly involved in shaping bioeconomy policy."

"Further assessment of the policy mix may provide guidance on how to adapt it and advance the transformation process."

"Narrowing down the vision to local strengths may help to identify targeted policy measures."

Introduction

"Integrating biological principles, such as metabolism, ecosystems and cycles, into urban planning and city life, and promoting a so-called "bio-principled city" was rated of very high relevance by participants in order to shape more sustainable urban development (Cuhls and others 124)."

“This leads to the following research questions, which we want to explore in more detail: To what extent are the urban bioeconomy and the idea of a bio-principled city considered in current policymaking?"

“This applies to the transformation towards a sustainable bioeconomy in general and a bio-principled city in particular.”

“By identifying and analysing the current policy mix supporting the development of a bio-principled city in Leipzig our findings shed light on the strengths and corresponding limitations of the current policy framework that arise for urban policy and practice to implement the vision of a bio-principled city.”

Theoretical anchoring

“To develop economically, socially, and environmentally compatible and resilient living environments for current and future urban residents, concepts for sustainable urban development have emerged in the past that aim to bring back nature into cities by integrating natural principles and materials into urban planning.”

“The concept has evolved since then and today describes more an economic system within cities and their rural surroundings in which bio-based processes, materials and services are used to promote sustainable urban development (Yang and Yang 125).”

“The Biocity adopts the abilities of forest ecosystems (i.e. reduces emissions of carbon dioxide and other greenhouse gases), is self-sufficient, reinforces the functions of existing city infrastructures, uses biodiversity for ecosystem services and promotes health living, makes use and develops local biobased recycled materials, promotes low mobility, harnesses regional bio-based value chains, provides publicly accessible nature, and is socially inclusive (Wilkes-Allemann and others 126).”

Methods

“By operationalising these visions, defining vision elements and defining lead questions guiding the analysis we aimed at elaborating in how far the current policy mix addresses the vision elements of a bio-principled city.”

“Our research revealed that policy documents differ in whether they address measures to promote vision elements of a bio-principled city, which is why we concentrated only on policy strategies that explicitly address such measures.”

“The resulting dataset provided the basis for a qualitative content analysis aiming at elaborating the extent to which the vision elements of a bio-principled city are currently promoted through proposed and implemented policy instruments.”

“This process resulted in a deductive-inductive coding system, providing information about which vision elements of a bio-principled city are currently addressed in policy documents and by which policy instruments they are promoted.”

Results

"Our desk research resulted in a dataset of 17 documents, including 11 policy strategies, four plans and programmes and two other records from 2018 to 2020 that explicitly address the bioeconomy and core vision elements of a bio-principled city at different policy levels."

"Market-based instruments promoting commercialisation and transformative governance are only infrequently mentioned in the policy documents analysed and are limited to technical assistance for project implementation and advisory services provided to municipal authorities, such as the European Investment Bank (EIB) Advisory Hub services and its Urban investment Advisory Platform (Urbis))."

"More specific bioeconomy-related vision elements, such as the integration of biological principles into urban planning and city life, and innovative sustainable building designs and construction techniques referring to biological principles and renewable resources were only rarely addressed and focused mostly on the use of wood and new composites in construction."

Discussion and conclusion

"The results of the study show that the current policy mix for a bio-principled city in Leipzig is complex, uncoordinated and lacks in definitive actions."

"Our case study has shown that the federal state and local policy levels would need to be even more explicitly involved in shaping bioeconomy policy in order to promote the transformation to a bio-principled city in Leipzig."

"It should be further interested as to whether these types of policy coordination adequately enable the implementation of cross-cutting policies and accelerate the transition to a bio-principled city, or whether new organizational structures and processes are needed to overcome issues such as departmental silos."

"We conclude that Leipzig's current extended policy mix, while not necessarily preventing a transition to a bio-principled city, does not explicitly promote it either."

"We see further potential for investigating how the current policy mix can be improved and how Leipzig's transition towards a bio-principled city can be advanced."

10.2.15 The Making of Sustainability: Ideological Strategies, the Materiality of Nature, and Biomass Use in the Bioeconomy [143]

This is a machine-generated summary of:

Boyer, Miriam; Kusche, Franziska; Hackfort, Sarah; Prause, Louisa; Engelbrecht-Bock, Friederike: The making of sustainability: ideological strategies, the materiality of nature, and biomass use in the bioeconomy [143]

Published in: Sustainability Science (2022)

Link to original: https://doi.org/10.1007/s11625-022-01.254-4.

If you want to cite the papers, please refer to the original.

For technical reasons we could not place the page where the original quote is coming from.

Abstract-Summary

"The bioeconomy, a recent addition to the political project of ecological modernization, is largely premised on the widespread use of biomass."

"Given this contradiction, we ask how the key institutions of the innovation system (government, science, and industry), construct and uphold the image of sustainability of biomass use in the bioeconomy."

"Through an analysis based on ideology critique, we look at the broad field of biomass policy in Germany, including official bioeconomy strategies and biomass potential calculations, expert portrayals of biomass use in the bioeconomy-themed Year of Science, and an iconic biomass-based commodity."

"We identify four central ideological strategies that uphold the image of sustainability and contribute to creating political consent for the political project of the German bioeconomy: seeking managerial solutions, relying on technological innovation, relegating solutions into the future, and obscuring the materiality of nature."

Introduction

"Despite the uncertainties and contradictions regarding the conditions that would make large-scale biomass use sustainable, the dominant policy discourse continues to create an image that portrays the large-scale substitution of fossil and mineral resources by biomass-based materials as a feasible goal that will facilitate the transition to a sustainable society: "With the expansion of the bioeconomy, the resource base of the economy is aligned towards sustainability and fossil raw materials are replaced" (BMBF and BMEL 127, p. 10)."

"Within the bioeconomy project, widespread criticism regarding the possibility of sustainably increasing the use of biomass is at odds with policy strategies and papers maintaining that biomass use is sustainable."

"We argue that proponents of the bioeconomy use distinct ideological strategies to uphold the image of an increase in biomass use as a sustainable option for future economies and thereby contribute to creating political consent for the bioeconomy."

"We ask how policy discourses in Germany sustain the image of sustainable biomass use, and how this contributes to creating political consent for the bioeconomy as part of the wider political project of ecological modernization."

Analytical framework and methods

"Despite the fact that biomass policy is largely structured around formal knowledge frameworks based on calculations, expert knowledge, and research results, we also expected to find common sense understandings across all documents we analyzed, since scientific, scholarly and other structured frameworks for understanding the world are not free of such suppositions (Gramsci 128, p. 1375)."

"For ideological strategies to 'work'—in our context, to create an image of sustainable biomass use for the bioeconomy—we assume that both formally structured and common sense understandings are reworked and contextualized, so as to appear plausible and appeal to particular social groups to create consent, i.e., active, "conscious and clear adoption" (Gramsci 129, p. 1671) of a collective will to move in a given political direction."

"This analysis revealed that many of the inductively derived codes referred to common sense notions as well as formally structured knowledge systems, which suggests that the two are strongly intertwined, together creating and stabilizing the ideological strategies that uphold the image of sustainable biomass use in the bioeconomy."

Results: ideological strategies regarding the sustainability of biomass use in the German bioeconomy

"Our analysis identified four principal ideological strategies employed to create and uphold the image of sustainable biomass use, namely: seeking managerial solutions, relying on technical innovation, relegating solutions to the future, and obscuring the materiality of nature."

"The emphasis is on formally structured knowledge frameworks such as quantitative calculations of biomass availability (e.g., residues of selected products of agriculture, forestry, and fisheries); economic evaluations and definitions of economic indicators; and systematic monitoring of the bioeconomy and its effects on land and resource use, demand for water, and CO_2 emissions."

"To what is suggested by policy statements such as, "with the expansion of the bioeconomy, the resource base of the economy will be sustainably oriented, and fossil raw materials replaced" (A1.6: 10), monitoring studies do not consider the availability of material that would be needed to transform all or most of the existing economy based on fossil raw materials to a biomass-based economy."

Discussion: ideological strategies, ecological modernization, and the materiality of nature

"Our results suggest that the four ideological strategies we identify uphold an image of sustainability in the discourse on biomass in the German bioeconomy."

"To the extent that the biomass-based bioeconomy and various kinds of other 'green' economies are seen as a plausible economic growth strategy, the materiality of nature is not only the object of an ideological strategy to create an image of sustainability, but also an increasingly large elephant in the room that may prove more and more difficult to obfuscate."

"Given the centrality of abstracting from the materiality of living nature as an ideological strategy for upholding the bioeconomy, we suggest that developing good sense notions will entail questioning current representations of nature and developing alterative understandings of nature—through, for example, practices such as farming or gardening, but also as a result of experiencing the everyday effects of the ecological crisis."

Conclusion

"We identify four ideological strategies, namely seeking managerial solutions, relying on technological innovation, relegating solutions into the future, and obscuring the materiality of nature, and describe their role in the creation of political consent for the political project of the bioeconomy as part of a larger project of ecological modernization."

"We show how all strategies contribute to creating consent for the larger political project of ecological modernization."

"A particular contribution of this paper to the literature on the bioeconomy and ecological modernization is the finding that 'obscuring the materiality of living nature' is a key component of these political projects."

"This is a strategy that merits more explicit analysis, since it not only suggests why current ideological strategies may cease to be effective in creating political consent, but also provides a starting point for countering these ideologies on the basis of new practical relationships with the materiality of living nature."

Bibliography

1. Stern T, Ploll U, Spies R, Schwarzbauer P, Hesser F, Ranacher L (2018) Understanding perceptions of the bioeconomy in Austria—an explorative case study. Sustain 10:4142. https://doi.org/10.3390/SU10114142
2. Macht J, Klink-Lehmann JL, Simons J (2022) German citizens' perception of the transition towards a sustainable bioeconomy: a glimpse into the Rheinische Revier. Sustain Prod Consum 31:175–189. https://doi.org/10.1016/J.SPC.2022.02.010
3. OECD (2009) The bioeconomy to 2030 -designing a policy agenda
4. Holmgren S, D'Amato D, Giurca A (2020) Bioeconomy imaginaries: a review of forest-related social science literature. Ambio 49:1860–1877. https://doi.org/10.1007/s13280-020-01398-6
5. European Commission (2018) A sustainable bioeconomy for Europe: strengthening the connection between economy, society and the environment. Luxembourg
6. Patermann C, Aguilar A (2018) The origins of the bioeconomy in the European Union. N Biotechnol 40:20–24. https://doi.org/10.1016/J.NBT.2017.04.002
7. Muscat A, de Olde EM, Ripoll-Bosch R, Van Zanten HHE, Metze TAP, Termeer CJAM, van Ittersum MK, de Boer IJM (2021) Principles, drivers and opportunities of a circular bioeconomy. Nat Food 2:561–566. https://doi.org/10.1038/s43016-021-00340-7
8. Salvador R, Barros MV, Donner M, Brito P, Halog A, De Francisco AC (2022) How to advance regional circular bioeconomy systems? Identifying barriers, challenges, drivers, and opportunities. Sustain Prod Consum 32:248–269. https://doi.org/10.1016/J.SPC.2022.04.025
9. Vivien FD, Nieddu M, Befort N, Debref R, Giampietro M (2019) The hijacking of the bioeconomy. Ecol Econ 159:189–197. https://doi.org/10.1016/J.ECOLECON.2019.01.027
10. Bryden J, Gezelius SS, Refsgaard K, Sutz J (2017) Innovation and development inclusive innovation in the bioeconomy: concepts and directions for research. Innov Dev. https://doi.org/10.1080/2157930X.2017.1281209
11. Giampietro M (2019) On the circular bioeconomy and decoupling: implications for sustainable growth. Ecol Econ 162:143–156. https://doi.org/10.1016/j.ecolecon.2019.05.001
12. Riemann L, Giurca A, Kleinschmit D (2022) Contesting the framing of bioeconomy policy in Germany : the NGO perspective. J Environ Policy Plan 1–17.https://doi.org/10.1080/1523908X.2022.2071689
13. Dupont-Inglis J, Borg A (2018) Destination bioeconomy – the path towards a smarter, more sustainable future. N Biotechnol 40:140–143. https://doi.org/10.1016/J.NBT.2017.05.010
14. Mont O (2004) Institutionalisation of sustainable consumption patterns based on shared use. Ecol Econ 50:135–153. https://doi.org/10.1016/J.ECOLECON.2004.03.030
15. Pyka A, Cardellini G, van Meijl H, Verkerk PJ (2022) Modelling the bioeconomy: emerging approaches to address policy needs. J Clean Prod 330:129801. https://doi.org/10.1016/J.JCLEPRO.2021.129801
16. Middlemiss L (2018) Sustainable consumption: key issues, 1st edn. Routledge. https://doi.org/10.4324/9781315628035

17. Allain S, Ruault JF, Moraine M, Madelrieux S (2022) The 'bioeconomics vs bioeconomy' debate: beyond criticism, Advancing Research Fronts. Environ Innov Soc Transitions 42:58–73. https://doi.org/10.1016/j.eist.2021.11.004
18. Agin S, Karlsson M (2021) Mapping the field of climate change communication 1993–2018: geographically biased, theoretically narrow, and methodologically limited 15:431–446. https://doi.org/10.1080/17524032.2021.1902363
19. Ballantyne AG (2016) Climate change communication: what can we learn from communication theory? Wiley Interdiscip Rev Clim Chang 7:329–344. https://doi.org/10.1002/WCC.392
20. Koteyko N, Nerlich B, Hellsten I (2015) Climate change communication and the internet: challenges and opportunities for research. Environ Commun 9:149–152. https://doi.org/10.1080/17524032.2015.1029297
21. Lamb WF, Steinberger JK (2017) Human well-being and climate change mitigation. Wiley Interdiscip Rev Clim Chang 8:e485. https://doi.org/10.1002/WCC.485
22. Stegmann P, Londo M, Junginger M (2020) The circular bioeconomy: its elements and role in European bioeconomy clusters. Resour Conserv Recycl X 6:100029. https://doi.org/10.1016/J.RCRX.2019.100029
23. Pfau SF, Hagens JE, Dankbaar B, Smits AJM (2014) Visions of sustainability in bioeconomy research. Sustainability 6:1222–1249. https://doi.org/10.3390/su6031222
24. Tan ECD, Lamers P (2021) Circular bioeconomy concepts—a perspective. Front Sustain 2:701509. https://doi.org/10.3389/frsus.2021.701509
25. Thompson, E.P. 1971. The moral economy of the English crowd in the eighteenth century. Past & Present 50 (1): 76–136.
26. Asdal, K. 2014a. From climate issue to oil issue: Offices of public administration, version of economics and the ordinary technologies of politics. Environment and Planning A 46 (9): 2110–2124.
27. EU. 2012a. Innovating for Sustainable Growth: A Bioeconomy for Europe. Directorate-General for Research and Innovation, European Commission, https://publications.europa.eu/en/publication-detail/-/publication/1f0d8515-8dc0-4435-ba53-9570e47dbd51 . Accessed 15 October 2020.
28. NFD. 2016. Kjente ressurser - uante muligheter. Regjeringens bioøkonomistrategi [Known resources, untold opportunities. The government's bioeconomy strategy]. Ministry of Trade, Industry and Fisheries, https://www.regjeringen.no/contentassets/32160cf211df4d3c8f3ab794f885d5be/nfd_biookonomi_strategi_uu.pdf . Accessed 15 October 2020.
29. EU. 2012b. Blue Growth: Opportunities for Marine and Maritime Sustainable Growth. Communication from the Commission to the European Parliament, the Council, the European Economic and Social Committee and the Committee of the Regions, COM(2012) 494 final, https://ec.europa.eu/maritimeaffairs/sites/maritimeaffairs/files/docs/publications/blue-growth_en.pdf . Accessed 15 October 2020.
30. SINTEF. 2012. Value Created from Productive Oceans in 2050. A report prepared by a working group appointed by the Royal Norwegian Society of Sciences and Letters (DKNVS) and the Norwegian Academy of Technological Sciences (NTVA), https://www.sintef.no/contentassets/f025260af6b8435394eced5e03939e11/value-created-from-productive-oceans-in-2050.pdf . Accessed 15 October 2020.
31. PNP. 1973b. Proceedings of the Norwegian Parliament. Stortingsforhandlinger 1972–73, Innst. O. nr. 57.
32. Ministry of Finance. 1974. Meld.st. nr. 25 (1973–74). Petroleumsvirksomhetens plass i det norske samfunn. [The petroleum industry's place in Norwegian society], https://www.regjeringen.no/contentassets/695b71fb157043998a9643f72b3ed843/stm197319740025000dddpdfs.pdf . Accessed 04 September 2020.
33. Bambo, T. L., & Pouris, A. (2020). Bibliometric analysis of bioeconomy research in South Africa. Scientometrics, 125(1), 29–51.
34. Louvel, R., & de Gromard, C. (2017). From biomass to bioeconomics, an energy strategy for Africa? Afrique Contemporaine, 1, 223–240.

35. Virgin, I., & Morris, E. J. (2016). Creating sustainable bioeconomies: The bioscience revolution in Europe and Africa. Taylor & Francis..
36. Bugge, M. M., Hansen, T., & Klitkou, A. (2016). What is the bioeconomy? A review of the literature. Sustainability, 8(7), 691.
37. European Commission. (2018a). A sustainable bioeconomy for Europe: Strengthening the connection between economy, society and the environment, update bioeconomy strategy. European Commission..
38. Gagolewski, M. (2011). Bibliometric impact assessment with R and the CITAN package. Journal of Informetrics, 5(4), 678–692.
39. Moral-Muñoz, J. A., Herrera-Viedma, E., Santisteban-Espejo, A., & Cobo, M. J. (2020). Software tools for conducting bibliometric analysis in science: An up-to-date review. El Profesional De La Información, 29(1), 4.
40. Saikia, K., Vallès, M., Fabregat, A., Saez, R., & Boer, D. (2020). A bibliometric analysis of trends in solar cooling technology. Solar Energy, 199, 100–114.
41. Aria, M., & Cuccurullo, C. (2017). Bibliometrix: An R-tool for comprehensive science mapping analysis. Journal of Informetrics, 11(4), 959–975.
42. McCormick, K., & Kautto, N. (2013). The bioeconomy in Europe: An overview. Sustainability, 5(6), 2589–2608.
43. Ikeda, K. (1979). Consumption and food utilization by individual larvae and the population of a wood borer Phymatodes maaki Kraatz (Coleoptera: Cerambycidae). Oecologia, 40(3), 287–298.
44. Cherubini, F. 2010. The biorefinery concept: Using biomass instead of oil for producing energy and chemicals. Energy Conversion and Management 51: 1412–1421. https://doi.org/10.1016/j.enconman.2010.01.015 .
45. Muscat, A., E.M. De Olde, I.J.M. De Boer, and R. Ripoll-Bosch. 2020. The battle for biomass: A systematic review of food-feed-fuel competition. Global Food Security 25: 1–11. https://doi.org/10.1016/j.gfs.2019.100330 .
46. D'Alessandro S, Cieplinski A, Distefano T, Dittmer K (2020) Feasible alternatives to green growth. Nat Sustain. https://doi.org/10.1038/s41893-020-0484-y
47. Priefer, C., J. Jörissen, and O. Frör. 2017. Pathways to shape the bioeconomy. Resources 6: 1–23. https://doi.org/10.3390/resources6010010 .
48. Bugge M, Hansen T, Klitkou A (2016) What is the bioeconomy? A review of the literature. Sustainability 8:691. https://doi.org/10.3390/su8070691
49. Kirchherr J, Reike D, Hekkert M (2017) Conceptualizing the circular economy: an analysis of 114 definitions. Resour Conserv Recycl 127:221–232. https://doi.org/10.1016/j.resconrec.2017.09.005
50. Bauer, F. 2018. Narratives of biorefinery innovation for the bioeconomy: Conflict, consensus or confusion? Environmental Innovation and Societal Transitions 28: 96–107. https://doi.org/10.1016/j.eist.2018.01.005 .
51. Van Der Minne, C., M. P. Hekkert, A. Nijhof, D. Loorbach, and C. J. A. M. Termeer. 2021. Handhold for sustainability innovators: Four perspectives on transitions thinking and doing. Den Haag (in Dutch).
52. Metze, T., and J. Dodge. 2016. Dynamic discourse coalitions on hydro-fracking in Europe and the United States. Environmental Communication 10: 365–379. https://doi.org/10.1080/17524032.2015.1133437 .
53. Van Hulst, M., and D. Yanow. 2016. From policy "Frames" to "Framing." The American Review of Public Administration 46: 92–112. https://doi.org/10.1177/0275074014533142 .
54. Van Eeten, M.J.G. 1999. Dialogues of the Deaf. Defining New Agendas for Environmental Deadlocks. Delft: Eburon.
55. Jakobsen, S.-E., A. Fløysand, and J. Overton. 2019. Expanding the field of Responsible Research and Innovation (RRI)—from responsible research to responsible innovation. European Planning Studies 27: 2329–2343. https://doi.org/10.1080/09654313.2019.1667617 .

56. D'Amato D, Droste N, Allen B, Kettunen M, Lähtinen K, Korhonen J, Leskinen P, Matthies BD, Toppinen A (2017) Green, circular, bio economy: a comparative analysis of sustainability avenues. J Clean Prod 168:716–734. https://doi.org/10.1016/J.JCLEPRO.2017.09.053
57. Lier, M. et al. Synthesis on Bioeconomy Monitoring Systems in the EU Member States—Indicators for Monitoring the Progress of Bioeconomy Natural resources and bioeconomy studies 38/2018 (Natural Resources Institute Finland (Luke), 2018).
58. Bracco, S., Calicioglu, O., Juan, M. G. S. & Flammini, A. Assessing the contribution of bioeconomy to the total economy: a review of national frameworks. Sustainability 10, 1698 (2018).
59. O'Brien M, Wechsler D, Bringezu S, Schaldach R (2017) Toward a systemic monitoring of the European bioeconomy: gaps, needs and the integration of sustainability indicators and targets for global land use. Land Use Policy 66:162–171. https://doi.org/10.1016/j.landusepol.2017.04.047
60. Egenolf, V. & Bringezu, S. Conceptualization of an indicator system for assessing the sustainability of the bioeconomy. Sustainability 11, 443 (2019).
61. Tukker, A. et al. Environmental and resource footprints in a global context: Europe's structural deficit in resource endowments. Glob. Environ. Change 40, 171–181 (2016).
62. Steinmann, Z. J. N. et al. Resource footprints are good proxies of environmental damage. Environ. Sci. Technol. 51, 6360–6366 (2017).
63. Oehmichen, K. et al. Die Alternativen WEHAM-Szenarien: Holzpräferenz, Naturschutzpräferenz und Trendfortschreibung. Szenarienentwicklung, Ergebnisse und Analyse (Thünen, 2018).
64. Natural Resource Use in the Group of 20: Status, Trends, and Solutions (International Resource Panel, 2019).
65. Mainar-Causapé, A. (2019). Análisis de los sectores de Bioeconomía a través de matrices de contabilidad social específicas: El caso de España. Investigaciones Regionales, 3(45), 273–282.
66. Philippidis, G., & Sanjuán, A. I. (2018). A re-examination of the structural diversity of biobased activities and regions across the EU. Sustainability (Switzerland), 10(11), 4325. https://doi.org/10.3390/su10114325
67. Philippidis, G., Sanjuán, A. I., Ferrari, E., & M'Barek, R. (2014). Employing social accounting matrix multipliers to profile the bioeconomy in the EU member states: Is there a structural pattern? Spanish Journal of Agricultural Research, 12(4), 913–926. https://doi.org/10.5424/sjar/2014124-6192
68. Sancho, F., & Cardenete, M. A. (2014). Multisectoral tools for key sectors analysis in regional analysis. Revista De Estudios Regionales, 100, 131–146.
69. Mainar-Causapé, A., Philippidis, G., & Sanjuán, A. I. (2020). Constructing an open access economy-wide database for bioeconomy impact assessment in the European Union member states. Economic Systems Research. https://doi.org/10.1080/09535314.2020.1785848
70. Iráizoz, B. (2006). Es determinante el método en la identificación de los sectores clave de una economía? Una aplicación al caso de las tables Input-Output de Navarra. Estadística Española, 48(163), 551–585.
71. Dietzenbacher, E., van der Linden, J., & Steenge, A. (1993). The regional extraction method: EC input-output comparisons. Economic Systems Research, 5(2), 185–206. https://doi.org/10.1080/09535319300000017
72. Ferreira V, Fabregat-Aibar L, Terceño A (2022) Research trends and hotspots in bioeconomy impact analysis: a study of economic, social and environmental impacts. Environ Impact Assess Rev 96:106842
73. European Commission (2018a) Communication COM/2018/673: a sustainable Bioeconomy for Europe: Strengthening the connection between economy, society and the environment. https://knowledge4policy.ec.europa.eu/publication/sustainable-bioeconomy-europestrengthening-connection-between-economy-society_en
74. European Commission (2018b) The European way to use our natural resources: action plan. https://knowledge4policy.ec.europa.eu/publication/bioeconomy-european-way-use-ournatural-resources-action-plan-2018_en

75. Martinez-Hernandez E, Sadhukhan J, Aburto J, Amezcua-Allieri MA, Morse S, Murphy R (2022) Modelling to analyse the process and sustainability performance of forestry-based bioenergy systems. Clean Techn Environ Policy ume 24:1709–1725. https://doi.org/10.1007/s10098-022-02278-1 . FebruaryPages
76. Marcone RD, Schmid M, Meylan G (2022) Closing the gap between EU-wide national bioeconomy monitoring frameworks and urban circular bioeconomy development. J Clean Prod 379:134563
77. Termeer CJAM, Metze TAP (2019) More than peanuts: transformation towards a circular economy through a small-wins governance framework. J Clean Prod 240:118272. https://doi.org/10.1016/j.jclepro.2019.118272
78. Bours SAMJV, Wanzenböck I, Frenken K, Wanzenböck I (2021) Small wins for grand challenges. A bottom-up governance approach to regional innovation policy. Eur Plan Stud 0(0):1–28. https://doi.org/10.1080/09654313.2021.1980502
79. Connor JO, Keil M (2017) The effects of construal level and small wins framing on an individual's commitment to an environmental initiative. J Environ Psychol 52:1–10. https://doi.org/10.1016/j.jenvp.2017.04.010
80. Termeer CJAM, Dewulf A (2019) A small wins framework to overcome the evaluation paradox of governing wicked problems. Policy Soc 38(2):298–314. https://doi.org/10.1080/14494035.2018.1497933
81. Termeer CJAM, Dewulf A, Biesbroek GR (2017) Transformational change: governance interventions for climate change adaptation from a continuous change perspective. J Environ Plan Manage 60(4):558–576. https://doi.org/10.1080/09640568.2016.1168288
82. von Wirth T, Fuenfschilling L, Frantzeskaki N, Coenen L (2019) Impacts of urban living labs on sustainability transitions: mechanisms and strategies for systemic change through experimentation. Eur Plan Stud 27(2):229–257. https://doi.org/10.1080/09654313.2018.1504895
83. Gorissen L, Spira F, Meynaerts E, Valkering P, Frantzeskaki N (2018) Moving towards systemic change? Investigating acceleration dynamics of urban sustainability transitions in the Belgian City of Genk. J Clean Prod 173:171–185. https://doi.org/10.1016/j.jclepro.2016.12.052
84. Lam DPM, Martín-López B, Wiek A, Bennett EM, Frantzeskaki N, Horcea-Milcu AI, Lang DJ (2020) Scaling the impact of sustainability initiatives: a typology of amplification processes. Urban Transform. https://doi.org/10.1186/s42854-020-00007-9
85. Loorbach D, Wittmayer J, Avelino F, von Wirth T, Frantzeskaki N (2020) Transformative innovation and translocal diffusion. Environ Innov Societal Transit. https://doi.org/10.1016/j.eist.2020.01.009
86. Weick KE (1984) Small wins: redefining the scale of social problems. Am Psychol 39(1):40–49. https://doi.org/10.1037/0003-066X.39.1.40
87. OECD (2009) The bioeconomy to 2030: designing a policy agenda. Main Findings and Policy Conclusions, Paris
88. Sillanpää, M. and Ncibi, C. (2017) A sustainable bioeconomy. The green industry revolution.
89. Lokesh K, Ladu L, Summerton L (2018) Bridging the gaps for a "circular" bioeconomy: selection criteria, bio-based value chain and stakeholder mapping. Sustainability (Switzerland) 10(6). https://doi.org/10.3390/su10061695
90. Salvador R, Puglieri FN, Halog A, Andrade FG, Piekarski CM, de Francisco AC (2021) Key aspects for designing business models for a circular bioeconomy. J Clean Prod 278:124341. https://doi.org/10.1016/j.jclepro.2020.124341
91. European Commission (2012) Innovating for sustainable growth: a bioeconomy for Europe. European Commission, Brussels
92. BMBF (2010) Nationale Forschungsstrategie BioÖkonomie 2030: Unser Weg zu einer bio-basierten Wirtschaft. BMBF, Berlin
93. Fund C, El-Chichakli B, Patermann C, Dieckhoff P (2015) Bioeconomy policy (part II): synopsis of national strategies around the world. A report from the German Bioeconomy Council. Bioökonomierat, Berlin

94. Lühmann M (2020) Whose European bioeconomy? Relations of forces in the shaping of an updated EU bioeconomy strategy. Environ Dev 35:100547
95. Levidow L, Nieddu M, Vivian F-D, Béfort N (2019) Transitions towards a European Bioeconomy: life sciences versus agroecology trajectories. In: Allaire G, Daviron B (eds) Ecology, capitalism and the new agricultural economy: the second great transformation. Routledge, London, pp 181–203
96. European Commission (2018) A sustainable Bioeconomy for Europe: strengthening the connection between economy, society and the environment. Updated Bioeconomy Strategy. European Commission, Brussels
97. Backhouse M, Lorenzen K, Lühmann M, Puder J, Rodríguez F, Tittor A (2017) Bioökonomie-Strategien im Vergleich. Gemeinsamkeiten, Widersprüche und Leerstellen, Working Paper Nr. 1 der BMBF-Nachwuchsgruppe 'Bioeconomy&Inequalities'. Friedrich-Schiller-Universität, Jena
98. European Commission (2018). A sustainable bioeconomy for Europe: Strengthening the connection between economy, society and the environment: Updated bioeconomy strategy. file://C:/Users/hayrym1/AppData/Local/Temp/KI0418806ENN.en.pdf
99. Stevenson, L., D. L. Haberman, P. Matthews Wright, & C. Witt (2017). Thirteen theories of human nature, 7th edition. Oxford University Press.
100. Hänninen, V., & E. Aaltola (Eds.) (2020). Ihminen kaleidoskoopissa: Ihmiskäsitysten kirjoa tutkimassa (Humanity in kaleidoscope: Investigating the diversity of views on human nature, in Finnish). Gaudeamus.
101. Takala, T., & Häyry, M. (2004). Is communitarian thinking altruistic? Trames, 8, 276–283.
102. Drexhage, J., & D. Murphy (2010). Sustainable development: From Brundtland to Rio 2012. United Nations Headquarters. http://www.surdurulebilirkalkinma.gov.tr/wp-content/uploads/2016/06/Background_on_Sustainable_Development.pdf
103. van Norren, D. (2020). The Sustainable Development Goals viewed through Gross National Happiness, Ubuntu, and Buen Vivir. International Environmental Agreements: Politics, Law and Economics, 20, 431–458. https://doi.org/10.1007/s10784-020-09487-3
104. United Nations (2015b). Transforming our world: The 2030 agenda for sustainable development. https://www.un.org/ga/search/view_doc.asp?symbol=A/RES/70/1&Lang=E
105. Hickel J (2020a) Less is more: how degrowth will save the world. Penguin Random House, London
106. Hickel J (2020b) Quantifying national responsibility for climate breakdown: an equality-based attribution approach for carbon dioxide emissions in excess of the planetary boundary. Lancet Planet Health 4(9):399-e404. https://doi.org/10.1016/S2542-5196(20)30196-0
107. Hickel J, Sullivan D, Zoomkawala H (2021) Plunder in the post-colonial era: quantifying drain from the global south through unequal exchange, 1960–2018. New Politi Econ. https://doi.org/10.1080/13563467.2021.1899153
108. Martínez-Alier J (2020) A global environmental justice movement: mapping ecological distribution conflicts Disjuntiva. Crítica De Les Ciències Socials 1(2):83–128. https://doi.org/10.14198/DISJUNTIVA2020.1.2.6
109. Büscher B, Feola G, Fischer A, Fletcher R, Gerber J-F, Harcourt W, Koster M, Schneider M, Scholtens J, Spierenburg M, Walstra V, Wiskerke H (2021) Planning for a world beyond COVID-19: five pillars for post-neoliberal development. World Dev 140:105357
110. Gills B, Morgan J (2020) Global climate emergency: after COP24, climate science, urgency, and the threat to humanity. Globalizations 17(6):885–902. https://doi.org/10.1080/14747731.2019.1669915
111. Martin A, Armijos T, Coolsaet B, Dawson N, Edwards GAS, Few R, Gross-Camp N, Rodríguez I, Schroeder H, Tebboth MLG, White CS (2020) Environmental justice and transformations to sustainability. Environ Sci Policy Sustain Dev 62(6):19–30. https://doi.org/10.1080/00139157.2020.1820294
112. Nightingale AS, Eriksen MT, Forsyth T, Pelling M, Newsham A, Boyd E, Brown K et al (2019) Beyond technical fixes: climate solutions and the great derangement. Clim Dev. https://doi.org/10.1080/17565529.2019.1624495

113. Temper L et al (2020) Movements shaping climate futures: a systematic mapping of protests against fossil fuel and low-carbon energy projects. Environ Res Lett 15:123004
114. 115. Birch K, Levidow L, Papaioannou T (2010) Sustainable capital? The neoliberalization of nature and knowledge in the European 'Knowledge-based Bio-economy.' Sustainability 2:2898–2918
115. 116. Tuck E, Yang KW (2012) Decolonization is not a metaphor. Decolon Indig Educ Soc 1(1):1–40
116. D'Alisa G, Demaria F, Kallis G (2015) Degrowth vocabulary for a new era. ISBN 9781138000773
117. Deschner C, Hust E (2018) Decolonisation and degrowth. Degrowth blog dated 1 Feb 2018. https://www.degrowth.info/blog/decolonisation-and-degrowth
118. 119. Feola G (2019) Degrowth and the unmaking of capitalism: beyond "decolonization of the imaginary". ACME Int J Crit Geograph 18(4):977–997
119. 120. Garcia-Arias J, Schöneberg J (2021) Urgencies and imperatives for revolutionary (environmental) transitions: from degrowth and postdevelopment towards the pluriverse? Environ Polit. https://doi.org/10.1080/09644016.2021.1911443
120. 121. Zografos C, Robbins P (2020) Green sacrifice zones, or why a green new deal cannot ignore the cost shifts of just transitions. One Earth. https://doi.org/10.1016/j.oneear.2020.10.012
121. Kruus K, Hakala T (2017) The making of bioeconomy transformation. VTT Technical Research Centre of Finland. https://cris.vtt.fi/en/publications/the-making-of-bioeconomy-transformation
122. UNECE (United Nations Economic Commission for Europe) (2012) From transition to transformation, sustainable and inclusive development in Europe and Central Asia. https://sustainabledevelopment.un.org/index.php?page=view&type=400&nr=795&menu=1515
123. Cuhls K, Kayser V, Grandt S, Hamm U, Reisch L, Daniel H, El-Chichakli B, Dieckhoff P. Global visions for the bioeconomy - an international Delphi-study. 2015. https://gbs2020.net/wp-content/uploads/2021/10/GBS2015_02_Delphi-Study.pdf . Accessed 4 Mar 2022.
124. Yang N-HN, Yang A. Urban bioeconomy: uncovering its components, impacts and the urban bio-symbiosis. Clean Prod Lett. 2022;3:100015. https://doi.org/10.1016/j.clpl.2022.100015 .
125. Wilkes-Allemann J, van der Velde R, Kopp M, Bernasconi A, Karaca E, Coleman Brantschen E, et al. Research agenda – biocities of the future. European Forest Institute; 2022. https://doi.org/10.36333/rs4 .
126. BMBF, BMEL (2020) Nationale Bioökonomiestrategie. Berlin
127. 128. Gramsci A (2012a) Gefängnishefte, Hefte 10 und 11. Argument Verlag, Hamburg
128. 129. Gramsci A (2012b) Gefängnishefte, Hefte 12 bis 15. Argument Verlag, Hamburg
129. Giurca, Alexandru Why is Communicating the Circular Bioeconomy so Challenging?. Circular Economy and Sustainability (2022). https://doi.org/10.1007/s43615-022-00229-y
130. Holden, Nicholas M.; Neill, Andrew M.; Stout, Jane C.; O'Brien, Derek; Morris, Michael A. Biocircularity: a Framework to Define Sustainable, Circular Bioeconomy. Circular Economy and Sustainability (2022). https://doi.org/10.1007/s43615-022-00180-y
131. Asdal, Kristin; Cointe, Béatrice; Hobæk, Bård; Reinertsen, Hilde; Huse, Tone; Morsman, Silje R.; Måløy, Tommas 'The good economy': a conceptual and empirical move for investigating how economies and versions of the good are entangled. BioSocieties (2021). https://doi.org/10.1057/s41292-021-00245-5
132. Mougenot, Benoit; Doussoulin, Jean-Pierre Conceptual evolution of the bioeconomy: a bibliometric analysis. Environment, Development and Sustainability (2021). https://doi.org/10.1007/s10668-021-01481-2
133. Starke, Jan R.; Metze, Tamara A. P.; Candel, Jeroen J. L.; Termeer, Catrien J. A. M. Conceptualizing controversies in the EU circular bioeconomy transition. Ambio (2022). https://doi.org/10.1007/s13280-022-01730-2
134. Bringezu, Stefan; Distelkamp, Martin; Lutz, Christian; Wimmer, Florian; Schaldach, Rüdiger; Hennenberg, Klaus Josef; Böttcher, Hannes; Egenolf, Vincent Environmental and socioeconomic footprints of the German bioeconomy. Nature Sustainability (2021). https://doi.org/10.1038/s41893-021-00725-3

135. Ferreira, Valeria; Pié, Laia; Mainar-Causapé, Alfredo; Terceño, Antonio The bioeconomy in Spain as a new economic paradigm: the role of key sectors with different approaches. Environment, Development and Sustainability (2023). https://doi.org/10.1007/s10668-022-02830-5
136. Dolge, Kristiāna; Balode, Lauma; Laktuka, Krista; Kirsanovs, Vladimirs; Barisa, Aiga; Kubule, Anna A Comparative Analysis of Bioeconomy Development in European Union Countries. Environmental Management (2022). https://doi.org/10.1007/s00267-022-01751-3
137. 138. Schagen, O. M.; Metze, T. A. P.; de Olde, E. M.; Termeer, C. J. A. M. Energizing a transformation to a circular bioeconomy: mechanisms to spread, deepen and broaden initiatives. Sustainability Science (2022). https://doi.org/10.1007/s11625-022-01249-1
138. Viaggi, Davide; Zavalloni, Matteo Bioeconomy and Circular Economy: Implications for Economic Evaluation in the Post-COVID Era. Circular Economy and Sustainability (2021). https://doi.org/10.1007/s43615-021-00113-1
139. Eversberg, Dennis; Holz, Jana; Pungas, Lilian The bioeconomy and its untenable growth promises: reality checks from research. Sustainability Science (2022). https://doi.org/10.1007/s11625-022-01237-5
140. Häyry, Matti; Laihonen, Maarit Situating a sustainable bioeconomy strategy on a map of justice: a solution and its problems. Environment, Development and Sustainability (2022). https://doi.org/10.1007/s10668-022-02720-w
141. Ramcilovic-Suominen, Sabaheta Envisioning just transformations in and beyond the EU bioeconomy: inspirations from decolonial environmental justice and degrowth. Sustainability Science (2022). https://doi.org/10.1007/s11625-022-01091-5
142. Boldt, Christin; Thrän, Daniela How to implement the urban bioeconomy: insights from Leipzig's current policy mix for transitioning towards a bio-principled city. Urban Transformations (2023). https://doi.org/10.1186/s42854-023-00052-0
143. Boyer, Miriam; Kusche, Franziska; Hackfort, Sarah; Prause, Louisa; Engelbrecht-Bock, Friederike The making of sustainability: ideological strategies, the materiality of nature, and biomass use in the bioeconomy. Sustainability Science (2022). https://doi.org/10.1007/s11625-022-01254-4

Chapter 11
Greening Urban Spaces

11.1 Introduction by Editor

The rapid urbanization of the twenty-first century has presented both challenges and opportunities for creating more livable and environmentally-friendly cities. United Nations Sustainable Development Goal number 11 is about Sustainable Cities and Communities. In this SDG the UN website observes that presently a majority of the global population resides in urban areas. They estimate that by the year 2050, almost 70% of the global population would reside in metropolitan regions. Urban areas play a key role in economic development and account for over 80% of the world's gross domestic product (GDP). On the other hand, UN observes that they contribute to about 70% of the total worldwide greenhouse gas emissions (Nations, Making Cities and Human settlements more inclusive, safe, resilient and sustainable, 2023). When urban development is carefully strategized and effectively administered, it has the potential to be sustainable and foster equitable economic growth.

As urban sprawls grow, they struggle to cope with the problem of sewage and solid waste. With rise in new technology, energy and water consumption is steadily growing in urban areas. However, there is a movement to greening cities. There is growing recognition of the importance of provisioning additional spaces to natural areas inside urban environments. This has led to the emergence of the idea of "urban green area" as a significant component of cities. Urban green areas play a significant role that has a positive influence on the quality of life.

Green cities are not just about green spaces. United Nations Environment Program (UNEP) advocates the use of Nature-based solutions (NbS) for urban spaces (UNEP, 2021). The World Bank (2008) and the International Union for Conservation of Nature (IUCN 2009) introduced the concept of Nature-based solutions around the late 2000s. NbS emphasises the significance of biodiversity conservation in addressing climate change mitigation and adaptation. Over time, NbS has evolved to expanding its applicability beyond climate-related issues to encompass a wider range of objectives and uses. NbS have a potential to effectively mitigate specific social challenges

G. Nulkar (ed.), *The Economics of Sustainable Development*,
https://doi.org/10.1007/978-981-99-7379-8_11

in urban areas related to climate change. By enhancing cities' resilience to climate change, NbS can contribute to improving the quality of life for urban residents and promoting biodiversity through the establishment of green spaces. There are studies to show that NbS contribute to a reduction in carbon emissions and improvement in carbon sequestration, and also to have valuable indirect effects and behavioural impacts (Nature-based solutions can pave the way to carbon–neutral cities in 2030, 2023).

Recognizing the ecological and psychological value of urban green spaces, several municipalities across the world are taking up activities like biophilic designs and nature-based solutions. There are several examples like the Singapore Green Plan 2030, that builds on existing climate adaptation and mitigation measures; or the Paris' 15-min city' that respond to the needs of greener and more walkable urban spaces. Several urban bodies in the USA and European Union are exploring nature-based solutions to address issues of biodiversity, climate resilience, and sustainable development.(Pincetl & Gearin, 2005)(Irvine et al., 2008)(Vázquez et al., 2020)(Irvine et al. 2008)(Vázquez et al., 2020)(Pincetl & Gearin, 2005).

Integrating green infrastructure into the urban landscape can provide a range of benefits, from regulating local temperatures and air quality to supporting diverse flora and fauna. As the COVID-19 pandemic has highlighted, access to nature and outdoor spaces is crucial for the well-being of city dwellers, especially during times of restricted mobility. Biophilic design principles, which seek to strengthen the connection between people and nature, are being incorporated into architectural practices to promote healthier, more sustainable built environments (Zhong et al., 2022). Studies show the positive impacts of biophilic design principles in cities. In the inner part of Los Angeles, a study looked at the local residents' attitudes and values towards a more inclusive green space strategy, and measured the potential value of nature's services in the city using GIS mapping (Pincetl & Gearin, 2005). The authors report strong community support for enhancing access to nature and the recognition of the multiple ecosystem services provided by urban greenery. Another study in London and Chicago provides evidence of positive impacts on human health and ecological benefits of incorporating biophilic elements into the built environment. By integrating natural materials, vegetation, and water features into the design of workplaces, neighborhoods, and public spaces, these studies found measurable improvements in air quality, thermal comfort, stress reduction, and community engagement. These cities are trying to overcome the challenges of climate change with nature-based solutions, and biophilic design promises a potential to improve the livability, sustainability, and resilience of urban environments (Andreucci et al., 2021)(Cabanek et al., 2020)(Panagopoulos et al., 2021)(Lee & Park, 2021).

However, the greening of urban spaces is not without its challenges. Equitable access to green amenities, the costs of implementation, and competing land use priorities must all be carefully considered. Innovative research and collaborative approaches are needed to overcome these barriers and realize the full potential of nature-based solutions in creating more sustainable and resilient cities.

Drawing on a research and case studies across the globe, this chapter examines the ecological and psychological benefits of urban greening, the key design paradigms and policy considerations, and emerging best practices from cities in the USA and European Union, and introduces readers to applying Nature-based solutions in urban spaces, circular economy, urban ecosystems, and green infrastructure. Readers will learn the outcomes of urban experiments like converting redundant parking spaces to green corners.

References

Andreucci, M B., Loder, A., Brown, M J., & Brajković, J. (2021, April 13). Exploring Challenges and Opportunities of Biophilic Urban Design: Evidence from Research and Experimentation. Multidisciplinary Digital Publishing Institute, 13(8), 4323–4323. https://doi.org/10.3390/su13084323.

Cabanek, A., Baro, M E Z D., & Newman, P. (2020, July 8). Biophilic streets: a design framework for creating multiple urban benefits. BioMed Central, 3(1). https://doi.org/10.1186/s42055-020-00.027-0.

Irvine, K N., Fuller, R A., Devine-Wright, P., Tratalos, J A., Payne, S R., Warren, P H., Lomas, K J., & Gaston, K J. (2008, January 1). Ecological and Psychological Value of Urban Green Space. Springer International Publishing, 215–237. https://doi.org/10.1007/978-1-4020-8647-2_10.

Lee, E J., & Park, S J. (2021, March 4). Toward the Biophilic Residential Regeneration for the Green New Deal. Multidisciplinary Digital Publishing Institute, 18(5), 2523–2523. https://doi.org/10.3390/ijerph18052523.

Nature-based solutions can pave the way to carbon–neutral cities in 2030. (2023). Nature Climate Change, 13, 774–775. https://doi.org/10.1038/s41558-023-01,750-0.

Panagopoulos, T., Sbarcea, M., & Herman, K. (2021, March 14). A biophilic mind-set for a restorative built environment., 17, 68–77. https://doi.org/10.22616/j.landarchart.2020.17.08.

Pincetl, S., & Gearin, E. (2005, August 1). The Reinvention of Public Green Space. Taylor & Francis, 26(5), 365–384. https://doi.org/10.2747/0272-3638.26.5.365.

UNEP. (2021, April 26). Nature-Based Solutions for Urban Challenges. Retrieved from United Nations Environment Programme: https://www.unep.org/resources/emerging-issues/nature-based-solutions-urban-challenges

United Nations (2023). Making Cities and Human settlements more inclusive, safe, resilient and sustainable. Retrieved from United Nations Statistics Division: https://unstats.un.org/sdgs/report/2022/Goal-11/

Vázquez, A V., Fernández-Pablos, E., Lozano-Diez, R V., & López-Zaldívar, O. (2020, July 1). Green space networks as natural infrastructures in PERI-URBAN areas. Springer Science + Business Media, 24(1), 187–204. https://doi.org/10.1007/s11252-020-01,019-w.

Zhong, W., Schröder, T., & Bekkering, J. (2022, February 1). Biophilic design in architecture and its contributions to health, well-being, and sustainability: A critical review. Elsevier BV, 11(1), 114–141. https://doi.org/10.1016/j.foar.2021.07.006.

11.2 Machine Generated Summaries

Disclaimer: The summaries in this chapter were generated from Springer Nature publications using extractive AI auto-summarization: An extraction-based summarizer aims to identify the most important sentences of a text using an algorithm and uses those original sentences to create the auto-summary (unlike generative AI). As the constituted sentences are machine selected, they may not fully reflect the body of the work, so we strongly advise that the original content is read and cited. The auto generated summaries were curated by the editor to meet Springer Nature publication standards. To cite this content, please refer to the original papers.

Machine generated keywords: urban, city, green, solution, green space, infrastructure, urban area, naturebase, ecosystem service, naturebase solution, ecosystem, space, service, biodiversity, European.

11.2.1 Reconnecting Cities to the Biosphere: Stewardship of Green Infrastructure and Urban Ecosystem Services [126]

This is a machine-generated summary of:

Andersson, Erik; Barthel, Stephan; Borgström, Sara; Colding, Johan; Elmqvist, Thomas; Folke, Carl; Gren, Åsa: Reconnecting Cities to the Biosphere: Stewardship of Green Infrastructure and Urban Ecosystem Services [126]

Published in: Ambio (2014).

Link to original: https://doi.org/10.1007/s13280-014-0506-y.

If you want to cite the papers, please refer to the original.

For technical reasons we could not place the page where the original quote is coming from.

Abstract-Summary

"Within-city green infrastructure can offer opportunities and new contexts for people to become stewards of ecosystem services."

"Key findings demonstrate that urban ecosystem services are generated by social–ecological systems and that local stewards are critically important."

"Land-use planning and management seldom account for their role in the generation of urban ecosystem services."

"The results highlight the importance and complexity of stewardship of urban biodiversity and ecosystem services and of the planning and governance of urban green infrastructure."

Introduction

"How do we ensure sustainable generation, management, and governance of ecosystem services for human well-being in cities, as well as ensure that cities contribute to incentives for better stewardship of distant landscapes and seascapes?."

"Providing but a fraction of the ecosystem services consumed, urban landscapes represent key arenas for learning about the way humans interact with the environment and what sustainable ecosystem stewardship might entail (Miller 1; Chapin and others 2)."

"We draw on more than 15 years of empirical work within cities in relation to ecosystem service generation in urban landscapes, particularly regulating ecosystem services, and their stewardship with enabling institutions (e.g., property rights), social networks and involvement of local user groups and civil society in green area management and governance."

"Our work in the Stockholm urban landscape, Sweden, has helped reveal green areas and ecosystem services not previously perceived in urban planning and clarified mismatches between institutions, governance, and urban ecosystems for human well-being (Electronic Supplementary Material, Appendix S1)."

Urban Social–Ecological Systems

"Comprehensive analyses of urban green spaces have shown that land uses such as private and public gardens, cemeteries, old brown-fields, and golf courses may contribute significantly to ecosystem services provided by the urban landscape (Colding and others 3; Goddard and others 4)."

"Many ecosystem services need to be locally provided in urban landscapes for easy access and use by a greater set of city-inhabitants, e.g., daily nature encounters, noise reduction, absorption of pollutants in water and air."

"As historical (Walker 5) and social movement research has indicated (Ansell 6; Ernstson and others 7), urban green areas attracting a high diversity of interest and user groups seem to have higher chances of being protected and creating a social environment that nurture stewardship of ecosystem services because of increased potential for effective collective action and combination of knowledge and skills."

Conclusion and Implications for Reconnecting Urban Areas to the Biosphere

"The length and diversity of borders, biophysical as well as administrative, call for careful thinking to make sure adjacency effects are positive and that regulating ecosystem services reach across the urban landscapes. (3) Cities hold unexplored potential for new urban spatial designs that integrate ecosystem services in the built environment, for restoring degraded ecosystems and for strengthening ecosystem functions through complementary designs of land use and urban green structures. (4) Mismatches between social and ecological boundaries are prevalent."

"With further research on the ecological underpinnings of ecosystem services, not least the cultural, future long-term urban social–ecological research must deepen our understanding of whether and how local stewardship and engagement in practical management of green infrastructures increase biodiversity and availability of ecosystem services in metropolitan landscapes, and if and how it actually stimulates a wider awareness and articulation of our global reliance on ecosystem services and results in an urban footprint both smaller and less detrimental to the resilience of the biosphere."

11.2.2 Principles for Urban Nature-Based Solutions [127]

This is a machine-generated summary of:

Kabisch, Nadja; Frantzeskaki, Niki; Hansen, Rieke: Principles for urban nature-based solutions [127]

Published in: Ambio (2022).

Link to original: https://doi.org/10.1007/s13280-021-01,685-w.

If you want to cite the papers, please refer to the original.

For technical reasons we could not place the page where the original quote is coming from.

Abstract-Summary

"Although principles for a common understanding and implementation of NBS were already developed on a landscape scale, specific principles are needed with regard to an application in urban areas."

"Given these challenges, in this perspective paper, we build upon and integrate knowledge from the most recent academic work on NBS in urban areas and introduce five distinct, integrated principles for urban NBS design, planning and implementation."

"To contribute to resilient urban futures, the design, planning, policy and governance of NBS should (1) consider the need for a systemic understanding, (2) contribute to benefiting people and biodiversity, (3) contribute to inclusive solutions for the long-term, (4) consider context conditions and (5) foster communication and learning."

Introduction

"Although the different sets of NBS principles and the IUCN standard for NBS are well-developed, we argue that specific principles are needed with regard to an application in urban areas—with the aim to provide a spatial translation and operationalisation of the existing knowledge on participatory planning and good governance of NBS to the urban scale."

"Integrative urban NBS principles are required that specifically address the five challenges considering i) potential development conflicts with urban systems nestedness; ii) specific urban biodiversity and related environmental conditions; iii) value plurality, multi-actor interdependencies and environmental injustices; iv) path-dependencies with cultural and planning legacies; and finally v) a potential misconception of cities as being artificial landscapes disconnected from natures."

"With a particular view on urban systems, these principles are intended to transcend governance gaps and stimulate the discussion around urban NBS towards more effective and sustainable designs that fit the urban morphology, social, ecological and institutional dynamics."

Five principles for urban nature-based solutions planning and governance

"The rationale on how our five principles for urban NBS are formulated and discussed is based on the five specific urban challenges as outlined above."

"The principles further build on knowledge from multiple urban disciplines (urban ecology, urban sociology, urban design, urban planning and governance)."

"The proposed urban principles can offer a new perspective to select and screen frameworks for designing, monitoring and evaluating planning and mainstreaming for NBS, and also to ensure that they contribute to resilient and inclusive cities on the short-term and in the long-term future of cities."

Urban NBS principle 1—Need for a systemic understanding: Urban NBS are integrated solutions and need to be based on a systems approach

"This interconnectedness of ecological, social and technical dimensions results in high systems complexity in which different kinds of knowledge is needed to the planning, design and management of NBS (Frantzeskaki and Kabisch 8; Keeler and others 9)."

"Considering systemic thinking at different spatial scales, urban planners need to ensure that core areas providing NBS such as urban forests, wetlands or large parks are protected and maintained for the whole city area, while at a neighbourhood scale street trees or other small private green patches might represent crucial NBS."

"In terms of governance for NBS, a systems approach would also translate openly in the need for cross-departmental, more intersectoral collaboration."

"Intersectoral collaborations means getting different urban planning departments such as transport and mobility, social policy, water infrastructure, green space planning and health with their particular expertise together to plan for integrating NBS to urban fabrics in an inclusive and multifunctional way (Kabisch and others 10; Bush 11; Frantzeskaki and others 12; Moosavi and others 13)."

Urban NBS principle 2—Benefiting people and biodiversity: urban NBS need to ensure a balanced delivery of multiple benefits for humans and non-humans

"Urban NBS should deliver and correspond to human needs and at the same time contribute to ecological flows and provide habitat for species diversity."

"While the ideal of NBS as functioning ecosystems need to be reconsidered for urban areas, NBS can and should be inspired by local natural ecosystems and designed as spaces that require little maintenance, are relatively stable in the long-term and fulfil both human- and nature-oriented purposes, i.e. prairie gardens in the US or xeric gardens in arid regions (Ignatieva and others 14)."

"We are here with Maller (15, p. 2) who discussed how NBS could be used as a lens through which urban areas could be designed or governed as"...places where multiple species and ecosystems are encouraged to flourish, including, but not limited to, humans."

Urban NBS principle 3—Inclusive solutions for the long-term: urban NBS need to be inclusively designed, planned, implemented and managed to appreciate long-term benefits

"Decisions for urban NBS implementation requires local knowledge and integration of cultural context in terms of inclusive design considering all dimensions of socio-environmental justice (distributive, procedural, interactional or recognition; see Kabisch and Haase 16; Baró and others 17; Pineda-Pinto and others 18)."

"The cultural benefits and contributions of nature need to be understood which in turn requires an integrated and inclusive approach to design and planning in terms of co-creation and co-design of NBS responding to the needs of citizens (Mattijssen and others 19; Buijs and others 20; van der Jagt and others 21; Frantzeskaki and others 22) while contributing to procedural and interactional justice."

"The long-term effectiveness or performance of NBS and its potential outcomes—be it synergies or eventually trade-offs—need to be considered and discussed openly and transparently (Gómez Martín and others 23)."

"An inclusive and thus, just approach to planning NBS will entail a holistic assessment and understanding of multiple benefits, potential trade-offs as well as an understanding of how their multifunctionality can be secured or hampered due to socio-economic and cultural contextual aspects throughout their lifecycle (Cousins 24; Giachino and others 25; Pineda-Pinto and others 18)."

Urban NBS principle 4—Context consideration: urban NBS should respect and be planned considering the local context

"In our understanding, urban NBS are an alternative type of infrastructure that helps adapting and mitigating societal challenges in a specific local, cultural and ecological context."

"Considering the socio-ecological context dynamics, NBS need to be designed to fit the context and produce (co-)benefits that restore local ecological flows and enrich biodiversity and the local community (Gomez Martin and others 23)."

"NBS need to encounter and be designed in such a way to 'enrich' if not fit in the cultural context of existing urban nature (Nagendra and Mundoli 26; Basu and Nagendra 27) which in turn not only benefits health and well-being but has a dedicated meaning to people."

"They also evidenced the historical trajectory for green spaces as urban commons in cities that is an enabling socio-cultural context for the development of new NBS as part of urban green commons."

Urban NBS principle 5—Communication and learning: urban NBS should support mutual learning for sustainability transitions in cities

"NBS in urban environments have the potential to (re)establish connections of people with nature and in this way may also contribute to pro-environmental behaviour and increased awareness of the significance of sustainability in everyday urban life (Soga and Gaston 28; West and others 29)."

"An increased sense of belonging and attachment might be related with aesthetical appeal of a NBS and support appreciation, recognition and awareness of the benefits of nature in cities (Frantzeskaki 30; Gómez Martín and others 23; Bayulken and others, 31)."

"In terms of making urban NBS part of sustainability transitions, they should be part of transdisciplinary and citizen-based environmental learning as well as awareness strategies and campaigns."

"Planning and implementation of NBS should be informed by citizens' concerns and preferences to support positive relations with urban nature."

Concluding remarks

"Acknowledging existing principles being important for a broader landscape scale, with this perspectives paper, we aim to stimulate discussions on the planning and governance of NBS in urban areas."

"We introduced five principles for urban NBS because cities come with specific challenges."

"Our five principles for urban NBS offer an extended, interlinked perspective when NBS are to be implemented in urban areas but may also guide future research on urban NBS and sustainability transitions."

"With the consideration of our five principles for urban NBS, we hope to contribute to an increased appreciation of nature in cities in which the beneficial contribution of nature is considered, given a proper weight in collaborative decisions and becomes a core solution, while still considering that NBS build on living organisms and ecological processes that require context-adaptation, specific conditions, appreciation of citizens and long-term care."

11.2.3 Moving Beyond the Nature-Based Solutions Discourse: Introducing Nature-Based Thinking [128]

This is a machine-generated summary of:

Randrup, Thomas B.; Buijs, Arjen; Konijnendijk, Cecil C.; Wild, Tom: Moving beyond the nature-based solutions discourse: introducing nature-based thinking [128]

Published in: Urban Ecosystems (2020).

Link to original: https://doi.org/10.1007/s11252-020-00.964-w.

Copyright of the summarized publication:

If you want to cite the papers, please refer to the original.

For technical reasons we could not place the page where the original quote is coming from.

Abstract-Summary

"Nature-based Solutions can be seen as a new conceptual approach to the human-ecological connection, and as an outcome of an evolutionary development of socio-ecological concepts."

"In this Communication, we argue for drawing up inspiration by nature as an outset for the development of more sustainable and inclusive cities, balancing anthropocentric and ecocentric values and acknowledging the importance of the social and governance dimensions in a more balanced socio-ecological perspective."

Introduction

"Nature-based solutions (NBS) (European Commission 32), the Intergovernmental Science-Policy Platform on Biodiversity and Ecosystem Services' (IPBES) note on Nature's Contribution to People (Escobedo and others 33), and the United Nations' 17 Sustainable Development Goals may be seen as efforts to address humans' dependency upon nature, in restoring and developing not only urban areas, but human life in general."

"We revisit the adage of being inspired by nature, and argue that acknowledging the interconnectedness of humans, nature and technology is a useful starting point to be inspired by natural processes in all aspects of urban development to create more holistic approaches to sustainable cities."

"By balancing anthropocentric and ecocentric values, as well as the relational values of nature, in combination with acknowledging the importance of the social and governance dimensions in a more balanced socio-ecological perspective, more healthy and long-term sustainable urban nature solutions may be developed."

Contemporary concepts are human-centered

"The ecosystem service (ESS) concept was a way of describing and organising the benefits that people derive from nature ('nature's gift to us')."

"The above concepts and approaches tend to deal with, and relate to finding ways to handle the urbanisation challenges by using and 'mobilising' nature."

"The development of these concepts shows a clear pattern: the naming and framing of our relationship with nature as well as the proposed solutions increasingly focused on instrumental values of nature: what good can it do for me and to us?."

"The concepts of ESS, Natural Capital and Nature's Contribution to People (ibid) have been critiqued for their technocratic ideals about knowledge, standardization, and science–society relations, for the way they combine nature and techniques, as well as for their quantification of these values (e.g., Turnhout and others 34; Schröter and others 35; Bekessy and others 36)."

Broadening the scope: nature-based thinking

"In order for planners, designers, managers, and citizens to really embrace nature, we argue that a more transformative 'turn' needs to be made beyond anthropocentric and solutions-based approaches."

"NBS (technological, scalable nature inspired solutions), GI (plans we make), ESS (systematic of values for humans benefit) etc. need to be placed in a broader framework, a new comprehensive paradigm on how we develop, design, and manage our cities."

"We argue that refocusing our thoughts and speech from Nature-based Solutions to Nature-based Thinking may stimulate a broader nature-based, but also social-ecological (inclusive) approach for greening our cities."

"We define NBT as an approach to urban inclusive planning, being inspired by nature to act socially, environmentally as well as economically in the transition towards sustainable cities."

"A new NBT framework will apply a holistic, nature-based approach to planning, design, construction, management and stewardship of urban nature."

Nature-based thinking in practice

"NBT capitalizes on the potential of cyclical process to plan, design, construct, and manage urban green spaces."

"The evolution of NBS is perhaps only one step in a transition towards deeper NBT which may require a turn away from traditional landscape planning processes, towards more gradual yet progressive and long-term approaches to restore the benefits and services of urban nature."

"NBT acknowledges the time dimension, by balancing strategic and visionary approaches with incremental planning processes."

"A technical approach to NBS may fall short of understanding the long-term context and interdependencies at the unpredictable and changing environmental, social and governance dimensions."

"We call for a new, cross-cutting, and interdisciplinary approach to planning, design, construction, and long-term maintenance of urban nature."

Three NBT dimensions

"By revisiting the ecological dimension we can create more room for nature also beyond human services and solutions, and especially in the urban areas, to build in more space for natural processes, ecosystem functioning, and long term unpredictability."

"By revisiting the community dimension we can create a new urban aesthetic, in which experiences with a diversity of natural elements may allow for more diversity in what nature (should) look like, rather than the opposite (because of limiting growing conditions, or limitations set by the economic system), and 3."

"The way new urban developments have often resulted in hostile growing conditions for urban nature will need to change."

"We propose that in NBT, urban vegetation including street trees are established in a nature inspired manner, e.g. locating trees strategically where there is room for growth, instead of systematically where technical solutions are needed."

Perspective

"There is a need to re-think the way we define concepts from a primarily anthropogenic perspective, and learn from the failure of contemporary governmental/technical approaches; we see the potential to learn and benefit from grassroots efforts to move beyond approaches that only address part of the larger need to fundamentally change our ways of shaping cities."

"We suggest a transition from perspectives and concepts which primarily recognise the performance of nature and the need to showcase via experimentation."

"The NBT perspective not only includes the environmental, social and economic dimensions of sustainability, but has an explicit focus on integrating these dimensions into three interconnected nexi; the ecological-governance nexus, where ecology becomes an integrative part of decision making, the ecological-community nexus, allowing for new urban aesthetics beyond manicured and sterile grasslands, and the community-governance nexus, recognising the agency and innovative potential of local communities and businesses."

11.2.4 Contribution of Prioritized Urban Nature-Based Solutions Allocation to Carbon Neutrality [129]

This is a machine-generated summary of:

Pan, Haozhi; Page, Jessica; Shi, Rui; Cong, Cong; Cai, Zipan; Barthel, Stephan; Thollander, Patrik; Colding, Johan; Kalantari, Zahra: Contribution of prioritized urban nature-based solutions allocation to carbon neutrality [129]

Published in: Nature Climate Change (2023).

Link to original: https://doi.org/10.1038/s41558-023-01,737-x.

If you want to cite the papers, please refer to the original.

For technical reasons we could not place the page where the original quote is coming from.

Abstract-Summary

"Carbon neutrality requires city-led climate action plans that incorporate both indirect and direct contributions of NBS."

"We assessed the carbon emissions mitigation potential of NBS in European cities, focusing particularly on commonly overlooked indirect pathways, for example, human behavioural interventions and resource savings."

"Spatially prioritizing different types of NBS in 54 major European Union cities could reduce anthropogenic carbon emissions by on average 17.4%."

"Coupling NBS with other existing measures in Representative Concentration Pathway scenarios could reduce total carbon emissions by 57.3% in 2030, with both indirect pathways and sequestration."

Main

"When all benefits are considered, NBS can contribute much more to urban climate neutrality goals than mere carbon sequestration effects."

"Due to the limited land resources in cities, the spatial allocation and configuration of urban NBS need to be optimized to achieve maximum carbon emissions reductions."

"In this Analysis, we assessed and quantified five potential carbon emissions reduction mechanisms for different types of NBS in EU cities."

"On the basis of sector-wise carbon emissions and the local context of 54 major European cities, we spatially allocated these five categories of NBS to each city and estimated the emissions reduction potential for each sector and city."

"We then compared the estimated emissions reduction potential against the 2030 climate neutrality targets of the 54 cities, to assess the potential contributions of prioritized NBS to these targets."

Multiple pathways of urban NBS to reduce carbon emissions

"From established definitions of NBS in the literature [37–39], we identified major mechanisms by which NBS can mitigate carbon emissions, namely saving resources and costs, reducing urban sprawl, promoting pro-environment behaviour, microclimate regulation and carbon sequestration."

"Carbon sequestration and saving resources and costs are more direct carbon emissions reduction mechanisms, for example, via carbon sequestration from urban trees and energy saving from using GI for stormwater treatment."

"Promoting pro-environmental behaviour generally reduces carbon emissions through indirect pathways, which are enabled by ecosystem services such as microclimate regulation, recreational opportunities and aesthetic nature experiences during daily routines."

"In the residential sector, urban parks and agriculture can have the strongest effects in promoting pro-environmental habits, including less driving, sustainable consumption, and reduced heating and cooling demand, thus reducing carbon emissions by up to 6.2% (ref [40])."

Spatial allocation of prioritized NBS implementation

"We selected 54 cities across Europe to model prioritized spatial allocation of NBS implementations in order to maximize the carbon emissions reductions."

"Although the allocation of prioritized NBS was mainly based on narratives of how different NBS fit with different socioeconomic contexts and their estimated emissions mitigation effects from the literature, some basic principles needed to be applied so that allocations for the 54 cities could be performed systematically."

"Road greening through street trees was allocated to the urban fringe of cities that are currently characterized by low-density development and high level of car-based commuting, including Rome, Athens, Naples, Valletta, Copenhagen, Helsinki and Stockholm."

"Many southern and eastern European cities with high residential emissions and high population density were co-allocated green access and green buildings."

Contribution of NBS to carbon emissions reduction in cities

"The highest carbon emissions reduction potential was found for eastern EU cities, where NBS^P could reduce total carbon emissions by 20.3%, closely followed by northern EU (18.2%)."

"The lowest carbon emissions reduction potential identified was in western European cities, closely followed by central EU cities, where NBS^P could reduce total carbon emissions by 13.0% and 13.2%, respectively."

"Among the remaining emissions reduction potential following NBS^P implementation in the three sectors, carbon sequestration could offset 47.6% and 10.7% of the carbon emissions in northern and eastern EU cities, but only 6.7% in central European cities."

"Western EU cities will need to rely on green–blue infrastructure to reduce industrial emissions, while northern EU cities will have to depend more on natural ecosystems for carbon sequestration and transport emissions reduction, as they were found to have the highest transportation emissions-saving potential (11.8%)."

Implications for pathways to climate-neutral cities by 2030

"NBS was least effective in SSP5, in which maximizing NBS implementations on all available land parcels in the RCP 8.5 scenario would reduce total carbon emissions by on average 16.2% (95% CI 2.7–29.7%) compared with the SSP5 baseline, with only 1.7% marginal reduction compared with the RCP 8.5 scenario."

"Taking carbon sequestration into consideration, maximizing NBS implementations and carbon sequestration on all available land parcels in the RCP 1.9 scenario would reduce total carbon emissions by on average 62.5% (95% CI 49.4–108.6%) compared with the SSP1 baseline, by capturing a further 13.3% of carbon emissions."

"Maximizing NBS implementations on all available land parcels in the RCP 8.5 scenario would reduce total carbon emissions by on average 19.9% (95% CI 0.0–65.3%) compared with the SSP5 baseline, by achieving a further 3.6% marginal reduction compared with the RCP 8.5 scenario."

Discussion and policy implications

"With prioritized NBS implementations, carbon sequestration was found to have a lower overall potential for meeting the goal of carbon neutrality in the European cities studied."

"Another policy implication relates to the 2030 policy goal of carbon–neutral cities, where more ambitious climate policies phasing out fossil fuels can seek to maximize NBS potential at the same time."

"More ambitious climate policies preserve more natural land, protect ecosystem quality, and promote high-density and compact urban development, which all present opportunities for spatial allocation of more implementations to maximize NBS potential."

"Pioneering cities that have created the best opportunities for NBS to be effective should extend their climate action plans to fully incorporate NBS implementations, which would maximize their chances of achieving city carbon neutrality."

"Inspired by how other environmental pollutants have been successfully handled in the past [41], we end by suggesting that the EU Commission should implement NBS-based spatial urban planning as a key strategy to achieve carbon–neutral cities."

Methods

"We conducted a systematic meta-analysis to estimate the effects of five categories of NBS implementations on carbon emissions reductions from three sectors (transport, residential and industrial), in addition to carbon sequestration."

"Spatial allocation of the most cost-effective NBS for emissions reduction was based on sectoral carbon emissions in each land use grid (30 m $\times$ 30 m) for the transport, residential and industrial sectors in each city."

"To carbon emissions at the scale of land use grid, we used socioeconomic variables (population density, building density, road networks and land use structure) and biophysical variables (ecosystem services and vegetative sequestration) to determine the allocation of NBS."

"To project how NBS would contribute to the climate neutrality goals of European cities in 2030, we conducted a comparative analysis of the percentage of carbon emissions mitigation in each city with its (emissions mitigation pathway) RCP for different SSPs."

11.2.5 Urban Living Labs, Circular Economy and Nature-Based Solutions: Ideation and Testing of a New Soil in the City of Turin Using a Multi-Stakeholder Perspective [130]

This is a machine-generated summary of:

Ascione, Grazia Sveva; Cuomo, Federico; Mariotti, Nicole; Corazza, Laura: Urban Living Labs, Circular Economy and Nature-Based Solutions: Ideation and Testing of a New Soil in the City of Turin Using a Multi-stakeholder Perspective [130]

Published in: Circular Economy and Sustainability (2021).

Link to original: https://doi.org/10.1007/s43615-021-00.011-6.

If you want to cite the papers, please refer to the original.

For technical reasons we could not place the page where the original quote is coming from.

Abstract-Summary

"In the attempt to foster circular economy (CE), cities are increasingly adopting urban living labs (ULLs) as sites of co-production aimed at testing alternative solutions based on the reuse of products, reduction of consumption and recycling of materials."

"The City of Turin joined proGIreg, a European project that entails the regeneration of former industrial districts by means of nature-based solutions (NBS)."

"The most promising is related to the production and test of the 'new soil', a blend obtained by mixing earth materials coming from construction sites with compost, zeolites and mycorrhizae."

"The case herein presented is interesting to analyse for the multi-stakeholder management setting used, where public institutions, private companies, research institutions, citizens and associations collaborated in the co-creation and testing phase of the NBS."

"The data collected through participant observation and direct interviews allow researchers to describe multi-stakeholders' dynamics and how they work."

"By consequence, the paper offers both theoretical contributions to the relational branch of stakeholder theory and practical evidence in demonstrating the importance of the relational branch of the theory over a more traditional transactional view."

Introduction

"Urban living labs (ULLs) are deemed to be innovative policy processes developed around the concept of multi-stakeholder engagement, where co-producers explore, examine, experiment, test and evaluate innovative ideas and alternative solutions in a real urban setting."

"ULLs, NBS and CE might be complementary in promoting policies and actions to regenerate and redesign the urban environment, but to be effective, they should be designed and tested to reach a shared consensus among the actors involved."

"In our paper, we present a case study of a systemic experience where a 'new soil' (artificial ground or technogenic soil) is tested as one of the most promising NBS for the urban requalification projects of the City of Turin (Italy), where the ULL is placed."

"We aim to shed light on how to examine the interplay between ULLs, CE and NBS, identifying in a multi-stakeholder approach a key driver of success and, lastly, providing a critique of the case tested."

Theoretical Framework

"Multi-stakeholder partnerships should be created to engage actors in the process of designing, developing, implementing, testing and evaluating an innovation [42]."

"Several studies acknowledge the importance of the relationship between urban regeneration, stakeholders' consideration and circular processes [43–45]; however, to our knowledge, no research links the idea of soil recovery with circular practices in the context of an urban regeneration process."

"In the case study presented in this research, new soil represents a NBS aimed at directly improving the condition of a post-industrialised brownfield area of Turin, named Mirafiori."

"Studies involving multi-stakeholder value creation in implementing a public policy around a CE issue are scarce [46, 47]."

"Studies involving public institutions are usually narrowed towards a stakeholder analysis to orient processes of communications, while less frequently the analysis has the strategic intent of interpreting relationships between actors to better frame future policies for co-management or co-design public initiatives."

Methods

"In the proGIreg framework, the artificial ground new soil has been formulated according to the following general composition: Building earth materials resulting from construction sites (particles with a diameter lower than 2 cm); Compost obtained from organic waste, in particular, plant waste; Natural zeolites, mainly chabazite; Natural mycorrhizae."

"Inside the proGIreg project, the innovative new soil has been tested in the ULL where an area of 2000 m^2, near the course of the Sangone River, has been dedicated to such an experiment."

"Researchers and public managers have started mixing up inert soil coming from construction sites with compost, zeolites and mycorrhizae, able to abide by the regulations and creating a new mixture that can be used to create new urban green areas."

"The Polytechnic of Turin played the role in supervising the coherence of the new soil experiment with the overall research project with many NBS and developing knowledge exchange dynamics between actors involved."

Results and Discussion

"During the interviews, researchers collected data about the future opportunities or constraints of new soil with the intent of developing a business model."

"In the case study of new soil, one of the most significant barriers in the development of a scalable business model has been the presence of strict environmental regulations that fix some limits in the presence of earth compounds when someone develops an earth composite."

"To guarantee that the outcome of the ULL for the NBS is effective, the business model of the new soil must be resilient, scalable and profitable."

"In the case herein presented, we give evidence to the fact that the relational view of stakeholder theory is playing and will play a pivotal role for the CE, and sooner or later, whoever wants to operate in developing sustainability innovation will not be exempted from adopting a multi-stakeholder management model."

Conclusion

"With the ULL, the city has reinvigorated the public interest towards the purpose of limiting the depletion of scarce resources, and it has played an active role in promoting a factual change in the regulation not only for the benefit of the project but also for all the private actors operating in the backfilling soil market."

"This entails that the new soil's composition is flexible so that it can be modulated according to specific needs; on the other hand, the possibility of exact replication is hampered by both chemical features of the earth materials and the requirements of diverse contexts of urban experimentation."

"With the hype of CE projects and ULL, experts of stakeholder theory and sustainability experts need to engage in studies that narrate how multi-stakeholder's arrangement happens in practice."

11.2.6 Natural Infrastructure in Sustaining Global Urban Freshwater Ecosystem Services [131]

This is a machine-generated summary of:

Chung, Min Gon; Frank, Kenneth A.; Pokhrel, Yadu; Dietz, Thomas; Liu, Jianguo: Natural infrastructure in sustaining global urban freshwater ecosystem services [131]

Published in: Nature Sustainability (2021).

Link to original: https://doi.org/10.1038/s41893-021-00.786-4.

If you want to cite the papers, please refer to the original.

For technical reasons we could not place the page where the original quote is coming from.

Abstract-Summary

"Rapid urbanization throughout the globe increases demand for fresh water and the ecosystem services associated with it."

"Natural infrastructure solutions have increased to provide freshwater ecosystem services, but little global research has examined the intricate relationships between built and natural infrastructure for providing freshwater ecosystem services to cities across the globe."

"Using network analysis, here we examine the interrelationships between built and natural infrastructure in 2113 watersheds for 317 cities worldwide, focusing on four key freshwater ecosystem services: freshwater provision, sediment regulation, flood mitigation and hydropower production."

"Improved understandings of the role of natural infrastructure in urban water networks must underpin strategic decision-making to sustainably provide freshwater ecosystem services to global cities."

Main

"With rapid increases in PAs and IWS worldwide, the networks of watershed conservation areas that act as a natural infrastructure may complement built infrastructure by providing various freshwater ES to cities."

"Our working hypotheses related to these questions are (1) that existing natural infrastructure in many cases helps enhance freshwater ES flows to cities by improving the capacity of built infrastructure, and (2) that environmental and socioeconomic factors in cities and source watersheds modulate such benefits, acting differently across different cities and source watersheds."

"The set of variables available to us at sending and receiving systems allows us to examine how built and natural infrastructure in source watersheds can be integrated to enhance the provision of freshwater ES for global cities (317 cities and 2,113 watersheds worldwide) while controlling for the net of geographic factors, watershed characteristics and city characteristics."

Results

"Increased urban populations had a positive relationship with the amounts of freshwater provision and sediment loads from source watersheds."

"New PA designations in such cities' freshwater source watersheds may be crucial to reduce sediment flows because high sediment levels in source watersheds create additional costs for urban water treatment [48–50]."

"Geological characteristics of watersheds also contributed to the flows of freshwater ES to cities."

"Watersheds with larger areas and greater distances between watersheds and cities had a positive relationship with more freshwater provisioning, sediment flows and hydropower production."

"We concentrated on freshwater source watersheds and hydropower watersheds, as both dams and PAs positively contributed to sediment reduction and hydropower production."

"PA designations are not the only type of natural infrastructure solution, and these alternative approaches could also act as a natural infrastructure in highly developed watersheds to sustain freshwater ES flows for cities."

Discussion

"Our network approach and results pointed out that natural infrastructure in source watersheds has been widely integrated with urban water supply networks by supporting the flows of freshwater ES to global cities."

"Our integration of natural infrastructure into urban water supply networks is a step to sustainably improve the supplies of freshwater ES to cities."

"The results indicate that natural infrastructure such as protected forests and wetlands already plays an important role in sustaining freshwater ES flows to cities as well as enhancing the performance of existing built infrastructure."

"Implementing natural infrastructure solutions requires examining causal relationships between freshwater ES supplies in source watersheds and their uses in cities, identifying how much natural infrastructure is needed to sustain freshwater ES flows during the most vulnerable seasons (for example, flooding, wildfire and drought seasons), and managing possible stakeholder conflicts between source watersheds and cities (for example, farmer–urban resident conflicts)."

Methods

"Freshwater source watersheds provide surface water to cities."

"Freshwater source watersheds and surface water intake points were obtained from the City Water Map database [51]."

"To examine which characteristics contribute to four freshwater ES flows from source watersheds to cities, we collected data regarding dams, watershed conservation activities (natural infrastructure solutions), environmental factors and socioeconomic factors in source watersheds and cities."

"We used multilevel models applied to egocentric network analysis to estimate the contribution of each independent variable to freshwater ES flows from source watersheds to cities [52]."

"At level 1, we modelled changes in freshwater ES flows as a function of forest cover in PAs (or non-PAs and both), wetland cover in PAs (or non-PAs and both), dam density (or reservoir storage capacity), irrigation area, watershed areas, distance from city to watershed, elevation and slope."

11.2.7 Constructing Smarter and More Sustainable Urban Ecosystems: A Dynamic Analysis of Challenges and Initiatives [132]

This is a machine-generated summary of:

Rodrigues, Maria I. R. P. R.; Ferreira, Fernando A. F.; Ferreira, Neuza C. M. Q. F.: Constructing smarter and more sustainable urban ecosystems: a dynamic analysis of challenges and initiatives [132]

Published in: Annals of Operations Research (2023).

Link to original: https://doi.org/10.1007/s10479-023-05,406-2.

If you want to cite the papers, please refer to the original.

For technical reasons we could not place the page where the original quote is coming from.

Abstract-Summary

"Cities need to become enablers of economic and social development and increased prosperity through the integration of technologies into projects promoting smarter and more sustainable urban ecosystems."

"Based on a constructivist and complementarity logic, this study sought to develop a multicriteria analysis model to support relevant decision-making processes in this study context."

"A panel of experts was recruited to create the analysis system using a combination of cognitive mapping and the decision-making trial and evaluation laboratory technique in an neutrosophic context."

"The entire procedure focused on enhancing the experts' learning through participation not only to structure the problem under analysis but also to identify and prioritize the factors and/or determinants of smartness and urban sustainability."

"The model provides a holistic, solid, and clear vision of the decision problem that fosters appropriate choices when creating and evaluating smart, sustainable urban ecosystems."

Introduction

"This approach creates smart, sustainable urban ecosystems, yet no clear consensus exists on which factors influence the development of smart, sustainable cities because of this topic's complexity and wide scope."

"To address these general limitations, the present study sought to strengthen decision makers' ability to understand which determining factors have to be considered while creating and/or analyzing smart, sustainable urban ecosystems."

"Two research questions were addressed: How can decision makers identify the key initiatives that contribute to and/or shape conceptualizations of smart, sustainable urban ecosystems?."

"The present research, therefore, combined DEMATEL and neutrosophic logic for the first time to facilitate a study of smart, sustainable urban ecosystems."

"The next section presents a literature review of the conceptualization of smart, sustainable urban ecosystems and some supporting approaches, after which the research gaps and theoretical framework are discussed."

Related literature and research gaps

"Smart, sustainable cities are urban entities capable of integrating social and digital technologies into processes and coordinating the existing knowledge to overcome economic, social, and environmental challenges that affect municipalities' long-term viability (Hara and others, 53)."

"Smart city initiatives are a long process that involves different components of urban areas (Khan and others, 54), but these efforts are extremely important to making urban environments smarter and more sustainable."

"Researchers failed to specify the criteria's significance for smart, sustainable urban ecosystem development."

"The present study sought to meet the need for a different analysis model that can overcome the aforementioned limitations and that will contribute to better practices and substantive developments in smart, sustainable urban ecosystem assessment."

"The aim was to identify concrete initiatives that help create and evaluate smart, sustainable urban ecosystems by using a multicriteria analysis system that addresses some of the limitations of previous investigations."

Methodological background

"Several methods have been developed previously to help decision makers appropriately structure complex issues, which produce graphical representations of problems and construct more innovative solutions (for examples and details, see Rosenhead (55) and Ackermann and Eden (56))."

"These maps help decision makers gain a fuller understanding of the relevant problem as they reflect on—and negotiate with each other over—which key criteria and/or determinants most influence the issue."

"Cognitive mapping is able to: (1) include qualitative variables; (2) structure complex decision situations; (3) support group work; and (4) facilitate the development and implementation of strategic solutions (Faria and others, 57)."

"Cognitive mapping is a process that encourages stakeholders to learn about the problem analyzed, understand the organization of decision makers' thought process, and obtain information about the issue at hand."

"DEMATEL is an extremely useful tool that converts factors' interdependence into cause-effect relationships and determines which criteria have a greater or lesser impact on the decision-support model (Dalvi-Esfahani and others, 58)."

Application and results

"SC19 is overall this cluster's most prominent factor in the present analysis system as its R + C value is the highest (i.e., 11.2726)."

"SC65, in contrast, is not only the most affected by the remaining SCs, with the highest C value of 7.3769, but also this cluster's factor of greatest importance to the analysis system, with an R + C value of 14.2719."

"This SC also is this cluster's most significant factor in the decision-support system, with an R + C value of 12.3205."

"SC118, in turn, receives the most effects from the remaining SCs, with the highest C value (9.6939), as well as being this cluster's most important SC in the analysis system, with an R + C value of 19.1565."

Conclusion

"The most important result in the present study was a model to support decisions regarding how to promote smart, sustainable urban ecosystems, which was produced by a combination of cognitive mapping and DEMATEL in a neutrosophic environment."

"The model simplifies decision-making processes as it identifies which areas to prioritize and what should be done to support the conceptualization and analysis of smart, sustainable urban ecosystems."

"The research included the definition of a set of measures essential to developing smart, sustainable urban ecosystems."

"This methodology considered the element of uncertainty in order to expand the conceptualization of smartness and urban sustainability and obtain more credible, transparent results."

"The construction of smarter and more sustainable urban ecosystems is clearly a topic with great potential for future research, and any new contributions that ensure more empirically robust findings will always offer significant benefits."

11.2.8 Finding Space for Nature in Cities: The Considerable Potential of Redundant Car Parking [133]

This is a machine-generated summary of:

Croeser, Thami; Garrard, Georgia E.; Visintin, Casey; Kirk, Holly; Ossola, Alessandro; Furlong, Casey; Clements, Rebecca; Butt, Andrew; Taylor, Elizabeth; Bekessy, Sarah A.: Finding space for nature in cities: the considerable potential of redundant car parking [133]

Published in: npj Urban Sustainability (2022).

Link to original: https://doi.org/10.1038/s42949-022-00.073-x.

If you want to cite the papers, please refer to the original.

For technical reasons we could not place the page where the original quote is coming from.

Abstract-Summary

"Delivering these benefits at scale will require large areas of scarce urban land to be converted into green space."

"We show an approach by which cities can make substantial progress towards their sustainability targets using NBS, by converting redundant street parking into biodiverse green space."

"We demonstrate that up to half of street parking in our case study municipality (The City of Melbourne) could be accommodated in garages within 200 m, freeing up large areas for greening."

"As many cities allocate extensive areas to both street parking and off-street garages, this approach to freeing up space for nature in cities is widely applicable."

Introduction

"Any urban land use change requires consideration of the practical trade-offs, and so identifying the most viable opportunities for large-scale, systematic change is an essential prerequisite for cities hoping to meet targets for NBS delivery."

"These traits make central Melbourne an ideal context to explore potential trade-offs between street parking and green space to deliver NBS at scale in highly urbanised areas."

"As public discourse around parking reform to date has often narrowly focused on the perceived loss of convenience of, and free access to, street parking for drivers [59–61], we use an interdisciplinary set of methods to explore an alternative side of the parking space trade-off, by demonstrating how multifunctional NBS could deliver a diverse set of potential ecosystem services for broader public benefit."

"We model a range of sustainability benefits delivered by replacing the redundant on-street parking with biodiverse green space."

Results.

"This study considers the reallocation of a portion of the City of Melbourne's 23,500 street parking spaces into vacant space in the 193,500 garage spaces within the municipality."

"11,668 spaces represent 47% of the 24,745 total on-street spaces in the city, which cover approximately 50 ha."

"Of this total area (municipality-wide), between 2.7 to 7.7 ha of de-paving opportunities exist within the flood-prone Elizabeth Street Catchment at the centre of Melbourne."

"To present this study's results in terms of the challenges cities seek to address using NBS, where possible, we compared our results to quantitative targets already established by City of Melbourne."

"The 2.7 to 7.7 ha of de-paving delivered in the flood-prone Elizabeth Street Catchment at the heart of the municipality represents between 4 and 12% of the 65 ha target for de-paving in this area, highlighting the need for complementary measures such as rooftop greening, permeable sidewalks and other de-paving solutions."

Discussion

"We examined the extent to which redundant street parking may be converted to biodiverse green space, quantifying the impacts of this change in terms of tree canopy, de-paving impervious surfaces, stormwater treatment and ecological connectivity."

"Our study highlights how a systematic reallocation of space in streetscapes can produce benefits at the scale that is required for cities to genuinely tackle significant urban sustainability challenges."

"The thousands of redundant car parking spaces in central Melbourne's streets represent an opportunity to replace up to 24 hectares of asphalt with biodiverse green space in the city's densest neighbourhoods."

"If the City of Melbourne were willing to replace parking at a reduced level—for example, by greening three street parking spaces for every two made available in parking garages—the scale of change would effectively be multiplied by that ratio."

Methods

"This part of the analysis required us to first establish how many potentially vacant off-street parking spaces exist in residential, commercial and other private garages."

"This analysis employed two additional datasets from the City of Melbourne Open Data Platform: a map of on-street public parking spaces, and a map of the street network."

"The overall canopy cover derived in each scenario was calculated by assuming that an equal proportion of each species was planted across the total number of viable parking spaces in each scenario."

"In each scenario, viable locations with existing tree canopy over parking lot centroids were excluded from the canopy analysis, assuming (conservatively) that trees would not be planted in these lots."

"To model the effect of parking space conversion on ecological connectivity we assumed that an area of species habitat corresponding to the spatial extent of each parking space would be being added to the landscape."

11.2.9 Urban Biodiversity: State of the Science and Future Directions [134]

This is a machine-generated summary of:

Rega-Brodsky, Christine C.; Aronson, Myla F. J.; Piana, Max R.; Carpenter, Ela-Sita; Hahs, Amy K.; Herrera-Montes, Adriana; Knapp, Sonja; Kotze, D. Johan; Lepczyk, Christopher A.; Moretti, Marco; Salisbury, Allyson B.; Williams, Nicholas S. G.; Jung, Kirsten; Katti, Madhusudan; MacGregor-Fors, Ian; MacIvor, J. Scott; La Sorte, Frank A.; Sheel, Vallari; Threfall, Caragh G.; Nilon, Charles H.: Urban biodiversity: State of the science and future directions [134]

Published in: Urban Ecosystems (2022).

Link to original: https://doi.org/10.1007/s11252-022-01,207-w.

If you want to cite the papers, please refer to the original.

For technical reasons we could not place the page where the original quote is coming from.

Abstract-Summary

"Recognition of urban biodiversity research has increased steadily."

"Knowledge of how ecological communities respond to urban pressures can assist in addressing global questions related to biodiversity."

"To assess the state of this research field in meeting this aim, we conducted a systematic review of the urban biodiversity literature published since 1990."

"While advances have been made in the field over the last 30 years, we found that urban biodiversity research has primarily been conducted in single cities within the Palearctic and Nearctic realms, within forest remnants and residential locations, and predominantly surveys plants and birds, with significant gaps in research within the Global South and little integration of multi-species and multi-trophic interactions."

"We advocate for future research that encompasses a greater diversity of taxonomic groups and urban systems, focusing on biodiversity hotspots."

"Implementing such research would enable researchers to move forward in an equitable and multidisciplinary way to tackle the complex issues facing global urban biodiversity."

Introduction

"The conservation and management of biodiversity in cities requires knowledge of the ecological patterns and processes that drive species' responses and adaptation."

"Urban biodiversity research is essential for understanding how intensive human activities affect the ecology and evolution of a region's species, which in turn can inform conservation initiatives designed to mitigate biodiversity loss (McKinney 62)."

"One recommendation to emerge from a workshop hosted by the Urban Biodiversity Network (UrBioNet) in March 2017 at Rutgers University, New Brunswick, New Jersey, was the need to assess the current state of urban biodiversity research in order to reflect on the work conducted since McDonnell and Pickett (63) and McKinney (62, 64) by identifying areas of saturation and gaps in the literature."

"In response to this recommendation, we performed a systematic review of the literature with the goal of addressing three objectives: (i) document patterns of geographic and taxonomic foci, and methodology used in urban biodiversity research since 1990, (ii) examine how ecosystem function, management, and restoration are addressed in urban biodiversity research, and (iii) identify critical knowledge gaps for future research."

Methods

"We reviewed the titles and abstracts of each for relevance regarding emphasis on biological communities, inclusion of multiple sites, and urban focus."

"We excluded review papers to avoid replicating any studies in our analyses; however, we kept meta-analyses as they presented new analyses regarding urban biodiversity trends over larger spatial or temporal dimensions compared to single studies."

"Studies were classified by publication year, journal, location(s) (city, country, biogeographic realm, or multiples of each), city of the lead author's institution, urban comparison type (urban only, gradient, urban vs. rural), how the urban area was defined (e.g., municipal boundary, land use, road density, population density), sampling effort (number of locations and duration) and methodology, land use and habitat types sampled, biodiversity metrics analyzed (e.g., species richness, taxonomic, functional, and/or phylogenetic diversity), data availability (e.g., species, traits, and/or coordinates identified), and linkages to restoration and management techniques and/or ecosystem service provision."

Results and Discussion

"As a local or "backyard" science, the study of urban biodiversity allows for opportunities for education and engagement with communities surrounding universities and other research institutions, which likely enhance conservation interest by urban residents, even for biodiversity beyond the city and in natural areas (e.g., Narango 65)."

"Forward for urban biodiversity research, studies that examine community patterns in multi-trophic interactions (e.g., pollination, predation, decomposition; Frey and others 66; Seibold and others 67; Tresch and others 68) and those that examine biodiversity of taxa closely associated with ecosystem function (e.g., soil microbial diversity), should be prioritized with biomonitoring surveys in taxa that are well studied (e.g., bees, birds)."

"This new direction in urban biodiversity research, where studies are conducted entirely inside city boundaries, highlights the diversity and complexity of urban habitats and land uses within the city itself, and the necessity to look at urban areas from a regional perspective."

The way forward

"The current geographic and study systems investigated (e.g., taxa, vegetation habitat, land use) bias our understanding of urban biodiversity towards birds and plants, forested ecosystems, the Global North, and areas of intense habitat management."

"We also recommend researchers and practitioners continue to broaden taxonomic representation in urban biodiversity research for understudied urban taxa (e.g., snails, spiders, reptiles, soil invertebrates, microbes) as these groups play important functional roles within urban ecosystems, as well as research that examines interactions of multiple taxa across trophic levels."

"Recent research demonstrates that urban biodiversity conservation initiatives are most successful when practitioners actively engage communities to understand people's needs and embrace the diversity of values the public hold towards biodiversity (Taylor and others 69)."

"We can commit to understanding biodiversity in cities and continuing to expand the scope, multidisciplinarity, equity, inclusivity, and rigor of urban ecological research within a rapidly urbanizing world, which could help pave the way to obtaining healthier, livable cities."

11.2.10 The "Sustainable" in Smart Cities: Ignoring the Importance of Urban Ecosystems [135]

This is a machine-generated summary of:

Mundoli, Seema; Unnikrishnan, Hita; Nagendra, Harini: The "Sustainable" in smart cities: ignoring the importance of urban ecosystems [135]

Published in: DECISION (2017).

Link to original: https://doi.org/10.1007/s40622-017-0152-x.

If you want to cite the papers, please refer to the original.

For technical reasons we could not place the page where the original quote is coming from.

Abstract-Summary

"Urban green spaces in cities are common pool resources (CPRs) or urban commons that provide provisioning, supporting, regulating and cultural ecosystem services."

"The smart cities model disregards the provisioning services of urban CPRs that are critical for livelihoods and subsistence of urban marginalised groups."

"The paper highlights the multiple uses of urban commons, with emphasis on the provisioning services that support lives and livelihoods of urban residents."

"We urge a systematic reconceptualisation of urban sustainability, and towards a multi-faceted use-value of urban ecosystems, to develop both ecologically and socio-culturally smart cities."

Introduction

"The proponents of the smart cities vision cite multiple advantages, and political acceptance for smart city projects is also much greater: after all who would oppose the idea of living in a smart city with technology providing solutions for urban problems (Hollands 70)."

"This popular vision of smart cities model has managed to find acceptance rapidly even in countries of the Global South, a region projected to witness rapid urbanisation and at the same time plagued by problems of social equity and environmental sustainability."

"At the same time that smart cities are presented as providing solutions to environmental problems such as climate change, the smart cities model to address urbanisation is itself seen as undermining environmental sustainability owing to its considerable requirements of energy and materials (Hollands 70, Vittanen and Kingston 71)."

Urban ecosystems and urban commons

"Green and open spaces in cities are also important providers of common pool resources (CPRs) or urban commons that serve a range of provisioning services to diverse groups of urban residents coming from different cultural, ethnic and socio-economic backgrounds."

"Provisioning ecosystem services are provided by urban CPRs like lakes, parks, wooded groves and avenue trees, as well as smaller vegetated spaces in slums and home gardens."

"Gourds, leafy vegetables, tomatoes, chillies, beans and a variety of flowers grown around water bodies like Chikkagubbi, and Varthur Lakes find their way into city markets and eventually into the homes of Bengaluru's residents."

"Keeping in mind the potential impacts of climate change that cities like Bengaluru will be exposed to in the coming decades, smaller trees provide reduced capacities for regulating services like absorbing pollution, providing shade and cooling, and sequestering carbon (Nagendra and Gopal 72)."

Ecosystem enclosures

"The very idea of a smart city could contribute to the conversion of these commons to public or private spaces, and along with it exclude access for certain groups."

"Fencing, entry fees and patrolling by guards of the PPP managed lakes actively contributed to this alienation, with those practising traditional livelihoods like grazers, fodder grass collectors, brick makers and dhobhis unable to access the lakes."

"While the PPP model for lake management has been discontinued in Bengaluru thanks to objections from citizens and a public interest litigation filed by the advocacy organisation, the Environmental Support Group, it does provide a useful lesson on how the transformation of urban commons to a public good results in exclusion of traditional users and marginalised communities."

"The collusion of the State in acquiring urban CPRs like lakes and gunda thopes as government property only to convert them to private property undermines the access to these commons for the larger population."

Conclusion

"There is an urgent need to develop a framework in planning for urban sustainability recognising the uniqueness of the ecological context of each city, and bridging the social and ecological systems in cities (Sassen 73, Yigitcanlar and Dizdaroglu 74)."

"Factoring in urban ecosystem services, especially the various provisioning services obtained from urban commons that are low cost and publicly utilised, is essential for city planning initiatives in the Indian context, providing a means to ensure the increased social and ecological resilience of the growing numbers of disempowered and disadvantaged in cities across India."

"Smart city initiatives cannot ignore this contested and dynamic nature of urban commons, even as they provide multiple ecosystem services."

"We urge a systematic reconceptualisation of urban sustainability to develop ecologically and socio-culturally smart cities: valuing and working with urban ecosystems to provide a richer, healthier and more fulfilling urban experience, one that provides avenues for the resilience for all urban residents."

11.2.11 Young People and Environmental Affordances in Urban Sustainable Development: Insights into Transport and Green and Public Space in Seven Cities [136]

This is a machine-generated summary of:

Nissen, Sylvia; Prendergast, Kate; Aoyagi, Midori; Burningham, Kate; Hasan, Mohammad Mehedi; Hayward, Bronwyn; Jackson, Tim; Jha, Vimlendu; Mattar, Helio; Schudel, Ingrid; Venn, Sue; Yoshida, Aya: Young people and environmental affordances in urban sustainable development: insights into transport and green and public space in seven cities [136]

Published in: Sustainable Earth (2020).

Link to original: https://doi.org/10.1186/s42055-020-00.039-w.

If you want to cite the papers, please refer to the original.

For technical reasons we could not place the page where the original quote is coming from.

Abstract-Summary

"Understanding young people's experiences in urban environments is therefore important as we strive to achieve both the Sustainable Development Goals and the Paris Agreement."

"Two amenities identified in the urban Sustainable Development Goal 11, transport and public and green space, are specifically recognised as applying to youth."

"There is little analysis that explicitly considers how youth experience these amenities across the Global North and South, and no current measures for understanding progress in youth experiences of green space and transport."

"This paper provides a comparative analysis of young people's experiences with local transport and green space in seven diverse urban communities (Christchurch, New Zealand; Dhaka, Bangladesh; Lambeth/London, UK; Makhanda, South Africa; New Delhi, India; São Paulo, Brazil; and Yokohama, Japan)."

"Drawing on focus groups and interviews conducted with 332 young people, we identify five affordances young people associate in relation to transport and public space across these diverse urban settings: (1) social inclusion and belonging; (2) autonomy; (3) physical comfort and security; (4) relaxation and reflection; and (5) health and fitness."

"In providing a comparative account of young people's experiences across diverse contexts, our discussion highlights how affordances in relation to transport or public and green space can help understand the multiple interconnections between the well-being of young people and sustainability."

"We argue that it is not merely the provision of transport or public and green space that matters, but the nuanced meaning of places and experiences as understood by local communities that needs to be recognised if we are to better support urban youth wellbeing and advance sustainable development goals."

Introduction

"There is surprisingly little analysis of how young people experience sustainable development in urban communities, especially in lower income countries where the vast majority of young people reside [75, 76]."

"We report on the urban affordances identified in focus group discussions conducted with 332 young people, aged 12 to 24 years, from seven urban communities of diverse size, in the Global South and Global North."

"Our focus in this paper is on young people's perspectives and experience of two amenities - transport and green and public space - which are the two areas under the Sustainable Development Goal 11 targets that explicitly specify age-related significance."

"Our comparative analysis of young people's experiences highlights how measuring affordances can extend our understanding of how young people's wellbeing might be supported in an urban century."

Young people in urban sustainable development

"and 11.7), is to consider young people's experiences of the affordances of these features of their urban environment."

"Youth and urban affordances research tended to focus on the functional aspects of the environment, and the opportunities and restrictions local places offer a young person."

"Clark and Uzzell [77] evaluated different local environments for what they can afford residents, including: a place to be active; enjoy; feel safe; avoid others; and meet new people."

"Examining affordances offers a distinct perspective on the relationship between young people and their local urban environments."

"Affordance theory helps challenge simple descriptive accounts of children's experiences in their environment."

"Other researchers have noted the way opportunities for youth to take advantage of urban affordances can be influenced by social inequality."

"Research also suggests that applying affordance theory effectively requires paying close attention to local experiences, rather than assuming a one-size-fits-all policy approach."

Methods

"This paper reports on the urban affordances discussed by 332 young people, aged 12 to 24 years, in seven urban communities of diverse size."

"Following on from a desk-based context study [78], a total of 45 focus groups and 32 individual interviews were conducted with young people between November 2017 and September 2018 across seven cities."

"The focus groups and individual interviews sought to gain in-depth, qualitative insight into the experiences of young people in their diverse local communities, what they value about their lives, and to explore their hopes and fears for the future [79]."

"Semi-structured discussion was used to ask young people what they liked most about living in their city; their everyday activities and experiences; what they would like to change about where they live; and who can make changes in their community."

Results

"Independently coding participants' discussions of transport and green space revealed strong resonances between the affordances young people described in relation to these two domains across all seven cities."

"Both transport and green space were described by the young people across all cities as offering a way to connect with other people and feel a sense of belonging (n = 61)."

"In Christchurch, young people spoke about how biking was "our mode of transport for independently getting around", while Participant U (Makhanda) showed us a picture of her bicycle, explaining "it takes me where my heart leads me, I go to town and back with it, it doesn't get tired.""

"In two cities, transport was described as a refuge: one participant in Lambeth spoke about a friend who rode public transport when he did not have a place to sleep, and in Christchurch young people in one focus group discussed families without homes living in cars."

Discussion

"Resonating with previous work by Clark & Uzzell [80] and Egli and others [81], this study identified how transport systems and public and green spaces can allow important affordances for young people's social inclusion, autonomy, physical comfort and security, relaxation and reflection, and health and fitness."

"The experiences of the young people we spoke to highlight the considerable challenges confronting the provision of sustainable transport and public space in cities."

"The concept of affordances makes an important additional contribution: that it is not merely the amenities or provision of transport or public and green space that matters, but the human interactions with it and the interconnections it enables with the natural world that need to be measured to understand how we can better support urban youth wellbeing in sustainable ways [82]."

"The synergies between the experiences of the young people in the seven cities – particularly the affordances offered by public transport and green and public space – is also suggestive of the utility of the affordances concept in a comparative context."

Conclusion

"In providing an analysis of young people's experiences in seven diverse cities spanning the Global North and Global South, this paper provides an important comparative contribution to analysis of progress towards the Sustainable Development Goal 11 for children and youth in urban environments."

"In listening to children and young people's perspectives, our study adds support to recent analysis [82, 83] that suggests understanding the nuanced local affordances offered by transport and green and public space is important for understanding child and youth wellbeing and well-becoming in low-carbon, sustainable ways."

"The paper identified five affordances that these amenities potentially provide for young people: social inclusion and belonging; autonomy; physical comfort and security; relaxation and reflection; and health and fitness."

"There is considerable scope to broaden targets and indicators to include more age-related analysis to understand what cities mean for children and young people, including in housing, air quality, waste and participatory planning processes."

11.2.12 Nature-Based Solutions in Cities—Contribution of the Portuguese National Association of Green Roofs to Urban Circularity [137]

This is a machine-generated summary of:

Pineda-Martos, Rocío; Calheiros, Cristina S. C.: Nature-Based Solutions in Cities—Contribution of the Portuguese National Association of Green Roofs to Urban Circularity [137]

Published in: Circular Economy and Sustainability (2021).

Link to original: https://doi.org/10.1007/s43615-021-00.070-9.

If you want to cite the papers, please refer to the original.

For technical reasons we could not place the page where the original quote is coming from

Abstract-Summary

"Green building–integrated systems and technologies (e.g., green roofs (GR) and walls (GW)) are classified as nature-based solutions (NBS) in the context of urban green infrastructure (GI), which contribute to add both natural elements and processes, as a result of locally designed, resource-efficient, and systemic interventions in cities."

"Sustainable and resilient societies under the challenge 'innovating with nature' are the leading aim of the EU Research and Innovation (R&I) policy agenda goals on NBS and re-naturing urban areas."

"The commitments of the foundations to encourage and promote the advanced adoption of green urban infrastructure practice and planning as part of the built environment, drive active efforts to support NBS innovation objectives and the transition from 'grey to green' infrastructure."

"The present manuscript aims at reflecting the crucial role of the Associations on GI/NBS, mainly GR and GW, to develop local frameworks applying innovative plans of action, and allocate R&I opportunities implementing relevant and inclusive urban regeneration solutions."

Introduction

"The challenges that contemporary European cities and societies will tackle (e.g. natural capital depletion, human health and well-being, food security, energy and water) are closely connected with environmental events together with the urbanisation and its new perspective from innovation and transformative opportunities with nature-based solutions (NBS) following environmental, social and economic demands [84–86]."

"The foreseeable future of cities in Europe will require tackling better knowledge, transversal research, public and private investments and policy strategies on NBS (e.g. the European Union (EU) Research and Innovation (R&I) policy agenda for NBS) to shape more sustainable and efficient urban conditions boosting the combined effect among nature, society and economy [84, 86–88]."

"The implementation and promotion of NBS or nature-based green and blue infrastructure solutions and technologies enhancing the natural capital would be target opportunities for innovation in a really smart way as sustainable and cost-effective ways and interactive services for creating a greener, competitive and resource-efficient ("resourceful') urban circular economy (CE) [84, 86, 89–91]."

Green Infrastructure, Biodiversity and Ecosystem Services in the Context of Nature-Based Solutions

"Assuring the services and benefits of healthy ecosystems, and preserving the long-term delivery of simultaneous ES linked to strategically planned (accessible and well-connected and managed) networks of GI at multiple spatial scales (i.e. spatial and temporal multifunctionality and connectivity aspects), coordinate and support numerous EU policy objectives, in areas such as agriculture, air pollution (air quality), biodiversity and nature, climate change and energy, environment and health, environmental technologies and earth observation (EO) [92], sustainable consumption and production, urban environment, waste and water among others; and fulfill variety of policy-maker and stakeholder group needs based on the best existing science and policy networks [93–95]."

"Nature-based solutions should aim at maintaining and enhancing natural capital, promoting biodiversity and human well-being while bringing other co-benefits (e.g. potential of GR and GW in providing holistic urban benefits), as being integrated into policies and planning [96]."

"Faivre and others (2017) [86] exemplified the strategic environmental, social and economic challenges faced by an urban GR as an innovative implemented NBS and suggested a three-scale (i.e. high, medium, low) classification grouping potential benefits as: high (biodiversity), medium–high (social cohesion, climate, health and green jobs), medium (water) or low-medium (disaster and food)."

Nature-Based Solutions and Major European Union Research Policy Priorities

"The active coordination and engagement of the EC in the direction of investing in NBS through the ambitious EU R&I policy (e.g. agenda on NBS and re-naturing cities) tackle cross-cutting societal and environmental challenges with aiming research on NBS for the benefit of biodiversity (i.e. habitat and species protection) and the support of urban-values ES delivery (e.g. ecosystem-based approaches) as drivers for innovation on smart growth, green economy and creating direct and indirect job opportunities in Europe [86, 96–99]."

"Current research initiatives and policies from existing environmental and NBS R&I agendas, plans and programmes in Europe, will assist the EU to accelerate steps forward to a circular, resource-efficient and low-carbon economy, besides smart, inclusive and sustainable development and growth [86, 100–102]."

Associations as Stakeholder Groups for Innovation with Nature in Europe

"The associations provide help in supporting current European and national policies' objectives and frameworks for urban NBS/GI implementation and offer in practice opportunities for NBS/GI upscaling and mainstreaming in cities."

"The associations may not be considered, in general, business stakeholders, but they are of great value having into consideration their insight on NBS, the mission that sets their foundations and, the members that represent."

"It is thus important that different categories of stakeholders from public to private sector, citizens and associations are engaged in the process of NBS implementation [103, 104]."

"The European Federation of Green Roof and Living Wall Associations (EFB) was established in 1997, thanks to the founding GR Associations from Austria, Germany and Switzerland, and, in 2019, had integrated fifteen-member associations from different European countries."

"GI/NBS organisations and associations are engaged in promoting urban greening and supporting their members in order to meet their interests."

Case study: the Portuguese National Association of Green Roofs

"The Portuguese National Association of Green Roofs (ANCV - Associação Nacional de Coberturas Verdes) is an NGO founded on the 19 of June 2015, which aims to promote GI/NBS in cities, especially those that can be installed in buildings (new or pre-existing) such as GR and GW, highlighting their importance and contributions to the possibility to create healthy, sustainable, biodiverse and resilient urban territories."

"Establishment of relationships and development of work with similar associations at international level, with a special focus on those sharing the typical Mediterranean climatic conditions v. Creation of a national database of GR and GW projects and an international database on municipal policies and strategies for GR and GW implementation vi."

"Membership dynamics: The opportunity to have detailed business information on the ANCV website is provided to companies' members, allowing the online visitors to have easy access when finding who can provide GR information, including project and consultancy; implementation and maintenance; thermal insulation; anti-root impermeabilisation and layers, blankets and filters; drainage; substrates; vegetation; irrigation systems; and GW."

Conclusion

"The European Union community is engaged with this alignment through policies that innovate with nature, highlighting the R&I agenda on nature-based solutions—inspired from the H2020 Expert Group report on 'Nature-Based Solutions and Re-Naturing Cities'—to promote the adoption of innovative business and investment models besides long-term financing institutional and legal frameworks for nature-based solutions in Europe; and the Biodiversity Strategy and its action plan that was recently adopted by the European Commission, representing a comprehensive long-term programme aspiring protecting nature and reversing the ecosystem degradation by 2030."

"In order that the policies and strategies give rise to applied action plans, every stakeholder plays a pivotal role acting in different dimensions to tackle the constraints and barriers of green infrastructure and nature-based solutions adoption."

"This paper delivers an overview on how association actions and dynamics can promote inclusion of nature-based solutions, although it would be interesting to look deeper into strategies to establish collaboration between stakeholders, integrate citizens' participation, and overcome barriers towards nature-based solutions implementation."

11.2.13 Are Cities Venturing Green? a Global Analysis of the Impact of Green Entrepreneurship on City Air Pollution [138]

This is a machine-generated summary of:

Cojoianu, Theodor F.; Hoepner, Andreas G. F.; Hu, Xi; Ramadan, Moustafa; Veneri, Paolo; Wójcik, Dariusz: Are cities venturing green? A global analysis of the impact of green entrepreneurship on city air pollution [138]

Published in: Small Business Economics (2023).

Link to original: https://doi.org/10.1007/s11187-023-00.764-4.

If you want to cite the papers, please refer to the original.

For technical reasons we could not place the page where the original quote is coming from.

Abstract-Summary

"Situated at the intersection between environmental entrepreneurship and urban sustainability, our paper seeks to investigate the links between city-level green venture ecosystems and the ability of urban centres to reduce air pollution."

"Looking closely at the subsectors that drive the results, we observe that the urban centres which hosted increased numbers of innovators in smart grid technologies, energy efficiency and wind energy generation (the low carbon energy sector overall) also experienced a decrease in air pollution over the 2010–2019 period."

Introduction

"Green entrepreneurs and venture capital investors are now at the forefront of solving many environmental problems, investing and developing new materials and clean technologies that in theory, it should have a positive environmental impact (Vedula and others, 105)."

"Situated at the intersection between environmental and urban entrepreneurship, our paper seeks to understand whether cities with VC-backed green are able to faster reduce their air pollution levels (proxied by the yearly average PM2.5 levels), for 12,834 urban centres around the world between 2010 and 2019."

"As sustainable urban entrepreneurship is a novel topic, and the previous literature does not fully allow us to form strong convictions on hypotheses, similarly to Cojoianu and others, (106, 107), Feldman and others (108) and Moeen and Agarwal (109), we employ an exploratory research design to unveil the impact of (green urban) entrepreneurship on an important and city-level relevant environmental indicator, air pollution."

Theoretical background and research questions

"Studies focusing on green entrepreneurship have yet to link the process of entrepreneurship with actual societal impact (e.g. air pollution reduction, climate change mitigation, biodiversity conservation at the city-region level) and have focused mostly on drivers of entry, growth and innovation of green solutions, regardless of whether these make an actual difference or not."

"We contend that both city-level and global green urban entrepreneurs can influence their host cities (i.e. the cities they have been founded in) for the interactions that they have with numerous stakeholders within the city, through the commercial solutions they offer to cities and municipalities as well as through the green knowledge spillovers they generate within the local economy (Cojoianu and others, 107)."

Data and methodology

"Our dependent variable which proxies the average air quality within a city over a year is the average population mean exposure to PM2.5, which was estimated by the OECD based on the Global Burden of Disease 2019 (GDB) project data."

"Using the two data providers above, and the green taxonomy provided by Cleantech-Group, using ArcGIS, we match each of our 167,256 start-ups and their addresses with the 12,834 urban centres in our database."

"For each urban centre, we build six key independent variables: the number of cumulative green start-ups, cumulative green start-up funding ($), the number of total cumulative green start-ups, cumulative total start-up funding ($), the number of new start-ups founded in a given urban centre (lagged 1 year) and the amount of new funding received by all start-ups in the urban centre (lagged 1 year)."

Synthesis of results

"Model 1 suggests that a 50% increase in the average cumulative green start-ups within a city is related to a 25% reduction in air quality over the 2010–2019 period."

"The method involves the following steps: First, we regress cumulative green entrepreneurship$_{i,t-1}$ on lagged air pollution$_{,t-2}$, to separate green entrepreneurship into two components, one which is driven by city air pollution, and one which is unrelated to city air pollution."

"We find that the effect of cumulative green entrepreneurship on urban air pollution retains its sign and significance as in our main models."

"We explore the differences in the air pollution reduction after the introduction of new legislation in a given country, between green VC cities and non-green VC cities."

Discussion and conclusions

"In responding to the call of Vedula and others (105) and Acs and others (2021) to further the research agenda on how entrepreneurship delivers environmental and/ or social returns, in particular, related to the sustainability of urban centres, our paper seeks to understand whether green VC-backed entrepreneurs have positively impacted air pollution levels in the urban centres they have been founded."

"Our study builds on the emerging literature on sustainable urban entrepreneurship, in particular on the framing of Cohen & Muñoz (110), by analysing how VC-backed green entrepreneurs, who operate both at the city level as well as the global city network level, influence the air pollution of their host urban centres."

"For entrepreneurship researchers and urban scholars, our study lays the foundation towards understanding the real environmental impacts of green entrepreneurship on their local environment."

11.2.14 Green and Digital Entrepreneurship in Smart Cities [139]

This is a machine-generated summary of:

Manjon, Miguel; Aouni, Zineb; Crutzen, Nathalie: Green and digital entrepreneurship in smart cities [139]

Published in: The Annals of Regional Science (2021).

Link to original: https://doi.org/10.1007/s00168-021-01,080-z.

If you want to cite the papers, please refer to the original.

For technical reasons we could not place the page where the original quote is coming from.

Abstract-Summary

"This paper investigates the relationship between the implementation of smart city initiatives and the number of new firms, paying special attention to the rates of green and digital entrepreneurship as smart cities tend to follow sustainable and/or digital orientations."

"We find evidence of a positive (causal) relation between smart city initiatives and entrepreneurship rates in a sample of Belgian municipalities, particularly when these initiatives follow a bottom-up approach and/or the level of implementation is high."

"Having sustainable and/or digital orientations in smart city initiatives does not generally make any difference in the rates of entrepreneurship, the exception being the digital rates in large municipalities."

Introduction

"Since one of the main purposes behind SC initiatives is to foster entrepreneurship, and the number of new businesses registered is one of the main indicators of the "competitiveness dimension" of the smart cities (Giffinger and others 111), we may intuitively expect to observe higher rates of entrepreneurship in the territories that launch SC initiatives (Richter and others 112; Montgomery 113)."

"This paper analyses whether there are statistically significant differences in the rates of (green and digital) entrepreneurship in local administrative units that develop SC initiatives (following a sustainable and/or technological orientation)."

"Our focus on green and digital entrepreneurship is motivated by the fact that SC initiatives mainly follow sustainable and/or technological orientations (Caragliu and others 114; Nam and Pardo 115)."

"This suggests that a link between SC initiatives following a sustainable and/or technological orientation and green and digital entrepreneurship exists, respectively (Sarma and Surmy 116)."

"In our case this is mostly motivated by the aim of analysing the link between a local event (the launch of SC initiatives and their orientation) and the creation of new firms in the local administrative units."

Entrepreneurship in smart cities

"This paper addresses this question by means of three hypotheses that, following the literature, capture the potential relationship between the rates of entrepreneurship and the implementation (Hypothesis 1), the level of implementation (Hypothesis 2) and the orientation of the SC initiatives (Hypothesis 3)."

"Since these hypotheses link a local event (the launch of SC initiatives and their orientation) with the creation of new firms in the territory (conditional on other determinants of the rates of entrepreneurship), we propose using data on municipalities and count data models to test them (Audretsch and Lehmann 117; Audretsch and Keilbach 118)."

"Although the evidence from case studies provides interesting insights on the relationship between SC and entrepreneurship (e.g. Vu and Hartley 119; Harrington 120; Sarma and Sunny 116), the availability of measures of SC initiatives and orientations for a large sample of municipalities is critical to empirically test the proposed hypotheses and provide findings that can be applied more generally."

Empirical analysis

"We assessed the role of the larger municipalities in shaping the relation between entrepreneurship and SC initiatives by considering an alternative specification that includes the product of the variables of interest (SC initiatives, SC implementation and SC orientations) with a dummy variable indicating those municipalities with more than 50,000 people between 20 and 60 years old in 2017."

"We considered an alternative model specification in which we included as additional regressors the products of the SC initiatives, implementation and orientation variables with a dummy variable identifying municipalities with more than 50,000 people between 20 and 60 (to assess the role of the larger municipalities in shaping the relationship between entrepreneurship and SC initiatives)."

"The ultimate row presents endogeneity tests rejecting the exogeneity of the SC initiatives variable and, for # new firms and # new digital firms, also that of the implementation variables (i.e. in these cases the residuals from the first-stage regression of the control function approach are statistically significant in the second-stage negative binomial model)."

Conclusions

"This suggests that smart cities may be acting as a local entrepreneurship-supporting policy that is not necessarily limited to initiatives emanating from the local government but from the whole local community."

"An analysis of the dynamics of the (causal) relationship between SC initiatives and entrepreneurship may provide interesting insights about the impact that these local initiatives have on the short- and long-term evolution of new business ventures."

"For if the key roles are the characteristics of the territory (e.g. the existence of a powerful entrepreneurship ecosystem), then local programmes oriented towards improving such characteristics should be implemented (programmes specifically oriented towards entrepreneurship or, perhaps, of a general nature if the empirical evidence shows that they perform well)."

"If the higher rates of entrepreneurship in smart territories actually have their origins in the entrepreneurs' characteristics (e.g. a strong background in STEM disciplines), then the policy recommendation would be to promote their acquisition by the local population of potential entrepreneurs."

11.2.15 The Socio-ecological Practice of Building Blue-Green Infrastructure in High-Density Cities: What Does the ABC Waters Program in Singapore Tell Us? [140]

This is a machine-generated summary of:

Liao, Kuei-Hsien: The socio-ecological practice of building blue-green infrastructure in high-density cities: what does the ABC Waters Program in Singapore tell us? [140]

Published in: Socio-Ecological Practice Research (2019).

Link to original: https://doi.org/10.1007/s42532-019-00.009-3.

If you want to cite the papers, please refer to the original.

For technical reasons we could not place the page where the original quote is coming from.

Abstract-Summary

"Blue-green infrastructure (BGI) is the network of green spaces that provides multiple water-related ecosystem services."

"This article explores the Active, Beautiful, Clean (ABC) Waters Program as a socio-ecological practice that builds BGI for the high-density city-state of Singapore."

"The ABC Waters Program provides valuable lessons for other high-density cities to overcome land scarcity as a constraint on BGI."

"The case study on the ABC Waters Program also reveals a research gap in the socio-ecological practice research on BGI planning and design."

Introduction

"This article focuses on a particular type of GI—blue-green infrastructure (BGI), defined here as the network of green spaces that provides multiple water-related ecosystem services, such as water supply, water quality treatment, flood hazard mitigation, and water-based recreation (see Haase 121, p. 5; Liao and others 122, p. 204)."

"Although neither GI, BGI, nor ecosystem services are mentioned, the ABC Waters Program exemplifies a shift from gray infrastructure to BGI in urban water management, showcasing a BGI-building program for a high-density city."

"Complementing the existing literature, this article examines the ABC Waters Program as a socio-ecological practice, which builds BGI for Singapore through retrofitting the existing terrestrial and aquatic green spaces."

"The objective is to gain insights into how high-density cities could build BGI with limited green spaces and to identify the research gap in BGI planning and design as a socio-ecological practice."

The ABC Waters Program

"Following the sustainable stormwater management practices in the West, e.g., low-impact develop (LID) and sustainable urban drainage system (SUDS), and drawing particularly from Water Sensitive Urban Design (WSUD) in Australia, the ABC Waters Program deploys nature-based solutions to enhance the purification (sedimentation, filtration, and biological uptake), detention, retention, and infiltration functions of terrestrial green spaces to manage stormwater runoff at source before it reaches the rivers/canals and eventually the reservoirs."

"To a lesser degree, the ABC Waters Program also addresses flood mitigation to cope with the increasing stormwater runoff associated increasing urban development and climate change through the "source-pathway-receptor" approach (PUB 123, p. 9)."

"The ABC Waters design features enhance the ecosystem services of the terrestrial green spaces by introducing the regulating services of water quality treatment and to a lesser degree, flood mitigation; and the ABC Waters design features are argued to also enhance the aesthetics and biodiversity of the urban landscape (PUB 123, p. 10)."

Lessons for other high-density cities

"While land availability seems like a constraint, Singapore has demonstrated that a high-density city can still build BGI by optimizing the existing green spaces and leveraging cultural ecosystem services to gain public support."

"313–314; Richards and others 124, p. 558), high-density cities can optimize the functionality of the available green spaces, making them as multifunctional as optimal to go beyond recreation and esthetics to provide more ecosystem services."

"Other high-density cities could also enhance the ecosystem services of their existing green spaces to "upgrade" them to BGI by waterway enhancement and even river restoration, as well as by nature-based stormwater facilities."

"Limited green spaces in high-density cities are inevitably expected to fulfill multiple social needs, cultural ecosystem services play an important role in the optimization of the existing green spaces to ensure a successful BGI program that requires public support."

A research gap in BGI planning and design as a socio-ecological practice

"Since green spaces should provide multiple water-related ecosystem services to be BGI, it would be a lack of rigor to call any urban green space or any preexisting collection of green spaces of a city BGI before its multifunctionality, in terms of ecosystem services, is assessed."

"Because ecosystem services are provided through ecosystem functions, BGI planning and design require explicit consideration of ecosystem functions."

"The social and economic functions of a BGI element (i.e., an individual green space) and BGI as a whole (i.e., the green space network) should be derived from ecosystem functions."

"To what degree should the ecosystem functions of a degraded river be restored for it to qualify as BGI?."

"Not all functions provided by an urban green space are ecosystem services."

"When a certain social function of an urban waterway is largely provided by engineering work, should they be considered ecosystem services?."

Conclusions

"The program tells us that a high-density city could systematically build BGI through optimizing the existing aquatic and terrestrial green spaces and leveraging cultural ecosystem services to gain public support."

"This article also enriches the literature on BGI planning and design by providing a case in high-density city in Asia, a context that is rather different from USA and Europe where most of the literature focus (Lennon 125, p. 4)."

"Against this backdrop, BGI planning and design will play an important role to move toward urban livability, resilience, and sustainability through the preservation, restoration, and creation of urban ecosystem services."

"Much more socio-ecological practice research on BGI is needed to better promote its planning and design for a high-density urban future in Asia."

Bibliography

1. Miller JR (2005) Biodiversity conservation and the extinction of experience. Trends Ecol Evol 20(8):430–434. https://doi.org/10.1016/j.tree.2005.05.013
2. Chapin, F.S., S.R. Carpenter, G.P. Kofinas, C. Folke, N. Abel, W.C. Clark, P. Olsson, D.M. Stafford Smith, et al. 2010. Ecosystem stewardship: Sustainability strategies for a rapidly changing planet. Trends in Ecology & Evolution 25: 241–249.
3. Colding, J., J. Lundberg, and C. Folke. 2006. Incorporating green-area user groups in urban ecosystem management. AMBIO 35: 237–244.
4. Goddard, M.A., A.J. Dougill, and T.G. Benton. 2010. Scaling up from gardens: Biodiversity conservation in urban environments. Trends in Ecology & Evolution 25: 90–98.
5. Walker, C. 2007. Redistributive land reform: For what and for whom? In The land question in South Africa: The challenge of transformation and redistribution, ed. L. Ntsebeza, and R. Hall, 132–151. Cape Town: HSRC Press.
6. Ansell, C.K. 2003. Community embeddedness and collaborative governance in the San Francisco Bay Area Environmental Movement. In Social movements and networks—Relational approaches to collective action, ed. M. Diani, and D. McAdam, 123–144. Oxford, UK: Oxford University Press.
7. Ernstson, H., S. Sörlin, and T. Elmqvist. 2008. Social movements and ecosystem services—The role of social network structure in protecting and managing urban green areas in Stockholm. Ecology and Society 13: 39.
8. Frantzeskaki, N., and N. Kabisch. 2016. Setting a knowledge co-production operating space for urban environmental governance Lessons from Rotterdam, Netherlands and Berlin, Germany. Environmental Science & Policy 62: 1–9. https://doi.org/10.1016/j.envsci.2016.01.010 .
9. Keeler, B.L., P. Hamel, T. McPhearson, M.H. Hamann, M.L. Donahue, K.A. Meza Prado, K.K. Arkema, G.N. Bratman, et al. 2019. Social-ecological and technological factors moderate the value of urban nature. Nature Sustainability 2: 29–38. https://doi.org/10.1038/s41893-018-0202-1 .
10. Kabisch N, Frantzeskaki N, Pauleit S, Naumann S, Davis M, Artmann M, et al. Nature-based solutions to climate change mitigation and adaptation in urban areas: perspectives on indicators, knowledge gaps, barriers, and opportunities for action. E&S. 2016;21 https://doi.org/10.5751/ES-08373-210239 .
11. Bush, J. 2020. The role of local government greening policies in the transition towards nature-based cities. Environmental Innovation and Societal Transitions 35: 35–44. https://doi.org/10.1016/j.eist.2020.01.015 .
12. Frantzeskaki, N., P. Vandergert, S. Connop, K. Schipper, I. Zwierzchowska, M. Collier, and M. Lodder. 2020. Examining the policy needs for implementing nature-based solutions in cities: Findings from city-wide transdisciplinary experiences in Glasgow (UK), Genk (Belgium) and Poznań (Poland). Land Use Policy 96: 104688. https://doi.org/10.1016/j.landusepol.2020.104688 .
13. Moosavi, S., G.R. Browne, and J. Bush. 2021. Perceptions of nature-based solutions for urban water challenges: Insights from Australian researchers and practitioners. Urban Forestry and Urban Greening 57: 126937. https://doi.org/10.1016/j.ufug.2020.126937 .
14. Ignatieva, M., D. Haase, D. Dushkova, and A. Haase. 2020. Lawns in cities: From a globalised urban green space phenomenon to sustainable nature-based solutions. Land. https://doi.org/10.3390/land9030073 .
15. Maller, C. 2021. Re-orienting nature-based solutions with more-than-human thinking. Cities 113: 103155. https://doi.org/10.1016/j.cities.2021.103155 .
16. Kabisch, N., and D. Haase. 2014. Green justice or just green? Provision of urban green spaces in Berlin, Germany. Landscape and Urban Planning 122: 129–139. https://doi.org/10.1016/j.landurbplan.2013.11.016 .

17. Baró, F., J. Langemeyer, E. Łaszkiewicz, and N. Kabisch. 2021. Editorial to the special issue "Advancing urban ecosystem service implementation and assessment considering different dimensions of environmental justice." Environmental Science and Policy 115: 43–46. https://doi.org/10.1016/j.envsci.2020.10.008 .
18. Pineda-Pinto, M., C.A. Nygaard, M. Chandrabose, and N. Frantzeskaki. 2021. Mapping social-ecological injustice in Melbourne, Australia: An innovative systematic methodology for planning just cities. Land Use Policy 104: 105361. https://doi.org/10.1016/j.landusepol.2021.105361 .
19. Mattijssen, T.J.M., A.P.N. van der Jagt, A.E. Buijs, B.H.M. Elands, S. Erlwein, and R. Lafortezza. 2017. The long-term prospects of citizens managing urban green space: From place making to place-keeping? Urban Forestry & Urban Greening 26: 78–84. https://doi.org/10.1016/j.ufug.2017.05.015 .
20. Buijs, A., R. Hansen, S. Van der Jagt, B. Ambrose-Oji, B. Elands, E. Lorance Rall, T. Mattijssen, S. Pauleit, et al. 2019. Mosaic governance for urban green infrastructure: Upscaling active citizenship from a local government perspective. Urban Forestry & Urban Greening 40: 53–62. https://doi.org/10.1016/j.ufug.2018.06.011 .
21. van der Jagt, A.P.N., M. Smith, B. Ambrose-Oji, C.C. Konijnendijk, V. Giannico, D. Haase, R. Lafortezza, M. Nastran, et al. 2019. Co-creating urban green infrastructure connecting people and nature: A guiding framework and approach. Journal of Environmental Management 233: 757–767. https://doi.org/10.1016/j.jenvman.2018.09.083 .
22. Frantzeskaki, N., T. Mcphearson, and N. Kabisch. 2021. Urban sustainability science: Prospects for innovations through a system's perspective, relational and transformations' approaches. This article belongs to Ambio's 50th Anniversary Collection. Theme: urbanization. Ambio. https://doi.org/10.1007/s13280-021-01521-1 .
23. Gómez Martín, E., R. Giordano, A. Pagano, P. van der Keur, and M. Máñez Costa. 2020. Using a system thinking approach to assess the contribution of nature based solutions to sustainable development goals. Science of the Total Environment 738: 139693. https://doi.org/10.1016/j.scitotenv.2020.139693 .
24. Cousins, J.J. 2021. Justice in nature-based solutions: Research and pathways. Ecological Economics 180: 106874. https://doi.org/10.1016/j.ecolecon.2020.106874 .
25. Giachino, C., G. Pattanaro, B. Bertoldi, L. Bollani, and A. Bonadonna. 2021. Nature-based solutions and their potential to attract the young generations. Land Use Policy 101: 105176. https://doi.org/10.1016/j.landusepol.2020.105176 .
26. Nagendra, H., and S. Mundoli. 2019. Cities and canopies. Penguin Random House, New Delhi: Trees in Indian cities.
27. Basu, S., and H. Nagendra. 2020. The street as workspace: Assessing street vendors' rights to trees in Hyderabad. India. Landscape and Urban Planning 199: 103818. https://doi.org/10.1016/j.landurbplan.2020.103818 .
28. Soga, M., and K.J. Gaston. 2020. The ecology of human–nature interactions. Proceedings of the Royal Society B: Biological Sciences 287: 20191882. https://doi.org/10.1098/rspb.2019.1882 .
29. West, S., L.J. Haider, S. Stålhammar, and S. Woroniecki. 2020. A relational turn for sustainability science? Relational thinking, leverage points and transformations. Ecosystems and People 16: 304–325. https://doi.org/10.1080/26395916.2020.1814417 .
30. Frantzeskaki, N. 2019. Seven lessons for planning nature-based solutions in cities. Environmental Science and Policy 93: 101–111. https://doi.org/10.1016/j.envsci.2018.12.033 .
31. Bayulken, B., D. Huisingh, and P.M.J. Fisher. 2021. How are nature based solutions helping in the greening of cities in the context of crises such as climate change and pandemics? A comprehensive review. Journal of Cleaner Production 288: 125569. https://doi.org/10.1016/j.jclepro.2020.125569 .
32. European Commission (2015) Nature-Based Solutions & Re-Naturing Cities. Final Report of the Horizon 2020 Expert group on 'nature-based solutions and re- Naturing cities'. Directorate-General for Research and Innovation–Climate Action, Environment, Resource

Efficiency and Raw Materials. https://op.europa.eu/en/publication-detail/-/publication/fb117980-d5aa-46df-8edc-af367cddc202/language-en
33. Escobedo, F.J., V. Giannico, C.Y. Jim, G. Sanesi, and R. Lafortezza. 2019. Urban forests, ecosystem services, green infrastructure and nature-based solutions: Nexus or evolving metaphors? Urban Forestry and Urban Greening 37: 3–12. https://doi.org/10.1016/j.ufug.2018.02.011 .
34. Turnhout E, Waterton C, Neves K, Buizer M (2013) Rethinking biodiversity: from goods and services to "living with". Conserv Lett 6:154–161
35. Schröter M, van der Zanden EH, van Oudenhoven APE, Remme RP, Serna-Chavez HM, de Groot RS, Opdam P (2014) Ecosystem services as a contested concept: a synthesis of critique and counter-arguments. Conserv Lett 6:514–523
36. Bekessy SA, Runge MC, Kusmanoff AM, Keith DA, Wintle BA (2018) Ask not what nature can do for you: a critique of ecosystem services as a communication strategy. Biol Conserv 224:71–74
37. Babí Almenar, J., T. Elliot, B. Rugani, B. Philippe, T. Navarrete Gutierrez, G. Sonnemann, and D. Geneletti. 2021. Nexus between nature-based solutions, ecosystem services and urban challenges. Land Use Policy 100: 104898. https://doi.org/10.1016/j.landusepol.2020.104898 .
38. Castellar, J. A. C. et al. Nature-based solutions in the urban context: terminology, classification and scoring for urban challenges and ecosystem services. Sci. Total Environ. 779, 146237 (2021).
39. Xie, L. & Bulkeley, H. Nature-based solutions for urban biodiversity governance. Environ. Sci. Policy 110, 77–87 (2020).
40. Vaccari, F. P., Gioli, B., Toscano, P. & Perrone, C. Carbon dioxide balance assessment of the city of Florence (Italy), and implications for urban planning. Landscape Urban Plan. 120, 138–146 (2013).
41. Jansson, Å., and P. Nohrstedt. 2001. Carbon sinks and human freshwater dependence in Stockholm County. Ecological Economics 39: 361–370.
42. Chroneer D, Stahlbrost A, Habibipour A (2019) Urban living labs: towards an integrated understanding of their key components. Technol Innov Manag Rev 9(3):50–62
43. Cerreta M, Savino V (2020) Circular enhancement of the cultural heritage: an adaptive reuse strategy for Ercolano Heritagescape. ICCSA. https://doi.org/10.1007/978-3-030-58808-3_72
44. De Medici S, Riganti P, Viola S (2018) Circular economy and the role of universities in urban regeneration: the case of Ortigia, Syracuse. Sustainability 10(11):4305 https://doi.org/10.3390/su10114305
45. Girard LF, Gravagnuolo A (2017) Circular economy and cultural heritage/landscape regeneration. https://doi.org/10.6092/2284-4732/5472
46. Casalegno C, Civera C, Mosca F, Freeman RE (2020) Circular economy and relationship-based view. Symphonya. Emerging Issues in Management 1:149–164. https://doi.org/10.4468/2020.1.12casalegno.civera.mosca.freeman
47. Wicks A, Elmore F A, Jonas D (2019) Connecting stakeholder theory to the law and public policy, in The Cambridge handbook of stakeholder theory edited by Jeffrey S. Harrison, Jay B. Barney, R. Edward Freeman, and Robert A. Phillips. Cambridge, United Kingdom, Cambridge University Press, pp. 97–116.
48. McDonald, R. I., Weber, K. F., Padowski, J., Boucher, T. & Shemie, D. Estimating watershed degradation over the last century and its impact on water-treatment costs for the world's large cities. Proc. Natl Acad. Sci. USA 113, 9117–9122 (2016).
49. Vercruysse, K., Grabowski, R. C. & Rickson, R. J. Suspended sediment transport dynamics in rivers: multi-scale drivers of temporal variation. Earth Sci. Rev. 166, 38–52 (2017).
50. Bilotta, G. S. & Brazier, R. E. Understanding the influence of suspended solids on water quality and aquatic biota. Water Res. 42, 2849–2861 (2008).
51. McDonald, R. I. et al. Water on an urban planet: urbanization and the reach of urban water infrastructure. Glob. Environ. Change 27, 96–105 (2014).

52. Wellman, B. & Frank, K. in Social Capital: Theory and Research (eds Lin, N. et al.) 233–273 (Routledge, 2001).
53. Hara, M., Nagao, T., Hannow, S., & Nakamura, J. (2016). New key performance indicators for a smart sustainable city. Sustainability, 8(3), 1–19.
54. Khan, H., Malik, M., Zafae, R., Goni, F., Chofreh, A., Klemeš, J., & Alotaibi, Y. (2020). Challenges for sustainable smart city development: A conceptual framework. Sustainable Development, 28(5), 1507–1518.
55. Rosenhead, J. (1996). What's the problem? An introduction to problem structuring methods. Interfaces, 26(6), 117–131.
56. Ackermann, F., & Eden, C. (2010). Strategic options development and analysis. In M. Reynolds & S. Holwell (Eds.), Systems approaches to managing change: A practical guide (pp. 135–190). Springer.
57. Faria, P., Ferreira, F., Jalali, M., & Bento, P. (2018). Combining cognitive mapping and MCDA for improving quality of life in urban areas. Cities: the International Journal of Urban Policy and Planning, 78, 116–127.
58. Dalvi-Esfahani, M., Niknafs, A., Nilashi, M., & Afrough, S. (2019). Social media addiction: Applying the DEMATEL approach. Telematics and Information, 43, 101250.
59. Guo, Z. Home parking convenience, household car usage, and implications to residential parking policies. Transp. Policy 29, 97–106 (2013).
60. Taylor, E. "Fight the towers! Or kiss your car park goodbye": How often do residents assert car parking rights in Melbourne planning appeals? Plan. Theory Pract. 15, 328–348 (2014).
61. Taylor, E. J. Journey into an immense heart of car parking. Plan. Theory Pract. 20, 448–455 (2019).
62. McKinney ML (2002) Urbanization, biodiversity, and conservation: The impacts of urbanization on native species are poorly studied, but educating a highly urbanized human population about these impacts can greatly improve species conservation in all ecosystems. Bioscience 52:883–890. https://doi.org/10.1641/0006-3568(2002)052[0883:ubac]2.0.co;2
63. McDonnell MJ, Pickett STA (1990) Ecosystem structure and function along urban-rural gradients: An unexploited opportunity for ecology. Ecology 71:1232–1237. https://doi.org/10.2307/1938259
64. McKinney ML (2008) Effects of urbanization on species richness: A review of plants and animals. Urban Ecosyst 11:161–176. https://doi.org/10.1007/s11252-007-0045-4
65. Narango DL (2020) Natural history in the city: Connecting people to the ecology of their plant and animal neighbors. Journal of Natural History Education & Experience 14:13–17
66. Frey D et al (2018) Predation risk shaped by habitat and landscape complexity in urban environments. J Appl Ecol 55:2343–2353. https://doi.org/10.1111/1365-2664.13189
67. Seibold S et al (2018) The necessity of multitrophic approaches in community ecology. Trends Ecol Evol 33:754–764. https://doi.org/10.1016/j.tree.2018.07.001
68. Tresch S et al (2019) Litter decomposition driven by soil fauna, plant diversity and soil management in urban gardens. Sci Total Environ 658:1614–1629. https://doi.org/10.1016/j.scitotenv.2018.12.235
69. Taylor L et al (2021) Enablers and challenges when engaging local communities for urban biodiversity conservation in Australian cities. Sustain Sci. https://doi.org/10.1007/s11625-021-01012-y
70. Hollands RG (2008) Will the real smart city please stand up? City 12(3):303–320
71. Vittanen J, Kingston R (2014) Smart cities and green growth: outsourcing democratic and environmental resilience to the global technology sector. Envrion Plan 46:803–819
72. Nagendra H, Gopal D (2011) Tree diversity, distribution, history and change in urban parks. Urban Ecosyst 14:211–223
73. Sassen S (2010) Cities are at the centre of our environmental future. S.A.P.I.E.N.S 2(3), 7 April 2010. https://sapiens.revues.org/948 . Accessed 17 Feb 2017
74. Yigitcanlar T, Dizdaroglu D (2015) Ecological approaches in planning for sustainable cities: a review of the literature. Glob J Enviorn Sci Manag 1(2):159–188

75. United Nations Human Settlements Programme. Cities of youth: Cities of prosperity. Nairobi: UN Habitat; 2013. Available from: https://unhabitat.org/books/cities-of-youth-cities-of-prosperity/ .
76. Huijsmans R, George S, Gigengack R, Evers S. Theorising age and generation in development. Eur J Dev Res. 2014;26:163–74.
77. Clark C, Uzzell D. The socio-environmental affordances of adolescents' environments. In: Spencer C, Blades M, editors. Children and their environments: learning, using and designing spaces. Cambridge: Cambridge University Press; 2006.
78. Shirtcliff B. Transformative power of city play: social media and place in a post-affordance world. Cities Health. 2018;3:1–14.
79. Burningham K, Venn S, Hayward B, Nissen S, Aoyagi M, Hasan MM, et al. Ethics in context: Essential flexibility in an international photo-elicitation project with children and young people. Int J Soc Res Method. 2019; In press.
80. Kyttä M. Children in outdoor contexts. Affordances and independent mobility in the assessment of environmental child friendliness. Helsinki: Helsinki University of Technology; 2008.
81. Kyttä M, Oliver M, Ikeda E, Ahmadi E, Omiya I, Laatikainen T. Children as urbanites: mapping the affordances and behavior settings of urban environments for Finnish and Japanese children. Children's Geographies. 2018;16(3):319–32.
82. Kaaronen RO. Affording sustainability: adopting a theory of affordances as a guiding heuristic for environmental policy. Front Psychol. 2017;8:1974.
83. Marcus L, Giusti M, Barthel S. Cognitive affordances in sustainable urbanism: contributions of space syntax and spatial cognition. J Urban Des. 2016;21(4):439–52.
84. European Commission (2021b) EU research policy, what nature-based solutions are, background, news and documents. Nature-based solutions research policy. European Commission website, In https://ec.europa.eu/info/research-and-innovation/research-area/environment/nature-based-solutions/research-policy_en .
85. Kolokotsa D, Lilli AA, Lilli MA, Nikolaidis NP (2020) On the impact of nature-based solutions on citizens' health & well being. Energ Buildings 229:110527. https://doi.org/10.1016/j.enbuild.2020.110527
86. Faivre, N., Fritz, M., Freitas, T., de Boissezon, B. & Vandewoestijne, S. Nature-based solutions in the EU: Innovating with nature to address social, economic and environmental challenges. Environ. Res. 159, 509–518 (2017).
87. Alberti V, Alonso Raposo M, Attardo C, Auteri D, Barranco R, Batista E, Silva F, Benczur P, Bertoldi P, Bono F, Bussolari I, Caldeira S, Carlsson J, Christidis P, Christodoulou A, Ciuffo B, Corrado S, Fioretti C, Galassi MC, Galbusera L, Gawlik B, Giusti F, Gomez J, Grosso M, Guimarães Pereira Â, Jacobs-Crisioni C, Kavalov B, Kompil M, Kucas A, Kona A, Lavalle C, Leip A, Lyons L, Manca AR, Melchiorri M, Monforti-Ferrario F, Montalto V, Mortara B, Natale F, Panella F, Pasi G, Perpiña C, Pertoldi M, Pisoni E, Polvora A, Rainoldi A, Rembges D, Rissola G, Sala S, Schade S, Serra N, Spirito L, Tsakalidis A, Schiavina M, Tintori G, Vaccari L, Vandyck T, Vanham D, Van Heerden S, Van Noordt C, Vespe M, Vetters N, Vilahur Chiaraviglio N, Vizcaino P, Von Estorff U, Zulian G (2019) In: Vandecasteele I, Baranzelli C, Siragusa A, Aurambout JP (eds) The future of cities – Opportunities, challenges and the way forward. Publications Office of the European Union, Luxembourg
88. Stefanakis AI, Calheiros CSC, Nikolaou I (2021) Nature-based solutions as a tool in the new circular economic model for climate change adaptation. Circ Econ Sust. https://doi.org/10.1007/s43615-021-00022-3
89. Atanasova N, Castellar JAC, Pineda-Martos R, Nika CE, Katsou E, Istenič D, Pucher B, Andreucci MB, Langergraber G (2021) Nature-based solutions and circularity in cities. Circ Econ Sust. https://doi.org/10.1007/s43615-021-00024-1
90. Langergraber G, Pucher B, Simperler L, Kisser J, Katsou E, Buehler D, Garcia Mateo MC, Atanasova N (2020) Implementing nature-based solutions for creating a resourceful circular city. Blue-Green Systems 2:173–185. https://doi.org/10.2166/bgs.2020.933

91. European Commission (2018). Infographic – Innovating with nature. https://ec.europa.eu/info/sites/info/files/research_and_innovation/research_by_area/images/naturebasedsolutions-w920px.png . Accessed 14 Jan 2021
92. Pineda-Martos R, Pérez-Urrestarazu L (2020) Using Earth observation satellites for identification and assessment of green roofs in European cities - Copernicus programme data and Google Earth Engine. In: Closed cycles and the circular society, The power of ecological engineering. Wädenswil, Switzerland, September 2–4, 2020. https://closedcycles2020.ch/index.php/program . Accessed 31 Dec 2020
93. Pozoukidou G (2020) Designing a green infrastructure network for metropolitan areas: a spatial planning approach. Euro-Mediterr J Environ Integr 5:40. https://doi.org/10.1007/s41207-020-00178-8
94. Madureira H, Andresen T (2014) Planning for multifunctional urban green infrastructures: Promises and challenges. Urban Des Int 19:38–49. https://doi.org/10.1057/udi.2013.11
95. Wamsler C, Pauleit S (2016) Making headway in climate policy mainstreaming and ecosystem-based adaptation: two pioneering countries, different pathways, one goal. Clim Chang 137:71–87. https://doi.org/10.1007/s10584-016-1660-y
96. Eggermont H, Balian E, Azevedo JMN, Beumer V, Brodin T, Claudet J, Fady B, Grube M, Keune H, Lamarque P, Reuter K, Smith M, van Ham C, Weisser WW, le Roux X (2015) Nature-based solutions: new influence for environmental management and research in Europe. GAIA 24(4):243–248. https://doi.org/10.14512/gaia.24.4.9
97. European Commission (2015a) Towards an EU research and innovation policy agenda for nature-based solutions & re-naturing cities – final report of the horizon 2020 Expert Group on 'Nature-Based Solutions and Re-Naturing Cities' (full version). Directorate-General for Research and Innovation. Publications Office of the European Union, Luxembourg. https://ec.europa.eu/newsroom/horizon2020/document.cfm?doc_id=10195 . https://op.europa.eu/en/publication-detail/-/publication/fb117980-d5aa-46df-8edc-af367cddc202 . Accessed 19 January 2021
98. European Commission (2021c) The EU and nature-based solutions. In: Nature-based solutions – nature-based solutions and how the Commission defines them, funding, collaboration and jobs, projects, results and publications. European Commission website. https://ec.europa.eu/info/research-and-innovation/research-area/environment/nature-based-solutions_en . Accessed 18 Jan 2021
99. European Commission (2016) Communication from the Commission to the European Parliament, the Council, the European Economic and Social Committee and the Committee of the Regions: next steps for a sustainable European future – European action for sustainability. COM(2016) 739 final. https://ec.europa.eu/transparency/regdoc/rep/1/2016/EN/COM-2016-739-F1-EN-MAIN.PDF . Accessed 18 January 2021
100. Chatzimentor A, Apostolopoulou E, Mazaris AD (2020) A review of green infrastructure research in Europe: challenges and opportunities. Landsc Urban Plan 198:103775. https://doi.org/10.1016/j.landurbplan.2020.103775
101. Davis M, Abhold K, Mederake L, Knoblauch D (2017) Nature-based solutions in European and national policy frameworks. Deliverable 1.5, NATURVATION (cities – nature – innovation). Horizon 2020 grant agreement No. 730243, European Commission. https://ec.europa.eu/research/participants/documents/downloadPublic?documentIds=080166e5b64f88d5&appId=PPGMS . Accessed 27 Jan 2021
102. Abad-Segura E, Batlles de la Fuente A, González-Zamar M-D, Belmonte-Ureña LJ (2020) Effects of circular economy policies on the environment and sustainable growth: worldwide research. Sustainability 12:5792. https://doi.org/10.3390/su12145792
103. Katsou E, Nika C-E, Buehler D, Marić B, Megyesi B, Mino E, Babí Almenar J, Bas B, Bećirović D, Bokal S, Đolić M, Elginöz N, Kalnis G, Garcia Mateo M-C, Milcusi M, Mousavi A, Rinčić I, Rizzo A, Rodriguez-Roda I, Rugani B, Šalaševičienė A, Sari R, Stanchev P, Topuz E, Atanasova N (2020) Transformation tools enabling the implementation of nature-based solutions for creating a resourceful circular city. Blue-Green Systems 2:188–213. https://doi.org/10.2166/bgs.2020.929

104. Cohen-Shacham, E., A. Andrade, J. Dalton, N. Dudley, M. Jones, C. Kumar, S. Maginnis, S. Maynard, et al. 2019. Core principles for successfully implementing and upscaling Nature-based Solutions. Environmental Science and Policy 98: 20–29. https://doi.org/10.1016/j.envsci.2019.04.014 .
105. Vedula, S., Doblinger, C., Pacheco, D., York, J., Bacq, S., Russo, M., & Dean, T. (2021). Entrepreneurship for the public good: A review, critique, and path forward for social and environmental entrepreneurship research. Academy of Management Annals, 16(1), 391–425.
106. Cojoianu, T., Clark, G.L., Hoepner, A.G.F., Pažitka, V., Wójcik, D., 2020a. Fin vs. tech: Are trust and knowledge creation key ingredients in fintech start-up emergence and financing? Small Business Economics https://doi.org/10.1007/s11187-020-00367-3
107. Cojoianu, T., Clark, G. L., Hoepner, A. G. F., Veneri, P., & Wójcik, D. (2020). Entrepreneurs for a low carbon world: How environmental knowledge and policy shape the creation and financing of green start-ups. Res. Policy, 49, 103988. https://doi.org/10.1016/j.respol.2020.103988
108. Feldman, M. P., Ozcan, S., & Reichstein, T. (2019). Falling not far from the tree: Entrepreneurs and organizational heritage. Organization Science. https://doi.org/10.1287/orsc.2018.1222
109. Moeen, M., & Agarwal, R. (2017). Incubation of an industry: Heterogeneous knowledge bases and modes of value capture. Strategic Management Journal. https://doi.org/10.1002/smj.2511
110. Cohen, B., & Muñoz, P. (2015). Toward a theory of purpose-driven urban entrepreneurship. Organization & Environment, 28, 264–285.
111. Giffinger R, Fertner C, Kramar H, Meijers E, Pichler-Milanović N (2007) Ranking of European medium-sized cities, Report of the Centre of Regional Science, Vienna. http://www.smart-cities.eu/download/smart_cities_final_report.pdf
112. Kraus S, Richter C, Papagiannidis S, Durst S (2015) Innovating and exploiting entrepreneurial opportunities in smart cities: evidence from Germany. Creat Innov Manag 24:601–616. https://doi.org/10.1111/caim.12154
113. Montgomery M (2018) Smart cities are smart places for entrepreneurs. Jul. 24. https://www.forbes.com/sites/mikemontgomery/2018/07/24/smart-cities-are-smart-places-for-entrepreneurs/#9452d1153214
114. Caragliu A, Del Bo C, Nijkamp P (2011) Smart cities in Europe. J Urban Technol 18:65–82. https://doi.org/10.1080/10630732.2011.601117
115. Nam T, Pardo TA (2011a) Smart city as urban innovation: focusing on management, policy, and context. In: Proceedings of the 5th international conference on theory and practice of electronic governance, pp 185–194. https://doi.org/10.1145/2072069.2072100
116. Sarma S, Sunny SA (2017) Civic entrepreneurial ecosystems: smart city emergence in Kansas City. Bus Horiz 60:843–853. https://doi.org/10.1016/j.bushor.2017.07.010
117. Audretsch DB, Lehmann EE (2005) Does the knowledge spillover theory of entrepreneurship hold for regions? Res Policy 34:1191–1202. https://doi.org/10.1016/j.respol.2005.03.012
118. Audretsch, D. B., & Keilbach, M. (2007). The theory of knowledge spillover entrepreneurship. Journal of Management Studies, 44, 1242–1254. https://doi.org/10.1111/j.1467-6486.2007.00722.x
119. Vu K, Hartley K (2018) Promoting smart cities in developing countries: policy insights from Vietnam. Telecommun Policy 42:845–859. https://doi.org/10.1016/j.telpol.2017.10.005
120. Harrington K (2017) Smart city leaders, champions, and entrepreneurs: The people part of vibrant smart cities. In: Kumar TMV (ed) Smart economy in smart cities. Springer, Berlin, pp 1005–1012
121. Haase D (2015) Reflections about blue-ecosystem services in cities. Sustain Water Qual Ecol 5:77–83
122. Liao KH, Deng SN, Tan PY (2017) Blue-green infrastructure: new frontier for sustainable urban stormwater management. In: Tan PY, Jim CY (eds) Greening cities: forms and functions. Springer, Singapore, pp 203–226
123. PUB (2018) Active, beautiful, clean waters design guidelines, 4th edn. PUB, Singapore
124. Richards DR, Passy P, Oh RRY (2017) Impacts of population density and wealth on the quantity and structure of urban green space in tropical Southeast Asia. Landsc Urban Plan 157:553–560

125. Lennon M (2014) Green infrastructure and planning policy: a critical assessment. Local Environ 20(8):1–24
126. Andersson, Erik; Barthel, Stephan; Borgström, Sara; Colding, Johan; Elmqvist, Thomas; Folke, Carl; Gren, Åsa Reconnecting Cities to the Biosphere: Stewardship of Green Infrastructure and Urban Ecosystem Services. Ambio (2014). https://doi.org/10.1007/s13280-014-0506-y
127. Kabisch, Nadja; Frantzeskaki, Niki; Hansen, Rieke Principles for urban nature-based solutions. Ambio (2022). https://doi.org/10.1007/s13280-021-01685-w
128. Randrup, Thomas B.; Buijs, Arjen; Konijnendijk, Cecil C.; Wild, Tom Moving beyond the nature-based solutions discourse: introducing nature-based thinking Urban Ecosystems (2020). https://doi.org/10.1007/s11252-020-00964-w
129. Pan, Haozhi; Page, Jessica; Shi, Rui; Cong, Cong; Cai, Zipan; Barthel, Stephan; Thollander, Patrik; Colding, Johan; Kalantari, Zahra Contribution of prioritized urban nature-based solutions allocation to carbon neutrality. Nature Climate Change (2023). https://doi.org/10.1038/s41558-023-01737-x
130. Ascione, Grazia Sveva; Cuomo, Federico; Mariotti, Nicole; Corazza, Laura Urban Living Labs, Circular Economy and Nature-Based Solutions: Ideation and Testing of a New Soil in the City of Turin Using a Multi-stakeholder Perspective. Circular Economy and Sustainability (2021). https://doi.org/10.1007/s43615-021-00011-6
131. Chung, Min Gon; Frank, Kenneth A.; Pokhrel, Yadu; Dietz, Thomas; Liu, Jianguo Natural infrastructure in sustaining global urban freshwater ecosystem services. Nature Sustainability (2021). https://doi.org/10.1038/s41893-021-00786-4
132. Rodrigues, Maria I. R. P. R.; Ferreira, Fernando A. F.; Ferreira, Neuza C. M Q. F. Constructing smarter and more sustainable urban ecosystems: a dynamic analysis of challenges and initiatives. Annals of Operations Research (2023). https://doi.org/10.1007/s10479-023-054 06-2
133. Croeser, Thami; Garrard, Georgia E.; Visintin, Casey; Kirk, Holly; Ossola, Alessandro; Furlong, Casey; Clements, Rebecca; Butt, Andrew; Taylor, Elizabeth; Bekessy, Sarah A. Finding space for nature in cities: the considerable potential of redundant car parking. npj Urban Sustainability (2022). https://doi.org/10.1038/s42949-022-00073-x
134. Rega-Brodsky, Christine C.; Aronson, Myla F. J.; Piana, Max R.; Carpenter, Ela-Sita; Hahs, Amy K.; Herrera-Montes, Adriana; Knapp, Sonja; Kotze, D. Johan; Lepczyk, Christopher A.; Moretti, Marco; Salisbury, Allyson B.; Williams, Nicholas S. G.; Jung, Kirsten; Katti, Madhusudan; MacGregor-Fors, Ian; MacIvor, J. Scott; La Sorte, Frank A.; Sheel, Vallari; Threfall, Caragh G.; Nilon, Charles H. Urban biodiversity: State of the science and future directions. Urban Ecosystems (2022). https://doi.org/10.1007/s11252-022-01207-w
135. Mundoli, Seema; Unnikrishnan, Hita; Nagendra, Harini The “Sustainable” in smart cities: ignoring the importance of urban ecosystems. DECISION (2017). https://doi.org/10.1007/s40622-017-0152-x
136. Nissen, Sylvia; Prendergast, Kate; Aoyagi, Midori; Burningham, Kate; Hasan, Mohammad Mehedi; Hayward, Bronwyn; Jackson, Tim; Jha, Vimlendu; Mattar, Helio Schudel, Ingrid; Venn, Sue; Yoshida, Aya Young people and environmental affordances in urban sustainable development: insights into transport and green and public space in seven cities. Sustainable Earth (2020). https://doi.org/10.1186/s42055-020-00039-w
137. Pineda-Martos, Rocío; Calheiros, Cristina S. C. Nature-Based Solutions in Cities—Contribution of the Portuguese National Association of Green Roofs to Urban Circularity. Circular Economy and Sustainability (2021). https://doi.org/10.1007/s43615-021-00070-9
138. Cojoianu, Theodor F.; Hoepner, Andreas G. F.; Hu, Xi; Ramadan, Moustafa; Veneri, Paolo; Wójcik, Dariusz Are cities venturing green? A global analysis of the impact of green entrepreneurship on city air pollution. Small Business Economics (2023). https://doi.org/10.1007/s11187-023-00764-4
139. Manjon, Miguel; Aouni, Zineb; Crutzen, Nathalie Green and digital entrepreneurship in smart cities. The Annals of Regional Science (2021). https://doi.org/10.1007/s00168-021-01080-z

140. Liao, Kuei-Hsien The socio-ecological practice of building blue-green infrastructure in high-density cities: what does the ABC Waters Program in Singapore tell us?. Socio-Ecological Practice Research (2019). https://doi.org/10.1007/s42532-019-00009-3

Chapter 12
The Pathway of Ecological Restoration

12.1 Introduction by Editor

Across the world, experiments have shown us the possibility of reversing ecological damage by employing restoration strategies to counter biodiversity loss and ecosystem degradation. The discipline of ecological restoration encompasses efforts aimed at enabling the restoration of the structure and function of ecosystems, thereby facilitating the provision of goods and services ensuing from the ecosystems. Ecological restoration has emerged as a critical strategy for addressing the pressing global challenges of climate change, land degradation, and the loss of vital ecosystem services. This comprehensive approach aims to revive the health, integrity, and sustainability of degraded ecosystems, ultimately enhancing their resilience to the impacts of a changing climate. This branch draws upon the foundations of ecological theory, and applies an interdisciplinary approach. The disciplines of earth sciences, life sciences and social sciences are employed to achieve ecological goals that benefit society.

Restoration on landscapes starts with establishing the baseline conditions, and asking practical questions, such as identifying suitable sites, understanding the nature of degradation, and establishing feasible restoration targets. The process includes technical considerations, and incorporates cultural, socioeconomic, and biogeographic perspectives to ensure the long-term viability and equitable distribution of restoration benefits.

Several countries including the European Union, the United States, and India have recognized the significant potential of ecological restoration to address pressing environmental and socioeconomic issues. In the EU, restoration efforts have focused on improving the resilience of ecosystems to invasive species and climate change, employing innovative engineering and community-based approaches (Tiền et al., 2021). Similarly, in the US, restoration initiatives are seen in riverine ecosystems, wetlands, and terrestrial ecosystems to enhance ecosystem services, such as reducing soil erosion, mitigating floods and carbon sequestration. They are also looked at

G. Nulkar (ed.), *The Economics of Sustainable Development*,
https://doi.org/10.1007/978-981-99-7379-8_12

as new employment opportunities in the emerging bioeconomy (Guo & Norman, 2013)(Zedler, 2005)(Zedler, 2005). India has also embraced ecological restoration as a means of reviving degraded lands and improving rural livelihoods, particularly through the integration of indigenous knowledge and community-driven approaches (Abhilash, 2021).

Degraded Lands and exhausted mines provide opportunities of restoration. The process begins with the identifying and removing the stresses, improving soil and abiotic conditions, removal of invasive vegetation, reintroduction of native species and the restoration of vital ecosystem processes. This allows the ecosystem to gradually recover its natural resilience and function. In some cases, more active interventions, such as soil amendments, water management, and the establishment of pioneering species, may be necessary to jumpstart the restoration process.

The integration of sustainable agricultural practices, such as agroforestry, conservation tillage, and the use of organic inputs, can further enhance the restoration of degraded lands by improving soil fertility, increasing carbon sequestration, and providing sustainable livelihoods for local communities (Wade et al., 2007).

Historically, ecological restoration initiatives have mostly focused on the implementation of conservation measures aimed at safeguarding endangered species and improving the quality of habitats within protected areas. However today, this discipline holds the promise of mitigating climate change. Moreover, restoration ecology explicitly includes societal values in its goals. This opens up opportunities of new job creation, which is dealt with in Chap. 9.

Ecological restoration requires multidisciplinary skills and could be challenging as many skills, such as ecological assessment, nurseries of native species, designing for ecosystem composition, structure, and function, and field techniques may need to be imparted. Restoration efforts extend beyond ecological considerations and encompass a broader social-ecological framework. The restoration practices offer such a vast potential that United Nations has declared this period as the decade of Ecological Restoration. The global initiative to envision, reconstruct, and rehabilitate ecosystems is being jointly spearheaded by the United Nations Environment Programme (UNEP) and the Food and Agriculture Organisation (FAO). This program assumes importance, especially as nations endeavour to recover from the impacts of the COVID-19 pandemic. The UN Decade on Ecosystem Restoration aims to prevent, halt and reverse the degradation of ecosystems on every continent and in every ocean. It can help to end poverty, combat climate change and prevent a mass extinction (Nations, Preventing, halting and reversing the loss of nature, 2023).

Ecological restoration has proven to be a powerful tool for the rehabilitation of degraded lands, including abandoned mines and other sites damaged by extractive industries. Through the integration of sustainable agricultural practices and the strategic deployment of restoration techniques, these efforts can not only revive the ecological functions of the land but also create new opportunities for sustainable livelihoods.

Ultimately, the success of ecological restoration in building climate resilience and supporting sustainable livelihoods hinges on a multifaceted approach that combines scientific knowledge, local expertise, and strategic policy frameworks. By fostering collaborative efforts between stakeholders, investing in research and innovation, and aligning restoration goals with broader sustainable development objectives, these regions can pave the way for a more resilient, equitable, and environmentally-sound future.

This chapter encompasses a collection of studies that examine the potential of ecological restoration in mitigating inequality, safeguarding communities against natural disasters, and rehabilitating various ecosystems such as wetlands, drylands, and forests. The chapter also presents a study on employing nature-based solutions to enhance circularity in economic operations.

References

Abhilash, P. (2021, February 17). Restoring the Unrestored: Strategies for Restoring Global Land during the UN Decade on Ecosystem Restoration (UN-DER). Multidisciplinary Digital Publishing Institute, 10(2), 201–201. https://doi.org/10.3390/land10020201.

Guo, Q., & Norman, S P. (2013, January 9). Improving restoration to control plant invasions under climate change. Informa, 203–216. https://doi.org/10.1201/b13865-16.

United Nations, (2023). Preventing, halting and reversing the loss of nature. Retrieved from Decade on Restoration: https://www.decadeonrestoration.org/

Tiền, H V., Anh, N T., Phong, N T., & Nhut, M L M. (2021, January 25). Ecological Engineering and Restoration of Eroded Muddy Coasts in South East Asia: Knowledge Gaps and Recommendations. Multidisciplinary Digital Publishing Institute, 13(3), 1248–1248. https://doi.org/10.3390/su13031248.

Wade, M R., Gurr, G M., & Wratten, S D. (2007, August 30). Ecological restoration of farmland: progress and prospects. Royal Society, 363(1492), 831–847. https://doi.org/10.1098/rstb.2007.2186.

Zedler, J B. (2005, September 2). Ecological Restoration: Guidance from Theory. eScholarship Publishing, University of California, 3(2). https://doi.org/10.15447/sfews.2005v3iss2art4 Nations, U. (2023). Preventing, halting and reversing the loss of nature. Retrieved from Decade on Restoration: https://www.decadeonrestoration.org/

12.1.1 Machine Generated Summaries

Disclaimer: The summaries in this chapter were generated from Springer Nature publications using extractive AI auto-summarization: An extraction-based summarizer aims to identify the most important sentences of a text using an algorithm and uses those original sentences to create the auto-summary (unlike generative AI). As the constituted sentences are machine selected, they may not fully reflect the body of the work, so we strongly advise that the original content is read and cited. The auto generated summaries were curated by the editor to meet Springer Nature publication standards. To cite this content, please refer to the original papers.

Machine generated keywords: restoration, forest, ecological, landscape, ecosystem, protection, habitat, property, ton, extensive, restore, area, program, poverty, decision.

12.1.2 Could Ecological Restoration Reduce Income Inequality? An Analysis of 290 Chinese Prefecture-Level Cities [91]

This is a machine-generated summary of:

Ma, Zihao; Tian, Xin; Zhang, Pingdan: Could ecological restoration reduce income inequality? An analysis of 290 Chinese prefecture-level cities [91].

Published in: Ambio (2023).

Link to original: https://doi.org/10.1007/s13280-022-01.815-y.

If you want to cite the papers, please refer to the original.

For technical reasons we could not place the page where the original quote is coming from.

Abstract-Summary

"To determine whether ecological restoration such as vegetation cover could affect the income gap, we used data for 290 prefecture-level cities in China from 2007 to 2018 and analyzed the effect of ecological restoration on income inequality in China."

"We found that ecological restoration can reduce income inequality in general, but this effect was not statistically significant until 2012."

"Due to some practical obstacles (e.g., employment opportunities, educational attainment, social discrimination), reducing income inequality through ecological restoration will be a time consuming process and requires constant effort from the Chinese government and local managers such as funding green industries, providing more targeted technical training for the poor and social services for the rural migrant workers."

Introduction

"Although Deng emphasized that the ultimate goal is to eliminate poverty and the polarization of wealth in China and achieve "common prosperity," relative poverty and the huge income gap still exist and restrict the development of the economic level of people living in less developed areas (Zhang 1)."

"Whether ecosystem restoration can promote relative poverty alleviation and achieve common prosperity (i.e., synergies between SDG 10 (Reduced Inequalities) and SDG 15) is not yet clear."

"We did not find any research mentioning whether or how ecological restoration, such as vegetation cover, could affect the income gap or common prosperity in China."

"Our aim is to test whether ecological restoration behavior can help reduce income inequality using econometric methods, provide a new path to alleviate inequality of income, and promote common prosperity not only in China but also with suitable modification to account for unique local conditions, in other developing countries around the world."

Materials and methods

"Since the way of implementing the policy of achieving common prosperity in China is to encourage some people and some regions to prosper prior to others to bring along the latter's prosperity, in time, to achieve the common prosperity, it is reasonable to divide the meaning of common prosperity into two steps: achieving "prosperity," which means the growth of the resident income of a relatively undeveloped area in a region (e.g., an undeveloped prefecture-level city in a province), and achieving "common," which means the elimination of the internal income gap in a region (e.g., urban–rural income gap)."

"We tested it by adding control variables to the model and observed whether the sign of the coefficient for the core explanatory variable ($lnndvi_{it}$) and its statistical significance changed significantly; if not, this supports our belief that our results are robust, which can prove that the effect of ecological restoration on common prosperity will not be affected by other control variables."

Results

"Our results showed that ecological restoration had a statistically significant negative effect on the municipal–provincial income gap and urban–rural income gap in China, suggesting that restoration could help achieve "prosperity" and "common prosperity.""

"Our robustness test showed that the sign of the coefficient for the core explanatory variable ($lnndvi_{it}$) and its statistical significance did not change significantly."

"Our heterogeneity analysis results showed that the effect of ecological restoration on the urban–rural income gap differed before and after 2012."

Discussion

"The implementation of ecological conservation projects and policies limits the use of natural resources, which forces rural livelihoods to change their traditional way of life because part of their income has been cut off (Wang and others 2), which may explain why ecological restoration could widen the urban–rural income gap in development zones to be optimized before 2012."

"The process of pursuing the common prosperity of urban and rural residents through ecological restoration may be challenging because of some practical obstacles (e.g., employment opportunities, educational attainment, social discrimination)."

"A well-functioning ecological environment may enhance the value of ecological products and increase the income of rural residents, which is a possible internal mechanism of such an effect."

"We will also test the change in the effect of ecological restoration on common prosperity in limited development zones and prohibited development zones over time when the relevant data are available to complete our research."

12.1.3 Identification of Priority Areas for Ecological Restoration Based on Ecological Security and Landscape Elements [92]

This is a machine-generated summary of:

Zhang, Jie; Chen, Hai; Ma, Yuhe; Liu, Di; Liang, Xiaoying; Chen, Wenting: Identification of priority areas for ecological restoration based on ecological security and landscape elements [92].

Published in: Environmental Science and Pollution Research (2022).

Link to original: https://doi.org/10.1007/s11356-022-24.711-0.

If you want to cite the papers, please refer to the original.

For technical reasons we could not place the page where the original quote is coming from.

Abstract-Summary

"Accurate identification of priority areas for ecological restoration is an important prerequisite for ecological protection and restoration, but it is a current challenge in landscape planning."

"A three-dimensional framework including natural potential, human disturbance, and landscape pattern factors was used to construct an ecological security evaluation index system, and spatial principal component analysis (SPCA) was used to quantitatively evaluate the ecological security levels of the study area."

"Priority areas for ecological restoration were identified by considering ecosystem security and the matching degree of landscape elements."

"The resulting area with low and moderately low security levels was 27,574.87 km^2 in size, accounting for 34.48% of the total study area, and the ecological security situation was not ideal."

"We identified seventeen ecological sources with an area of 5789.36 km^2, and the important ecological sources were mainly distributed in the south of the study area."

"We found that the high matching degree of landscape elements included four watersheds with an area of 7571.17 km^2, mainly distributed in the southern part of the study area."

"Fifty-one basins with a low matching degree of landscape elements were identified, covering an area of 50,399.44 km^2 and mainly distributed in the west and north of the study area."

"We identified three levels of areas to be restored, of which the level I ecological restoration priority area was the smallest, at 7047.61 km^2."

"We discussed ecological restoration strategies for different levels of ecological restoration priority areas and provided new methods for identifying priority ecological restoration areas in the future."

Introduction

"Finding an approach that considers both regional ecological security and the matching degree of landscape elements for ecological restoration has become a popular research topic (Fang and others 3; Wang and others 4; Richardson and Davidson 5; Zhao and others 6)."

"Most previous research has identified priority areas for ecological restoration based on ecological security assessment or landscape element identification (Ni and others 7; Xie and others 8)."

"The change of landscape pattern will degrade ecosystem services and cause adverse ecological effects, which is one of the important factors affecting regional ecological security (Wang and others 9)."

"Their research showed that landscape patterns have an important impact on regional ecological security."

"Comprehensively considering ecosystem security and the matching degree of landscape elements is necessary to effectively identify ecological restoration priority areas."

"Priority areas for ecological restoration were identified by considering ecosystem security and the matching degree of landscape elements, and practical restoration strategies were provided."

Study area and data sources

"This provides a good research platform for us to construct an index system based on the three dimensions of natural potential, human disturbance, and landscape pattern to evaluate the ecological security situation and identify ecological restoration priority areas in northern Shaanxi."

"The land use data of the study area in 2020 were obtained from the 1:100,000 land use vector dataset of Shaanxi Province from the Western China Environmental and Ecological Science Data Center (http://westdc.westgis.ac.cn/)."

"The land use could be divided into six first-class land types, including cultivated land, forestland, grassland, water area, construction land and unused land, and 22 s-class land types, such as paddy fields, dry land, urban land, and rural residential areas."

"The industrial and mining land, urban land, residential area, and road data were extracted based on the land use data."

Methods

"This study comprehensively considered socioeconomic factors, natural factors, and landscape pattern factors to construct an ecological security index system, which can comprehensively and accurately evaluate ecological security."

"The four indicators of distance from industrial and mining land, distance from urban land, distance from residential areas, and distance from roads were selected to characterize the impact of human activities on the ecological security of the study area (Li and others 10; Huang and others 11; Wang and others 12)."

"CONTAG and SHEI were selected to characterize the impact of the spatial heterogeneity of landscape units on ecological security (Yan and others 13; Zhang and others 14)."

"The specific formula was as follows: In this formula, ESI represents the comprehensive evaluation results of ecological security; a_{ij} represents the corresponding j principal component of the i grid; F_j represents the variance contribution rate of the jth principal component; and m and n represent the number of grids and principal components, respectively."

Identification of landscape elements

"The ecological security evaluation results in this paper comprehensively considered the factors of natural potential, human disturbance, and landscape pattern in the study area, which can well reflect the resistance encountered in the process of ecological source expansion."

"The MCR model formula was as follows: In this formula, MCR represents the minimum cumulative resistance value between the ecological source j and any grid i; f_{min} represents the positive correlation function between MCR and an ecological process; and D_{ij} denotes the distance between the surface i grid of the landscape pattern resistance and the j ecological source."

"In the field of ecology, the gravity model can quantify the strength of interactions between patches of different sources to determine the relative importance of potential ecological corridors within a region."

"The advantage of this model is the ability to quantify interactions between ecological sources, providing a practical way to assess landscape connectivity and determine the location of important ecological corridors."

Identification of priority areas for ecological restoration

"Previous studies have only focused on the quantitative relationship of landscape elements, ignoring the impact of spatial characteristics and the matching degree of landscape elements on ecological security."

"A higher matching degree of landscape elements showed that the landscape elements in the basin were more important to the ecological process of the whole study area."

"We should protect areas with a low ecological security and a high matching degree of landscape elements (Peng and others 15; Wang and others 16)."

"We initially set the priority of the ecosystem security to be higher than the matching degree of landscape elements, mainly to divide the order of level II and level III ecological restoration priority areas."

"Under the premise of a high level of overall ecological security in the study area, a high matching degree of landscape elements can play a better role (Ran and others 17)."

Results

"The area with the high ecological security level in the study area was 9485.06 km^2, accounting for 11.86% of the total study area."

"The area with the moderately high ecological security level was 19,302.27 km^2, accounting for 24.13% of the total study area."

"The medium level ecological security area was the largest, at 23,619.69 km^2, accounting for 29.53% of the total study area."

"The landscape elements had a low degree of matching in the study area, which had a certain impact on the ecological security of the watershed."

"Priority areas for ecological restoration should be identified by comprehensively considering the ecological system security and the matching degree of landscape elements to ensure the regional ecological security and the sustainability of the role of landscape elements."

"This area was characterized by a good matching degree of landscape elements but a poor ecological security level."

Discussion

"In terms of this study area, landscape pattern factors had a negative effect on ecological security, which meant that if these factors are ignored, the ecological security evaluation results may be overestimated."

"This paper proposed a new idea to quantitatively identify priority areas for ecological restoration considering ecosystem security and the matching degree of landscape element."

"We considered not only the quantitative relationship of landscape elements, such as the area of ecological sources, the length of ecological corridors, and the number of ecological nodes, but also considered the landscape pattern index of ecological sources to characterize the influence of the shape of source on the degree of matching of ecological elements."

"Integrating ecological corridors and ecological nodes into the evaluation of the matching degree of landscape elements based on the actual situation of the study area will be a focus of future research."

Conclusions

"Three levels of ecological restoration priority areas were identified based on a comprehensive consideration of ecosystem security and the matching degree of ecological elements."

"There were four watersheds with a high matching degree of landscape elements, with an area of 7571.17 km^2, accounting for 9.47% of the total area of the study area, mainly distributed in the south of the study area."

"There were 51 watersheds with a low matching degree of landscape elements, with an area of 50,399.44 km^2, accounting for 63.01% of the total area of the study area, mainly distributed in the west and north of the study area."

"The focus of restoration of the level III restoration area is to improve the matching degree of landscape elements in these watersheds."

"Integrating ecological corridors and ecological nodes into the evaluation of the matching degree of landscape elements based on the actual situation of the study area will be a focus of future research."

12.1.4 Protecting People and Property While Restoring Coastal Wetland Habitats [93]

This is a machine-generated summary of:

Weinstein, Michael P.; Guo, Qizhong; Santasieri, Colette: Protecting People and Property While Restoring Coastal Wetland Habitats [93].

Published in: Estuaries and Coasts (2021).

Link to original: https://doi.org/10.1007/s12237-021-00.900-x.

If you want to cite the papers, please refer to the original.

For technical reasons we could not place the page where the original quote is coming from.

Abstract-Summary

"Regulating and provisioning roles of coastal habitats have increasingly prompted policy makers to consider the value of ecosystem goods and services in these same decisions, broadly defined as "the benefits people obtain from ecosystems.""

"If unrestricted tidal flow were reestablished with mobile gates or similar devices, up to 226 ha of tidal salt marsh would be potentially restored to Spartina spp."

"Using existing literature and a value transfer approach, the estimated total economic value (TEV) of goods and services provided annually by these 226 ha of restored wetlands ranged from \$2,058,182 to \$2,390,854 y^{-1}."

"The associated annual engineering cost for including a mobile gate system to fully restore tidal flows to the upstream degraded wetlands was about \$1,925,614 y^{-1}

resulting in a benefit–cost ratio range of 0.98–1.14 over 50 years (assuming no wetland benefits realized during the first 4 years)."

"Inclusion of a cost-effective mobile gate system in any engineering design to improve long-term flood resilience in the region would produce dual benefits of protecting people and property from major storms, while preserving and enhancing ecosystem values."

Introduction

"Included are the protection of wetlands that contribute to the success of regional fisheries and coastal ecosystem processes and functions such as flood mitigation, carbon sequestration, and groundwater recharge that warrant future protection (ICCATF 18; zu Ermgassen, and others 19)."

"Striking a balance between protection of people and property, and protection of the structural and functional characteristics of ecosystems, is essential to ensuring the persistence of ecosystem services that are derived from tidal wetlands into the future (ICCATF 18)."

"We address a single element of any proposed "blue print" for flood abatement at Greenwich or elsewhwere; whatever the cost of providing protection against the 100-y storm and including the effects of sea level rise, what would be the element cost of including a mobile flood gate in the designs that will allow unimpeded tidal flow upstream of the levees, and consequently allow restoration of these degraded wetlands?"

Study Location and Existing Hydraulic Conditions

"Levees surrounding Greenwich are located approximately 0.3 to 1.3 mi (0.5 to 2.1 km) from the village center and have various degrees of wetland disturbance upstream of their footprints."

"With the exception of the eroded area below levee 48, the tidal marshes downstream of the three levees are relatively undisturbed and are dominated by extensive coverage by Spartina spp."

Ecosystem Valuation Estimation

"We supplemented this datum with an estimate for New Jersey wetlands based on Costanza and Mates (20) who used 100 studies to calculate a median TEV of \$6847 ha^{-1} y^{-1} (2004 USD), or \$9107 ha^{-1} y^{-1}, in 2018 USD."

"This led to an underlying assumption that a reasonable estimate for economic value of ecosystem goods or services at Greenwich could be inferred from the analysis of existing valuation studies at other sites."

"The area of wetland to be restored was assumed to be equivalent to the tidal inundation areas upstream of the three levees."

"If unimpeded upstream tidal flow were reestablished through each levee, the potential wetland area restored would be about 181 ha."

"These are likely to be underestimates because not all goods and services for tidal wetlands at Greenwich can be valued."

Proposed Engineering Design for Mobile Tide Gates

"Design parameters for the gates were obtained from existing data: mean high water (MHW), mean higher high water (MHHW), and mean tide level (MTL)."

"Tidal creek widths required for restoration of full tidal flow at Levees 48 and 50 were estimated as 152.4 m and 9.14 m, respectively."

"The water head was determined as the difference between water surface elevations at upstream and downstream sides of the gates."

"The downstream water surface elevation was assumed to be at the top elevation of the gate itself, and the upstream water surface elevation was assumed to be at the mean tide level (MTL)."

"The estimated heights required for the mobile gates at the four levees were thus estimated to be 3.8 m (Levee 48), 5.4 m (Levee 50); 5.7 m (Levee 51—Mill Creek), and 5.7 m (Levee 51—Mounce Creek)."

Proposed Installation and Maintence Costs for Mobile Tide Gates, and Benefit/ Cost

"Estimating construction costs based on the actual design of a storm surge barrier for each levee location would be difficult and would require detailed knowledge of the general characteristics and dimensions of each component, including levees, closure structures, gates, and gate monoliths."

"We applied an approach proposed by the US Army Corps of Engineers (USACOE 21), that considers average construction costs for storm surge barriers in various parts of the world."

"At Greenwich, costs were estimated from the barrier width, barrier height, and head (water level differential) acting on the barrier, as described above."

"From maintenance numbers of three large barriers in the world (Thames Barrier, Maeslant Barrier, Eastern Scheldt Barrier), it has been estimated that the annual maintenance costs are approximately 0.5% of the first construction costs (van Ledden and others 22)."

"Total annual costs are estimated using a 50-year project life and annual interest rate of 3.5%."

Discussion

"We previously conducted extensive studies of the flow of nutrients from dominant marsh vegetation including not only P. australis and S. alterniflora, but also microphytobenthos and phytoplankton, which attests to the relative value of tidal salt marshes in supporting the estuarine food web and secondary production."

"And others cited above, we believe that the wetland functions and values of the restored Spartina-dominated marsh upstream of Levee 50 (< 5% of the total area to be restored) will substantially exceed the near zero values of the extant Phragmites-dominated marsh currently at the site, and the former values have consequently been used in our TEV estimates."

"The application of a literature-based approximation to "total economic value (TEV)" for tidal salt marsh goods and services at the Greenwich study sites should be considered with some caution."

Conclusions

"We provide a concise case study which synthesizes ecosystem service valuation with engineering considerations to estimate the value-added benefits of including mobile tide gates that restore unimpeded tidal flows upstream in flood mitigation enhancement works, against their costs."

"Four decades, ecosystem goods and services' valuation has been one of fastest-growing areas of research in environmental and ecological economics (Schaefer and others 23)."

"It is being shown that sustainable, multi-functional use of an ecosystem is usually not only ecologically sound, but also economically beneficial, both to local communities and to society as a whole (Balmford and others 24)."

"Monetary valuation should always be seen as a supplement to, not as a replacement for, ecological, social, and cultural values under consideration in the decision-making process."

12.1.5 Nature-Based Solutions as a Tool in the New Circular Economic Model for Climate Change Adaptation [94]

This is a machine-generated summary of:

Stefanakis, Alexandros I.; Calheiros, Cristina S.C.; Nikolaou, Ioannis: Nature-Based Solutions as a Tool in the New Circular Economic Model for Climate Change Adaptation [94].

Published in: Circular Economy and Sustainability (2021).

Link to original: https://doi.org/10.1007/s43615-021-00.022-3.

If you want to cite the papers, please refer to the original.

For technical reasons we could not place the page where the original quote is coming from.

Abstract-Summary

"While our modern global society enjoys the benefits of an economic growth that was never seen before, at the same time, it is facing the existential threat of climate change."

"We look back to nature to find the solutions we need to increase the resiliency of our societies, to protect and restore the ecosystems, and to maintain the necessary economic growth without further undermining the planet's boundaries."

"This opinion paper discusses the potential of nature-based solutions to stimulate a circular model of economic growth, their contribution to new circular strategies for climate change adaptation, the various benefits of such concepts, and the key actions needed to increase the awareness and attract more investments on nature-based approaches in the near future."

The Global Environment Crisis

"Circular economy target is to improve the way we meet our current needs but through the use of less raw materials and natural resources and the reduction of the environmental impacts such as greenhouse gas emissions."

"Circular economy has the potential to significantly contribute to climate change mitigation by improving material management, dematerialization, decarbonization, and systematic change with long-term prospects."

"Climate change and infrastructure development make disaster-prone areas more vulnerable to extreme weather events and natural disasters (e.g., floods, landslides, forest fires) that cause loss of life and result in billions of euros of damage."

"Green infrastructure solutions or also known as nature-based solutions (NBS) can, however, limit the impacts of such events on human society and the environment and support the wider vision of circular economy for increased resilience and effectively addressing the challenges posed by climate change."

Defining Nature-Based Solutions

"The fundamental requirement for NBS to work effectively is first to understand the natural environment and the natural processes in order to harness the power of ecosystems and the sophistication of nature as infrastructure to provide natural services to benefit society and the environment and to turn environmental, social, and economic challenges into innovation opportunities."

"At the 2016 World Conservation Congress and members' assembly, the International Union for Conservation of Nature's (IUCN) members adopted a resolution (WCC-2016-Res-069-EN) that defined for the first time the use NBS as "actions to protect, sustainably manage and restore natural or modified ecosystems that address societal challenges effectively and adaptively, simultaneously providing human well-being and biodiversity benefits" (available at https://portals.iucn.org/library/sites/library/files/resrecfiles/WCC_2016_RES_069_EN.pdf)."

"Another definition is given by the European Commission as viewed from a social aspect of NBS implementation: "Nature-based solutions to societal challenges are solutions that are inspired and supported by nature, which are cost-effective, simultaneously provide environmental, social and economic benefits and help build resilience."

Nature-Based Solutions as a New Tool for Economic Growth

"Infrastructures that add more carbon into the atmosphere will eventually undermine the growth of the economies, and poor decisions in water and land-use infrastructure will not meet the adaptation needs particularly important for the developing countries."

"Investment decisions for major infrastructures (i.e., which sectors will receive funding) that are currently being made will determine the ability of developing countries to achieve their growth targets and development aspirations while avoiding the catastrophic results of climate change [25]."

"There is a need to develop a wider framework for infrastructure that promotes a sustainable model of economic growth, the circular model."

"The adoption of green infrastructure and an integrated operation and management plan of urban infrastructures, urban water, and waste and urban green is viewed as a critical tool for sustainable economic growth."

Why We Need Nature-Based Solutions

"The European Commission identified four key areas for NBS implementation [26]: (i) supporting sustainable urbanization to stimulate economic growth and enhance human well-being, while making the urban area more attractive, (ii) restoring degraded ecosystems by improving their resilience and increasing the ecosystem services they offer, (iii) climate change adaptation and mitigation with focus on carbon storage, and (iv) improving risk management and resilience."

"NBS for climate change mitigation are seen primarily as measures (e.g., reforestation, forest conservation and management, agroforestry, cropland nutrient management, conservation agriculture, coastal wetland restoration, peatland conservation and restoration) that can either reduce CO_2 emissions or remove CO_2 from the atmosphere, which is why the capacity of NBS to sustain or enhance carbon storage and carbon sequestration is highly attractive within the circular economy context [27]."

How to Spread the Word About Nature-Based Solutions

"In this frame, for NBS to play their role in the new circular approach, financial institutions should develop new funding streams and models that can support long-term investment in NBS for adaptation, including private sector actors."

"Integrate the nature-based approach in financial conditions Procurement, financing conditions, industry standards, and other policies should be improved to ensure that NBS are included and evaluated among the various adaptation solutions and their benefits are assessed for all options under consideration."

"Only a slow shift is visible towards NBS integration, although few global financial institutions such as the World Bank and the Asian Development Bank (ADB) are increasingly including the nature-based approach in their financing conditions."

"To help increase the impact of science on the design and implementation of NBS for climate change adaptation, more and up-to-date evidence-based information needs to be supplied to the policymakers."

Conclusion

"Nature-based solutions are facilitators towards the transition to circular economy, supporting a more sustainable management of the built environment."

"Considering their multiple benefits that go beyond climate adaptation, NBS can further establish ecosystems in the urban environment, forests, and coastal wetlands that support the economy and the society and provide fuel, food, livelihoods, critical habitat for biodiversity, health, jobs, education, and recreation benefits, and sustain cultural traditions."

"To stimulate greater commitment and attention to the undervalued role of nature in climate adaptation, it is necessary to better and more effectively highlight successful NBS implementation projects."

"The circular economy approach and targets create the necessary framework and favorable conditions to attract more investments for NBS scaling up and to further exploit the inherent potential of nature for climate adaptation and protect those most likely to be affected by climate change."

12.1.6 *Evaluating the Evolution of Forest Restoration Research in a Changing World: A "Big Literature" Review [95]*

This is a machine-generated summary of:

Nunez-Mir, Gabriela C.; Iannone, Basil V.; Curtis, Keeli; Fei, Songlin: Evaluating the evolution of forest restoration research in a changing world: a "big literature" review [95].

Published in: New Forests (2015).

Link to original: https://doi.org/10.1007/s11056-015-9503-7.

If you want to cite the papers, please refer to the original.

For technical reasons we could not place the page where the original quote is coming from.

Abstract-Summary

"In a rapidly changing world, characterized by novel ecological scenarios and fluctuating socioeconomic and ecological demands, ecological restoration practices must constantly adapt to emerging issues and circumstances."

"To describe and evaluate the evolution of research pertaining to the ecological restoration of forests within the field of forestry, we performed a review of all 29,766 abstracts published over the last 35 years (1980–2014) in 15 leading forestry journals using automated content analysis, a machine learning-based tool for automated review of large volumes of literature ("big literature")."

"We found not only a 50-fold increase in the prominence of restoration ecology in forestry literature from 2000 to 2013, but also an evolution in the focus of forest restoration research."

"This evolution is likely in response to emerging topics and issues affecting forest restoration, such as exotic species, altered disturbance regimes, degraded tropical forests, and ecosystem services."

Introduction

"Changing demands at the social (e.g., public concern for ecological sustainability), economic (e.g., cost-effectiveness of forest restoration and management) and ecological (e.g., emerging need to broaden the pool of species for restoration) level

(DellaSala and others 28; Menz and others 29; Millar and others 30; Oliet and Jacobs 31), along with emerging challenges presented by accelerated global change and increasingly unpredictable extreme climatic events (Harris and others 32; Hobbs and Cramer 33; Jalili and others 34; Stanturf and others 35; Vitousek and others 36), require the simultaneous and reciprocate evolution of these scientific disciplines."

"The objectives of this review were: (1) to assess the importance of the concept of restoration ecology in forestry research, (2) to identify the major concepts associated with forest restoration, and (3) to explore the progression of these concepts over time."

"We intend to explore and describe the evolution of restoration ecology in forest science in order to assess if the discipline and practice of forest restoration has been progressing in a direction that responds to current social and ecological demands in a changing world."

Methods

"ACA is an innovative content analysis technique for the synthesis of large volumes of literature, in which a body of text is processed by text-parsing software and classified at a high resolution (i.e. text segments of 2–3 lines) into categories named concepts (Smith and Humphreys 37)."

"To assess the importance of restoration over time, we first used ACA to identify the proportion of the literature analyzed for each decade classified under the concept of restoration ecology (ACA concepts are italicized to distinguish from ordinary words)."

"To determine the focus of restoration research in the forestry science literature and to assess the evolution of this focus, we identified the top 30 concepts having the strongest relationship with restoration ecology in each decade."

"The prominent concepts identified through the profiled analysis were then ranked by the strength of their association to restoration ecology (i.e. the percentage of text segments that were classified under both the particular concept and restoration ecology)."

Results and discussion

"Unlike frequency trends, the majority of concepts displayed a steady increase in the strength of their association to restoration ecology over time, possibly suggesting that as the field of restoration ecology continues to grow within the field of forestry science, these concepts have become more strongly associated to restoration in the literature."

"Our review of the forest restoration literature from recent decades has evidenced that restoration ecology is indeed not only rising in prominence, but also evolving and expanding in its focus to address emerging issues and challenges."

"These findings coincide with trends previously recognized in the field of restoration ecology in general (i.e. beyond forests), in which basic and applied restoration research focuses mostly on established ecological principles and concepts (e.g. competition and physiological limits) (Young and others 38) and how they apply to the restoration of site-level conditions (Brudvig 39)."

Concluding remarks

"Our "big literature" review clearly shows that the study of restoration ecology within the context of forestry science is not only growing in importance, but is also expanding in focus to adapt to the new challenges and demands emerging in an increasingly changing world."

"Our comparison of the literature from recent decades revealed shifts in the focus of research, suggesting an expansion of the field to explore established concepts and remaining issues from new perspectives, as well as undertake new concepts and emerging issues."

"At this moment, restoration ecology in the context of forest science appears to be going through the stage of expansion, marking an ideal time for the exploration of new ideas, concepts and perspectives."

"As it continues to grow, the field of restoration ecology, both in general and in the specific context of forests, has the potential to provide unique contributions to the research, conservation, and management of forest ecosystems."

12.1.7 Climate Resilient Development Pathways in Global Drylands [96]

This is a machine-generated summary of:

Stringer, Lindsay C.; Simpson, Nicholas P.; Schipper, E. Lisa F.; Eriksen, Siri H.: Climate Resilient Development Pathways in Global Drylands [96].

Published in: Anthropocene Science (2022).

Link to original: https://doi.org/10.1007/s44177-022-00.027-z.

If you want to cite the papers, please refer to the original.

For technical reasons we could not place the page where the original quote is coming from.

Abstract-Summary

"The Intergovernmental Panel on Climate Change (IPCC) Working Group II Report on Adaptation, Impacts and Vulnerability identifies the urgent need to embark upon Climate Resilient Development Pathways."

"Climate Resilient Development acknowledges that adaptation needs to be undertaken together with mitigation and development, in joined-up, inclusive, just and equitable ways, across multiple arenas of engagement."

"In highly vulnerable systems with complex development contexts, such as drylands, where globally, warming is already exceeding that of humid areas, urgent action is vital, as the window of opportunity for Climate Resilient Development is rapidly closing."

"This paper considers challenges and opportunities in charting Climate Resilient Development Pathways in the world's drylands."

"Adequate assessment of trade-offs in Climate Resilient Development actions in drylands remains vital to the framing of Climate Resilient Development Pathways for different groups."

Introduction

"The world needs to embark upon Climate Resilient Development Pathways: development trajectories that integrate adaptation, mitigation and sustainable development to advance towards high resilience and low risk futures to secure a liveable future for all this decade, to have the greatest likelihood of keeping global average temperature rise to below 1.5 °C, as well as avoiding getting locked into long-term pathways toward less sustainable futures."

"To achieve Climate Resilient Development at a global scale requires the design and navigation of diverse Climate Resilient Development Pathways at smaller scales, with different pathways being appropriate in different social–ecological systems, including different climate-sensitive environments, such as mountains, drylands, small islands and polar regions."

"This paper examines some of the major findings in the IPCC report and highlights some of the contextual challenges and opportunities in determining Climate Resilient Development Pathways in drylands."

Climate Change and Development in the Drylands

"A wide variety of traditional and emerging responses to climate-related water scarcity across these areas typify the challenges of living in drylands (Leal Filho and others 40), while viable alternatives to natural resource-based livelihoods are largely lacking."

"Six percent of the global dryland area experienced desertification over the period 1982–2015, due to a combination of unsustainable land use and land management practices, and climate change (Mirzabaev and others 41)."

"Rural agricultural and natural resource dependent livelihoods are presented with very different challenges to and opportunities for Climate Resilient Development compared to the livelihoods of people living in dryland megacities in complex, diversified economies, such as Cairo, Tehran, Beijing, Mexico City and Los Angeles (Stringer and others 42)."

"Climate Resilient Development that integrates adaptation, mitigation and development in the drylands needs to tackle these multiple challenges through the societal transformations that ensue, while also negotiating an inevitable diversity of complex trade-offs that operate across different temporal, spatial and social scales."

Climate Resilient Development Pathways in the Drylands

"The IPCC report challenges popular dryland policy actions, such as sedentarisation of mobile pastoralists and land privatisation, leading to exclusion and inequity (Mirzabaev and others 41) and notes that restoration needs to be undertaken in ways that support achievement of higher levels of Climate Resilient Development."

"Reforestation and afforestation programmes in drylands require full impact assessments to be carried out given concerns for water availability and wider impacts on biodiversity and ecosystem services (Bond and others 43; Veldman and others 44) to ensure mitigation does not undermine adaptation and development and lead towards lower levels of Climate Resilient Development."

"Training land users (particularly women) on how to care for the stumps and revive them supported climate change mitigation goals, and enabled adaptation and development as trees recovered, while new business models were created following local processing of non-timber tree products."

Research Gaps

"Alongside this, knowledge gaps remain as to what specific actions are needed, where and by whom, to support system transformations, and how costs and benefits can be balanced over appropriate and acceptable time frames, particularly given the urgency to act now."

"While business as usual is taking us towards increasing economic and non-economic costs, and inaction is exacerbating the challenge, there are currently no reliable estimates of the magnitude of investments needed to ensure we are embarking on Climate Resilient Development pathways."

"Robust approaches for monitoring, evaluation and learning are needed, to assess our progress along Climate Resilient Development Pathways in the drylands."

"If warming can be limited to 1.5 °C, the magnitude of climate risks that need to be adapted to can be reduced and the chances of a sustainable future for all, including those in the drylands, are increased."

Conclusions

"The recent IPCC report highlighted the urgent need to embark on Climate Resilient Development Pathways and that climate action that includes fundamental system transformations needs to begin this decade."

"This commentary has set out some of the challenges and opportunities for dryland systems to move toward higher levels of Climate Resilient Development."

"Inclusivity and equity need to be at the heart of actions to reduce emissions, adapt effectively and improve human livelihoods and wellbeing."

"All stakeholders at all levels, drawing on all kinds of knowledge, need to engage across all arenas (political, economic and financial, ecological, socio-cultural, knowledge and technology, and community) if Climate Resilient Development and the SDGs are to be achieved."

12.1.8 The Science and Practice of Ecological Restoration: A Mental Models Analysis of Restoration Practitioners [97]

This is a machine-generated summary of:

Walpole, Emily H.; Toman, Eric; Stidham, Melanie; Wilson, Robyn: The science and practice of ecological restoration: a mental models analysis of restoration practitioners [97].

Published in: Environment Systems and Decisions (2020).

Link to original: https://doi.org/10.1007/s10669-020-09.768-x.

If you want to cite the papers, please refer to the original.

For technical reasons we could not place the page where the original quote is coming from.

Abstract-Summary

"While substantial resources have been dedicated to advance ecological restoration efforts, limited research has investigated how ecological restoration is conceptualized by restoration practitioners."

"These conceptualizations are important because they likely inform goal-setting and selected management approaches, while different conceptualizations may contribute to conflicts that inhibit restoration efforts."

"Despite broad recognition of social and economic constraints, participant responses suggested limited consideration of how restoration may incorporate and achieve social and economic goals."

"Literature and history both suggest that these components are critical for the long-term success of ecological restoration efforts, and so their lack of salience to participants point to these as potential areas of improvement."

"Our management and policy recommendations include taking steps to more effectively integrate social scientific input in restoration planning, which may be supported by incorporating decision-aiding tools such as Structured Decision Making to guide restoration-planning efforts."

Introduction

"The manner in which land managers and other decision-makers define and think about forest restoration will have significant implications for the development of restoration goals, the level of agreement or conflict among participants, and how restoration activities are carried out on the ground (Vining and others 45; Hjerpe and others 46; Egan and others 47)."

"Walpole and others (48) observed differences in restoration goals and perceptions between conservation group and agency-based decision-makers in CFLRP efforts, including disagreements about the need for action in riparian and old-growth conifer systems."

"Given the range of potential conditions that could be emphasized through restoration efforts, developing a shared conceptualization, or mental model, among decision-makers regarding the need, goals, process, and priorities of restoration activities will likely contribute to their success over time."

"We applied a mental models approach to capture conceptualizations of forest restoration as a concept and process among land managers and other decision-makers engaged in three CFLRP locations."

Research methods

"Following Morgan and others's mental models methodology (49), we followed an iterative process of literature review and expert feedback to develop an initial influence diagram, which was finalized and condensed into a technical model illustrating the key concepts and practices associated with forest restoration."

"These responses were useful in distinguishing between concepts that participants were truly unaware of (e.g. responded "I don't know" to a direct prompt) versus those that were not salient enough to come up unprompted during interviews. (See Morgan and others 49 for a more in-depth description of the interview methodology used, and the electronic supplementary material for the interview protocol used.) All interviews were transcribed verbatim and imported into MaxQDA V11 (VERBI) for analysis; coding was guided by a 62-item code manual based on a summary of the topics identified in the literature and expert review process (i.e. the technical model)."

Results

"While every participant readily described ecologically-based desired future conditions, fewer than half mentioned social or economic goals unprompted; suggesting that these areas are less of a focus overall."

"These comments most commonly described the specific management actions that could be taken to achieve restoration objectives; indeed, the code "Management actions" was the most frequent code used by participants in the study, with an average of five total (i.e. prompted and unprompted) comments per interview."

"Despite the high salience in this area however, fewer than a third of participants (32%) discussed unprompted how the collaborative sought to actually engage the relevant public to participate in forest restoration decisions or incorporate their feedback into management activities, and "Public input" was also one of our least frequently discussed codes of the study overall."

Discussion

"There were also notable differences between the technical model and our participant's mental models with regards to the low salience of many social concepts identified as relevant to forest restoration, as well as some evidence of a focus on more quantifiable and near-term management planning topics related to restoration efforts."

"The collaboration and restoration literature emphasizes that providing benefits to diverse stakeholders and the public (i.e. economic and social as well as ecological benefit) is important for the long-term viability of most land management and restoration projects (Gobster and Hull 50; Wondolleck and Yaffee 51)."

"The key decision-makers we identified through CFLRP project coordinators and snowball sampling primarily had natural science backgrounds, it is possible that mandated project involvement or cross-training with social scientific experts, or continuing education in human dimensions of resource management, may provide useful skills to more fully consider how to incorporate social and economic goals into projects."

Summary and conclusions

"These results provide important insight into efforts to implement restoration goals through mental model-style interviews with decision-makers at three USDA Forest Service CFLRP locations."

"Our results also found that the ecological aspects of restoration were mentally emphasized among participants to the detriment of many social topics and goals."

"Collaborative restoration efforts may benefit from pioneering new methods of public involvement and increased social scientific input to more evenly balance the ecological, economic, and social benefits of restoration and achieve long-term program support."

"We suggest that this may be accomplished by more proactive integration of social and economic goals into project planning, cross-training and new incentive structures for USDA Forest Service employees, and incorporating new methods of public outreach through Structured Decision Making (SDM) processes or input from Forest Service or other social scientists."

12.1.9 Assessing Restoration Potential of Semi-natural Grasslands by Landscape Change Trajectories [98]

This is a machine-generated summary of:

Pitkänen, Timo P.; Mussaari, Maija; Käyhkö, Niina: Assessing Restoration Potential of Semi-natural Grasslands by Landscape Change Trajectories [98].

Published in: Environmental Management (2014).

Link to original: https://doi.org/10.1007/s00267-014-0242-x.

If you want to cite the papers, please refer to the original.

For technical reasons we could not place the page where the original quote is coming from.

Abstract-Summary

"We present a conceptual framework for evaluating the restoration potential of partially overgrown and selectively managed semi-natural grasslands in a moderately transformed agricultural environment in south-western Finland."

"On the basis of the spatio-temporal landscape trajectory analysis, we construct potential restoration scenarios based on expected semi-natural grassland characteristics that are derived from land productivity, detected grassland continuum, and date of overgrowth."

"These scenarios are evaluated using landscape metrics, their feasibility is discussed and the effects of potential restoration are compared to the present extent of open semi-natural grasslands."

"Our results show that landscape trajectory analysis and scenario construction can be valuable tools for the restoration planning of semi-natural grasslands with limited resources."

Introduction

"Restoration of the former semi-natural grasslands is the main tool to increase the habitat area, relying mostly on plants that have survived in a seed bank or refuge habitats, or are being spontaneously dispersed from nearby populations (Helm and others 52; Kuussaari and others 53; Dahlström and others 54; Auffret and Cousins 55)."

"To maximize the benefits of grassland management, practical planning solutions have focused on the spatial and temporal optimization of grazing and/or mowing, to sustain a sufficient amount of semi-natural grassland habitats of various characteristics (Vessby and others 56; van Teeffelen and others 57; Öckinger and others 58)."

"On the basis of the spatio-temporal landscape trajectory analysis, we construct potential restoration scenarios based on the expected semi-natural grassland characteristics that are derived from land productivity, detected grassland continuum and date of overgrowth."

"These scenarios are evaluated using landscape metrics, their feasibility is discussed and the effects of potential restoration are compared to the present extent of open semi-natural grasslands."

Materials and Methods

"In class 4A, including present or overgrown grasslands on high-productive ground, this classification was based on the assumption of more intensive management, thus reducing species richness and impeding potential restoration actions (Berendse and others 59; Eriksson and others 60)."

"The extended core scenario includes, in addition to trajectories 1A–B, recently overgrown areas (trajectory class 2) that can be expected to possess a significant number of semi-natural grassland characteristics and have the highest potential for restoration."

"The maximum limits scenario consisted of trajectory classes 1A–B, 2, and 3, and reflects the largest potential area of semi-natural grasslands, provided that restoration is also extended to long overgrown habitats."

"Landscape configuration of management and restoration scenarios was measured using five widely applied landscape indices: total area (TA), percentage of landscape (PL), number of patches (NP), effective mesh size (EMS) and connectance index (CI)."

Results

"Mälhamn, located approximately 750 m away from Berghamn as measured by the closest shore-to-shore distance, shares many characteristics with the village island: both high- and low-productive grasslands have occupied substantial areas, coniferous coverage has always been virtually negligible and bedrock-dominated areas have remained unchanged."

"Areas of overgrown grassland, resulting mainly from recent succession to scrubland or deciduous forest (class 2), appear rather frequently on Berghamn while their proportion on Mälhamn is smaller."

"Restoration actions are targeted to the areas of extended core habitats, including recently overgrown semi-natural grasslands, all the islands are able to gain new areas of semi-natural grassland but the increase on Berghamn is more substantial than that on Mälhamn and Boskär."

"The maximum limits scenario, extending restoration actions to semi-natural grasslands with more prolonged overgrowth, indicates evidence of further enlargement of grassland area on Boskär while the other islands remain close to the level of the previous scenario."

Discussion

"In the Berghamn study area, trajectory analysis indicates a polarization of the agricultural scenery where the previous landscape, characterized by well-connected grazed grasslands and transitional components between the grasslands and forests, has turned into more separated patches of still open or restored grasslands and rather thickly overgrown forests."

"One fact which has been assumed, but not fully supported by the study materials, is the long-term continuity of the grasslands prior to 1939, forming the basis for finding valuable and restorable semi-natural grassland habitats."

"While the importance of change trajectory analysis and historical knowledge of agrarian landscapes is often highlighted (Eriksson and others 60; Bruun and others 61; Lindborg and Eriksson 62; Lunt and Spooner 63; Reitalu and others 64), far less attention has been given to the question of how to support functional habitat networks by increasing the area of high-quality semi-natural grasslands."

"Provided that remnants of a functional semi-natural grassland network still exist and overgrowth has not demolished all the previous habitat characteristics, restoration of non-intensively managed semi-natural grasslands is to be prioritized instead of former fields."

Conclusions

"Increasing the area of managed semi-natural grasslands should be considered as one of the priorities to reach the goals set for biodiversity preservation, both at national and international levels."

"Landscape trajectory analysis and scenario construction, in a similar manner to that presented above, can be valuable for the restoration planning of semi-natural grasslands and should be considered as an essential tool in the pre-stratification of restoration potentialities."

"Several matters of concern are highlighted above but when the analysis is carefully performed and its practical limitations are understood, the results can prove beneficial for restoration action."

"We encourage the readers to assimilate the message while taking the fine details of our case study as curiosities and make an improved version of the methodology that will be suitable for semi-natural grasslands elsewhere."

12.1.10 Mud, Muddle and Models in the Knowledge Value-Chain to Action on Tropical Peatland Conservation [99]

This is a machine-generated summary of:

van Noordwijk, Meine; Matthews, Robin; Agus, Fahmuddin; Farmer, Jenny; Verchot, Louis; Hergoualc'h, Kristell; Persch, Sebastian; Tata, Hesti Lestari; Lusiana, Betha; Widayati, Atiek; Dewi, Sonya: Mud, muddle and models in the knowledge value-chain to action on tropical peatland conservation [99].

Published in: Mitigation and Adaptation Strategies for Global Change (2014).

Link to original: https://doi.org/10.1007/s11027-014-9576-1.

If you want to cite the papers, please refer to the original.

For technical reasons we could not place the page where the original quote is coming from.

Abstract-Summary

"Tropical peatlands are known not only for their high, area-based, carbon emissions in response to land-use change but also as hot spots of debate about associated data uncertainties."

"Debate includes the ways of reducing emissions through rewetting, reforestation and agroforestry."

"A knowledge value-chain that is long and complex links (a) fundamental understanding of peat and peatland processes leading to sciencebased quantification and default values, (b) willingness and (c) ability to act towards emission reduction, and ultimately (d) to local, national and global actions that effectively provide rules, incentives and motivation to conserve peat and reduce emissions."

"We conclude that, to shorten the denial and conspiracy-theory stages of debate that otherwise slow down steps B and C, networks of international and national scientists have to be involved at the early stage of identifying policysensitive environmental issues."

"While drainage of peatlands triggers landscape-scale increases in emissions, factors beyond drainage depth, including nutrient supply, may have a major influence on decomposition rates."

"With progress on A leading to new internationally accepted defaults and with resistance on step B reduced, the reality of C and lack of working solutions for D is currently constraining further progress."

Introduction

"Emission estimates from tropical peatlands converted from their wet natural forest condition to a drier form of human land use are large numbers with wide confidence intervals (Agus and others, 65; IPCC, 66)."

"A fundamental understanding of peat and peatland dynamics is certainly needed as the basis for appropriate policy responses, but the current debate on policy and economic incentives to reduce emissions are several steps beyond understanding of the basic processes."

"From the perspective of reducing global emissions from all land uses, uncertainty over emission rates from peatlands hardly affects the conclusions about which land uses are desirable, but it does matter if reductions are to be quantified as a basis for economic incentives."

"The Reducing Emissions from Deforestation and Degradation through Alternative Land Uses in Rainforests of the Tropics (REDD-ALERT) project (Matthews and others 67) has engaged with the science as well as with policy responses on this issue, and some lessons have been learnt at the science–policy interface."

Methods: a knowledge value chain

"Similar to the stepwise conversion of raw materials to tradable commodities, branded products, consumer-oriented services and highly valued experiences that are described in economic value chains with a focus on the value added in each step, we here describe a knowledge value (addition) chain, that potentially leads from raw data to positive impacts on lives and landscapes."

"A and D are specific to peatlands, B and C are more generic aspects of land-based emission control."

"It is a value chain, in the sense that unless step A provides a credible account of the issue, it can be easily dismissed as irrelevant."

"The chain concept does not imply that all efforts have to go into the first step before the others become relevant."

Knowledge chain transitions

"Accounting units shift from C stock in a unit volume of peat anywhere in the profile (mg C cm^{-3}) to a gaseous flux per unit soil mass or volume (μg CO_2 g^{-1} (dry weight) hr^{-1}, or μg CO_2 cm^{-3} h^{-1}) to area-based expressions (mg CO_2 m^{-2} h^{-1} or Mg CO_2 ha^{-1} $year^{-1}$) at measurement point, field, enterprise and landscape scales and, further, through product-based ones for footprints per unit of harvested product (Mg CO_2 Mg^{-1} product; Davis and others 68) to emission-based units for assessments of opportunity costs (USD Mg^{-1} CO_2) and emission-reduction strategies."

"The fundamental understanding of gaseous emission processes usually represents an hourly change dynamic and the basic unit of flux data collection is less than an hour; area-based emission estimates, however, are typically expressed per year (Groffman and others, 69), while economic assessments require a life-cycle analysis, which may well mean a 25 year period for plantation crops."

Results: progress on issues along the knowledge value-chain

"As the distinctions between peat- and plant-based respiration are key to the policy-relevant interpretation of surface flux for annual carbon budgets as emissions or recycling, we analyzed the underlying science and concepts."

"While there is no reason to assume that peat- and plant-based respiration are directly interacting or respond in similar ways to external variation, the use of fixed fractions to split measured surface flux into peat- and plantbased respiration is an assumption that may fail beyond calibration ranges."

"Murdiyarso and others (70) and Hergoualc'h and Verchot (71, 72) used measurements of peat respiration, litterfall rate and fine root turnover as the basis for their correction factors to convert flux data to emission estimates."

"A method used by Husnain and others (73) to keep areas of peat free from vegetation to directly measure peat-based respiration may provide an overestimate because decomposition of root necromass left after removal of aboveground biomass probably contributed to the surface flux."

Discussion

"Our quick tour of the issues associated with reducing emissions from tropical peatlands suggests that there is some progress in all four requirements for mitigation action, basic understanding (a), willingness (b) and ability (c) to act and effective alternative land uses (d), but that progress has been uneven."

"The knowledge to action framework (Cash and others 74; Clark and others 75) on salience, credibility and legitimacy of scientific knowledge, appears only to focus on steps A and B of the knowledge-value chain as described here."

"Given the progress achieved on a, b and c, it now seems that stage D has become the limiting step of the knowledge-to-action value-chain."

"The knowledge to action value chain involves multiple transformations."

"Mitigation of and adaptation to climate change require synergy between multiple layers where action is needed, the knowledge value chain depends on multiple steps of boundary work, without losing its core content and identity."

12.1.11 An Extensive Review on Restoration Technologies for Mining Tailings [100]

This is a machine-generated summary of:

Sun, Wei; Ji, Bin; Khoso, Sultan Ahmed; Tang, Honghu; Liu, Runqing; Wang, Li; Hu, Yuehua: An extensive review on restoration technologies for mining tailings [100].

Published in: Environmental Science and Pollution Research (2018).

Link to original: https://doi.org/10.1007/s11356-018-3423-y.

If you want to cite the papers, please refer to the original.

For technical reasons we could not place the page where the original quote is coming from.

Abstract-Summary

"The ecological restoration of mining tailings has become one of the urgent tasks for mine workers and environmental engineers all over the world."

"Aim of the present paper is to highlight the previous restoration techniques and the challenges encountered during the restoration of mine tailings."

"As it is a common practice that, before restoring of tailings, the site should be evaluated carefully."

"Studies showed that the mine tailings' adverse properties, including excessive heavy metal concentration, acidification, improper pH value, salinization and alkalization, poor physical structure and inadequate nutrition, etc., are the major challenges of their restoration."

"Four restoration technologies, including physical, chemical, phytoremediation, and bioremediation, are used to restore the mining tailings."

"The ecosystem restoration is perceived to be a more promising technology for mine tailings."

Introduction

"Tailing, in context to mining, is defined as the waste or non-economic by-product generated during mining activities, processing of minerals, and other materials that contain small amounts of residual valuable minerals, chemicals, water, and heavy metals (Adiansyah and others 76)."

"Studies indicated that the process of mineral exploitation and utilization generated a large amount of waste rock and tailings, which, when not disposed properly, potentially polluted to soil, air, and water and created many health issues (Bhattacharya and others 77; Schoenberger 78)."

"One example is the failure of a sulfide mining tailing storage facility that resulted in an environmental disaster in Spain on 25 April 1998, with large amounts of toxic mud and acidic water released into the Agrio River and significant kills the freshwater organisms and pollutes the water sources (Grimalt and others 79)."

Challenges encountering during restoration of mine tailings

"Acid production decreases the pH value of the tailings soil and increases the solubility of heavy metals."

"The low pH and high heavy metal contents in the tailings soil can cause plant growth disorders, the so-called biotoxicity."

"These two factors are likely to have synergistic effects, because low pH values increase the bioactivity of heavy metals, and high levels of heavy metals and strong acidity are usually the greatest limiting factors for re-establishing plant growth in the tailings area (Glaser and Birk 80; Liu and others 81), thus causing great difficulties in the restoration of tailings."

"To leading to a higher concentration of heavy metals in the soil, the acidification of tailings can also lead to the activation of aluminum ions in the soil, inducing aluminum toxicity."

"Many studies have shown that if the nutrient content in the soil is sufficiently high, the adverse effect of excessive heavy metal contents on plant growth can be offset to some extent (Johnson and others 82; Rah and Bradshaw 83)."

Techniques for mine tailing restoration

"Principle, chemical stabilization technology promotes the transformation of heavy metals from more bioavailable forms such as F1 or F2, to less bioavailable forms by applying immobilized reagents, thus reducing the toxic effects of heavy metals on soil plants and microorganisms."

"Because the physicochemical properties of tailings are not suitable for the growth and reproduction of plants, due to high concentrations of heavy metals, high acidity, poor nutrition, salinization and alkalization, and physical limitations, it is necessary to improve the soil to resolve the problems existing in different tailings (Lee and others 84)."

"Plants that are used for restoration of tailings can tolerate high concentration of heavy metals and harsh growth environment, such as high acidity, poor nutrition, salinization and alkalization, and physical limitation (Li 85)."

"Among graminaceous plants, Vetiveria zizanioides found to have a strong resistance to acidic, barren soil and heavy metals, hence suitable for tailing restoration (Rotkittikhun and others 86)."

Comparison of different restoration technologies for their applicability and selection

"There are several factors contribute to the cost involved for restoration of mining tailings including those related to tailing site (such as type and depth of soil, depth of groundwater, possible migration pathways and purpose of restoration), the heavy metals (such as type of metal, and concentration of metal in tailings) and the restoration technology (various technologies have different requirements and cost of action)."

"Chemical restoration is a cost-effective technique compared to physical remediation methods."

"Compared to physical restoration, phytoremediation is a low-cost restoration method."

"Physical restoration is considered most effective method with regard to the time required for complete restoration of mining tailings."

"As for phytoremediation, there are several factors influence the time required for restoration of mining tailing sites, including metal levels found in tailings, heavy metal mobility, target metal final concentrations, plant biomass, and rates of plant growth (Shahid and others 87)."

Ecosystem restoration

"This is because many of the mining sites have been severely damaged or contaminated, and even the restoration techniques above are unlikely to make them suitable for the survival of vegetation or ecosystems."

"Some mines have been mined too deep, piled too high or otherwise altered such that the landscape has changed so much that it cannot be restored to its original form."

"In the more developed mines in Australia, the mining areas have been successfully restored, and many have returned to local ecosystems dominated by Eucalyptus marginata and E. tetrodonta."

"The growth of fynbos forest after mining of many gold deposits in South Africa, prepared by certain measures including the restoration of topsoil, rational fertilization, and strict control of alien species invasion, eventually restored the local vegetation, which shows that taking some measures can restore the original vegetation type (Holmes and Richardson 88; Holmes 89; Keeley and Bond 90)."

Conclusions

"Working mechanism, advantages and disadvantages available restoration technologies, including physical, chemical, phytoremediation and bioremediation, for mining tailings have been described in detail."

"Economic considerations and time required can affect the selection and applicability of available mining tailing restoration technologies."

"The ecosystem restoration is suggested to be a more favorable technology for mine tailings."

"The restoration of tailings is a systematic project, and it is difficult to achieve the desired results simply by relying on a single restoration technology."

Bibliography

1. Zhang, W. 2016. The evolution of China's pay inequality from 1987 to 2012. Journal of Current Chinese Affairs 45: 183–217.
2. Wang, C., Y. Wen, W. Duan, and F. Han. 2013. Coupling relationship analysis on households' production behaviors, protection attitudes, resource endowment and family wealth in nature reserves: A structural equation model. Chinese Geographical Science 23: 506–518.
3. Fang Y, Wang J, Huang KY, Zhai TL (2020) Determining and identifying key areas of ecosystem preservation and restoration for territorial spatial planning based on ecological security patterns: a case study of Yantai city. J Nat Resour 35:190–203
4. Wang CX, Yu CY, Chen TQ, Feng Z, Hu YC, Wu KN (2020) Can the establishment of ecological security patterns improve ecological protection? An example of Nanchang. China Sci Total Environ 740:140051
5. Richardson BJ, Davidson NJ (2021) Financing and governing ecological restoration projects: the Tasmanian Island Ark. Ecol Manag Reser 22:36–46
6. Zhao WK, Liao YP, Zhou SQ, Zhou B (2022) Ecological remediation strategy for urban brownfield renewal in Sichuan Province, China: a health risk evaluation perspective. Sci Rep-Uk 12:4300
7. Ni QL, Hou HP, Ding ZY, Li YB, Li JR (2020) Ecological remediation zoning of territory based on the ecological security pattern recognition: taking Jiawang district of Xuzhou city as an example. J Nat Resour 35:204–216
8. Xie LX, Bai YP, Che L, Qiao FW, Sun SS, Yang XD (2021) Construction of ecological zone based on value-risk ecological function area in the Upper Yellow River. J Nat Resour 36:196–207
9. Wang HR, Zhang MD, Wang CY, Wang KY, Wang C, Li Y, Bai XL, Zhou YK (2022a) Spatial and temporal changes of landscape patterns and their effects on ecosystem services in the Huaihe River Basin. China Land 11:513
10. Li QP, Zhang ZD, Wan LW, Yang CX, Zhang J, Ye C, Chen YC (2019) Landscape pattern optimization in Ningjiang River Basin based on landscape ecological risk assessment. Acta Geogr Sin 74:1420–1437
11. Huang MY, Yue WZ, Feng SR, Cai JJ (2019) Analysis of spatial heterogeneity of ecological security based on MCR model and ecological pattern optimization in the Yuexi county of the Dabie Mountain Area. J Nat Resour 34:771–784
12. Wang ZY, Shi PJ, Zhang XB, Tong HL, Zhang WP, Liu Y (2021) Research on landscape pattern construction and ecological restoration of Jiuquan City based on ecological security evaluation. Sustainability 13:5732
13. Yan YC, Ju HR, Zhang SR, Chen GK (2021) The construction of ecological security patterns in coastal areas based on landscape ecological risk assessment: a case study of Jiaodong Peninsula, China. Int J Environ Res Public Health 18:12249
14. Zhang XM, Du HM, Wang Y, Chen Y, Ma L, Dong TX (2021a) Watershed landscape ecological risk assessment and landscape pattern optimization: take Fujiang River Basin as an example. Hum Ecol Risk Assess 27:2254–2276
15. Peng J, Yang Y, Liu YX, Hu YN, Du YY, Meersmans J, Qiu SJ (2018) Linking ecosystem services and circuit theory to identify ecological security patterns. Sci Total Environ 644:781–790
16. Wang ZF, Liu Y, Xie XQ, Wang XK, Lin H, Xie HL, Liu XZ (2022b) Identifying key areas of green space for ecological restoration based on ecological security patterns in Fujian Province. China Land 11:1496
17. Ran YJ, Lei DM, Li J, Gao LP, Mo JX, Liu X (2022) Identification of crucial areas of territorial ecological restoration based on ecological security pattern: a case study of the central Yunnan urban agglomeration. China Ecol Indic 143:109318
18. Interagency Climate Change Adaptation Task Force (ICCATF). 2011. Federal actions for a climate resilient nation. DC: Washington.

19. zu Ermgassen, et al. 2021. Valuation, social and human dimensions in tidal marsh ecology. Estuaries and Coasts (this issue).
20. Costanza, R., and W. Mates. 2007. Valuing New Jersey's natural capital: An assessment of the economic value of the State's natural resources. Final Report: New Jersey Department of Environmental Protection, Trenton, NJ.
21. United States Army Corps of Engineers (USACOE). 2015. North Atlantic coast comprehensive study: Resilient adaption to increasing risk, final report. Planning Analyses, January: Appendix C.
22. Van Ledden, M., H.J. Lansen, Ridder, et al. 2012. Reconnaissance level study Mississippi storm surge barrier. ASCE, Proceedings Coastal Engineering Conference, Santander, Spain.
23. Schaefer, M., E. Goldman, A.M. Bartuska, A. Sutton-Grier, and J. Lubchenco. 2015. Nature as capital: Advancing and incorporating ecosystem services in United States federal policies and programs. Proceedings of the National Academy of Sciences 112 (24): 7383–7389.
24. Balmford, A., A. Bruner, P. Cooper, et al. 2002. Economic reasons for conserving wild nature. Science 297 (5583): 950–953.
25. Brown M, Kim Y, Romani M (2015) Infrastructure finance in the developing world. Green infrastructure: definition and needs. The Global Green Growth Institute, Seoul, Korea & The Intergovernmental Group of Twenty Four on Monetary Affairs and Development, Washington
26. EC (European Commission) (2015) Towards an EU research and innovation policy agenda for nature-based solutions & re-naturing cities. Final report. Directorate-General for Research and Innovation Climate Action, Environment, Resource Efficiency and Raw Materials, Publications Office of the European Union, Luxembourg
27. Bulkeley H, (2020). Nature based solutions towards sustainable communities. Analysis of EU-funded projects. Directorate-General for Research and Innovation, European Commission, Publications Office of the European Union, Luxembourg
28. DellaSala D et al (2003) A citizen's call for ecological forest restoration: forest restoration principles and criteria. Ecol Restor 21:15
29. Menz MHM, Dixon KW, Hobbs RJ (2013) Hurdles and opportunities for landscape-scale restoration. Science 339:526–527
30. Millar CI, Stephenson NL, Stephens SL (2007) Climate change and forests of the future: managing in the face of uncertainty. Ecol Appl 17:2145–2151
31. Oliet J, Jacobs D (2012) Restoring forests: advances in techniques and theory. New For 43:535–541. https://doi.org/10.1007/s11056-012-9354-4
32. Harris JA, Hobbs RJ, Higgs E, Aronson J (2006) Ecological restoration and global climate change. Restor Ecol 14:170–176
33. Hobbs RJ, Cramer VA (2008) Restoration ecology: interventionist approaches for restoring and maintaining ecosystem function in the face of rapid environmental change. Annu Rev Environ Resour 33:39–61
34. Jalili A et al (2010) Climate change, unpredictable cold waves and possible brakes on plant migration. Global Ecol Biogeogr 19:642–648
35. Stanturf JA, Palik BJ, Dumroese RK (2014) Contemporary forest restoration: a review emphasizing function. For Ecol Manage 331:292–323
36. Vitousek PM, Mooney HA, Lubchenco J, Melillo JM (1997) Human domination of Earth's ecosystems. Science 277:494–499
37. Smith AE, Humphreys MS (2006) Evaluation of unsupervised semantic mapping of natural language with Leximancer concept mapping. Behav Res Methods 38:262–279
38. Young TP, Petersen DA, Clary JJ (2005) The ecology of restoration: historical links, emerging issues and unexplored realms. Ecol Lett 8:662–673. https://doi.org/10.1111/j.1461-0248.2005.00764.x
39. Brudvig LA (2011) The restoration of biodiversity: where has research been and where does it need to go? Am J Bot 98:549–558
40. Leal Filho W, Totin E, Franke JA, Andrew SM, Abubakar IR, Azadi H, Nunn PD, Ouweneel B, Williams PA, Simpson NP (2022) Understanding responses to climate-related water scarcity in Africa. Sci Total Environ 806:150420. https://doi.org/10.1016/j.scitotenv.2021.150420

41. Mirzabaev A, Stringer LC, Benjaminsen TA, Gonzalez P, Harris R, Jafari M, Stevens N, Tirado CM, Zakieldeen S (2022) Cross-chapter paper 3: deserts, semi-arid areas and desertification. In: Pörtner H-O, Roberts DC, Tignor M, Poloczanska ES, Mintenbeck K, Alegría A, Craig M, Langsdorf S, Löschke S, Möller V, Okem A, Rama B (eds) IPCC, 2022: Climate change 2022: impacts, adaptation, and vulnerability. Contribution of Working Group II to the sixth assessment report of the intergovernmental panel on climate change. Cambridge University Press, Cambridge (in press)
42. Stringer LC, Mirzabaev A, Benjaminsen TA, Harris RMB, Jafari M, Lissner TK, Stevens N, Tirado-von der Pahlen C (2021) Climate change impacts on water security in global drylands. One Earth 4(6):851–864. https://doi.org/10.1016/j.oneear.2021.05.010
43. Bond WJ, Stevens N, Midgley GF, Lehmann CER (2019) The trouble with trees: afforestation plans for Africa. Trends Ecol Evol 34(11):963–965. https://doi.org/10.1016/j.tree.2019.08.003
44. Veldman JW, Overbeck GE, Negreiros D, Mahy G, Le Stradic S, Fernandes GW, Durigan G, Buisson E, Putz FE, Bond WJ (2015) Where tree planting and forest expansion are bad for biodiversity and ecosystem services. Bioscience 65(10):1011–1018. https://doi.org/10.1093/biosci/biv118
45. Vining J, Tyler E, Kweon BS (2000) Public values, opinions, and emotions in restoration controversies. In: Gobster P, Hull B (eds) Restoring nature: perspectives from the social sciences and humanities. Island Press, Washington DC, pp 143–161
46. Hjerpe E, Abrams J, Becker DR (2009) Socioeconomic barriers and the role of biomass utilization in southwestern ponderosa pine restoration. Ecol Restor 27:169–177. https://doi.org/10.3368/er.27.2.169
47. Egan D, Hjerpe EE, Abrams J (2011) Human dimensions of ecological restoration. Island Press, Washington DC
48. Walpole EH, Toman E, Wilson RS, Stidham M (2017) Shared visions, future challenges: a case study of three Collaborative Forest Landscape Restoration Program locations. Ecol Soc 22(2):35. https://doi.org/10.5751/ES-09248-220235
49. Morgan MG, Fischhoff B, Bostrom A, Atman CJ (2002) Risk communication: a mental models approach. Cambridge University Press, New York
50. Gobster PH, Hull RB (2000) Restoring nature: perspectives from the Social Sciences and Humanities. Island Press, Washington DC
51. Wondolleck JM, Yaffee SL (2000) Making collaboration work: lessons from innovation in natural resource managment. Island Press, Washington DC
52. Helm A, Hanski I, Pärtel M (2006) Slow response of plant species richness to habitat loss and fragmentation. Ecol Lett 9:72–77. https://doi.org/10.1111/j.1461-0248.2005.00841.x
53. Kuussaari M, Bommarco R, Heikkinen RK, Helm A, Krauss J, Lindborg R, Öckinger E, Pärtel M, Pino J, Rodà F, Stefanescu C, Teder T, Zobel M, Steffan-Dewenter I (2009) Extinction debt: a challenge for biodiversity conservation. Trends Ecol Evol 24:564–571. https://doi.org/10.1016/j.tree.2009.04.011
54. Dahlström A, Rydin H, Borgegård S-O (2010) Remnant habitats for grassland species in an abandoned Swedish agricultural landscape. Appl Veg Sci 13:305–314. https://doi.org/10.1111/j.1654-109X.2009.01068.x
55. Auffret AG, Cousins SAO (2011) Past and present management influences the seed bank and seed rain in a rural landscape mosaic. J Appl Ecol 48:1278–1285. https://doi.org/10.1111/j.1365-2664.2011.02019.x
56. Vessby K, Söderström B, Glimskär A, Svensson B (2002) Species-richness correlations of six different taxa in swedish seminatural grasslands. Conserv Biol 16:430–439. https://doi.org/10.1046/j.1523-1739.2002.00198.x
57. van Teeffelen AJA, Cabeza M, Pöyry J, Raatikainen K, Kuussaari M (2008) Maximising conservation benefit for grassland species with contrasting management requirements. J Appl Ecol 45:1401–1409. https://doi.org/10.1111/j.1365-2664.2008.01514.x
58. Öckinger E, Lindborg R, Sjödin NE, Bommarco R (2012) Landscape matrix modifies richness of plants and insects in grassland fragments. Ecography 35:259–267. https://doi.org/10.1111/j.1600-0587.2011.06870.x

59. Berendse F, Oomes MJM, Altena HJ, Elberse WTh (1992) Experiments on the restoration of species-rich meadows in The Netherlands. Biol Conserv 62:59–65. https://doi.org/10.1016/0006-3207(92)91152-I
60. Eriksson Å, Eriksson O, Berglund H (1995) Species abundance patterns in Swedish semi-natural pastures. Ecography 18:310–317. https://doi.org/10.1111/j.1600-0587.1995.tb00133.x
61. Bruun HH, Fritzbøger B, Rindel PO, Hansen UL (2001) Plant species richness in grasslands: the relative importance of contemporary environment and land-use history since the Iron Age. Ecography 24:569–578. https://doi.org/10.1111/j.1600-0587.2001.tb00491.x
62. Lindborg R, Eriksson O (2004) Historical landscape connectivity affects present plant species richness. Ecology 85:1840–1845. https://doi.org/10.1890/04-0367
63. Lunt ID, Spooner PG (2005) Using historical ecology to understand patterns of biodiversity in fragmented agricultural landscapes. J Biogeogr 32:1859–1873. https://doi.org/10.1111/j.1365-2699.2005.01296.x
64. Reitalu T, Johansson LJ, Sykes MT, Hall K, Prentice HC (2010) History matters: village distances, grazing and grassland species diversity. J Appl Ecol 47:1216–1224. https://doi.org/10.1111/j.1365-2664.2010.01875.x
65. Agus F, Gunarso P, Sahardjo BH, Harris N, van Noordwijk M, Killeen TJ (2013a) Historical CO2 Emissions from Land Use and Land Use Change from the Oil Palm Industry in Indonesia, Malaysia and Papua New Guinea. Roundtable on Sustainable Palm Oil, Kuala Lumpur
66. IPCC (2014) 2013 Supplement to the 2006 IPCC Guidelines for National Greenhouse Gas Inventories: Wetlands. In: Hiraishi T, Krug T, Tanabe K, Srivastava N, Baasansuren J, Fukuda M, Troxler TG (eds). Methodological Guidance on Lands with Wet and Drained Soils, and Constructed Wetlands for Wastewater Treatment. IPCC, Switzerland
67. Matthews, RB, van Noordwijk M, Lambin E, Meyfroidt P, Gupta J, Verchot L, Hegouac'h K, Veldkamp E (2014) Implementing REDD + (Reducing Emissions from Deforestation and Degradation): case study evidence on governance, evaluation and impacts. Mitig Adapt Strateg Clim Chang -- This issue (MITI-D-14-00055)
68. Davis SC, Boddey RM, Alves BJR, Cowie A, Davies C, George B, Ogle SM, Smith P, van Noordwijk M, van Wijk M (2013) Management swing potential for bioenergy crops. Glob Chang Biol Bioenergy. https://doi.org/10.1111/gcbb.12042
69. Groffman PM, Venterea RT, Verchot LV, Potter CS et al (2006) Landscape and regional scale studies of nitrogen gas fluxes. In: Scaling and Uncertainty Analysis in Ecology: Methods and Applications. Springer, Dordrecht, pp 191–203
70. Murdiyarso D, Hergoualc'h K, Verchot LV (2010) Opportunities for reducing greenhouse gas emission in tropical peatlands. Proc Natl Acad Sci U S A 107(46):19655–19660
71. Hergoualc'h K, Verchot LV (2011) Stocks and fluxes of carbon associated with land-use change in Southeast Asian tropical peatlands: a review. Glob Biochem Cy 25. https://doi.org/10.1029/2009GB003718
72. Hergoualc'h K, Verchot LV (2013) Greenhouse gas emission factors for land use and land-use change in Southeast Asian peatlands. Mitig Adapt Strateg Glob Chang. https://doi.org/10.1007/s11027-013-9511-x. This issue
73. Husnain H, Agus F, Wigena IP, Dariah A, Marwanto S (2014) Peat CO2 emissions from several land use types in Indonesia. Mitig Adap Strateg Glob Chang. https://doi.org/10.1007/s11027-014-9550-y. This issue
74. Cash DW, Clark WC, Alcock F, Dickson NM, Eckley N, Guston DH, Jager J, Mitchell RB (2003) Knowledge systems for sustainable development. Proc Natl Acad Sci U S A. https://doi.org/10.1073/pnas.1231332100
75. Clark WC, Tomich TP, van Noordwijk M, Guston D, Catacutan D, Dickson NM, McNie E (2011) Boundary work for sustainable development: natural resource management at the Consultative Group on International Agricultural Research (CGIAR). Proc Natl Acad Sci U S A. https://doi.org/10.1073/pnas.0900231108
76. Adiansyah JS, Rosano M, Vink S, Keir G (2015) A framework for a sustainable approach to mine tailings management: disposal strategies. J Clean Prod 108:1050–1062

77. Bhattacharya A, Routh J, Jacks G, Bhattacharya P, Mörth M (2006) Environmental assessment of abandoned mine tailings in Adak, Västerbotten district (northern Sweden). Appl Geochem 21:1760–1780
78. Schoenberger E (2016) Environmentally sustainable mining: the case of tailings storage facilities. Res Policy 49:119–128
79. Grimalt JO, Ferrer M, Macpherson E (1999) The mine tailing accident in Aznalcollar. Sci Total Environ 242:3–11
80. Glaser B, Birk JJ (2012) State of the scientific knowledge on properties and genesis of Anthropogenic Dark Earths in Central Amazonia (terra preta de Índio). Geochim Cosmochim Acta 82:39–51
81. Liu X, Zhang A, Ji C, Joseph S, Bian R, Li L, Pan G, Paz-Ferreiro J (2013) Biochar's effect on crop productivity and the dependence on experimental conditions—a meta-analysis of literature data. Plant Soil 373:583–594
82. Johnson MS, McNeilly T, Putwain PD (1977) Revegetation of metalliferous mine spoil contaminated by lead and zinc. Environ Pollut 12:261–277
83. Rah S, Bradshaw AD (1979) The use of metal tolerant plant populations for the reclamation of metalliferous wastes. J Appl Ecol 16:595–612
84. Lee SH, Ji W, Lee WS, Koo N, Koh IH, Kim MS, Park JS (2014) Influence of amendments and aided phytostabilization on metal availability and mobility in Pb/Zn mine tailings. J Environ Manag 139:15–21
85. Li MS (2006) Ecological restoration of mineland with particular reference to the metalliferous mine wasteland in China: a review of research and practice. Sci Total Environ 357:38–53
86. Rotkittikhun P, Chaiyarat R, Kruatrachue M, Pokethitiyook P, Baker AJ (2007) Growth and lead accumulation by the grasses Vetiveria zizanioides and Thysanolaena maxima in lead-contaminated soil amended with pig manure and fertilizer: a glasshouse study. Chemosphere 66:45–53
87. Shahid M, Pinelli E, Dumat C (2012) Review of Pb availability and toxicity to plants in relation with metal speciation; role of synthetic and natural organic ligands. J Hazard Mater 219-220:1–12
88. Holmes PM, Richardson DM (1999) Protocols for restoration based on recruitment dynamics, community structure, and ecosystem function: perspectives from South African fynbos. Restor Ecol 7:215–230
89. Holmes PM (2001) Shrubland restoration following woody alien invasion and mining: effects of topsoil depth, seed source, and fertilizer addition. Restor Ecol 9:71–84
90. Keeley JE, Bond WJ (1997) Convergent seed germination in South African fynbos and Californian chaparral. Plant Ecol 133:153–167
91. Ma, Zihao; Tian, Xin; Zhang, Pingdan Could ecological restoration reduce income inequality? An analysis of 290 Chinese prefecture-level cities. Ambio (2023). https://doi.org/10.1007/s13280-022-01815-y
92. Zhang, Jie; Chen, Hai; Ma, Yuhe; Liu, Di; Liang, Xiaoying; Chen, Wenting Identification of priority areas for ecological restoration based on ecological security and landscape elements. Environmental Science and Pollution Research (2022). https://doi.org/10.1007/s11356-022-24711-0
93. Weinstein, Michael P.; Guo, Qizhong; Santasieri, Colette Protecting People and Property While Restoring Coastal Wetland Habitats. Estuaries and Coasts (2021). https://doi.org/10.1007/s12237-021-00900-x
94. Stefanakis, Alexandros I.; Calheiros, Cristina S.C.; Nikolaou, Ioannis Nature-Based Solutions as a Tool in the New Circular Economic Model for Climate Change Adaptation. Circular Economy and Sustainability (2021). https://doi.org/10.1007/s43615-021-00022-3
95. Nunez-Mir, Gabriela C.; Iannone, Basil V.; Curtis, Keeli; Fei, Songlin Evaluating the evolution of forest restoration research in a changing world: a "big literature" review. New Forests (2015). https://doi.org/10.1007/s11056-015-9503-7
96. Stringer, Lindsay C.; Simpson, Nicholas P.; Schipper, E. Lisa F.; Eriksen, Siri H. Climate Resilient Development Pathways in Global Drylands. Anthropocene Science (2022). https://doi.org/10.1007/s44177-022-00027-z

97. Walpole, Emily H.; Toman, Eric; Stidham, Melanie; Wilson, Robyn The science and practice of ecological restoration: a mental models analysis of restoration practitioners. Environment Systems and Decisions (2020). https://doi.org/10.1007/s10669-020-09768-x
98. Pitkänen, Timo P.; Mussaari, Maija; Käyhkö, Niina Assessing Restoration Potential of Semi-natural Grasslands by Landscape Change Trajectories. Environmental Management (2014). https://doi.org/10.1007/s00267-014-0242-x
99. van Noordwijk, Meine; Matthews, Robin; Agus, Fahmuddin; Farmer, Jenny; Verchot, Louis; Hergoualc'h, Kristell; Persch, Sebastian; Tata, Hesti Lestari; Lusiana, Betha; Widayati, Atiek; Dewi, Sonya Mud, muddle and models in the knowledge value-chain to action on tropical peatland conservation. Mitigation and Adaptation Strategies for Global Change (2014). https://doi.org/10.1007/s11027-014-9576-1
100. Sun, Wei; Ji, Bin; Khoso, Sultan Ahmed; Tang, Honghu; Liu, Runqing; Wang, Li; Hu, Yuehua An extensive review on restoration technologies for mining tailings. Environmental Science and Pollution Research (2018). https://doi.org/10.1007/s11356-018-3423-y

Chapter 13
Measuring Development and Well-Being

13.1 Introduction by Editor

The conventional measure of a nation's economic growth has long been the Gross Domestic Product (GDP), which quantifies the total economic output of a nation. However, ever since it was proposed by Simon Kuznets, the Gross Domestic Product (GDP) has been subject to criticism. Former President John F. Kennedy made a notable statement regarding the measurement of Gross Domestic Product (GDP). He said "*…it (GDP) measures everything in short, except that which makes life worthwhile.*' Many scholars have argued that GDP is falls short in assessing the overall well-being and quality of life within a society (Costanza et al., 2009)(Leon & Boris, 2010)(England, 1998). The shortcomings of GDP have triggered notable efforts in development of alternative frameworks and indices that could capture human development and well-being more holistically.

One such metric is the Human Development Index (HDI), developed by the United Nations Development Program (UNDP). The HDI combines measures of life expectancy, education, and per capita income to give a more comprehensive assessment of a country's level of development. The OECD's Well-Being Framework is another comprehensive approach that examines a range of factors beyond just economic output, including health, education, environment, and work-life balance. These alternatives reflect a growing concurrence that economic growth does not necessarily translate to improved quality of life and well-being (Giannetti et al., 2015).

In face of global competition, rapid economic growth has triggered extensive depletion of natural resources, increased the rift between the rich and the poor, and widened the gap between some of the developed and developing nations. Measures that go beyond GDP can capture factors that directly impact people's experiences, such as poverty, employment, and the state of the natural environment (Giannetti et al., 2015)(Costanza et al., 2009).

G. Nulkar (ed.), *The Economics of Sustainable Development*,
https://doi.org/10.1007/978-981-99-7379-8_13

While GDP remains an important economic metric, policymakers and researchers have increasingly advocated for the adoption of more nuanced, multidimensional frameworks to evaluate societal progress. These new approaches can provide valuable insights to guide decision-making and ensure that development efforts truly enhance human well-being. Improving the quality of life, as opposed to economic growth, is the core tenet of human development. It emphasizes the freedom that individuals have. Health is one of its important dimensions.

According to the World Health Organisation "Health is a state of complete physical, mental, and social well-being, not just the absence of disease or infirmity". Human development and well-being have multiple dimensions that be broken down into two key areas: objective well-being and subjective well-being. Different scales and indices have been made to measure both areas as a result. To measure the well-being of individuals in society, it is imperative to extend our measures beyond the mere functioning of the economic system. Instead, we must take into account living conditions of individuals and households. Across the world, several organisations are involved in developing guidelines and frameworks for improving the measurement of well-being.

The UNDP-HDI emphasizes the fact that economic growth alone is insufficient to determine whether or not a country is developing. The index has three dimensions: Long and healthy life, Knowledge, and A decent standard of living. Each dimension has indicators and a dimension index. The measure captures the overall attainment in fundamental aspects of human development. The OECD Well-being Framework has 11 dimensions that cover outcomes at the individual, household or community level. They include material conditions like housing, work and job quality; quality-of-life factors that encompass how well people are and how well they feel they are; how healthy and safe their places of living are; how connected and engaged people are, and their work-life balance, social connections, and civic engagement.

The story of Bhutan is well known. Having shunned metrics of economic growth, Bhutan embraced the measure of Gross National Happiness (GNH). The evolution of the GNH goes back to 1972 when the king of Bhutan, Jigme Singye Wangchuck, chose the pursuit of citizens happiness over economic measures (Forum, 2016). The GNH mandates four dimensions – good governance, conservation of the environment, preservation of cultures, and sustainable and equitable socio-economic development.

As nations across the world embrace the UN Sustainable Development Goals, the GDP loses not just its glamour, but also its relevance to modern conditions. Nations must inevitably move to a new measure that assesses the health of the economy and its citizen. In this chapter we present studies that look at the multiple dimensions of human development, discuss well-being, examine how to adjust ecological deficits in GDP, and green development efficiency.

References

Costanza, R., Hart, M., Posner, S., & Talberth, J. (2009, January 1). Beyond GDP: The Need for New Measures of Progress. https://open.bu.edu/bitstream/2144/22665/1/PP-004-GDP.pdf

England, R. (1998, April 1). Measurement of social well-being: alternatives to gross domestic product. Elsevier BV, 25(1), 89–103. https://doi.org/10.1016/s0921-8009(97)00.098-0.

Fan, V Y., Bloom, D E., Ogbuoji, O., Prettner, K., & Yamey, G. (2018, October 23). Valuing health as development: going beyond gross domestic product. BMJ, k4371-k4371. https://doi.org/10.1136/bmj.k4371.

Giannetti, B F., Agostinho, F., Almeida, C., & Huisingh, D. (2015, January 1). A review of limitations of GDP and alternative indices to monitor human wellbeing and to manage eco-system functionality. Elsevier BV, 87, 11–25. https://doi.org/10.1016/j.jclepro.2014.10.051.

Leon, E D., & Boris, E T. (2010, January 1). The state of society: Measuring economic success and human well-being. https://doi.org/10.1037/e725342011-001.

13.2 Machine Generated Summaries

Disclaimer: The summaries in this chapter were generated from Springer Nature publications using extractive AI auto-summarization: An extraction-based summarizer aims to identify the most important sentences of a text using an algorithm and uses those original sentences to create the auto-summary (unlike generative AI). As the constituted sentences are machine selected, they may not fully reflect the body of the work, so we strongly advise that the original content is read and cited. The auto generated summaries were curated by the editor to meet Springer Nature publication standards. To cite this content, please refer to the original papers.

Machine generated keywords: wellbee, individual, measure, wellbeing, life, indicator, subjective, measurement, gdp, human development, domain, good, dimension, index, social.

13.2.1 *Putting the Biophysical (Back) in Economics: A Taxonomic Review of Modeling the Earth-Bound Economy [107]*

This is a machine-generated summary of:

Sherwood, John; Carbajales-Dale, Michael; Haney, Becky Roselius: Putting the Biophysical (Back) in Economics: A Taxonomic Review of Modeling the Earth-Bound Economy [107]

Published in: Biophysical Economics and Sustainability (2020).

Link to original: https://doi.org/10.1007/s41247-020-00.069-0.

If you want to cite the papers, please refer to the original.

For technical reasons we could not place the page where the original quote is coming from.

Abstract-Summary

"Within the last 10 years, a wide variety of research papers have been published that include some biophysical aspects in a model of the economy."

"BPE models of the economy originate from a variety of disciplines, and thus BPE research articles are published across a wide spectrum of academic journals."

"As inter-disciplinary researchers ourselves, we want to understand what BPE modeling approaches have been used so far."

"We examine and classify over one hundred published articles that use biophysical models of the economy."

"Although BPE modeling approaches are quite varied, grouping the research by common characteristics reveals several active research areas."

"Several modeling approaches have not been used in published works yet."

"We conclude by suggesting which BPE modeling approach might be particularly appropriate for a variety of research questions."

Introduction

"To be fair, instances of natural resource depletion leading to major economic contraction are rare in the modern era and early attempts by economists to model or account for biophysical limits on the economy were unsuccessful."

"Most of the biophysical models of the economy have been developed outside of the economics discipline because ecosystem services are not traded in the market, nor are ecological systems necessarily well-understood by economists."

"The contribution of this paper is to develop a taxonomy for biophysical models of the economy and classify a sample of over 100 BPE articles published in the last 10 years."

"Classifying 110 BPE articles according to the taxonomy allows researchers interested in biophysical economics an opportunity to see the broad landscape of modeling approaches that have been published so far."

What Makes a Model of the Economy a Biophysical Model?

"Mainstream environmental economic models examine the markets and market failures related to natural resources, energy, and waste."

"Biophysical models of the economy also account for the physical flows of natural resources, energy, and waste through the economy."

"Biophysical economics is the study of the ways and means by which human societies procure and use energy and other biological and physical resources to produce, distribute, consume and exchange goods and services, while generating various types of waste and environmental impacts."

"The first holistic biophysical model of the economy in the current era is the seminal, Limits to Growth work led by Donella Meadows (Meadows et al. 1)."

"Voudouris et al. (2); Sherwood et al. (3) and others have been developing Agent-Based models to study biophysical economics."

"BPE models also appear in macroeconomics, energy economics, natural resource economics, and complexity economics."

The Biophysical Economic Modeling Landscape.

"Paper, we have selected 110 articles containing biophysical economic models to analyze based on their topics addressed and model characteristics."

"For each journal, search with keywords: Model, input–output Model, systems dynamics Model, production function Agent-based model 3."

"The categories are Individual-based Model, Agent-based Model, Input–Output Model, Systems Dynamics, and Aggregate Production Function."

"A category of 2 represents agent-based models."

"A category of 1 for the framework characteristic represents true individual-based models, where an agent is a specific person or entity, rather than an industry or country (a country might often be the "agent" in ABMs, such as Voudouris et al. (2)) The Rai and Robinson (4) example would be categorized as an individual-based model, as the "agent" is a specific household, the lowest level of analysis for household solar adoption."

"A category of 4 represents mainstream economic models such as aggregate production functions."

Using the Taxonomy.

"We chose one model for each level of the model framework characteristic (excluding individual-based modeling), to show how a specific combination of characteristics applies to a research topic."

"These combinations of characteristics are deliberate, and gaps between potential combinations and actualized combinations may point towards topics or research questions that may not have been modeled yet."

"This study extends a line of research that seeks to incorporate energy parameters into a standard macroeconomic model (Ayres and Warr 5; Heun et al. 6)."

"This model demonstrates the viability of using ABM to study biophysical economics."

"The nature of the modeling framework allowed for deep insight into indicators that are normally highly aggregated, such as production and consumption of oil for every country."

"Palmer and many others have recently begun to utilize environmentally-extended input–output models to study biophysical economics indicators (King et al. 7; King 8; Brand-Correa et al. 9; Heun et al. 10)."

"These examples represent a range of biophysical economic modeling frameworks, objectives, and mechanisms."

Qualitative Analysis of Modeling Space

"The top-left box shows that there are two individual-based models in the dataset."

"Box A shows that there are no individual-based models at the country, continent, or world scale."

"Box B indicates that individual-based and agent-based models tend to extend across a long term time horizon."

"Few agent-based models occur in short- or medium term timeframes."

"Box C indicates that individual-based and agent-based models are usually theoretical."

"Box D indicates that agent-based models are grounded in behavioral economics or social science."

"Box E shows that an agent-based model by definition is a simulation."

"Box F shows that there are no short term systems dynamics models."

"Similar to agent-based models, this gap appears to be a byproduct of the systems dynamics framework."

"The majority of time horizon gaps occur for the short term (less than 5 years) time horizon models."

Conclusion and Future Research

"Because of the multi-disciplinary nature, biophysical economics modeling approaches are published across a large spectrum of academic journals."

"The primary contribution of this study has been to develop a taxonomy of biophysical economics models as a way to understand the current landscape of this diverse body of work."

"Several gaps in the taxonomy suggest that some modeling approaches have yet to be explored in the published literature."

"The Ethos gaps are a natural result of the context of how models are developed."

13.2.2 Adjusting GDP for Ecological Deficit: The Index of Debt to the Future (IDF) [108]

This is a machine-generated summary of:

Lianos, Theodore P.; Pseiridis, Anastasia: Adjusting GDP for ecological deficit: the Index of Debt to the Future (IDF) [108]

Published in: SN Business & Economics (2021).

Link to original: https://doi.org/10.1007/s43546-021-00041-0.

If you want to cite the papers, please refer to the original.

For technical reasons we could not place the page where the original quote is coming from.

Abstract-Summary

"Many criticisms have been raised against the use of GDP as a measure of economic welfare."

"The Index of Debt to the Future (IDF) proposed in this paper adjusts the measured GDP by subtracting the part of GDP that creates the ecological deficit as estimated by the difference between biocapacity and ecological footprint."

"That part of GDP is a measure of the debt of the present generation to the future generations."

"The adjusted GDP, defined as the difference between GDP and the implied debt, is the net economic magnitude which the present generation should consider as its own."

Introduction

"It is now generally accepted that the Gross Domestic Product (GDP) is not a perfect measure of economic welfare."

"There is production taking place which is not included in the measured GDP as, for example, housework, unpaid voluntary work, and unregistered transactions."

"The other two examples fall in the category of what (Nordhaus and Tobin 11) call instrumental expenditures and there is no reason to exclude them from the measurement of GDP."

"Intended transactions create some utility, otherwise they would not take place, and they should be included in the measurement of GDP."

"In spite of the above criticisms, all countries continue to estimate GDP and use it for various purposes instead of the other proposed measures of economic welfare (e.g., ISEW, GPI, see below)."

"GDP is clearly defined as a measure based on market transactions and therefore dubious imputations are avoided."

Literature review: measuring economic activity

"Another way of measuring aggregate production while taking account of the environmental impact of production is that suggested by the Environmentally Sustainable National Income (eSNI) suggested by Roefie Hueting, which is defined as "the maximal attainable production by which vital environmental functions remain available for future generations, based on the technology available at the time."

"Another index of measuring well-being is the National Accounts of Well-being which uses information of subjective judgments about satisfying life, vitality, self-esteem, optimism, etc. The results of these indices, as valuable as they may be for other purposes, are not directly connected to economic welfare and therefore seem to fall outside the area of our present interest."

"There is the ambitious and apparently expensive project of the United Nations (12) for a System of Environmental Economic Accounts (SEEA)."

Research approach: adjusting GDP for debts to the future

"This can be done by combining two widely used measures – the GDP and the ecological footprint – into a new index: the Index of Debt to the Future (IDF)."

"The ecological footprint by definition underestimates the impact that economic activity has on nature."

"This measure has the benefit that is based on readily available data: GDP, biocapacity, and ecological footprint."

"Using world data for the global biocapacity and ecological footprint for 2016, the world IDF for 2016 can be calculated as follows: An alternative, more intuitive, derivation of IDF would be which yields The above IDF means that 40.5% of the gross world product produced is in excess of what would be produced in ecological balance; this excess GDP is achieved using "excess" (borrowed) resources that should be used in the future."

Comparison of GDP_{EDF} with other indices of sustainable welfare

"The Index of Debt to the Future (IDF) suggested here attempts to correct the traditionally measured GDP by reducing it by the extent to which the production of goods causes ecological deficit."

"The resulting Ecological Debt–Free GDP (GDP_{EDF}) calculated with the use of the IDF does not contradict other indices of sustainable economic welfare as, for example, the ISEW and GPI."

"The IDF is based on the idea that the satisfaction we derive from our present consumption level is less (or should be less) because we know that our consumption creates problems for the present and the future generations."

"Our situation is similar to that of the individual who consumes 100 euros every day knowing that every day he/she creates a debt of 40 euros to support their present consumption level, and this growing debt negatively affects the level of utility derived from future consumption."

Results: calculating the ecological debt-free GDP using the IDF

"The Index of Debts to the Future can be very easily estimated for individual countries and for the world as a whole, since data for GDP, ecological footprint, and biocapacity are readily available."

"It is interesting to compare the ratio of GDP to IDF–adjusted GDP (GDP_{EDF}) with the ratio of GDP to ISEW for some countries."

"In the case of the world GDP (i.e., GWP), the ratio GWP/GDP_{EDF} is smaller than the GDP/GDP_{EDF} and GDP/ISEW ratios in all cases."

"For Australia the GDP/GDP_{EDF} ratio is less than one, implying that biocapacity is greater than the ecological footprint, and therefore no adjustment is needed for the estimate of the Ecological Debt-Free GDP, as Australia is an ecological creditor country."

"This follows from the fact that the adjustment made to GDP by ISEW is less than that required for ecological equilibrium as shown by the adjustment made by the IDF that is suggested in this paper."

Discussion

"During the course of writing this article, the COVID pandemic emerged; the pandemic shattered our collective certainties that the threats to humanity will unfold in the near or far future; the pandemic showed that the threat to our lives and way of living is closer in time and space than we thought, revealed the vulnerability of our current well-being, and diverted resources from existing problems to a new and pressing one."

"Almost every recent policy document or scientific article acknowledges the vast, massive, or otherwise dramatic behavioural changes that have to be achieved in the next few years, if humanity is to stand a chance of survival from climate change, from biodiversity loss, and from a cascading and interacting set of problems created by humanity's economic activity."

"The authors hope that the simple idea provided in this paper will help visualize and disseminate the message of urgency and facilitate the behavioural change that is so urgently needed."

Conclusions

"The Index of Debt to the Future (IDF) proposed in this paper adjusts the measured GDP by subtracting the part of GDP that creates the ecological deficit."

"That part of GDP is a measure of the debt of the present generation to the future generations."

"The IDF-adjusted GDP (GDP_{EDF}), defined as the difference between GDP and the implied debt to the future, is the net economic magnitude which the present generation should consider as its own."

"The comparison between the IDF-adjusted GDP and those of ISEW and GPI indicates that the ISEW and GPI indices underestimate the needed adjustment."

"It is clear that the necessary adjustment to GDP is so large that governments and international organisations cannot ignore the debt that past and existing generations have accumulated on the shoulders of the future ones."

13.2.3 The Gross Developmental Potential (GDP2): A New Approach for Measuring Human Potential and Wellbeing [109]

This is a machine-generated summary of:

Halfon, Neal; Chandra, Anita; Cannon, Jill S.; Gardner, William; Forrest, Christopher B.: The Gross Developmental Potential (GDP2): a new approach for measuring human potential and wellbeing [109]

Published in: BMC Public Health (2022).

Link to original: https://doi.org/10.1186/s12889-022-14.030-x.

If you want to cite the papers, please refer to the original.

For technical reasons we could not place the page where the original quote is coming from.

Abstract-Summary

"The Gross Developmental Potential (GDP2) is proposed as a new capabilities-based framework for assessing threats to thriving and understanding progress in achieving lifelong health and wellbeing."

"Moving beyond the Gross Domestic Product's (GDP) focus on economic productivity as a measure of progress, the GDP2 focuses on seven essential developmental capabilities for lifelong health and wellbeing."

"The GDP2 capability domains include Health -living a healthy life; Needs-satisfying basic human requirements; Communication-expressing and understanding thoughts and feelings; Learning-lifelong learning; Adaption -adapting to change; Connections -connecting with others; and Community -engaging in the community."

"The framework was then vetted in focus groups of community leaders from three diverse settings."

"This prototype GDP2 framework will next be used to develop specific measures and indices and guide the development of community-level GDP2 dashboards for local sense-making, learning, and application."

Background

"There are concerns now about social and emotional health impacts (e.g., school anxiety, depression) and learning loss, which will affect current youths' health and educational outcomes, and through them, the social and economic progress of the nation [13]."

"Although lagging measures of economic output such as the Gross Domestic Product (GDP) are correlated with other measures of social progress, they do not directly measure the human capabilities necessary to achieve optimal levels of health and well-being."

"This paper reports on an effort to build a new framework for a new measure of social progress – gross developmental potential (GDP2) – that could complement the GDP."

"GDP2 attempts to capture the potential of a nation by assessing the developmental capabilities that promise to define their young people's life course trajectories, enabling them to thrive as youth and adults."

Methods

"Once the initial set of developmental capabilities was identified, we solicited additional subject-matter expert input, conducted a focused literature review, and interviewed community leaders to refine our approach further."

"The second meeting reviewed the ultimate framework for GDP2 to vet core developmental capabilities that would be used in the pilot communities."

"This framework was also reviewed through interviews with four experts on subjects relevant to our capabilities set and on measuring well-being more generally."

"We shared the final capability set with stakeholder groups from three diverse communities to obtain their reactions to the capabilities, identify capabilities that may be missing, and finally hear how they might use a sentinel measure such as GDP2."

"Communities were presented the broad descriptions of the capabilities to develop community-informed definitions of capabilities, learn which capabilities were considered priorities by communities, and help us understand what potential indicators we might seek to measure."

Results

"Having the Health capability does not necessarily mean that a person will realize a long, healthy life because that capability refers to an ability."

"The Learning capability is the ability of a person to learn across the life course as an individual and in groups and apply those learnings to new challenges, both inside and outside of formal education systems."

"The leaders noted that this capability also refers to abilities that enable individuals to participate effectively in group learning to obtain knowledge and solve problems, including efforts that employ combinations of human and machine intelligence [14]."

"The Adaptation capability is the ability of a person to adapt to life's challenges, self-regulate behavior, and agilely adjust to rapid social, cultural, and technological change."

"Community leaders underscored this capability, noting that individuals must understand that they are part of a larger social fabric and acknowledge their responsibility to address community problems."

Discussion

"We have proposed and presented a new capabilities-based framework for measuring human potential that cities, counties, states, and the nation can use to monitor social progress."

"Through extensive literature review and expert input, we developed a set of seven capabilities that community leaders affirmed as important for human potential measurement."

"The new GDP2 measure must accurately represent the development of the seven capabilities and do it in a useful way to communities and the wide range of place-based health improvement initiatives."

"Developing new measures of each capability allows us to capture precisely the data needed in the most parsimonious way."

"The other option is to further develop the capabilities framework in collaboration with several communities and work with those communities to create a dashboard of existing indicators and indices."

Conclusions

"We need to rethink how society can help children and youth adapt in productive ways for a more promising future."

"Community leaders indicated that the GDP2 could elevate the importance and attention paid to positive health development pathways, which are leading indicators of future economic and social progress."

"It provides a robust way to consider community and national investments in positive health and well-being."

13.2.4 The Indicator of a Happy, Long and Sustainable Life [110]

This is a machine-generated summary of:

Gaucher, Renaud; Dialga, Issaka; Vennin, Coralie: The indicator of a happy, long and sustainable life [110]

Published in: Social Indicators Research (2021).

Link to original: https://doi.org/10.1007/s11205-021-02.727-2.

If you want to cite the papers, please refer to the original.

For technical reasons we could not place the page where the original quote is coming from.

Abstract-Summary

"Many indicators are used to measure the quality of life of people and, consequently, these can be used to inform us about the quality of public policies."

"This question can be asked to anyone and we could build a main indicator for public policies using this question and the responses of people."

"Our personal response is the following: What is really important for us as an individual is to live a happy and long life, and a life that does not prevent following generations from having lives as happy and long as ours, that is a sustainable life."

"As we are negative utilitarianists and we consider that reducing premature deaths is more important than lengthening the duration of life, the indicator follows these two approaches."

Introduction

"There are a lot of single and composite indicators that can be used to measure the quality of life, and it is easy to design new composite indicators using new combinations of several single indicators and to consider that these combinations are better than the previous ones."

"There are two ways to develop the indicator of a happy, long and sustainable life."

"The first way is to have three single indicators, one used to measure happiness, one used to measure the duration of life, and one used to measure sustainability."

"As the indicator of a happy, long and sustainable life, as a composite indicator, is our personal answer to the question of what is really important for us as an individual, the chosen weights, which reflect our opinion, are obviously subjective, and we choose equal weighting."

Happiness

"Two, Veenhoven's definition of happiness can be summarized in a few simple words, which make it easy to use to empower human beings in their lives."

"The distinction between experienced happiness and remembered happiness is very powerful and it would be a good idea to use the concept of experienced happiness when thinking about public policies, however, it is difficult to measure experienced happiness over a long period of time."

"As happiness is defined by Veenhoven, (15) as liking the life one leads, the best item can be 'How much do you like the life you lead?'."

"Three, improving the degree of happiness of the less happy, the unhappiest, this is the negative utilitarianist approach (Popper, 16)."

"Our negative utilitarianist measure of happiness using a 1 to 10 scale is:where i and n denote respectively the degree of happiness, numerical response, assigned by the respondent and n number of respondents."

Long life

"Veenhoven (17, 18) has developed an indicator that links happiness and duration of life through life expectancy at birth."

"Veenhoven (17, 18) uses life expectancy at birth to operationalize life duration."

"It is not possible to measure the duration of life of a person when he or she is alive, because duration can only be calculated after death."

"Life expectancy at birth is calculated using the average age at death in a population age and, based on this calculation, an assumption of life expectancy is made for the living, however, the living conditions of the people can be very different from the conditions those now dead lived under."

"Healthy life expectancy (Robine & Ritchie, 19; Sanders, 20) might have been another option to use as proxy variable in the life duration calculation."

"This is why we choose to use Potential Years of Life Lost (PYLL) as a proxy variable of duration of life."

Sustainability

"The concept of ecological deficit/reserve fits the narrowest sense of definition of sustainability of Brown et al., (21), survival of human beings."

"The concept of ecological deficit/reserve partially fits the definition of sustainability as life above biological survival."

"The concept of ecological deficit/reserve partially fits the definition of sustainability as persistence of all components of the biosphere, even those with no apparent benefit to humanity."

"Global Footprint Network operationalizes the concept of ecological deficit/ reserve by subtracting the ecological footprint of a region or country from the biocapacity of the region or country."

"In order to implement a measure of sustainability in our indicator, we choose to operationalize the concept of ecological deficit/ reserve differently and to divide the biocapacity of a country by the ecological footprint of the same country."

"A space without any human activity would have a 0 ecological footprint and an infinite sustainability ratio."

First results of the indicator of a happy, long and sustainable life as a composite indicator

"Since the indicator of a happy, long and sustainable life is based on a negative utilitarianist approach, the geometric aggregation method is more appropriate because this method values (overestimates) lower values more than higher values."

"The developed form of the composite indicator is: where i the degree of happiness assigned by the respondent, j the country, k the dimension of the indicator, happiness, long life and sustainability, PYLL the value of the Potential Years of Life Lost, BIO the biocapacity of the country j, and EF the ecological footprint of the country j. When a characteristic is measured by country, countries are often ranked from the most successful country to the least."

Discussion

"Many indicators of a happy, long and sustainable life are possible depending on social preferences and scientific debates."

"Each dimension of the indicator of a happy, long and sustainable life can also be measured differently, and this measurement could be relevant."

"We would like to discuss two limitations of the indicator of a happy, long and sustainable life and the way it might be used."

"Two, a public policy that could have a positive impact on the indicator of a happy, long and sustainable in the near future may have a negative impact in the distant future, and vice versa."

"Can we find a better definition of what is an indicator of a happy, long and sustainable life?."

"To what extent does the indicator of a happy, long and sustainable life tap something different from other composite indicators such as GDP per capita, HDI or SDI?."

13.2.5 Measuring Well-Being and Lives Worth Living [111]

This is a machine-generated summary of:

Fleurbaey, Marc; Ponthiere, Gregory: Measuring well-being and lives worth living [111]

Published in: Economic Theory (2022).

Link to original: https://doi.org/10.1007/s00199-022-01,446-0.

Copyright of the summarized publication:

The Author(s), under exclusive licence to Springer-Verlag GmbH Germany, part of Springer Nature 2022.

If you want to cite the papers, please refer to the original.

For technical reasons we could not place the page where the original quote is coming from.

Abstract-Summary

"We study the measurement of well-being when individuals have heterogeneous preferences, including different conceptions of a life worth living."

"When individuals differ in the conception of a life worth living, the equivalent income can regard an individual whose life is not worth living as being better off than an individual whose life is worth living."

"In order to avoid this paradoxical result, we reexamine the ethical foundations of well-being measures in such a way as to take into account heterogeneity in the conception of a life worth living."

13.2.6 Beyond Geography: Social Quality Environments and Health [112]

This is a machine-generated summary of:

Li, Yang; Spini, Dario; Lampropoulos, Dimitrios: Beyond Geography: Social Quality Environments and Health [112]

Published in: Social Indicators Research (2023).

Link to original: https://doi.org/10.1007/s11205-023-03,073-1.

If you want to cite the papers, please refer to the original.

For technical reasons we could not place the page where the original quote is coming from.

Abstract-Summary

"Prior research suggests four domains of social quality: socio-economic security, social cohesion, social inclusion, and social empowerment, based on the assumption that these domains influence health and well-being."

"We investigate whether and to what extent social quality environments defined with reference to the cross-cutting social quality domains reliably predict various types of health, using data collected in a municipality in Switzerland as part of a participatory action research project."

"Results indicate interaction among various domains of social quality for a subset of health measures."

"Findings suggest that environments defined as combinations of social quality domains effectively distinguish between population segments with varying levels of health."

Introduction

"Social quality deals with the interdependencies between the social environment and the individuals who reside and carry out daily activities within such environment, capturing the ecological connections between multi-level contexts that shape people's lives and health across time and space (Abbott & Wallace, 22; Holman & Walker, 23)."

"We examine the predictive power of meso-contexts of social quality for various health outcomes based on the social locations shared by people in specific social quality environments defined with reference to the cross-cutting domains of social quality."

"This research addresses two focal questions: (1) Whether and to what extent are inequalities in health outcomes explained by social quality environments that cut across socio-economic security, social cohesion, social inclusion, and social empowerment? (2) How much relative predictive power does each of the aforementioned social quality dimensions have, relative to an analysis that does not account for any intersecting social quality dimension?."

Methods.

"Within the context of PAR, the Cause Commune project was developed to better understand social problems and to identify intervention pathways in Chavannes-près-Renens, a municipality in the Swiss canton of Vaud, bordering Lake Geneva to the south and France to the west."

"Chavannes-près-Renens has a diverse population: 52% of its 8060 inhabitants (December, 2019) are of non-Swiss origins, representing nearly 100 nationalities spanning Africa, Asia, North and South Americas, the Middle East, and other parts of Europe (Plattet & Spini, 24)."

"The Cause Commune project involved the entire adult resident population of Chavannes-près-Renens."

"With support from the municipality administration, all 6220 adults who were eligible for the survey were contacted, and 1492 individuals participated by responding to the questionnaire, yielding a participation rate of 24% (Plattet & Spini, 24)."

"The final analytic sample consisted of 1401 adult residents aged 18 or older in Chavannes-près-Renens."

Measures

"We measured all four domains of social quality using survey questions already established in prior literature (van der Maesen & Walker, 25)."

"A detailed questionnaire containing specific question wording and response options for all measures of social quality can be found in Appendix 2 (supplementary material)."

"To operationalize the social quality "environments" conceptualized in this paper based on respondents' cross-classified attributes, all measures of social quality were coded as tertiles (high, mid, low) so that the respondents were roughly evenly distributed across the tertiles for each domain of social quality."

"This novel and innovation technique is uniquely suited to address our research questions because it allows for an ecological examination of health inequities by partitioning the total variance into micro- and meso- levels, making it possible to evaluate the interlocking dimensions of social quality (Evans and others, 26; Li, 27)."

Results

"In terms of social quality domains (1 = worst to 5 = best), the mean level of social inclusion was about 3.7, and the mean levels of social support, empowerment, and economic security were approximately 3.1."

"Results showed that the predictive power of the social quality domains differed by measures of health."

"This suggests that a portion of the between-strata variation remains unexplained by the additive effects of the social quality domains, but was captured by intersectional interaction."

"This implies that domains of social quality have interacted to shape health for people at unique intersections of social locations who share a common set of social exposures such as norms, opportunities, and constraints relating to health."

Discussion

"Framed within the social quality model (van der Maesen & Walker, 25), and drawing insights from ecological theories (Bronfenbrenner, 28), this study investigated whether and to what extent social quality environments defined with reference to the cross-cutting domains of social quality reliably predict physical, mental, and functional health, utilizing data collected as part of a participatory action research project."

"The study also found that the various domains of social quality had differing predictive power for health, largely concurring with prior research (Abbott and others, 29, 22; Lin, 30; Yuan & Golpelwar, 31)."

"While the present study focused on social quality environments defined with reference to the cross-cutting social quality domains independent of geographic location, it is possible that a combination of social and spatial locations—i.e. socio-spatial units—would have even higher predictive power for health."

"Social quality environments, defined according to the cross-cutting domains of economic security, social support, social inclusion, and social empowerment, effectively distinguish between population segments in terms of their physical, functional, and mental health."

13.2.7 A Development-Based Approach to Global Climate Policy [113]

This is a machine-generated summary of:

Clark, Susan Spierre; Seager, Thomas P.; Selinger, Evan: A development-based approach to global climate policy [113]

Published in: Environment Systems and Decisions (2015).

Link to original: https://doi.org/10.1007/s10669-015-9533-x.

If you want to cite the papers, please refer to the original.

For technical reasons we could not place the page where the original quote is coming from.

Abstract-Summary

“A common approach for addressing climate change is to reduce the cost of greenhouse gas mitigation through market-based mechanisms, which enable an economically efficient allocation of emissions.”

“From a human development perspective, a market-based approach to emission allocation might not be appropriate, considering the value it places on social well-being.”

“The framework enables a critical examination of policy prescriptions that employ market-based methods for emission allocation due to their potential for unintended consequences for underdeveloped nations.”

Introduction

“These interdependencies are particularly problematic for developing nations, given that efforts to stabilize the climate system will undoubtedly place limits on CO_2 emissions, a byproduct of energy production under current technology constraints, but also a major greenhouse gas (GHG) contributing to global warming.”

“Simply put, energy consumption and its concomitant emissions are highly correlated with achieved HDI, as it is with economic development, but the former is distinct because it indicates that as nations increase their per capita CO_2 emissions, those emissions are resulting in less and less HDI improvements.”

“Economic losses from GHG mitigation in developed nations may not be as critical to human development progress (as defined by the HDI) as an economic analysis would predict, which questions the viability of economic-driven mechanisms to effectively address current human development inequities around the world.”

Development equity and climate policy background.

“Allocating GHG emissions based on development equity has been acknowledged across the climate ethics literature (e.g., Jamieson 32; Singer 33; Gardiner 34; Shue 35; Lomborg 36; Baer and others 37; Moellendorf 38).”

“Small island nations and the least developed countries (i.e., group of 77) call for large emission cuts by big emitters, including a “loss and damage mechanism” to help alleviate impacts of climate change (Morgan and Waskow 39; Sethi 40).”

> “Global climate policy is currently led by United Nations Framework Convention on Climate Change (UNFCCC), which does recognize development in its principle of “common but differentiated responsibilities.””

"The UNFCCC recognizes development as a priority, yet the major finding of a review article in Climate Change finds that the market-based CDM does not significantly contribute to sustainable development in developing countries, since sustainable benefits are not monetized and therefore do not drive investments (Olsen 41)."

Empirical evidence.

"Prior, many studies recognize that per capita CO_2 emissions and HDI level of nations are highly correlated (Alam and others 42, 43; Pasternak 44; Goldenberg and others 45; Dias and others 46; Moran and others 47)."

"Studies also recognize the diminishing returns to HDI as nations transition to higher levels of per capita emissions (Pasternak 44; Martínez and Ebenhack 48; Mechtenberg and others 49; Spierre and others 50)."

"Steinberger and Roberts (51) found that HDI levels were achieved at decreasing levels of energy and carbon emissions among many nations from 1975 to 2005, suggesting a decoupling of energy and carbon."

"Consistent advances in HDI are found among many of the least developed nations (HDI below 0.5) as emissions increase over time (Spierre and others 50)."

"This indicates that the ability to emit CO_2 is at least a co-requisite in the export-oriented development strategies of nations that attained the greatest HDI improvements (Spierre and others 50)."

A theoretical framework.

"From a capabilities perspective, current technology limitations make energy and CO_2 important co-requisites for transforming resources into achieved human development because they enable modern forms of communication, transportation, education, and health care."

"According to our theory, the emissions generated to build basic infrastructure (such as roads, communication, and energy systems) and provide essential services (e.g., hospitals, schools, banks) directly work to improve human development, which would explain why CO_2 and HDI remain highly correlated within the least developed nations of the world."

"This framework elucidates the notion that income and resources do not always lead to human development improvements if a society lacks the capacity to transform them into valuable achievements."

"Income in more wealthy nations can be used to purchase "luxury" goods and services (e.g., vacations and recreational vehicles) that do not directly lead to improvements in human development indicators."

A critique of an unlimited CO_2 market.

"Underdeveloped nations may not invest the proceeds of emissions permit sales in effective local human development when they can gain greater risk-adjusted returns in more technologically sophisticated countries."

"If the ability of a developed country to pay for CO_2 emissions exceeds that of the local populace in an underdeveloped country, then it may be in the best short-term interest for the underdeveloped country (especially during times of extreme drought) to sell their extra emission rights to the developed nation in return for money that could allow them to import needed water (or food)."

"We reason that unlimited emissions trading is an insufficient mechanism for enhancing the capabilities of people living in developing countries because it results in a transfer of financial resources to less-developed nations and tends to neglect the importance of transferring knowledge, empowerment, or longer-term development tools needed for sustainable development improvements."

Policy proposal.

"Considering existing empirical evidence and the established theoretical framework, as well as the arguments against unlimited CO_2 permit markets, we highlight three components of an international carbon market that we think is least-likely to negatively impact international human development progress, including (1) placing the greatest CO_2 restrictions on the most developed nations, whose human development is least likely to be impacted, (2) restrictions placed on underdeveloped nations should be based on future human development need, since CO_2 is at least a co-requisite for achieving human development, and (3) emissions trading should be prohibited between countries with large development disparities to protect impoverished nations from coercion (as explained by Sandel)."

"The latter may be achieved by limiting emissions trading to nations with similar human development levels."

"Limiting the carbon market to parties with similar development levels would result in the least redeveloped nations retaining more of their emission rights, where we observe the greatest returns to human development from increasing per capita CO_2 emissions."

Conclusion.

"Article 1 of the United Nation's Declaration on the Right to Develop (United Nations 1986) states, "The right to development is an inalienable human right by virtue of which every human person and all peoples are entitled to participate in, contribute to, and enjoy economic, social, cultural, and political development, in which all human rights and fundamental freedoms can be fully realized.""

"This declaration shows the UN's commitment to advance development throughout the world."

"With the aid of existing empirical evidence and development theory, we argue that an unlimited emission market that includes both developed and developing nations will result in perverse human development consequences."

13.2.8 Wellbeing Economics Narratives for a Sustainable Future [114]

This is a machine-generated summary of:

Waddock, Sandra: Wellbeing Economics Narratives for a Sustainable Future [114]

Published in: Humanistic Management Journal (2021).

Link to original: https://doi.org/10.1007/s41463-021-00.107-z.

If you want to cite the papers, please refer to the original.

For technical reasons we could not place the page where the original quote is coming from.

Abstract-Summary

"There is increasing attention to the idea of bringing about what is termed a wellbeing economy, and recognition that a coherent story or narrative is important in countering the strength of today's dominant economic narrative–neoliberalism."

"Many of these initiatives have allied with an aggregator called WEAll, the Wellbeing Economy Alliance."

"In an effort to determine what new economics/next economies initiatives mean by wellbeing economy, this study assesses the statements provided by WEAll allies to WEAll when seeking membership to see whether there is a consistent narrative about wellbeing economy that emerges."

"Four nuanced, yet related, versions of the wellbeing economy narrative emerged. "‘‘

"Transformation" is mainly a critique of business as usual, emphasizing transformation towards a wellbeing economy that critiques neoliberalism and emphasizes wellbeing of all people."

Telling a Different Story: What Are Practitioners Saying?

"The early part of the twenty-first century has seen a proliferation of initiatives oriented towards creating and implementing what can broadly be called a new economics, one that the Wellbeing Economy Alliance (WEAll, https://wellbeing economy.org/) broadly characterizes as wellbeing economics."

"These think tanks, institutes, networks, coalitions, development authorities, and other initiatives, some global and others more locally or regionally oriented, are generally geared, as the WEAll name suggests, towards the emergence of wellbeing for all (where all also includes non-human beings)."

"These efforts put WEAll in the center of a large coalition of loosely-coupled allies who identify with the concept of transforming economics towards wellbeing."

"The purpose of the present study is to determine from their own statements in joining WEAll what these allies understand a wellbeing economy to be—and to simultaneously determine whether there are any common or shared memes emerging from their efforts that can help shape a new economics narrative."

Methods

"The present study is a descriptive and integrative effort to determine what the current messaging (narrative emergence) about wellbeing economics that WEAll is attempting to foster actually is."

"Of particular interest were a series of questions that allies answered: What does the wellbeing economy mean to your organization/group (Meaning)?."

"What is your organization/group already doing to advance the wellbeing economy (Doing)?."

"The answers to each question were loaded separately into the qualitative data analysis program NVivo, where raw word counts and associated WordClouds for each question were developed, to provide first cut insight into what key words (memes) were emerging from the data."

"Following that step a more qualitative approach seeking themes from each code category (references in NVivo) highlighting the ways in which narratives being used within each category of coded data are expressed provided more insight into the specific ways that new economic narratives are emerging."

What Is Wellbeing Economics?

"That noted, the wordcloud and word count for Meaning of Wellbeing Economy also highlight several other words that are used in responses to that question 28 or more times (where a natural break point seems to occur): people, planet, works, well, human, system, community, society, and social."

"Memes in the good life or people-centric narrative emphasize the sustainability of people-oriented economics where basic needs are met and a good life for all people is possible through economic wellbeing, care, reduced inequalities, poverty reduction, and health lives lived sustainable."

"Memes for the integrated or both people- and nature-centric view of a wellbeing economy present a life-centered, holistic, and integrated narrative that emphasizes the value of all beings, the integral nature of the web of life, and the interconnectedness of social, economic, and ecological issues."

Limitations

"There are many other initiatives and entities, not part of WEAll, that might have something to say about wellbeing economics."

"Additional studies could seek out these initiatives, or gather data from WEAll members' websites to assess the more formal versions of their statements about wellbeing economy, though one benefit of the present study is that allies were explicitly asked to state their understanding of what a wellbeing economy is—which may or may not appear on their websites."

"Would be to interview key figures in these initiatives for more elaborated perspectives on the wellbeing economy."

"This study attempts to provide a cohesive perspective on emerging definitions of wellbeing economy and, hence, hopefully provides good insight into that construct."

Discussion and Conclusions

"A core question underpinning this study is whether a narrative of a wellbeing economy can give voice to as powerful, consistent, and emphatic an expression of what a wellbeing economy is, needs to be, and does as neoliberalism has done for itself."

"The narratives represent differently nuanced approaches and possibly are coalescing in what has been labeled an integrated framing that brings the needs of humans, other beings, and planetary considerations together holistically."

"Another effort, very much aligned with these six tenets, which they are meant to support, and which provides a holistic perspective on achieving a wellbeing economy is offered by Raworth's (52) Doughnut Economics, which integrates the planetary boundaries perspective with meeting a specific set of core human needs in an integrated way—much as the Integrated perspective on wellbeing economics offered above suggests."

[Section 6]

"Some observers argue that in very real ways this economic narrative of neoliberalism has resulted in many of today's most significant problems—climate change, species extinction, and growing inequality, among them (e.g., Lovins and others 53; Korten 54; Korten 55)."

"Such myths are powerful because we humans are storytelling creatures, who use memes, tropes, and ideas to constitute the stories and narratives that guide our behaviors, attitudes, and practices (e.g., Riedy 56, 57; Blackmore 58)."

"Despite neoliberalism's dominance, multiple attempts to tell a different story or create a new narrative around economics have emerged, both academically (e.g., Waddock 59, for a synthesis) and in practice."

"Others look at broad patterns over time to see the emergence of the story or narrative about neoliberalism (along with the industrial revolution) as decidedly needing to change its core ideas or memes if major problems are to be resolved (e.g., Lent 60; Harari 61)."

13.2.9 The Low Dimensionality of Development [115]

This is a machine-generated summary of:

Kraemer, Guido; Reichstein, Markus; Camps-Valls, Gustau; Smits, Jeroen; Mahecha, Miguel D.: The Low Dimensionality of Development [115]

Published in: Social Indicators Research (2020).

Link to original: https://doi.org/10.1007/s11205-020-02,349-0.

If you want to cite the papers, please refer to the original.

For technical reasons we could not place the page where the original quote is coming from.

Abstract-Summary

"The general question which independent dimensions are essential to capture all aspects of development still remains open."

"Using a nonlinear dimensionality reduction approach we aim to extract the core dimensions of development in a highly efficient way."

"We find that more than 90% of variance in the WDIs can be represented by solely five uncorrelated dimensions."

"The first dimension, explaining 74% of variance, represents the state of education, health, income, infrastructure, trade, population, and pollution."

"Although this dimension resembles the HDI, it explains much more variance."

"The second dimension (explaining 10% of variance) differentiates countries by gender ratios, labor market, and energy production patterns."

"The data-driven nature of the extracted dimensions complements classical indicator approaches, allowing a broader exploration of global development space."

"The extracted independent dimensions represent different aspects of development that need to be considered when proposing new metric indices."

Introduction

"The UNDP's most prominent indicator is the Human Development Index (HDI), which is the geometric mean of indicators describing life expectancy, education, and income (UNDP 62)."

"The rationale for dimensionality reduction methods like PCA is that often the intrinsic dimension of a dataset is much lower than the number of variables describing it."

"To monitor the SDGs, 169 specific targets have been developed which are measured using 232 different indicators included in the WDIs (The World Bank 63; United Nations General Assembly 64), leading to substantial interactions across and within the targets that need to be analyzed (Costanza et al. 65)."

"If the intrinsic dimensionality of development proves to be high, one would indeed need to track many indicators synchronously to understand the interplay of different aspects of development."

"In the case of a low-dimensional development space, it would be sufficient to track either the emerging dimensions, or the closely related variables to monitor development across countries and time."

Data and Methods

"To make our analyses possible, we therefore had to select a subset of indicators, countries and years with few missing observations and to fill in the remaining missing observations using gapfilling techniques (see next section)."

"We develop an approach to find subsets of the WDI database which we can gapfill and use for estimating distances among data points."

"Dimensionality reduction describes a family of multivariate methods that find alternative representations of data by constructing linear or, in our case, nonlinear combinations of the original variables so that important properties are maintained in as few dimensions as possible."

"We developed "Ensemble Isometric Feature Mapping" (e-Isomap) to produce the final nonlinear embedding based on the different gapfilled subsets of data."

"To assess the influence of single variables on the final e-Isomap dimensions, we calculated the distance correlation (dcor, Székely et al. 66), which is a measure of dependence between variables that takes nonlinearities into account."

Results

"Our results suggest that the dimensions resulting from the e-Isomap can be indeed interpreted analogously to traditional indicators of development."

"Other indicators that strongly correlate with this dimension are related to pollution and primary production and include tariffs and imports as well as trade, the climate impact of GDP (gross domestic product), and development aid received."

"Dimension 1 has a strong nonlinear correlation with the HDI (dcor = 0.93), and can be interpreted as a measure of development sensu HDI, even though it includes much more than the aspects measured by the HDI."

"Dimension 2 (9.9% of the variance) is strongly related to gender ratios in the general population and the labor market, as well as primary energy production and consumption and the fraction of 25–29 year old people."

"Dimension 3 shows mostly employment/unemployment ratios, but there are no really strong general trends observable."

Discussion

"The large number of measured variables and derived indicators that are used today to describe development could suggest that global development is a high dimensional process requiring many indicators to describe it accurately."

"To put in other terms, the fact that many properties of development are highly correlated (Ghislandi et al. 67) also means that one can summarize them efficiently in very few dimensions."

"The fact that these factors co-vary in a way that we can represent them in a single dimension can guide the development of novel metric indices."

"Collapsing many aspects of development into a single dimension, which in turn forms the main gradient along which countries move over time, essentially expresses (nonlinear) covariations that should not be studied in isolation."

"The proposed approach can help in the planning of adding measures of development and testing their redundancy with respect to the existing indicators, simplifying e.g. reporting of complementary dimensions of development."

Conclusions

"Exploring the meaning of the emerging dimensions allows us to understand which aspects of development are underrepresented in current databases."

"The second insight is that development as described by the dimensional space, remains to be a highly complex process that involves strong nonlinear interactions."

"We have elaborated some of these aspects, but a more profound exploration of the five-dimensional development space is still needed."

"The third insight is that single countries' trajectories in the low dimensional space show abrupt changes that coincide with major environmental hazards or socioeconomic anomalies."

"Our analysis gives new insights into the general structure of development which is of low dimensionality, but highly nonlinear and interconnected."

13.2.10 Looking Good or Feeling Well? Understanding the Combinations of Well-Being Indicators Using Insights from the Philosophy of Well-Being [116]

This is a machine-generated summary of:

Cohen Kaminitz, Shiri: Looking Good or Feeling Well? Understanding the Combinations of Well-Being Indicators Using Insights from the Philosophy of Well-Being [116]

Published in: Social Indicators Research (2020).

Link to original: https://doi.org/10.1007/s11205-020-02.289-9.

If you want to cite the papers, please refer to the original.

For technical reasons we could not place the page where the original quote is coming from.

Abstract-Summary

"In the context of the broad question of the combination of different indicators of social well-being, this paper points to a distinction between four normative approaches to social well-being."

"The paper analyses these four approaches and their significance for the issue at hand, thereby establishing grounds for a non-arbitrary interpretation of the data represented by indicators and non-arbitrary choice between indices."

"A particular contribution of the paper is defining and exemplifying the QSH approach, at the heart of which is the normative conviction that a society is not better off unless both subjective and objective aspects have been improved."

Bridging Conceptions of Well-Being and Well-Being Indicators

"The categories of indicators are divided on the objective versus subjective question in the sense of what is being measured."

"Building on these two sets of divisions, scholars connected subjective-indicators of social well-being with subjective philosophical approach to well-being, and objective-indicators with objective approach to well-being."

"The crucial question to ask, becomes: what kind of a concept of social well-being is an indicator or an index—be it subjective, objective or mixed—meant to represent?."

"The subjective approach to social well-being sees the well-being of a society as dependent on individuals' attitudes to the things and events achieved or occur in their life."

"The objective approach to social well-being does not depend on such attitudes, and thus can build on indicators that do not represent individuals' attitudes."

"If we accept that it matters independently, we refer to this indicator as representing an objective element of social well-being."

Pure Cases

"This would be extremely odd, since by definition the social objectivist does not take personal attitudes as sufficient or even necessary conditions of well-being, so why would she count on subjective indicators alone to represent it?."

"Advocating the use of subjective indicators alone in order to represent social well-being is strongly linked to the social subjectivist approach."

"In what sense are objective indicators linked with objective approaches to social well-being?."

"The complexity is due to the possibility that one could hold a subjective approach to social well-being and yet advocate the use of an objective indicator."

"This means that a social subjectivist could prefer this objective indicator over a subjective indicator (such as Life-Satisfaction)."

"Regarding the indicator of GDP per capita, however, this conclusion is problematic, and, if representing social well-being, is better taken as an indicator representing an objective element of social well-being (i.e. attitude-independent)."

Combining Subjective and Objective Indicators

"Combining objective and subjective indicators can be grounded on different approaches."

"This program considers objective indicators as relevant to measuring social well-being because they represent opportunities to meet human needs, and those opportunities have the potential to transform into subjective well-being when perceived (Costanza and others 68: 18)."

"This index is a combination of a subjective indicator ('average happiness' based on surveys) and an objective one (standard life-expectancy in years), when: Veenhoven criticizes other indices for confusing means with ends, with the objective conditions considered as means (also as 'inputs') and individuals' happiness/satisfaction as ends or as 'output' (Veenhoven 69: 9–10)."

"Even though combining objective and subjective indicators, the index (implicitly) builds on a subjective approach to social well-being."

"These last examples show that combining subjective and objective indicators does not necessarily indicate any renouncing of the philosophical positions of either the social subjectivist or the social objectivist."

Other Approaches to Social Well-Being

"If we hold to a social pluralistic approach, and ascribe to objective and subjective aspects such and such relevancy to social well-being (the weights representing the proportion), then, first, we should decide which indicators represent for us the subjective elements and which indicators represent for us the objective ones."

"This interpretation is distinctive because, unlike the other approaches, it takes the objective and subjective aspects together to provide evidence for a significant change in social well-being (again, this does not mean that we necessarily need both subjective and objective indicators in order to represent the two aspects)."

"A QSH interpreter could justify the claim that it is society C that is the most advantageous because it has the higher joint performance of both objective and subjective aspects (represented by the scores of the two indices)."

Conclusion

"This paper sets out a path that connects different indicators and indices of social well-being with philosophical conceptions/aspects of well-being."

"Since indicators and their compositions have different possible meanings underlying them, avoiding the discussion of these meanings may lead to confusion, and to lack of correspondence between the object we measure and the concept we attempt to represent by measurement."

"This effort is nevertheless limited, though, because, important as it may be, deciding between approaches of social well-being far from exhausts the many normative and practical decisions bound up in deliberately developing or choosing and weighing indicators."

"A distinct question concerns who is to decide between indices (or between approaches to social well-being) and who is to interpret them."

"Regardless of the procedure selected, the point made by this paper is that the concept of social well-being underlying the index decided upon should not be overlooked."

[Section 6]

"This paper provides a conceptual framework for weighing and combining different well-being indicators in this context."

"This paper nevertheless assumes that the development of a deliberative manner to combine different indicators and weigh them is essential in order to meet particular political goals, for it opens a path to comparing across societies and assessing improvements in social well-being."

"Within the broad question of how to select aggregate measures, or how to combine different indicators, the present paper singles out one particularly important issue, namely, how to select aggregate measures of subjective and objective indicators."

"The paper will not address the many methodological difficulties and shortcomings of objective and subjective indicators."

"The paper will not address the debate over the selection and combination of different indicators within the two categories."

13.2.11 Can the Market Economy Deal with Sustainability? [117]

This is a machine-generated summary of:

Schoenmaker, Dirk; Stegeman, Hans: Can the Market Economy Deal with Sustainability? [117]

Published in: De Economist (2022).

Link to original: https://doi.org/10.1007/s10645-022-09,416-6.

If you want to cite the papers, please refer to the original.

For technical reasons we could not place the page where the original quote is coming from.

Abstract-Summary

"A system analysis of the current neoclassical theory shows that the system components (goal function, interaction mechanisms, actors and outcomes) are predominantly defined in terms of economic growth and facilitated by market exchange."

"The one size fits all 'economic mechanism design' cannot deliver social outcomes regarding sustainability."

"The explicit recognition that an economy has different domains (ecological, social, economic) broadens the options for incorporating sustainability within the economic system."

"We show that the alternative for market economics does not only have to be government intervention but can also include private forms of collective decision-making."

Introduction

"The follow-up question is, what can we do if the market economy cannot deal with sustainability: which changes are needed to fit sustainability within an economic system?."

"We analyse the mainstream theoretical market economy to highlight the underlying assumptions and how they relate to sustainability."

"Sustainability is not substantially visible in the different institutional setups of market economies (Magnin, 70), where sustainability is partly seen as a public good (clean air) that the government should provide."

"The standard approach in economics to achieve sustainability is to define property rights, correct market failures (e.g., via Pigouvian taxes) and implement redistribution policies (Besley & Ghatak, 71)."

"The explicit recognition that an economy has different, interdependent, domains (ecological, social, economic) broadens the options for incorporating sustainability within an economy with still a role for market economics."

A Non-Sustainable Market Economy

"The violations of the invisible hand conditions increasingly lead to unsustainable market outcomes, because it is still assumed that they hold in practice: (1) Imperfect competition: There is extensive empirical evidence that there are increasing concentrations in market power, driven by globalisation and technology, resulting in a 'winner-take-all' economy, reducing economic growth, and suppressing careers and wages for workers on the one hand and raising profits and hence wealth for top-earners and shareholders on the other side (Autor and others, 72; Eeckhout, 73; Philippon, 74)."

"Given the increasing scale of the externalities associated with private property and the allocative efficiencies (underutilisation of assets) that result from absolute ownership, instead of private property as a solution, private ownership becomes more and more of an impediment for efficient markets. (6) Finite substitutability between capitals: Perfect substitutability is one of the core concepts of market economics: different forms of capital (human capital, social capital, natural capital, financial capital and produced capital) can be substituted for one another: manual labour can be replaced by produced capital (machines) in combination with natural capital (fossil fuel)."

An Encompassing Framework for Sustainability Economics

"The focus is on provisioning and appropriation by various social groups, thereby relegating market exchange to a specific feature of particular societies and opening the door to other ways to create and exchange value, such as collective decision-making, either in social groups (households, clubs, religious institutions) or in politics."

"Economics should relate to these different fields and make clear how these different 'domains' (economy, society and ecology) interact and contribute to the well-being of humans now and in the future."

"Without a flourishing planet and inclusiveness, efficient market interactions (the goal of the economic domain) make no sense."

"The value expression is biophysical boundaries achieved through the output of common and public goods, mostly related to the governance of ecological sustainability (preservation), but can also entail the production of nature services (regeneration)."

Use the Full Range for Optimal Well-Being/Real-Life Conceptual Economics

"Introducing different nested and interdependent domains, goal functions, interaction mechanisms, and criteria broadens economics."

"The application of a particular goal function and interaction mechanism depends on the current state, values, rules, and conditions."

"Although GDP is an imperfect measure of well-being, replacing the goal is a necessary but not sufficient condition to reach a sustainable economy: it should address all elements of the system, from its goal, interaction mechanism and role of actors to criteria of evaluation when the interaction is a success."

"The solution to market failures and the omission of sustainability in market economics is to do what implicitly already happens, but what is theoretically missing: put the different goals of the planet and society explicitly in the economic domain."

"It is possible to use collective decision-making as an interaction mechanism, use different criteria or set different goals."

Conclusion

"Because GFTS (globalisation, financialisation, technology and scale) increasingly violates invisible hand assumptions, the outcomes are increasingly unsustainable."

"To address this problem, we suggest using a nested domain approach: the planet (the ecological side of sustainable development) and the society (the social side of sustainable development) are added to the economic domain."

"Ecological, social or economic single-domain approaches cannot address the multiple-domain inquiries."

"Acknowledging that the economy is a nested part of a social system and a subset of ecosystems means that we need no longer solve sustainability problems in a reductionist way as 'externalities' within the economic system."

"The main challenge of the new governance framework for sustainable development is balancing the different goals."

"Different goals and interaction mechanisms for different domains provide a richer toolbox for policymakers to address the sustainability challenges."

13.2.12 Income Inequality, Carbon Emissions, and Green Development Efficiency [118]

This is a machine-generated summary of:

Sun, Yongchun: Income inequality, carbon emissions, and green development efficiency [118]

Published in: Environmental Science and Pollution Research (2022).

Link to original: https://doi.org/10.1007/s11356-022-23.583-8.

If you want to cite the papers, please refer to the original.

For technical reasons we could not place the page where the original quote is coming from.

Abstract-Summary

"Income inequality affects not only social well-being, health, and carbon emissions but also the strategy for green development in China."

"Based on the panel data of 205 cities in China from 2010 to 2020, a panel model with partial linear functional coefficient is used to analyze and test the relationship between income inequality, carbon emissions, and green development efficiency under different regional economic development levels."

"The empirical results show that the impact of income inequality on carbon emissions and green development efficiency is significant."

"The number of cities where income inequality has an inhibitory effect on carbon emissions and green development efficiency has increased over time, and the impact of income inequality in a few cities on green development efficiency is not significant."

Introduction

"This paper uses green development efficiency to measure economic development more comprehensively and also provides a new perspective for the impact of income inequality."

"The impact of income inequality on carbon emissions and green development efficiency could be restricted by the level of regional economic development and show heterogeneity."

"This paper aims to analyze the specific impact of income inequality on carbon emissions and green development efficiency considering different levels of the economic development in Chinese cities."

"The indicator of economic growth is simply measured by GDP when the existing literature studies the link between income inequality and economic growth, while this paper uses the indicator of green development efficiency, which can reflect high-quality economic development."

"We test the hypothesis that the impact of income inequality on carbon emissions and green development efficiency depends on the economic development level."

Literature review

"Although many literatures examined the relationship between income inequality and carbon emissions (Uddin and others 76), they did not have an agreed point of view."

"The results of the relationship between income inequality and economic development remain controversial."

"The existing literature examines the relationship between income inequality, environmental issues, and economic development extensively from multiple perspectives and methods."

"The relationship between income inequality, carbon emissions, and green development efficiency is rarely included in a framework to explore, so there is a lack of an accurate grasp of the relationship between the three and the exploration of the impact mechanism."

"This paper incorporates green development efficiency into the research on income inequality and carbon emissions and considers the differences in regional economic development levels to study the impact of income inequality on carbon emissions and green development efficiency."

Methodology

"Considering that the impact of income inequality on carbon emissions and green development efficiency may be affected by the levels of regional economic development, some studies may construct an interaction term between the level of regional economic development and income inequality, but this strategy may lead to estimation bias (Du and others 77, 78)."

"Based on the panel data, An and others (75) extended a partially linear variable coefficient model with fixed effects and used a series method to estimate the model."

"The coefficient function $\gamma(\cdot)$ is estimated as: Therefore, the final settings of the estimated PLFC model in this paper are as follows: where the two explained variables in this paper are green development efficiency (GTFP) and carbon emissions (CO_2)."

Data

"The three key variables in this paper, including green development efficiency, carbon emissions, and income inequality, cannot be obtained directly, and need to be obtained through measurement."

"Based on the DEA-SBM model, this paper employs the GML index to measure the green total factor productivity (GTFP) of 205 cities in China."

"The input indicators used in this paper to measure urban GTFP of the high-quality development include labor, capital stock, and energy inputs."

"Economic output is GDP at constant price with 2010 as the base period and the influence of price factors removed, while green output is measured by the green coverage rate of each city, which represents the quality of life of residents."

"This indicator can indirectly affect carbon emissions and the high-quality development of urban green economy by optimizing the industrial structure and promoting the level of technological innovation (Juuti 79)."

Empirical results

"For cities with a moderate level of regional economic development and most cities with the high economic development, the increase in income inequality could increase carbon emissions and green development efficiency."

"For a few cities with a high level of regional economic development, income inequality could inhibit carbon emissions and the efficiency of green economic development."

"For cities with the moderate regional economic development, the effect of income inequality on carbon emissions and green development efficiency is gradually increasing."

"In cities with the moderate economic development and a few high-level cities, the effect of income inequality on carbon emissions is relatively moderate and has no influence on the efficiency of green development."

"For cities with the high regional economic development, the number of cities where income inequality promotes carbon emissions and green development efficiency is decreasing, which is inseparable from China's environmental policies in recent years."

Conclusions and policy implications

"Taking into account the heterogeneity of regional economic development levels, this paper adopts a PLFC panel model to study the impact of income inequality on carbon emissions and green development efficiency."

"While cities with higher economic development, the number of cities where income inequality promotes carbon emissions and green development efficiency is decreasing."

"The empirical results suggest that in the case of different economic development levels, income inequality has different effects on CO_2 emissions and urban green development efficiency."

"It is necessary to promulgate relevant supporting policies to adapt to regional economic development, continuously improve the income distribution mechanism, promote common prosperity, and improve the ecological environment, which are the requirements for high-quality urban economic development."

"There are many channels through which income inequality affects carbon emissions and green development efficiency."

13.2.13 Sustainable Development Goals and Inclusive Development [119]

This is a machine-generated summary of:

Gupta, Joyeeta; Vegelin, Courtney: Sustainable development goals and inclusive development [119]

Published in: International Environmental Agreements: Politics, Law and Economics (2016).

Link to original: https://doi.org/10.1007/s10784-016-9323-z.

If you want to cite the papers, please refer to the original.

For technical reasons we could not place the page where the original quote is coming from.

Abstract-Summary

"The concept of inclusive development emphasizes the social, ecological and political dimensions of development."

"It presents inclusive development as having three key dimensions (social, ecological, and relational inclusiveness) with five principles each."

"The paper concludes that while the text on the SDGs fares quite well on social inclusiveness, it fares less well in respect to ecological and relational inclusiveness."

"This implies that there is a risk that implementation processes also focus more on social inclusiveness rather than on ecological and relational inclusiveness."

Introduction

"The literature on, and politics of, sustainable development shows that achieving strong sustainability, which implies no trade-offs between the economic, social and ecological goals, is rare; politics tend to make trade-offs in favour of the economy at the cost of social and ecological issues (Lorek and Spangenberg 80)."

"They have led to the adoption of a document entitled 'Transforming our World: The 2030 Agenda for Sustainable Development' which includes the sustainable development goals (UNGA 81) under the auspices of the United Nations General Assembly (for background see Chasek and Wagner and Gellers in this issue)."

"The sustainable development goals (SDGs) set goals to ensure a 'social floor' á la Raworth (82), and 'planetary boundaries' á la Rockström and others (83) and suggest some of the means that should help in achieving these goals."

"Its central argument is that without a deep commitment to inclusive development, the SDGs run the risk of not directing the substantive transformation needed to achieve strong sustainable development."

Inclusive development

"This translates into the following principles: (a) develop specific rules on public and merit goods and ensure that these do not become privatized or securitized directly or indirectly; (b) address all drivers at all levels, and combat 'offshoring' and 'virtualization' of the economy by large corporations and the wealthy to bypass tax and other rules; tax the accumulation of wealth and check when deregulation affects the rights of the poorest; (c) question the discourses that lead to exclusive growth and ensure that markets function within the constraints of inclusive development rules and that these rules are made legally binding; (d) test instruments for inclusiveness and ensure downward accountability; and (e) promote the rule of law and constitutionalism at the global and national levels as a way to combat power politics at both the global and national level to ensure that no country or powerful entity is above the law."

Applying ID to the SDGs

"While these goals clearly meet two of the five principles—namely that of (a) enhancing the level of protection for the poor/marginalized and small-scale sector; and (b) enhancing development opportunities for all to a limited extent—there is little emphasis on including the knowledge of all, or engaging all, or promoting effective capacity building to enable better participation."

"They set targets on production and consumption patterns (Goal 12) including sustainable and resilient agricultural practices, maintaining ecosystem services and adaptation to climate change, genetic diversity and securing seed and plant banks (Goal 2), demand side management through energy efficiency and supply side management through promoting renewable energy (Goal 7), resource efficiency, decoupling growth from pollution (Goal 8) and sustainable industrialization (Goal 9); and on enhancing the quality of life by providing access to open and green spaces for all, sustainable transport systems, sustainable urbanization, sustainable human settlement planning, and improving air quality and waste management within sustainable and resilient cities (Goal 11)."

Conclusion

"This paper argues that an inclusive development analysis of the SDGs may help to assess whether the textual design truly aims at ensuring that development focuses both on social and ecological issues, as well as on the political tools for achieving the transformation to which it aims."

"Recognizing the risk that the SDG agenda may go the same route as the sustainable development discourse—that priority will go to economic growth over social and economic goals—this paper has surveyed the 17 SDGs to assess how well they represent inclusive development."

"From the perspective of inclusive development, successful implementation of the SDGs would not only mean that poverty and marginalization are reduced, ecological sustainability is enhanced, and the gap between the powerful and less powerful is minimized."

"The methods for implementation would need to embrace the principles of inclusive development in themselves to overcome unequal power relations in order to genuinely address the goals."

13.2.14 The Role of Financial Fragility and Financial Control for Well-Being [120]

This is a machine-generated summary of:

Bialowolski, Piotr; Weziak-Bialowolska, Dorota; McNeely, Eileen: The Role of Financial Fragility and Financial Control for Well-Being [120]

Published in: Social Indicators Research (2021).

Link to original: https://doi.org/10.1007/s11205-021-02,627-5.

If you want to cite the papers, please refer to the original.

For technical reasons we could not place the page where the original quote is coming from.

Abstract-Summary

"Financial fragility is recognized as a substantial issue for individual well-being."

"Various estimates show that between 46 and 59% of American adults are financially fragile and thus vulnerable in terms of their well-being."

"Impacts of financial fragility and financial control on 17 well-being outcomes were examined, including emotional well-being (nine outcomes), physical well-being (four outcomes), social well-being (two outcomes), in addition to an unhealthy days summary measure and the flourishing index."

"Financial fragility was shown to be on average less influential for the well-being outcomes than financial control."

"Less evidence in support of a harmful role of financial fragility for well-being is provided."

"Tests for moderation effects revealed no interaction between financial control and financial fragility within our sample, indicating that financial control did not modify the relationship between financial fragility and well-being."

Introduction

"Evidence for the financial control and well-being link is ambiguous."

"More empirical evidence on the role of financial control is needed."

"This article aims to provide a broader perspective and examine how financial fragility and financial control influence emotional, physical, and social well-being."

"Striving to evaluate earlier findings, within our sample we assess the extent to which financial fragility has a detrimental effect on well-being [research question 1 (RQ1)], and what the scale (if any) of positive effect of financial control on well-being can be identified (RQ2)."

"We also examine within our sample whether financial control reduces the detrimental impact of financial fragility on well-being (RQ3)."

"It is the first to investigate the impact of financial fragility on physical, emotional, and social well-being among a sample of U.S. residents, also considering a potential moderating effect of financial control."

Theoretical Background

"Among five domains of financial capability identified by Atkinson and others (86), four relate entirely to financial control (i.e., oversight over financial issues, planning for the future, choosing financial products wisely, and seeking information), while only one partially relates to financial distress and fragility (i.e., being able to make ends meet)."

"The financial locus of control adds an additional dimension to the traditional concept of locus of control that, in turn, represents someone's generalized attitudes, beliefs, or expectations regarding the nature of the causal relationship between her behavior and its consequences (Buddelmeyer and Powdthavee 87; Rotter 88)."

"This might suggest that locus of control in the financial domain has a strong linkage with well-being."

"Although the evidence of the role of financial control for well-being is limited, importance of the locus of control for financial behaviors has already been established."

Material and Methods

"Invitation letter was sent in June 2018 to all members of a credit union residing in one of the North Carolina counties in the U.S. 4083 persons responded to the survey."

"In September 2019 they were invited to participate in the second wave of the survey."

"The sample selection process was not different than in most of the state-of-the-art surveys with the exception that instead of selecting a random sample, we approached the whole population of customers of the federal credit union (Bialowolski and others 89)."

"The sample was relatively overrepresented by females, older, white and well-educated participants."

Measures

"The temporal association between financial fragility and financial control and well-being outcomes was modeled using three lagged linear regression models: M1 (financial fragility): M2 (perception of financial control): M3 (moderation effect): In all specifications subscript i represents an individual, subscript j represents one out of

17 outcomes, variables $FF_{i,2017}$ and $FC_{i,2017}$ indicate exposure related to financial fragility and perception of financial control, respectively."

"β_0 represents a constant, β_1 shows the impact of individual level characteristics (control variables) on a well-being outcome, β_2 represents the strength with which a well-being measure in 2017 is related to well-being in 2018, γ_1 shows the effect of financial situation on a well-being outcome, γ_2—the effect of financial control and γ_3—the effect of moderation (interaction)."

"Since we used a sample that precludes statistical inference for any population beyond ours, we focused on deriving plausible values for the temporal associations between well-being outcomes and either financial control or financial fragility."

Results

"Financial control was also found to limit the subsequent number of days with poor physical health and poor mental health (effect sizes: $\beta = -0.18$ and $\beta = -0.16$, respectively)."

"We also found that financial control was associated with moderately high effect sizes (above 0.1) with the following three well-being measures: (1) days in which one feels sad or depressed ($\beta = -0.12$), (2) days in which pain makes it difficult to perform everyday activities ($\beta = -0.11$), and (3) days with insufficient rest or sleep ($\beta = -0.11$)."

"Financial control also revealed higher effect sizes in its temporal association with subsequent the unhealthy days summary measure—one of two composite measures used in the analysis."

Discussion

"Using longitudinal observational data from the SHINE Well-Being Survey completed by 1448 members of a federal credit union, we examined the prevalence of financial fragility and financial control, their influence on emotional, physical, and social well-being, as well as their impact on composite measures of well-being."

"We also investigated the role that financial control plays in modifying the relationship between financial fragility and well-being, since it has been shown that people who feel in control are less likely to have excessive debt (Tokunaga 90; Wahlund and Gunnarsson 91), demonstrate higher savings (Cobb-Clark and others 92), and are also more likely to incur financial risks (Kesavayuth and others 93)."

"It can be also argued that financial control translates into long lasting effects (which can be identified in a longitudinal study like ours) and financial fragility has only very immediate consequences and thus does not have a long-term detrimental effect on well-being."

"Financial control and financial fragility are highly correlated which may reduce the potential to study the interaction between them."

13.2.15 *Measuring Well-Being: Trial of the Neighbourhood Thriving Scale for Social Well-Being Among Pro-social Individuals [121]*

This is a machine-generated summary of:

Baldwin, Cathy; Vincent, Penny; Anderson, Jamie; Rawstorne, Patrick: Measuring Well-Being: Trial of the Neighbourhood Thriving Scale for Social Well-Being Among Pro-Social Individuals [121]

Published in: International Journal of Community Well-Being (2020).

Link to original: https://doi.org/10.1007/s42413-020-00.067-6.

If you want to cite the papers, please refer to the original.

For technical reasons we could not place the page where the original quote is coming from.

Abstract-Summary

"An online questionnaire was used to measure neighbourhood thriving (NT) among 212 pro-social volunteers involved in revitalising neighbourhoods in the UK city of Stoke-on-Trent between May and October 2018."

"The 11 sub-scales of NT showed satisfactory internal consistency reliability and preliminary evidence of construct validity."

"The sub-scales were used tentatively to examine NT among the volunteer sample, which showed the highest sub-scale score for Positive Regard and the lowest score for Celebration."

"Further validation work is needed before the NT scales can be used with confidence."

"Validated scales offer potential benefits including: measuring NT pre- and -post project implementation; establishing which dimensions of NT are, and are not, working well in a community and need strengthening through further initiatives, and establishing which specific groups of people are experiencing lower levels of NT and designing projects that meet their needs."

Introduction: Neighbourhood Flourishing

"We report on a trial of a conceptual framework measuring social well-being called the 'neighbourhood thriving framework' (originally outlined under the name 'neighbourhood flourishing framework' in an earlier paper by authors Anderson and Baldwin (94) in the English city of Stoke-on-Trent."

"The neighbourhood thriving framework is based on an original framework called the 'neighbourhood flourishing framework' (NFF) that was designed to measure the effects of urban design on neighbourhood flourishing (NF)."

"In a conceptual departure from the original NFF framework, this specific target group was selected due to there being three well-established social epidemiological causal pathways to well-being anchored in pro-social social structures and action: networks, participation and pro-social behaviours."

"In using our adapted version of the NFF (known here as the neighbourhood thriving framework for differential purposes) to collect data from pro-social volunteers, we sought to perform exploratory factor analysis, a statistical analysis technique, on the data to see to what extent the resulting factor structure reflected these three pathways."

Original Framework

"It comprised a preliminary indicator set of 14 features centred on the combined notion of social feelings and functionings at the group, not individual level, in relation to geographic or non-spatial communities."

"It evolved from developments in psychology where individual subjective well-being is conceived of as split between momentary experiences of pleasure or positive emotions (hedonic well-being) and more important, long-term and active processes of 'well-doing' (Aristotle's eudaimonia) (see, e.g. Keyes 95; Ryan and Deci 96; Huppert and So 97)."

"Huppert and So (97) combined these two types of wellbeing - feeling good (hedonic well-being) and functioning effectively (eudaimonic well-being) - into an operational definition of individual mental well-being increasingly referred to as flourishing."

"Whilst there are other indicator sets that combine hedonia and eudaimonia, they are not designed to measure neighbourhood-based flourishing, as shaped by the particular social experiences that people accrue by living in a geographically defined neighbourhood."

Measuring Well-Being on the Group/Social Level

"With the exception of one item (positive relationships), Huppert and So's scale did not address social well-being – well-being at the group or community level, i.e. how the individual responds to experiences of the social environment which can affect their health (Larson 98; Keyes 99)."

"Going beyond individual psychological functioning, the presence of both social feelings (e.g. I feel close to my community) and social functioning (e.g. I feel supported by my community) reach beyond perceptions of one's self and represent a key part of positive social health."

"They explore pathways between individual and group experiences of processes described by sociological concepts such as social integration (as defined by Keyes); social networks - the structures of society/community; social cohesion - the quality of individual and group-level relationships within networks; social capital - the ways in which networks operate to provide members with benefits; the provision of social support - a particular kind of network benefit; and individual and group-level health and well-being."

Social Epidemiological Pathways/Mechanisms Underpinning Social Well-Being

"Social networks are linked to health outcomes through four sub-casual paths: through their functions in providing social support, the social influence of network members on other members, e.g. the provision of information on maintaining good health, social engagement and attachment, and the access that networks provide to resources and material goods (Berkman and Glass 100, p.143)."

"It involves people participating in a community context in social activities in networks, as shaped by local issues, the geographic location, and local 'culture, norms, values and institutions' (Keyes and others 101, p. 98)."

"Berkman and Glass (100, p. 146–7) elaborated on an unclear pathway between social engagement and participation and positive health status."

"When people participate in social networks, a third pathway emerges between the actual prosocial behaviours and actions that each individual engages in, and the influence of these on their health and well-being, and that of an aggregation of their fellow community members'."

Devising the Neighbourhood Thriving Framework for Testing on pro-Social Volunteers

"To achieve this, Baldwin and Rawstorne devised 5–7 proxy indicators per dimension (sub-scale) in anticipation of the potential loss of proxy indicators during Exploratory Factor Analysis (EFA) and reliability analysis."

"Possible, proxy indicators were adapted from the ESS (2006; 2012), and existing national UK surveys, the Citizenship Survey (102; UK Data Archive Study No 7111), UK Household UK Household Longitudinal Study (Institute for Social and Economic

Research 103), ONS Social Capital Indicators Review (Foxton and Jones 104) and surveys from environmental psychology (e.g. Williams and Vaske 105) and social epidemiology (e.g. Cobb 106)."

"Original proxy indicators were devised through discussion/refinement between the authors, with disagreement rare and discussed until agreement was reached."

"Face validity (i.e. proxy indicators appearing to measure what they are intended to measure) and content validity (i.e. proxy indicators measuring the full breadth of each construct) were built into the design of the proxy indicators measuring each sub-scale."

Materials and Methods

"We sought volunteers involved in pro-social/community activities at neighbourhood level, via a request circulated to public health authorities/community development organisations in UK cities/regions using our networks and online mailing list Community Empowerment Evidence Network CEEN@JISCMAIL.AC.UK Interest was received from Aylesbury, Sheffield, Exeter and Stoke-on-Trent, and South Wales."

"A downloadable sheet listing the contact details of the authors, community development workers from 1000 Lives and MCM, and local mental health support services for respondents to ask questions or seek support should they experience any discomfort after completing the questionnaire was clearly marked beside the consent form."

"The limitations of a convenience sample were accepted, and the following survey inclusion criteria were set for respondents: they should be aged 18 or over, live or work in Stoke, volunteer with MCM, 1000 Lives or another local community action group, and be able to complete the survey online."

Results

"For the statement that their neighbourhood had got better, agreement with the statement was associated with higher NT scale scores while disagreement with the statement was associated with lower scores (Collective positive effort, $F(1, 204) = 47.00$, $p < 0.001$; Participation, $F(1, 204) = 17.72$, $p < 0.001$; Celebration, $F(1, 204) = 24.61$, $p < 0.001$; Social network pathway, $F(1, 204) = 24.81$, $p < 0.001$; Optimism about the community, $F(1, 204) = 71.78$, $p < 0.001$; Social cohesion, $F(1, 204) = 57.41$, $p < 0.001$; Engagement pathway, $F(1, 204) = 28.19$, $p < 0.001$; Safety, $F(1, 204) = 34.94$, $p < 0.001$; Autonomous citizenship, $F(1, 204) = 41.64$, $p < 0.001$; Positive regard, $F(1, 204) = 20.36$, $p < 0.001$; Low resilience, $F(1, 204) = 33.11$, $p < 0.001$)."

Discussion

"Other validation work that is needed is in showing whether the scales can be used to predict (at least correlate with) behaviours consistent with a person's scores on each construct, and also examining whether a total NT score is predictive of behaviours."

"It will also be important for future studies to apply and validate the scales in populations that are comprised of people other than pro-social volunteers, particularly in random samples of local communities."

"With a mid-point of 3 on the 5-point Likert scale, and based on higher scores for each of the scales (except for Low Resilience) indicating higher levels of NT, we can observe that the mean of each of the scales was equal to or above the mid-point, likely indicative of NT."

"It may surprise some readers that the sample of pro-social volunteers in this study did not confirm their neighbourhood had become a better place to live and/or work in over the previous 2 years."

Policy Recommendations

"Since the current study did not set out to evaluate the two community development projects that were commissioned by Stoke-on-Trent City Council Public Health Department in 2012 to reduce health inequalities through social action, and because the current sample comprised pro-social volunteers and not a broad cross-section of the community, we are unable to ascertain the specific benefits of these programmes to their communities."

"Future community development programmes should aim to address these conditions: the first three of which can be cultivated through local action: social cohesion (through structured mixing of people from different backgrounds in collaborative activities), safety (through design efforts to transform urban environments with poor street lighting, poor traffic control, and no eyes on the street, and the targeted reduction both crime and violence, and their risk factors), and resilience (through cultivating social capital, social cohesion, well-being and the eradication of inequalities)."

Conclusions

"More validation work is needed on these scales, particularly with non pro-social volunteers, before they can be used with confidence to measure the 11 dimensions identified in the study."

"When these measures were applied to a sample of pro-social community volunteers in Stoke, we observed the highest scale score for Positive Regard and the lowest scale score for Celebration."

"The scale may develop into a useful tool for evaluating the success of community projects when administered pre and post project implementation."

"It may also be useful for gauging which members of a community are experiencing low levels of NT and to tailor the design of projects at different groups of community members and their needs."

Bibliography

1. Meadows DH, Meadows DL, Randers J, Behrens WW (1972) The limits to growth. Club of Rome, New York
2. Voudouris V, Stasinopoulos D, Rigby R, Di Maio C (2011) The ACEGES laboratory for energy policy: exploring the production of crude oil. Energy Policy 39(9):5480–5489. https://doi.org/10.1016/j.enpol.2011.05.014
3. Sherwood J, Ditta A, Haney B, Haarsma L, Carbajales-Dale M (2017) Resource criticality in modern economies: agent-based model demonstrates vulnerabilities from technological interdependence. BioPhys Econ Res Qual 2(3):9. https://doi.org/10.1007/s41247-017-0026-z
4. Rai V, Robinson SA (2015) Agent-based modeling of energy technology adoption: empirical integration of social, behavioral, economic, and environmental factors. Environ Model Softw 70:163–177. https://doi.org/10.1016/j.envsoft.2015.04.014
5. Ayres RU, Warr B (2010) The economic growth engine: how energy and work drive material prosperity. Edward Elgar Publishing, Cheltenham
6. Heun MK, Santos J, Brockway PE, Pruim R, Domingos T, Sakai M, Heun MK, Santos J, Brockway PE, Pruim R, Domingos T, Sakai M (2017) From theory to econometrics to energy policy: cautionary tales for policymaking using aggregate production functions. Energies 10(2):203. https://doi.org/10.3390/en10020203 . http://www.mdpi.com/1996-1073/10/2/203
7. King C, Maxwell J, Donovan A, King CW, Maxwell JP, Donovan A (2015) Comparing world economic and net energy metrics, part 2: total economy expenditure perspective. Energies 8(11):12,975–12,996, https://doi.org/10.3390/en81112347 . http://www.mdpi.com/1996-1073/8/11/12347
8. King CW (2016) Information theory to assess relations between energy and structure of the US economy over time. BioPhys Econ Res Qual 1(2):10. https://doi.org/10.1007/s41247-016-0011-y
9. Brand-Correa LI, Brockway PE, Copeland CL, Foxon TJ, Owen A, Taylor PG, Brand-Correa LI, Brockway PE, Copeland CL, Foxon TJ, Owen A, Taylor PG (2017) Developing an input-output based method to estimate a national-level energy return on investment (EROI). Energies 10(4):534. https://doi.org/10.3390/en10040534 , http://www.mdpi.com/1996-1073/10/4/534
10. Heun MK, Owen A, Brockway PE (2018) A physical supply-use table framework for energy analysis on the energy conversion chain. Appl Energy 226:1134–1162
11. Nordhaus W, Tobin J (1972) “Is growth obsolete?” In economic research: retrospect and prospect. NBER. https://www.nber.org/chapters/c7620 . Accessed 28 Sept 2020
12. United Nations (ed.) (2014) System of environmental-economic accounting 2012: Central framework. United Nations, New York
13. Branje S, Morris AS. The impact of the COVID-19 pandemic on adolescent emotional, social, and academic adjustment. J Res Adolesc. 2021;31(3):486–99.
14. Mulgan G. Big mind: How collective intelligence can change our world. Princeton: Princeton University Press; 2018.
15. Veenhoven, R. (1984). Conditions of happiness. Dordrecht/Boston: D.Reidel.
16. Popper, K. R. (1952). The open society and Its enemies. Routledge & Kegan Paul.
17. Veenhoven, R. (1996). Happy life expectancy. Social Indicators Research, 39(1), 1–58.
18. Veenhoven, R. (2004). Happy life years: A measure of gross national happiness. In K. Ura & K. Galay (Eds.), Gross national happiness and development, Proceedings of the first international seminar on Operationalization of Gross National Happiness. (pp. 287–318). Thimphu, Bhutan: Center of Bhutan Studies.
19. Robine, J. M., & Ritchie, K. (1991). Healthy life expectancy: evaluation of global indicator of change in population health. British Medical Journal, 302(6774), 457–460.
20. Sanders, B. S. (1964). Measuring community health levels. American Journal of Public Health and the Nations Health, 54(7), 1063–1070.
21. Brown, B. J., Hanson, M. E., Liverman, D. M., & Merideth, R. W. (1987). Global sustainability: toward definition. Environmental Management, 11(6), 713–719.

22. Abbott, P., & Wallace, C. (2012). Social quality: A way to measure the quality of society. Social Indicators Research, 108(1), 153–167.
23. Holman, D., & Walker, A. (2018). Social quality and health: Examining individual and neighbourhood contextual effects using a multilevel modelling approach. Social Indicators Research, 138(1), 245–270.
24. Plattet, A., & Spini, D. (2021). Cause Commune. Méthodologie d'un projet d'action-recherche sociale et participative à Chavannes-près-Renens. LIVES Working Paper, 89(1), 1–52. https://doi.org/10.12682/LIVES.2296-1658.2021.89.1
25. van der Maesen, L. J., & Walker, A. C. (2005). Indicators of social quality: Outcomes of the European scientific network. European Journal of Social Quality, 5(1–2), 8–24.
26. Evans, C. R., Williams, D. R., Onnela, J. P., & Subramanian, S. V. (2018). A multilevel approach to modeling health inequalities at the intersection of multiple social identities. Social Science & Medicine, 203, 64–73.
27. Li, Y. (2022). Physically constructed and socially shaped: Sociomaterial environment and walking for transportation in later life. Journal of Aging and Physical Activity, 1, 1–8.
28. Bronfenbrenner, U. (1979). The ecology of human development: Experiments by nature and design. Harvard University Press.
29. Abbott, P., Wallace, C., & Sapsford, R. (2010). Surviving the transformation: Social quality in Central Asia and the Caucuses. Journal of Happiness Studies, 12, 199–223.
30. Lin, K. (2014). Social quality and happiness—An analysis of the survey data from three Chinese cities. Applied Research in Quality of Life, 11, 23–40.
31. Yuan, H., & Golpelwar, M. (2013). Testing subjective well-being from the perspective of social quality: Quantile regression evidence from Shanghai, China. Social Indicators Research, 113(1), 257–276.
32. Jamieson D (2001) Climate change and global environmental justice. In: Miller C, Edwards P (eds) Changing the atmosphere: expert knowledge and environmental governance. MIT Press, Cambridge, MA
33. Singer P (2004) One atmosphere. In: One world: the ethics of globalization. Yale University Press
34. Gardiner S (2004) Ethics and global climate change. Ethics 114:555–600
35. Shue H (1993) Subsistence emissions and luxury emissions. Law & Policy 15(1):39–58
36. Lomborg B (2001) Global warming. In: Lomborg B (ed) The sceptical environmentalist. Cambridge University Press, Cambridge, p 258–324
37. Baer P, Anthanasiou T, Kartha S, Kemp-Benedict E (2009) Greenhouse development rights: a proposal for a fair global climate treaty. Ethics, Place and Environ 12(3):267–281
38. Moellendorf D (2011) A normative account of dangerous climate change. Clim Change 108:57–72
39. Morgan J, Waskow D (2013) A new look at climate equity in the UNFCCC. Clim Policy 14(1):17–22. Available at: http://www.tandfonline.com/doi/abs/https://doi.org/10.1080/14693062.2014.848096#.VLL7KivF-AU
40. Sethi N (2015) BASIC nations to discuss climate change. Live Mint. Available at: http://www.livemint.com/Politics/T2bUYB9daeJalW9bOMwMbP/BASIC-nations-to-discuss-climate-change.html
41. Olsen KH (2007) The clean development mechanism's contribution to sustainable development: a review of the literature. Clim Change 84:59–73. https://doi.org/10.1007/s10584-007-9267-y
42. Alam MS, Bala BK, Huq AMZ, Matin MA (1991) A model for the quality-of-life as a function of electrical energy-consumption. Energy 16(4):739–745
43. Alam MS, Roychowdhury A, Islam KK, Huq AMZ (1998) A revisited model for the physical quality of life (PQL) as a function of electrical energy consumption. Energy 23(9):791–801
44. Pasternak AD (2000) Global energy futures and human development: a framework for analysis. US Department of Energy, Oak Ridge
45. Goldenberg J, La Rovere EL, Coelho ST (2004) Expanding access to electricity in Brazil. Energy Sustain Dev 8(4):86–94

46. Dias RA, Mattos CR, Balestieri JAP (2006) The limits of human development and the use of energy and natural resources. Energy Policy 34(9):1026–1031
47. Moran DD, Wackernagel M, Kitzes JA, Goldfinger SH, Boutaud A (2008) Measuring sustainable development—nation by nation. Ecol Econ 64(3):470–474. https://doi.org/10.1016/j.ecolecon.2007.08.017
48. Martínez DM, Ebenhack BW (2008) Understanding the role of energy consumption in human development through the use of saturation phenomena. Energy Policy 36:1430–1435. https://doi.org/10.1016/j.enpol.2007.12.016
49. Mechtenberg A, Borchers K, Miyingo EW, Hormasji F, Hariharan A, Makanda JV, Musaazi MK (2012) Human power (HP) as a viable electricity portfolio option below 20 W/Capita. Energy Sustain Dev 16:125–145. https://doi.org/10.1016/j.esd.2011.12.006
50. Spierre SG, Seager TP, Selinger E (2013) The diminishing returns to the 2010 human development index. J Sustain Dev 6(6):34
51. Steinberger J, Roberts T (2010) From constraint to sufficiency: the decoupling of energy and carbon from human needs, 1975–2005. Ecol Econ, 425–433.https://doi.org/10.1016/j.ecolecon.2010.09.014
52. Raworth, K. 2017. Doughnut economics: Seven ways to think like a 21st-century economist. White River Junction: Chelsea Green Publishing.
53. Lovins, L.H., S. Wallis, A. Wijkman, and J. Fullerton. 2018. A finer future: Creating an economy in service to life, a report to the club of Rome. Gabriola Island: New Society Publishers.
54. Korten, D. 2018. Ecological civilization: The vision. Posted at: https://davidkorten.org/home/ecological-civilization/ .
55. Korten, David C. 2015. Change the story, change the future: A living economy for a living earth. San Francisco: Berrett-Koehler Publishers.
56. Riedy, C. 2020a. Discourse coalitions for sustainability transformations: Common ground and conflict beyond neoliberalism. Current Opinion in Environmental Sustainability. 45: 100–112.
57. Riedy, C. 2020b. Storytelling the future: Storytelling practice in transformative systems. In Higher education: An Educator's handbook, 71–87. Abington: Routledge.
58. Blackmore, S. 2000. The meme machine. Vol. 25. Oxford: Oxford Paperbacks.
59. Waddock, S. 2020. Reframing and transforming economics around life. Sustainability, m 12, 7553, 1–16, https://www.mdpi.com/2071-1050/12/18/7553/pdf .
60. Lent, J. 2017. The patterning instinct: A cultural history of Humanity's search for meaning. Amherst: Prometheus Books.
61. Harari, Y.N. 2014. Sapiens: A brief history of humankind. New York: Random House.
62. UNDP. (2019). Human Development Report 2019 Beyond income, beyond averages, beyond today: Inequalities in human development in the 21st century. United Nations Development Programme, New York, NY, USA: Human Development Reports.
63. The World Bank.(2018a). Sustainable development goals (SDG)|data catalog. Retrieved May 3, 2018, from https://datacatalog.worldbank.org/dataset/sustainable-development-goals .
64. United Nations General Assembly. (2017a). Global indicator framework for the Sustainable Development Goals and targets of the 2030 Agenda for Sustainable Development. https://doi.org/10.1891/9780826190123.0013
65. Costanza, R., Fioramonti, L., & Kubiszewski, I. (2016). The UN sustainable development goals and the dynamics of well-being. Frontiers in Ecology and the Environment, 14(2), 59–59. https://doi.org/10.1002/fee.1231 .
66. Székely, G. J., Rizzo, M. L., & Bakirov, N. K. (2007). Measuring and testing dependence by correlation of distances. The Annals of Statistics, 35(6), 2769–2794. https://doi.org/10.1214/009053607000000505 .
67. Ghislandi, S., Sanderson, W. C., & Scherbov, S. (2019). A simple measure of human development: The human life indicator. Population and Development Review, 45(1), 219.
68. Costanza, R., Fisher, B., Ali, S., Beer, C., Bond, L., Boumans, R., & Gayer, D. E. (2008). An integrative approach to quality of life measurement, research, and policy. SAPI EN. S. Surveys and Perspectives Integrating Environment and Society, 1.1.

69. Veenhoven, R. (1996). Developments in Satisfaction Research. Social Indicators Research,37(1), 1–46.
70. Magnin, E. (2018). Varieties of capitalism and sustainable development: Institutional complementarity dynamics or radical change in the hierarchy of institutions? Journal of Economic Issues, 52(4), 1143–1158.
71. Allen, C. R., Angeler, D. G., Garmestani, A. S., Gunderson, L. H., & Holling, C. S. (2014). Panarchy: Theory and application. Ecosystems, 17(4), 578–589
72. Autor, D., Dorn, D., Katz, L. F., Patterson, C., & van Reenen, J. (2020). The fall of the labor share and the rise of superstar firms. The Quarterly Journal of Economics, 135(2), 645–709.
73. Eeckhout, J. (2021). The Profit Paradox: How Thriving Firms Threaten the Future of Work. (1st edition). Princeton University Press.
74. Philippon, T. (2019). The Great Reversal: How America Gave Up on Free Markets. Harvard University.
75. An Y, Hsiao C, Li D (2016) Semiparametric estimation of partially linear varying coefficient panel data models. In: Gonz´alez-Rivera G, Hill RC, Lee T-H (ed) Essays in Honor of Aman Ullah (Advances in Econometrics, Volume 36). Bingley, UK: Emerald
76. Uddin MM, Mishra V, Smyth R (2020) Income inequality and CO2 emissions in the G7, 1870–2014: evidence from non-parametric modelling. Energy Econ 88:104780
77. Du K, Cheng Y, Yao X (2021a) Environmental regulation, green technology innovation, and industrial structure upgrading: the road to the green transformation of Chinese cities. Energy Econ 98:105247
78. Du M, Liao L, Wang B, Chen Z (2021b) Evaluating the effectiveness of the water-saving society construction in China: a quasi-natural experiment. J Environ Manag 277:111394
79. Juuti T (2022) The role of financial development in the relationship between income inequality and economic growth: an empirical approach using cross-country panel data. Qual Quant 56(3):985–1021
80. Lorek, S., & Spangenberg, J. H. (2014). Sustainable consumption within a sustainable economy: Beyond green growth and green economies. Journal of Cleaner Production, 62, 33–44.
81. UNGA (United Nations General Assembly). (2014). Report of the Open Working Group of the General Assembly on Sustainable Development Goals, UNGA a/68/970.
82. Raworth, K. (2012). A safe and just space for humanity: Can we live within the doughnut. Oxfam Policy and Practice: Climate Change and Resilience, 8(1), 1–26.
83. Rockström J, Steffen W, Noone K, Persson Å, Chapin FS, Lambin EF, Nykvist B et al (2009) A safe operating space for humanity. Nature 461(7263):472–475
84. Gupta, R., Hasler, A., & Lusardi, A. (2018). Financial fragility in the US: Evidence and implications. Global Financial Literacy Excellence Center.
85. Report on the Economic Well-Being of U.S. Households in 2017 (Issue May). (2018).
86. Atkinson, A., McKay, S., Kempson, E., & Collard, S. (2006). Levels of financial capability in the UK: Results of a baseline survey. Consumer Research, 46, 1–150.
87. Buddelmeyer, H., & Powdthavee, N. (2016). Can having internal locus of control insure against negative shocks? Psychological evidence from panel data. Journal of Economic Behavior and Organization, 122, 88–109. https://doi.org/10.1016/j.jebo.2015.11.014 .
88. Rotter, J. B. (1966). Generalized expectancies for internal versus external control of reinforcement. Psychological Monographs (General & Applied), 80, 1–28
89. Bialowolski, P., Weziak-Bialowolska, D., & McNeely, E. (2021). A socially responsible financial institution – The bumpy road to improving consumer well-being. Evaluation and Program Planning. https://doi.org/10.1016/j.evalprogplan.2021.101908 .
90. Tokunaga, H. (1993). The use and abuse of consumer credit: Application of psychological theory and research. Journal of Economic Psychology, 14(2), 285–316. https://doi.org/10.1016/0167-4870(93)90004-5 .
91. Wahlund, R., & Gunnarsson, J. (1996). Mental discounting and financial strategies. Journal of Economic Psychology, 17(6), 709–730. https://doi.org/10.1016/S0167-4870(96)00041-4 .

92. Cobb-Clark, D. A., Kassenboehmer, S. C., & Sinning, M. G. (2016). Locus of control and savings. Journal of Banking and Finance, 73, 113–130. https://doi.org/10.1016/j.jbankfin.2016.06.013 .
93. Kesavayuth, D., Ko, K. M., & Zikos, V. (2018). Locus of control and financial risk attitudes. Economic Modelling, 72, 122–131. https://doi.org/10.1016/j.econmod.2018.01.010 .
94. Anderson, J. and Baldwin, C. (2017). 'Building wellbeing: Neighbourhood flourishing and approaches for participatory Urban Design intervention', In R. Phillips, C. Wong (eds.), Handbook of community well-being research, International Handbooks of Quality-of-Life, Springer Publications. https://doi.org/10.1007/978-94-024-0878-2_17 .
95. Keyes, C. L. M. (2002). The mental health continuum: From languishing to flourishing in life. Journal of Health and Social Behavior, 43, 207–222.
96. Ryan, R. M., & Deci, E. L. (2001). On happiness and human potentials: A review of research on hedonic and eudaimonic well-being. Annual Review of Psychology, 52, 141–166.
97. Huppert, F. A., & So, T. T. (2013). Flourishing across Europe: Application of a new conceptual framework for defining well-being. Social Indicators Research, 110(3), 837–861.
98. Larson, J. S. (1993). The measurement of social wellbeing. Social Indicators Research, 28, 285–289.
99. Keyes, C. L. M. (1998). Social well-being. Social Psychology Quarterly, 61, 121–140.
100. Berkman, L. and Glass, T. (2000). Social Integration, Social Networks, Social Support and Health. In Berkman LF and Kawachi I (eds), Social Epidemiology. USA: Oxford University Press, pp.137–173.
101. Keyes, C., Cicognani, E., Pirini, C., Joshanloo, M., Rostami, R., & Nosratabadi. (2007). Social participation, sense of community and social well-being: Study on American, Italian and Iranian University students. Social Indicators Research, 89(1), 97–112.
102. UK Data Archive Study Number 7111. (2011). Citizenship Survey, 2010–2011: Communities Study 2010/11 Questionnaire Q2 (clean version).
103. Institute for Social and Economic Research. (2014). Understanding society: UK household longitudinal study 2013–2014. University of Essex/ESRC: https://www.understandingsociety.ac.uk/
104. Foxton, F., & Jones, R. (2011). Social capital indicators review. London: Office for National Statistics (ONS) Available at: www.ons.gov.uk .
105. Williams, D. R., & Vaske, J. J. (2003). The measurement of place attachment: Validity and generalizability of a psychometric approach. Forest Science, 49, 830–840.
106. Cobb, S. (1976). Social support as a moderator of life stress. Psychosomatic Medicine, 38(5), 300–314.
107. Sherwood, John; Carbajales-Dale, Michael; Haney, Becky Roselius Putting the Biophysical (Back) in Economics: A Taxonomic Review of Modeling the Earth-Bound Economy. Biophysical Economics and Sustainability (2020). https://doi.org/10.1007/s41247-020-00069-0
108. Lianos, Theodore P.; Pseiridis, Anastasia Adjusting GDP for ecological deficit: the Index of Debt to the Future (IDF). SN Business & Economics (2021). https://doi.org/10.1007/s43546-021-00041-0
109. Halfon, Neal; Chandra, Anita; Cannon, Jill S.; Gardner, William; Forrest, Christopher B. The Gross Developmental Potential (GDP2): a new approach for measuring human potential and wellbeing. BMC Public Health (2022). https://doi.org/10.1186/s12889-022-14030-x
110. Gaucher, Renaud; Dialga, Issaka; Vennin, Coralie The indicator of a happy, long and sustainable life. Social Indicators Research (2021). https://doi.org/10.1007/s11205-021-02727-2
111. Fleurbaey, Marc; Ponthiere, Gregory Measuring well-being and lives worth living. Economic Theory (2022). https://doi.org/10.1007/s00199-022-01446-0
112. Li, Yang; Spini, Dario; Lampropoulos, Dimitrios Beyond Geography: Social Quality Environments and Health. Social Indicators Research (2023). https://doi.org/10.1007/s11205-023-03073-1

113. Clark, Susan Spierre; Seager, Thomas P.; Selinger, Evan A development-based approach to global climate policy. Environment Systems and Decisions (2015). https://doi.org/10.1007/s10669-015-9533-x
114. Waddock, Sandra Wellbeing Economics Narratives for a Sustainable Future. Humanistic Management Journal (2021). https://doi.org/10.1007/s41463-021-00107-z
115. Kraemer, Guido; Reichstein, Markus; Camps-Valls, Gustau; Smits, Jeroen; Mahecha, Miguel D. The Low Dimensionality of Development. Social Indicators Research (2020). https://doi.org/10.1007/s11205-020-02349-0
116. Cohen Kaminitz, Shiri Looking Good or Feeling Well? Understanding the Combinations of Well-Being Indicators Using Insights from the Philosophy of Well-Being. Social Indicators Research (2020). https://doi.org/10.1007/s11205-020-02289-9
117. Schoenmaker, Dirk; Stegeman, Hans Can the Market Economy Deal with Sustainability?. De Economist (2022). https://doi.org/10.1007/s10645-022-09416-6
118. Sun, Yongchun Income inequality, carbon emissions, and green development efficiency. Environmental Science and Pollution Research (2022). https://doi.org/10.1007/s11356-022-23583-8
119. Gupta, Joyeeta; Vegelin, Courtney Sustainable development goals and inclusive development. International Environmental Agreements: Politics, Law and Economics (2016). https://doi.org/10.1007/s10784-016-9323-z
120. Bialowolski, Piotr; Weziak-Bialowolska, Dorota; McNeely, Eileen The Role of Financial Fragility and Financial Control for Well-Being. Social Indicators Research (2021). https://doi.org/10.1007/s11205-021-02627-5
121. Baldwin, Cathy; Vincent, Penny; Anderson, Jamie; Rawstorne, Patrick Measuring Well-Being: Trial of the Neighbourhood Thriving Scale for Social Well-Being Among Pro-Social Individuals. International Journal of Community Well-Being (2020). https://doi.org/10.1007/s42413-020-00067-6

The manufacturer's authorised representative in the EU is Springer Nature Customer Service Centre GmbH, Europaplatz 3, 69115 Heidelberg, Germany. If you have any concerns regarding our products, please contact ProductSafety@springernature.com

Printed and bound by CPI Group (UK) Ltd, Croydon, CR0 4YY
01/10/2026
02225857-0018